Mastering the TOGAF® Standard

A Practical Translation of the World's Leading Architecture Framework

Eric Jager

Mastering the TOGAF® Standard: A Practical Translation of the World's Leading Architecture Framework

Eric Jager
Almere, Flevoland, The Netherlands

ISBN-13 (pbk): 979-8-8688-1813-4   ISBN-13 (electronic): 979-8-8688-1814-1
https://doi.org/10.1007/979-8-8688-1814-1

Copyright © 2025 by Eric Jager

This work is subject to copyright. All rights are reserved by the Publisher, whether the whole or part of the material is concerned, specifically the rights of translation, reprinting, reuse of illustrations, recitation, broadcasting, reproduction on microfilms or in any other physical way, and transmission or information storage and retrieval, electronic adaptation, computer software, or by similar or dissimilar methodology now known or hereafter developed.

Trademarked names, logos, and images may appear in this book. Rather than use a trademark symbol with every occurrence of a trademarked name, logo, or image we use the names, logos, and images only in an editorial fashion and to the benefit of the trademark owner, with no intention of infringement of the trademark.

Figures mentioned in Appendix A are reprinted and reproduced in electronic form from the TOGAF® Standard, 10th Edition with permission granted by The Open Group, L.L.C. The official standard can be found at https://www.opengroup.org/library/c220, and remains the authoritative version for all purposes. ©The Open Group. All rights reserved. To explore how the TOGAF Standard can be used for commercial or non-commercial purposes, please visit The Open Group Licensing page at https://www.opengroup.org/licensing-commercial-and-non-commercial.

This publication is an independent guide designed to support the understanding and application of the TOGAF® Standard. While it is not formally endorsed by The Open Group, it has been developed with respect for the principles of the standard and the contributions of its global community. The perspectives shared reflect the author's interpretation and are not intended to represent the official views of The Open Group or its member organizations.

The use in this publication of trade names, trademarks, service marks, and similar terms, even if they are not identified as such, is not to be taken as an expression of opinion as to whether or not they are subject to proprietary rights.

While the advice and information in this book are believed to be true and accurate at the date of publication, neither the authors nor the editors nor the publisher can accept any legal responsibility for any errors or omissions that may be made. The publisher makes no warranty, express or implied, with respect to the material contained herein.

Managing Director, Apress Media LLC: Welmoed Spahr
Acquisitions Editor: Aditee Mirashi
Editorial Assistant: Jacob Shmulewitz

Cover designed by eStudioCalamar

Cover image designed by Eric Jager

Distributed to the book trade worldwide by Springer Science+Business Media New York, 1 New York Plaza, New York, NY 10004. Phone 1-800-SPRINGER, fax (201) 348-4505, e-mail orders-ny@springer-sbm.com, or visit www.springeronline.com. Apress Media, LLC is a Delaware LLC and the sole member (owner) is Springer Science + Business Media Finance Inc (SSBM Finance Inc). SSBM Finance Inc is a **Delaware** corporation.

For information on translations, please e-mail booktranslations@springernature.com; for reprint, paperback, or audio rights, please e-mail bookpermissions@springernature.com.

Apress titles may be purchased in bulk for academic, corporate, or promotional use. eBook versions and licenses are also available for most titles. For more information, reference our Print and eBook Bulk Sales web page at http://www.apress.com/bulk-sales.

Any source code or other supplementary material referenced by the author in this book is available to readers on GitHub. For more detailed information, please visit https://www.apress.com/gp/services/source-code.

If disposing of this product, please recycle the paper

Table of Contents

About the Author xiii

About the Reviewer xv

Acknowledgments xvii

Chapter 1: Introduction..1

    1.1. Why This Book 5

    1.2. Who This Book Is For 7

    1.3. How to Read This Book 9

Chapter 2: Preliminary Notes 13

    2.1. A Brief History 13

    2.2. The Evolution of the TOGAF Standard...16

        2.2.1. The Early Versions 17

        2.2.2. The More Recent Versions 18

        2.2.3. Comparison of TAFIM and TOGAF 26

    2.3. Summary..29

Chapter 3: The TOGAF Standard...31

    3.1. What Is the TOGAF Standard 32

    3.2. Fundamental Content..36

    3.3. Series Guides 40

    3.4. The TOGAF Library 43

    3.5. Key Advantages of the TOGAF Standard.......................................46

    3.6. Summary..50

TABLE OF CONTENTS

Chapter 4: Architecture Definition 53
4.1. What Is Enterprise Architecture 54
4.2. Confusion Over Definitions 57
4.3. Multiple Definitions 58
4.4. Summary 61

Chapter 5: Enterprise Metamodel 63
5.1. Sections of the Enterprise Metamodel 63
5.1.1. General Entities 65
5.1.2. Business Architecture 74
5.1.3. Data Architecture 92
5.1.4. Application Architecture 96
5.1.5. Technology Architecture 98
5.2. Tailoring the Enterprise Metamodel 101
5.3. Summary 107

Chapter 6: Content Framework 111
6.1. A Categorization Framework 111
6.2. Advantages of the Content Framework 114
6.3. Mapping to the Architecture Development Method 117
6.4. Tailoring the Content Framework 123
6.5. Summary 125

Chapter 7: Architecture Repository 127
7.1. The Central Hub 128
7.2. Key Components of the Architecture Repository 129
7.3. The Three Continua 136
7.3.1. The Enterprise Continuum 139
7.3.2. The Architecture Continuum 141
7.3.3. The Solutions Continuum 147

TABLE OF CONTENTS

7.4. Catalogs, Diagrams, Maps, and Matrices 150
    7.4.1. Catalogs 151
    7.4.2. Diagrams and Maps 151
    7.4.3. Matrices 153
7.5. Architecture Deliverables 154
    7.5.1. Architecture Artifacts 155
    7.5.2. Building Blocks 163
7.6. Views and Viewpoints 176
    7.6.1. Relationship Between Views and Viewpoints 177
7.7. Using the Architecture Repository 179
    7.7.1. Deciding on a Tool 180
7.8. Summary 183

Chapter 8: Architecture Governance 189
8.1. Managing the Architecture 190
8.2. Implementing Architecture Governance 193
8.3. Architecture Board 204
    8.3.1. Diverse Leadership 206
    8.3.2. When to Set Up an Architecture Board 209
    8.3.3. Key Responsibilities 211
    8.3.4. Composition and Size 213
    8.3.5. How to Set Up an Architecture Board 215
8.4. Summary 224

Chapter 9: Architecture Capability 229
9.1. The Ability to Practice Architecture 230
9.2. Creating the Architecture Capability 230
9.3. Positioning the Architecture Capability 236
9.4. Summary 241

TABLE OF CONTENTS

Chapter 10: Architecture Roles, Skills, and Competencies 247
10.1. Difference Between Jobs, Roles, and Assignments 248
10.2. Core Architecture Roles 249
10.3. The Role of the Segment Architect 255
    10.3.1. The Horizontal Perspective 258
    10.3.2. The Vertical Perspective 260
10.4. Skills and Competencies 263
10.5. Summary 277

Chapter 11: Architecture Domains 279
11.1. Four Architecture Domains 280
11.2. Difference in Domains and Layers 286
11.3. Mapping Roles to Domains 289
11.4. Summary 292

Chapter 12: Architecture Development Method 295
12.1. The Heart of the TOGAF Standard 296
12.2. Architecture Document Creation 300
12.3. Preliminary Phase 305
    12.3.1 Determining and Establishing the Architecture Capability 305
    12.3.2 Preliminary Phase Artifacts 320
12.4. Phase A: Architecture Vision 324
    12.4.1 Creating the Architecture Vision 324
    12.4.2 Architecture Vision Phase Artifacts 336
12.5. Phase B: Business Architecture 347
    12.5.1 Developing Business Architecture 348
    12.5.2 Business Architecture Methods and Modeling Techniques 360
    12.5.3 Phase B (Business Architecture) Artifacts 366

TABLE OF CONTENTS

- 12.6. Phase C: Information Systems Architectures 398
  - 12.6.1 Developing Data Architecture 400
  - 12.6.2 Phase C (Data Architecture) Artifacts 410
  - 12.6.3 Developing Application Architecture 418
  - 12.6.4 Phase C (Application Architecture) Artifacts 427
- 12.7. Phase D: Technology Architecture 441
  - 12.7.1 Developing Technology Architecture 442
  - 12.7.2 Phase D (Technology Architecture) Artifacts 453
- 12.8. Phase E: Opportunities and Solutions 459
  - 12.8.1 Defining Opportunities and Solutions 460
  - 12.8.2 Phase E (Opportunities and Solutions) Artifacts 475
  - 12.8.3 Creating an Architecture Roadmap 485
- 12.9. Phase F: Migration Planning 495
  - 12.9.1 Creating a Migration Planning 495
  - 12.9.2 Phase F (Migration Planning) Artifacts 507
- 12.10. Phase G: Implementation Governance 513
  - 12.10.1 Establishing Architecture Governance 514
  - 12.10.2 Architecture Performance Monitoring 522
- 12.11. Phase H: Architecture Change Management 528
  - 12.11.1 Managing Architecture Changes 528
- 12.12. Requirements Management 539
  - 12.12.1 Capture, Manage, and Refine Requirements 540
  - 12.12.2 Requirements Management Artifacts 545
- 12.13. The Architecture Development Method in Short 547
- 12.14. Summary 557

TABLE OF CONTENTS

Chapter 13: Applying the ADM 561
13.1. Various Architecture Development Method Approaches 562
13.2. Using the TOGAF Framework with Different Architecture Styles..............563
13.3. Applying Iteration to the Architecture Development Method 564
13.3.1. Different Ways to Apply Iteration 566
13.3.2. How to Apply Iteration 573
13.4. Architecture Partitioning 580
13.4.1. Why Use Architecture Partitioning 580
13.4.2. Supporting Architecture Partitioning 583
13.5. Summary...583

Chapter 14: ADM Techniques 589
14.1. Technique or No Technique 589
14.2. Principles 592
14.2.1 The Added Value of Principles 593
14.2.2 The Structure of a Principle 594
14.2.3 Enterprise Principles 597
14.2.4 Architecture Principles 599
14.3. Stakeholder Management..601
14.3.1 Stakeholder Analysis 605
14.3.2 Stakeholder Mapping 609
14.3.3 Communications Plan..611
14.3.4 Stakeholder Engagement 615
14.4. Architecture Patterns 616
14.5. Gap Analysis...617
14.6. Interoperability Requirements...623
14.7. Business Transformation Readiness Assessment....................................627
14.8. Risk Management 632

TABLE OF CONTENTS

    14.8.1 Risk Classification 638
    14.8.2 Risk Identification 642
    14.8.3 Initial Risk Assessment 644
    14.8.4 Risk Mitigation and Residual Risk Assessment 647
    14.8.5 Risk Monitoring 650
    14.8.6 Visualizing Risks 651
  14.9. Architecture Alternatives and Trade-Offs 656
  14.10. Summary 663

Chapter 15: Architecture Maturity Models 669

  15.1. Determining Architecture Maturity 670
    15.1.1. Architecture Capability Maturity Model 670
    15.1.2. Capability Maturity Model Integration 681
  15.2. Maturing the Architecture Capability 690
  15.3. Alignment with Competencies 695
  15.4. Summary 699

Chapter 16: Alignment with Other Frameworks 701

  16.1. Mapping Existing Frameworks and Methods 701
  16.2. Enterprise Architecture Implementation Wheel 703
    16.3.3. Mapping to the Architecture Development Method 707
  16.4. Summary 711

Chapter 17: Tailoring the TOGAF Standard 713

  17.1. Designed for Customization 714
  17.2. Tailoring the TOGAF Standard for the Healthcare Industry 719
  17.3. Summary 722

ix

TABLE OF CONTENTS

Chapter 18: Agile Architecture 725
18.1. What Is Agile Architecture 726
18.1.1. The Basic Concepts Explained 727
18.2. Three Levels of Architecture Granularity 731
18.3. Architecture Partitions and Agile Sprints 735
18.4. Developing an Agile Architecture 739
18.4.1. Mapping the ADM to Agile Concepts 740
18.4.2. ADM Phases and Agile Sprints 744
18.5. Summary 754

Chapter 19: Digital Transformation 757
19.1. What Is Digital Transformation? 758
19.2. Digital Transformation and Enterprise Architecture 759
19.3. A Structured Approach to Digital Transformation 763
19.4. Summary 777

Chapter 20: Putting the TOGAF Standard into Practice 781
20.1. A Case Scenario 782
20.1.1 Scoping the Work 782
20.1.2 Envisioning the Architecture 784
20.1.3 Starting the Development 787
20.1.4 Setting the Direction 794
20.1.5 Getting Ready for Implementation 798
20.1.6 Overseeing the Architecture Work 805
20.1.7 Maintaining Control 806
20.2. Summary 808

Chapter 21: The TOGAF Standard on a Page **817**
   21.1 Connecting the Concepts 817

Chapter 22: Closing Remarks **823**

Appendix A: List of Copyrighted Figures **831**

References **835**

Index **839**

About the Author

Eric Jager is the author of the bestselling book *Getting Started with Enterprise Architecture* and a Certified Master Architect in the field of Enterprise Architecture. He is also a Certified TOGAF Enterprise Architecture Practitioner, Certified Business Architect, and ArchiMate Practitioner. Eric is familiar with various architecture methodologies including the TOGAF Standard and the Zachman Framework.

With over 18 years of experience practicing Enterprise Architecture and extensive knowledge of its development and application, Eric is considered a thought leader in the field. He has worked for many organizations, ranging from government agencies to healthcare institutions.

Drawing from his years of practical experience, Eric created the *Enterprise Architecture Implementation Wheel*, a method for implementing Enterprise Architecture that can be used by novice and experienced architects alike. Eric lectures on architecture at Eindhoven University of Technology and speaks at various conferences and seminars. He focuses on the practical and pragmatic application of Enterprise Architecture and writes about his experiences on his blog: https://eawheel.com/blog/.

About the Reviewer

Daniël van Winsum is an Enterprise Architect at Solventa B.V. with more than 15 years of experience in the field. He has built up a strong track record in various sectors, including higher education, municipalities, and housing associations.

Daniël is certified in the TOGAF Standard, ArchiMate, and BIZBOK and is well versed in various architecture methodologies such as the TOGAF Standard, Dynamic Enterprise Architecture (DYA), and General Enterprise Architecture (GEA).

After studying Business Administration at the University of Groningen, he specialized in the interaction between IT and business strategy with an Executive Master in Information Management at the TIAS Business School.

Daniël has played a key role in introducing architecture-driven thinking to organizations and has built teams from the ground up. He has made significant contributions to strategic digital transformation efforts, both in developing enterprise architecture content and transforming IT business models. Daniël was also instrumental in the development of the first versions of the Dutch Higher Education Reference Architecture (HORA).

Acknowledgments

First and foremost, I would like to thank The Open Group for granting me permission to use a selection of their illustrations. While most of the figures in this book are my own, some are copyrighted by The Open Group. See Appendix A – List of Copyrighted Figures.

Special thanks go to Daniël van Winsum, a former colleague, for agreeing to review this book. His keen insight and attention to detail were essential in ensuring the material was presented clearly.

I would also like to thank the editorial and copywriting staff at Apress Media for guiding me through the writing and editing process with good-natured professionalism. They have produced another fine product.

Thanks also to my wife, Nienke, whose trust and support made this book possible.

Finally, I would like to thank Vasileios Lessis. He generously offered valuable suggestions for visualizing the alternative reading tours included in this book, as well as insightful feedback.

CHAPTER 1

Introduction

 This chapter is part of the PINK reading tour. See Figure 1-1 for alternative reading tours.

Architecture is a fantastic profession. But I am not talking about the architecture of buildings or cities (which is also great, by the way). I am talking about the profession of Enterprise Architecture.

With the tools we have, we as architects can help organizations translate their strategy into an executable whole. We can help them by giving them insight into the implications of implementing such a strategy. We can also interpret those implications in a way that is understandable to a wide range of stakeholders. For each level of an organization, we can create a visualization that gives those stakeholders a clear picture of the impact that proposed changes have or will have. We can do all of this because we have several tools at our disposal.

One such tool is architecture frameworks. For us as architects, these are reference works. They describe the outlines of how we can map the changes needed for and by an organization and how to interpret them. Architecture frameworks help us determine what impact these changes will have on achieving the desired outcomes and thus the intended goals. These frameworks provide us with a great tool, and because of them, we don't have to reinvent the proverbial wheel. We can tap into the knowledge and expertise of countless people who have gone through the trouble of

CHAPTER 1 INTRODUCTION

figuring things out for us, trying them out, testing them, and capturing them in the frameworks. All we have to do is follow them and adapt them to our needs and specific situations or challenges.

The TOGAF Standard is known as *the world's leading architecture framework* and has been used by tens of thousands of architects for decades [1]. This is not surprising because it is a very complete and comprehensive framework. All the ins and outs that we as architects could possibly face are described and explained in the framework.

The reason that I wanted to write a book about the TOGAF Standard is not because I thought that the TOGAF Standard was incomplete or lacking something. Quite the contrary. It may actually seem overly complete in some areas. Perhaps a bit cumbersome in its practical application. And maybe even a little inconsistent in some respects. No, the main reason for writing this book is that I believe that the TOGAF Standard is a very applicable framework that, in my opinion, has not yet been properly applied by the profession. I believe this is because not everyone knows or understands exactly *how to apply* the framework. Partly because of the amount of information in the framework and the sometimes fuzzy (or even academic and stale) way in which that information is explained, many architects give up and do not use the approach described in the TOGAF Standard. And that's a shame.

I hope to present an approach - a *practical* approach - that clarifies the methodology in such a way that more architects will start using (at least parts of) the framework. Because that is what the TOGAF Standard is all about: using the parts that apply to your own situation or organization and applying the method in a way that makes it work for the architectural challenges.

This book is certainly not a one-to-one description of everything that's in the TOGAF Standard. It covers the most important parts of the framework, supplemented by additional topics that I felt could help paint a complete enough picture of how to use the framework. I have also included certain areas that I believed needed more context and

CHAPTER 1   INTRODUCTION

explanation. If you're looking for a complete listing of every possible topic, I suggest you read the TOGAF Standard itself, which is a good idea on its own. The primary purpose of this book is to help you master the most important parts of the world's leading architecture framework.

The book focuses on the *practical application* of the TOGAF *Framework*. This framework consists of a substantial number of fundamental components that are important for building a solid Enterprise Architecture. Series Guides highlight specific topics or present additional aspects of the architecture.

The TOGAF Framework and its fundamental components are designed to be durable and stable. The Series Guides offer industry-specific, architectural-style, purpose-specific, and problem-specific guidance on how to use the framework. They also contain additional information on a variety of architecture-related topics. The Series Guides are expected to evolve the fastest of all the TOGAF documentation. Due to their rapid growth, The Open Group has decided to no longer include the supporting guides as a permanent part of the TOGAF Framework. With the release of *The TOGAF Standard, 10th Edition*[1], The Open Group divided the Standard into two sections: *Fundamental Content* (see Chapter 3, Section 3.2) and supplementary information on architecture-specific topics, known as *Series Guides* (see Chapter 3, Section 3.3).

The TOGAF Framework (consisting of the aforementioned Fundamental Content) remains the essential scaffolding across industries, domains, and styles. It is therefore the focus of this book. Some of the topics covered in the Series Guides are also covered in this book, but as The Open Group itself states:

Not all the TOGAF Series Guides will be relevant in every situation [2].

---

[1] *The TOGAF Standard, 10th Edition*, was released in April 2022.

CHAPTER 1   INTRODUCTION

However, Enterprise Architects who are planning the deployment of the TOGAF Standard should be aware of the guidance available.

As I said earlier, this book will focus primarily on the TOGAF Framework, with the addition of those components and/or Series Guides that have proven to be essential in guiding architects in the implementation of Enterprise Architecture.

If you were to summarize the TOGAF Standard in a nutshell, it would boil down to the *establishment and implementation* of the Architecture Capability. This capability is implemented by one or more *architecture roles*, which in turn use architecture *skills* and *competencies*.

Architecture work begins with a thorough understanding of the organization's goals, such as improving customer experience, accelerating digital transformation, or optimizing operations. This foundational step, *understanding business needs*, involves collecting and analyzing both business and technology objectives.

The next step is to *design the architecture*. With a clear understanding of business needs, the architecture design process begins. This includes data management strategies, application integration methods, and the selection of appropriate technologies, all aimed at supporting rapid development and efficient digital delivery cycles.

The TOGAF Standard can then be used to develop a flexible *implementation roadmap* that emphasizes agility and iterative progress. Tasks are prioritized to ensure a seamless and adaptable transition, with a focus on *delivering value* early and consistently. This is called *implementation planning*.

Finally, there is *governance and execution*. This involves implementing governance mechanisms to maintain alignment with the *architectural vision*. The TOGAF Standard recognizes the need for continuous evolution of the architecture to quickly adapt to changing business needs.

By following the structured phases of the Architecture Development Method, organizations can effectively develop and implement an Enterprise Architecture that is responsive to dynamic business environments. It's important to note that not everything in the TOGAF Standard needs to be used in its entirety. Tailoring the framework is a must.

Perhaps a little too bluntly stated for some, but that is what it comes down to. If you keep this in mind when applying the TOGAF Standard, the framework suddenly feels a lot less overwhelming.

1.1. Why This Book

As I mentioned at the beginning of this chapter, the TOGAF Standard is the most widely used and applied Enterprise Architecture framework in the world.

> The TOGAF Framework is and remains the essential scaffolding across industries, domains, and styles [3].

Despite the fact that the TOGAF Standard is considered the *de facto* standard worldwide, there is also criticism of the framework. Criticism that, in my opinion, is mostly unfounded.

Some quotes collected over the years:

> "Real examples demonstrating the actual practical usage of TOGAF's recommendations are missing."
>
> "TOGAF's prescriptions are vague and inarticulate since it only states that the ADM should be adapted without specifying how."
>
> "Using TOGAF can be best explained as studying TOGAF and then doing something else instead."

CHAPTER 1   INTRODUCTION

The main criticisms of the framework are its *lack of real added value* and its *vague prescriptions*, which make practitioners feel more or less compelled (or forced) not to use the framework for the reasons mentioned.

The people who created the framework invested a great deal of time and effort into developing a useful tool from which others can benefit.

In this book, I aim to refute these arguments for not using the framework. I therefore provide examples of the *practical* application of the TOGAF Standard. This will demonstrate the added value of the framework and how to interpret the so-called vague prescriptions mentioned above. These examples come from daily practice and are easy to understand and use.

I am not saying that there is nothing left to be desired in terms of the framework. I also see room for improvement in some areas. For example, the argument that I myself can agree with is that the TOGAF Standard mostly describes the *what* and not the *how*.

The description of the phases of the Architecture Development Method focuses on recommendations on *what* may be done to define and deploy an Enterprise Architecture.

Guidance on *how* to do what is specified can be found in the TOGAF Series Guides [2].

A lot of time and attention is spent in the framework interpreting all sorts of things that *must* or *can* be done to achieve an Enterprise Architecture implementation. However, describing *how* implementers of the framework should actually go about doing this is less well articulated, even in the aforementioned Series Guides. An omission, in my opinion.

CHAPTER 1   INTRODUCTION

I have mentioned this on a number of occasions when I had the opportunity to have conversations with people from The Open Group. So far, it has not led to a change in the framework's approach.

This is one of the reasons why I decided to write this book. With this book I want to show *how* the TOGAF Standard can be used to achieve an Enterprise Architecture implementation. Practical and pragmatic. With powerful examples.

By doing so, the third quote above can also be debunked. After reading this book, everyone will be able to apply the theory of the TOGAF Standard in a practical and pragmatic way. There will no longer be any reason to *study TOGAF and then do something else instead*.

With this book, I enable the reader to use the world's leading architecture framework as it was intended to be used: *practical* and *pragmatic*, but also *tailored* to the situation.

A lot of care and thought has gone into this book. I realize that nothing is ever perfect, and this book is no different. If you find any errors or typos, please report them. You can find instructions on how to do this on my website: `https://eawheel.com/books/errata/`.

1.2. Who This Book Is For

This book is first and foremost for architects. People who have been practicing the profession of architecture for many years. It's written for people who know TOGAF or for those who have at least heard of the framework. It's for architects with strong opinions about the TOGAF Standard, for those that say they don't use the framework but fail to give a *solid argument* as to why they don't use it.

CHAPTER 1 INTRODUCTION

This book is really for *two kinds of people*:

- Those who **have** reviewed or read the TOGAF Standard
- Those who **have not** reviewed or read the TOGAF Standard

For those who have: In all likelihood, this group of people does not apply the TOGAF Standard in their daily work. Or at the very least, they think they do not. This may be because the theory described in the framework is not sufficiently supported by meaningful examples or situations from practice. Or it may be because the theory is too vague in some areas, making it difficult to apply in real life. To this group of people I would like to say: this is *definitely* a book for you!

For those who haven't: For the group of people who have not yet had the time and opportunity to familiarize themselves with the TOGAF Standard, this book is an excellent starting point. It explains the theory of the framework in a practical and pragmatic way, using clear examples and real-world scenarios. The group of people who have not yet embraced the framework may be put off by the amount of theory or by the fact that in some situations the framework is talked about in a less positive way. With this book, I hope to address both of these concerns.

The theory laid out in the TOGAF Standard is extensive, to say the least. And it needs to be, since it is the *blueprint* for Enterprise Architecture. The scope of Enterprise Architecture is broad and has many different aspects. All of these aspects must be described in detail in a framework in such a way that there is little (if any) ambiguity. With this book, I offer the reader the opportunity to absorb this theory in an accessible way. I translate the theory from the TOGAF Standard in a practical and pragmatic way so that it becomes highly accessible to everyone.

Mastering the TOGAF® Standard is a book for anyone who wants to learn how to use the world's leading architecture framework.

1.3. How to Read This Book

The book provides a *practical translation* of the major topics that exist in the TOGAF Standard. This wealth of information is complemented by a brief history of the framework and its evolution over time. An explanation of the framework's composition is also included in this book.

My goal with each topic is not just to paint a picture of a certain approach that can be taken. I wanted to describe not only the *what*, as is already expertly done in the TOGAF Standard itself, but more importantly the *how*. Therefore, I have tried to provide each major topic in the framework with clear steps to support its implementation or to clarify exactly what is meant by a particular topic. I also packed the book with the necessary illustrations to support the topics. I felt it was important to not only describe the theory, but to bring that theory to life with easy-to-understand examples.

In addition to the standard route of moving through the book in its entirety, there are a number of alternative routes you can take. These alternative routes allow you to focus on specific content:

- Learn about **the very basics** of the TOGAF Standard.
- Learn about **the major concepts** of the TOGAF Standard.
- Dive **into the details**.
- Content you **won't find easily elsewhere**.

CHAPTER 1   INTRODUCTION

Figure 1-1 illustrates the different reading tours and their content.

For example, if you are one of the people who **has** reviewed or read the TOGAF Standard, you may decide to read the book by following the blue or gold reading tour as shown in Figure 1-1. In this way, you can dive into the rich details of the framework or catch up on content that you may not easily find elsewhere.

Another way to read the book is to follow the green reading tour. This will allow you to learn about the most important concepts contained in the TOGAF Standard. If you are just interested in brushing up on the basics, then the pink reading tour is the way to go.

Each chapter is preceded by the following reference to the reading tour(s) of which the chapter is a part.

 This chapter is part of the PINK reading tour. See Figure 1-1 for alternative reading tours.

However, if you are one of those people who **have not** reviewed or read the TOGAF Standard, I recommend that you read the book *in its entirety*. Reading the entire book from cover to cover will reveal all the ingenious relationships between the various parts of the framework. This is, of course, the best way to read *Mastering the TOGAF Standard*.

CHAPTER 1 INTRODUCTION

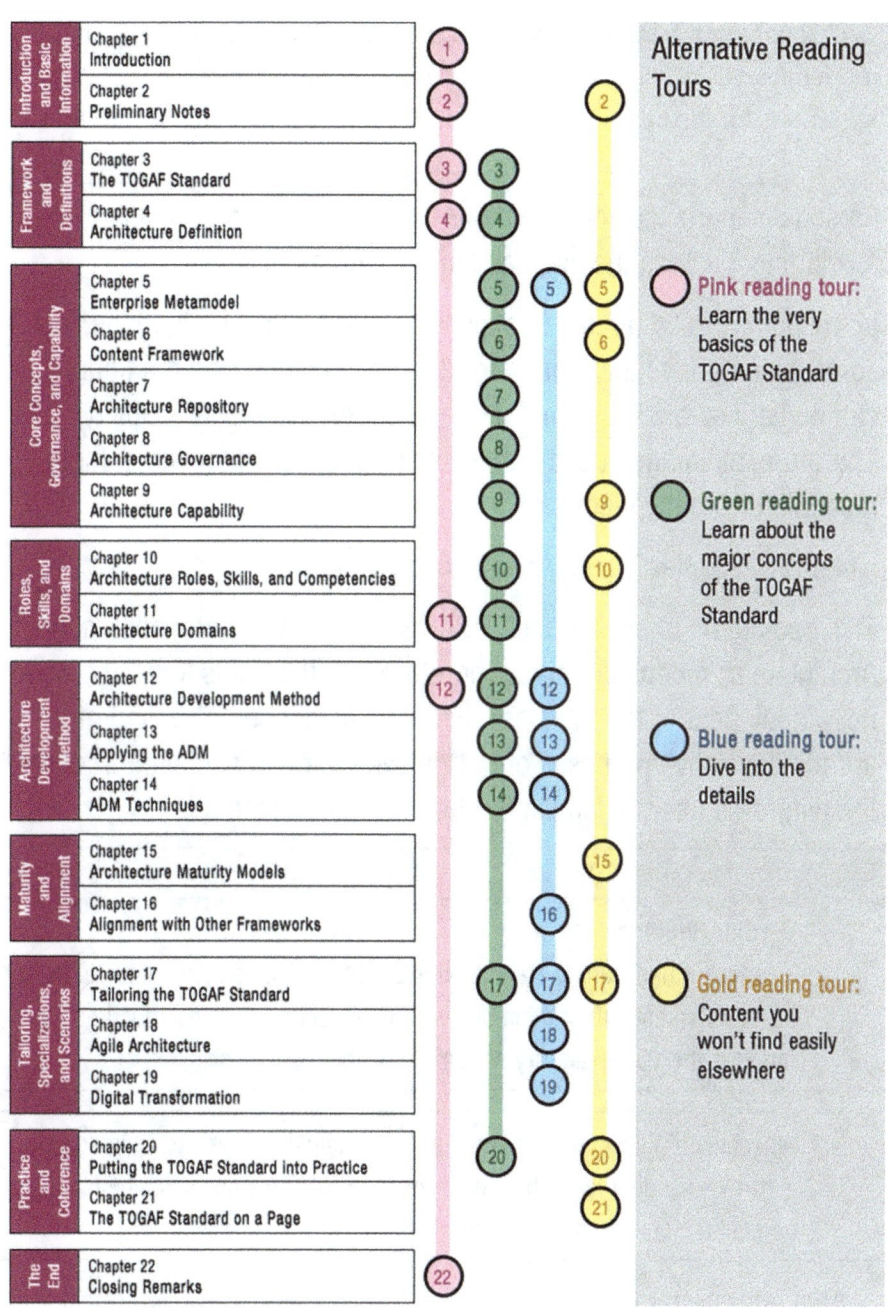

Figure 1-1. Alternative reading tours

CHAPTER 1   INTRODUCTION

Throughout the book, I have used quotes and definitions, personal experiences, references (indicated by a number between brackets: [13]), and suggestions for further reading to paint as complete a picture as possible.

Personal experiences or comments are indicated by a specific type of formatting, such as the one used in this sentence.

In the book, I used figures to explain certain topics or situations. In some cases, the amount of information contained in such a figure can be very extensive, so it may be at the expense of readability. Therefore, all images used in this book can be downloaded in high resolution from my website:

`https://eawheel.com/books/media/`

This book aims to explain the TOGAF Standard. It is not a book about ArchiMate or modeling ethics. Therefore, the diagrams in this book do not strictly adhere to ArchiMate modeling guidelines. The diagrams are for illustrative purposes only. However, to maintain recognizability, the diagrams use the familiar ArchiMate color scheme.

Further Reading
Tips for further reading are indicated by a book icon and are listed in these Further Reading information boxes. The tips either reference documents found in the TOGAF Library or point to online resources.

Whenever a piece of information needs (or could use) an abbreviated step-by-step approach, the steps icon is used. The steps described in this particular block are short and to the point.

All names used in examples, artifacts, and case scenarios in this book are fictitious and do not refer to actual persons or organizations.

CHAPTER 2

Preliminary Notes

 This chapter is part of the PINK and GOLD reading tours. See Figure 1-1 in Chapter 1, Section 1.3, for alternative reading tours.

To fully understand and appreciate the TOGAF Standard, some background information is necessary. It is essential to know the origins of the world's leading architecture framework as this helps to form a complete picture of the evolution that the TOGAF Standard has gone through and the moments in time that determined the growth of the framework.

The emergence of architecture frameworks has been the catalyst for the creation and evolution of the TOGAF Standard as we know it today.

This chapter discusses the development of the TOGAF Standard and its subsequent evolution in detail.

2.1. A Brief History

Enterprise Architecture emerged in the 1960s and 1970s as large organizations began to see the need for more formal methods of managing their complex (IT) systems and aligning them with business goals and intended outcomes. At that time, several attempts were made to develop system architectures and information models.

CHAPTER 2 PRELIMINARY NOTES

In the 1980s, the term Enterprise Architecture began to take hold. At that time, the focus was on defining and documenting the structure and components of an organization's information systems. The primary focus lay at the intersection of IT and information delivery. During this decade, methodologies such as John Zachman's Framework for Enterprise Architecture emerged. Enterprise Architecture has evolved in response to the increasing complexity of IT environments and related business processes.

The Zachman Framework (actually an *ontology*) was published by John Zachman in 1987 [4]. It was the first framework to focus on structuring IT infrastructure and business processes.

During the 1990s, the Zachman Framework became increasingly popular as an Enterprise Architecture methodology. The framework was used by many organizations, including government agencies and other large corporations. The framework underwent several revisions to better meet the changing needs of organizations.

The 1990s marked a period of increasing interest and growth in the field of Enterprise Architecture. More and more organizations recognized the importance of aligning IT with their business goals. In 1995, The Open Group introduced what is now called the TOGAF Standard, which provided a comprehensive approach to Enterprise Architecture.

In the early 2000s, more emphasis was placed on the integration of IT and business strategies, leading to the adoption of Enterprise Architecture as a strategic management discipline. Enterprise Architecture frameworks and methodologies, such as the Zachman Framework and the TOGAF Standard, became more widely accepted and used. In the mid- to late 2000s, Enterprise Architecture evolved to address the complexity of globalized and networked organizations. The focus shifted to a more holistic approach to Enterprise Architecture that included not only IT systems, but also business processes, people, and organizational structures. This broader perspective was necessary to adapt to rapidly changing market dynamics and technological innovations.

The evolution of Enterprise Architecture brought a new level of focus to the strategic importance of IT within organizations. It became a way to align IT infrastructure and business processes with business goals, thereby increasing the value of IT. Enterprise Architecture also offered a way to manage and reduce the complexity of IT environments, thereby reducing the cost and risk of IT projects.

The evolution of today's leading architectural frameworks, starting with the introduction of the Zachman Framework, is shown in Figure 2-1.

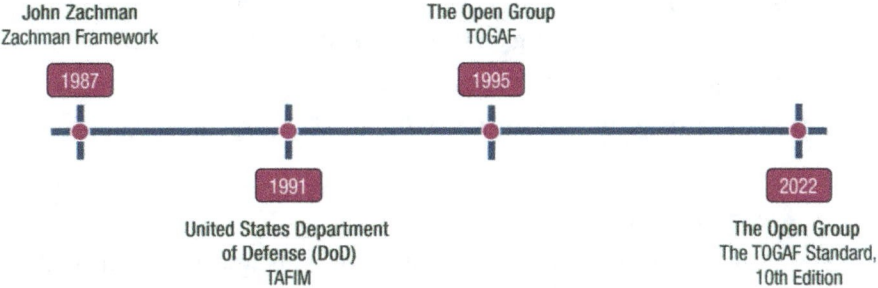

Figure 2-1. *The evolution of today's leading architecture frameworks*

Shortly after 2000, the Zachman Framework was further developed, extended, and adapted to the changing needs of organizations. More attention was given to the relationship between Enterprise Architecture and digital transformation, and new tools and techniques were developed to support Enterprise Architecture modeling. Of course, the Zachman Framework has also been criticized, and the applicability and practical value of the framework and its role in an ever-changing IT environment have been debated.

Throughout the 2010s, Enterprise Architecture became increasingly aligned with other strategic management disciplines, such as business process management, data management, and enterprise security. Enterprise Architecture evolved into an essential tool for guiding digital transformation initiatives, cloud adoption, and agile development practices.

CHAPTER 2 PRELIMINARY NOTES

In recent years, the profession has evolved in response to the increasing significance of digitization, data-driven decision-making, and emerging technologies, including artificial intelligence, the Internet of Things (IoT), and blockchain. Enterprise Architecture has become more adaptive and agile in order to address the dynamic and rapidly changing business landscape.

Today, organizations use this discipline to optimize business processes and IT infrastructure, accelerate digital transformation, and strengthen their competitive position. Enterprise Architecture continues to evolve and adapt to the ever-changing technological environment and business needs. Throughout its history, the discipline has evolved from primarily IT-focused to *strategic and business-centric*.

Further Reading

Additional information about the Zachman Framework in the form of the document "A Framework for Information Systems Architecture" can be found at the following URL:

https://www.academia.edu/30793320/A_framework_for_information_systems_architecture

The development of the Zachman Framework is detailed in a document available at this URL:

https://zachman-feac.com/the-zachman-framework-evolution

2.2. The Evolution of the TOGAF Standard

Today, the TOGAF Standard – the world's leading architecture framework – consists of several components, including an Architecture Development Method (ADM), a set of standards, and a set of tools and techniques. The

TOGAF Standard is designed to help organizations develop a holistic and integrated Enterprise Architecture that is aligned with their business objectives.

Looking at the evolution of the TOGAF Standard over the years, the framework has evolved from an enhanced version of the Technical Architecture Framework for Information Management (TAFIM) framework to a full-fledged Enterprise Architecture framework.

2.2.1. The Early Versions

Back in 1995, the first edition of what is now known as the TOGAF Standard was published under the name of the *Open Architecture Framework*. This name was initially changed to The Open Group Architecture Framework with the same speed with which it was coined. Later, it was decided to use only the original acronym in conjunction with the word *standard*: the TOGAF Standard. The very first version of the framework (see Figure 2-2) served as a *proof of concept*. It was the result of a *proof of need* from a year earlier (1994).

Because The Open Group was committed to an annual release cycle in the early years, TOGAF 2.0 appeared in 1996. The first edition's proof of concept was transformed into a *proof of application*.

In 1997, prompted by the annual release schedule, version 3.0 appeared. The purpose of this release was to reaffirm its relevance to practical architectures. The following years saw the release of TOGAF 4 and 5 in 1998 and 1999, respectively.

The fourth edition of the architecture framework offered the ability to access content via the Internet, which at the time led to a significant increase in the number of users of the Standard.

CHAPTER 2   PRELIMINARY NOTES

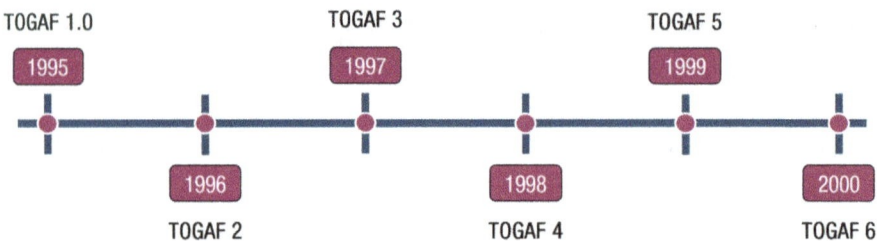

Figure 2-2. *The early versions of the TOGAF Standard throughout the years*

TOGAF 5 was largely rewritten to focus more on the refined and expanded Architecture Development Method.

In 2000, TOGAF 6 was released, with The Open Group doing a lot of work to incorporate building blocks into the framework (see Chapter 7, Section 7.5.2).

2.2.2. The More Recent Versions

When TOGAF 7 was released in 2001, it was given a subtitle. The term *Technical Edition* was added to the name of the framework.

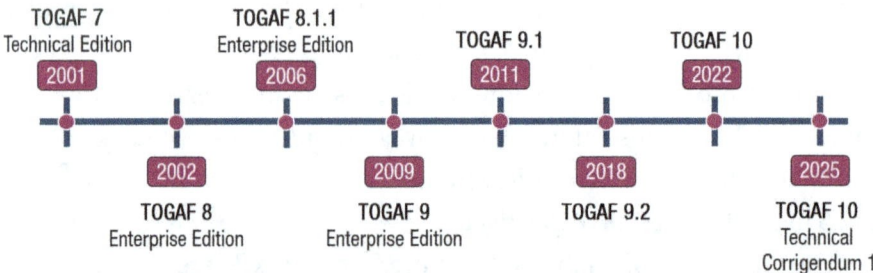

Figure 2-3. *The more recent versions of the TOGAF Standard throughout the years*

CHAPTER 2   PRELIMINARY NOTES

> Given that there are still many architects (and non-architects) who deliberately refer to the TOGAF Standard as an IT framework, it may well be that this idea originated with the seventh edition of TOGAF.

The seventh edition of this well-known framework had been extensively updated (see Figure 2-3). This version provided a more structured approach to Enterprise Architecture, including guidance on architecture development, the Architecture Development Method, and architecture views. The Business Scenarios topic also received a significant update with additional material. Finally, a section comparing the framework to other architectural frameworks had been included. Meanwhile, the TOGAF Framework consisted of a more elaborate Architecture Development Method, a Foundation Architecture, and a Resource Base.

The Foundation Architecture, shown in Figure 2-4, consisted of a Technical Reference Model, a Standards Information Base, and a Building Block Information Base.

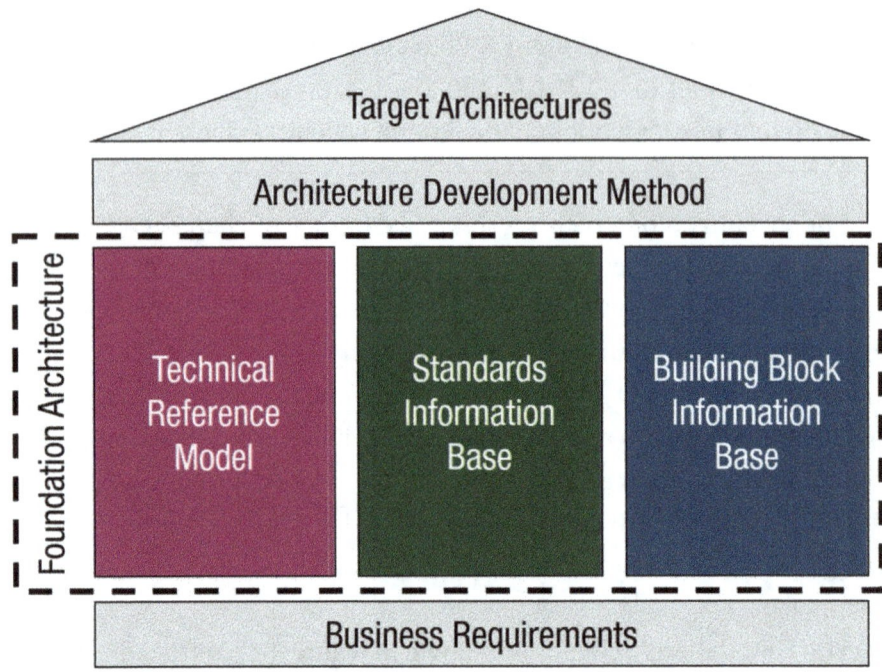

Figure 2-4. *The Foundation Architecture in context of the framework*

The Resource Base included advice on architecture views, architecture governance, business scenarios, Architecture Principles, and additional reference models. Today, the Resource Base can be compared to the TOGAF Library (see Chapter 3, Section 3.4).

The revised version of the Architecture Development Method played an increasingly important role in the development of the framework. Where the very first version of the TOGAF Standard used the TAFIM framework, slowly but surely the contours of the Architecture Development Method as we know it today began to emerge (see Figure 2-5).

CHAPTER 2 PRELIMINARY NOTES

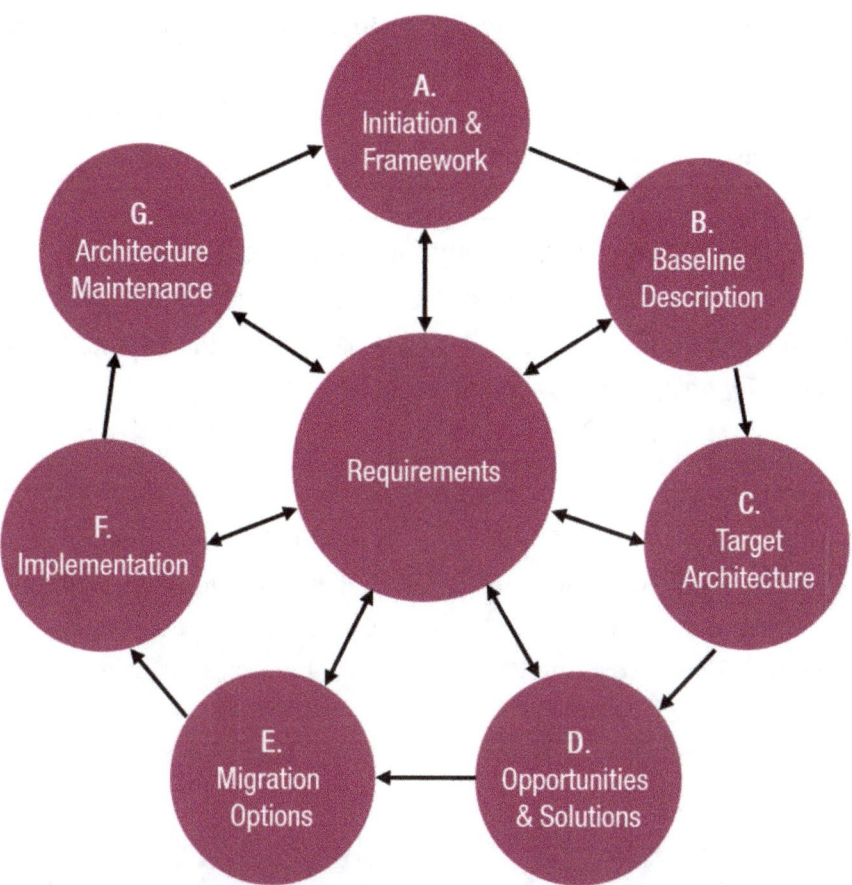

Figure 2-5. *The Architecture Development Method in TOGAF 7 (2001)*

One year after the publication of the seventh edition, another edition was released. TOGAF 8 saw the light of day in 2002. This version expanded on the framework's capabilities for broader Enterprise Architecture applications.

CHAPTER 2   PRELIMINARY NOTES

Remarkably, after a year of working with the Technical Edition subtitle, it was already abandoned. The subtitle *Enterprise Edition* was quickly added, hoping to rapidly wipe out the memory of the Technical Edition. This turned out to be a futile hope.

Today, more than two decades after the release of TOGAF 8, the framework is still often associated with the technical origins with which it began. The question, of course, is whether the addition of the words "*Technical Edition*" actually caused this.

Undoubtedly, it did not help.

In the eighth edition of the TOGAF Standard, the now familiar Preliminary Phase was added to the Architecture Development Method. This is illustrated in Figure 2-6. At the time, this phase was called Preliminary Framework and Principles. The other phases were given the names they still carry today.

CHAPTER 2 PRELIMINARY NOTES

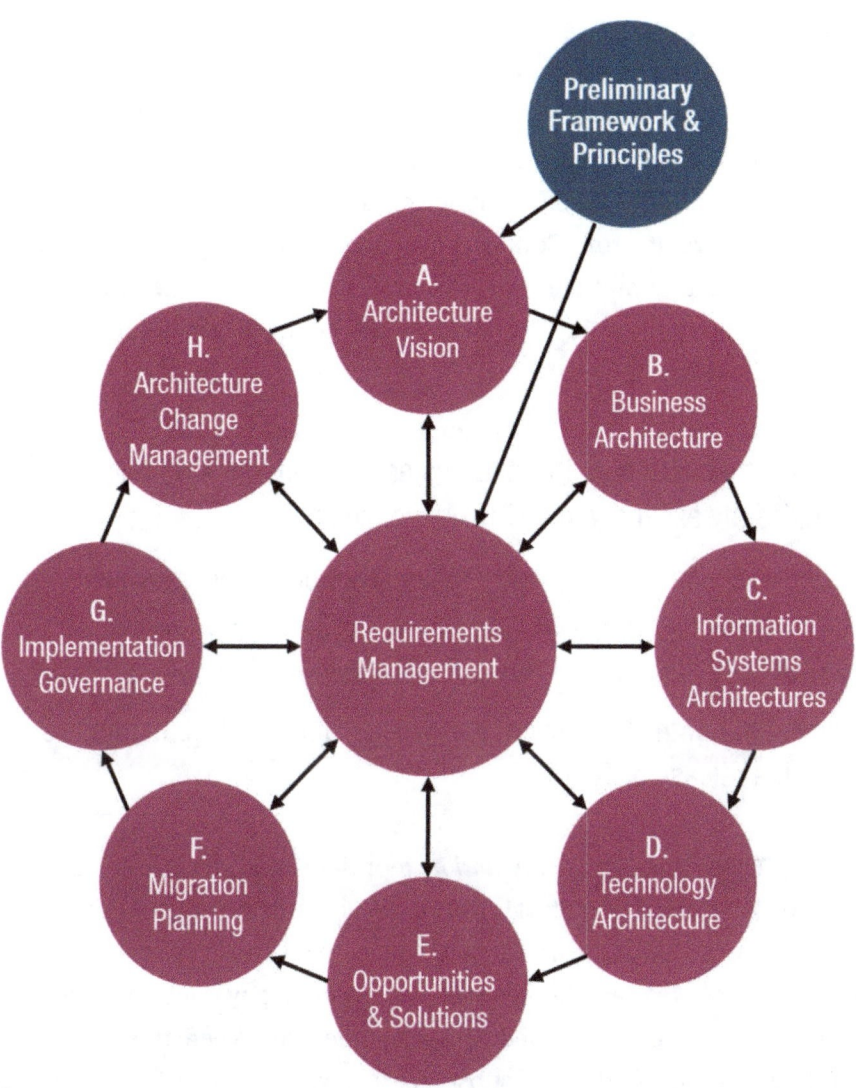

Figure 2-6. *The Architecture Development Method in TOGAF 8 (2002)*

A year before the release of version 8.1.1, TOGAF became a registered trademark of The Open Group. TOGAF 8.1.1 became available in 2006, and this release was also subtitled Enterprise Edition. The last version to carry the suffix was version 9, released in 2009. All versions of the framework released after version 9 had to do without a subtitle.

The transition from TOGAF 8 to TOGAF 9 was considered revolutionary. It brought significant updates, including a modular structure and more detailed guidance on the Architecture Development Method and Architecture Content Framework. Starting with version 9, the Architecture Content Framework was introduced, connecting all the artifacts in the framework. The Architecture Repository also saw the light of day with version 9.

It is interesting to note that the Business Architecture domain was not significantly expanded until TOGAF version 9.2.

Prior to version 9.2, this domain was described very briefly, and architects had to make do with references to other bodies of knowledge.

Version 9.2 of the framework was released in 2018, some nine years after the release of TOGAF 9.

The TOGAF Standard, 10th Edition, represents a significant update to the framework. Released in 2022, it is the first version of the TOGAF Framework to have the word *Standard* added to it. The tenth edition leverages other available frameworks and bodies of knowledge on many fronts. A good example of this is in the area of Business Architecture. Where the TOGAF Standard made an effort to add substance to it with version 9.2, and attempted to incorporate the domain of Business Architecture, the publication of version 10 has again made reference to other bodies of knowledge. For example, the framework now refers to the Business Architecture Body of Knowledge (BIZBOK) Guide [5] - maintained by the Business Architecture Guild - for information on and application of the Business Architecture domain. These references appear in a number of Series Guides.

CHAPTER 2 PRELIMINARY NOTES

A Technical Corrigendum was published on May 15, 2025. This document included updates to the TOGAF Standard, consisting mostly of textual adjustments and further clarifications. One significant change brought about by the Technical Corrigendum was the removal of all arrowheads from the Architecture Development Method (see Figure 2-7). This change was intended to further emphasize the iterative nature of the approach.

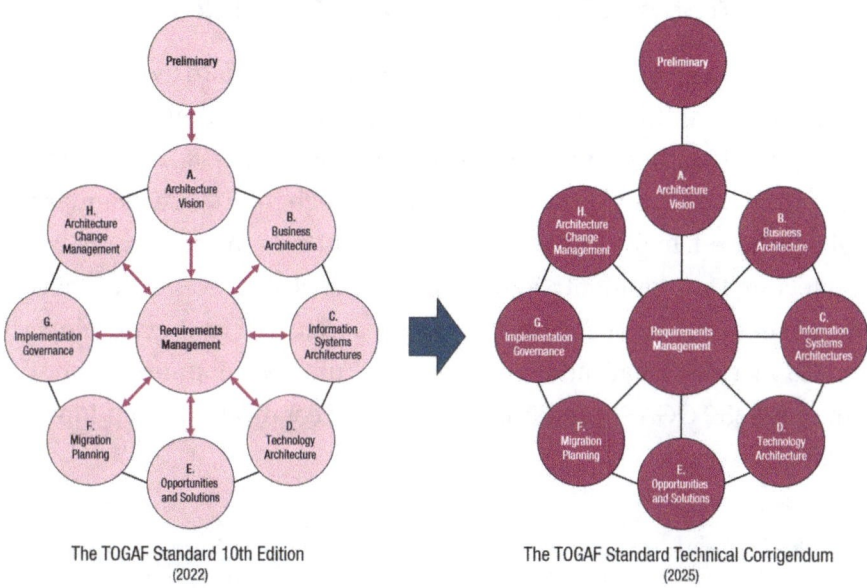

Figure 2-7. *The Architecture Development Method with and without arrowheads*

The TOGAF Standard aims to enhance the flexibility, usability, and accessibility of the framework for Enterprise Architects. It continues to evolve, with newer versions being developed to address modern Enterprise Architecture challenges. The focus has shifted to ensuring alignment between business and IT, incorporating agile methodologies, and supporting digital transformation initiatives.

> **Further Reading**
> Additional information about the TOGAF Standard version 7 can be found at the following URL:
>
> ```
> https://www.opengroup.org/architecture/togaf7-doc/arch/
> ```

2.2.3. Comparison of TAFIM and TOGAF

As noted in the previous paragraphs, the TOGAF Standard owes its origins to the TAFIM framework. A very early version of the Architecture Development Method can be seen on the left of Figure 2-8. This left image comes from the 1996 TAFIM framework on which the TOGAF Standard is originally based. In the TAFIM model, all the phases that now make up the Architecture Development Method are already visible, even if they have different names. The phases have been expanded and further developed over the years. A visualization of the Architecture Development Method from the 2025 TOGAF Standard, 10th Edition Technical Corrigendum 1, is shown on the right.

CHAPTER 2 PRELIMINARY NOTES

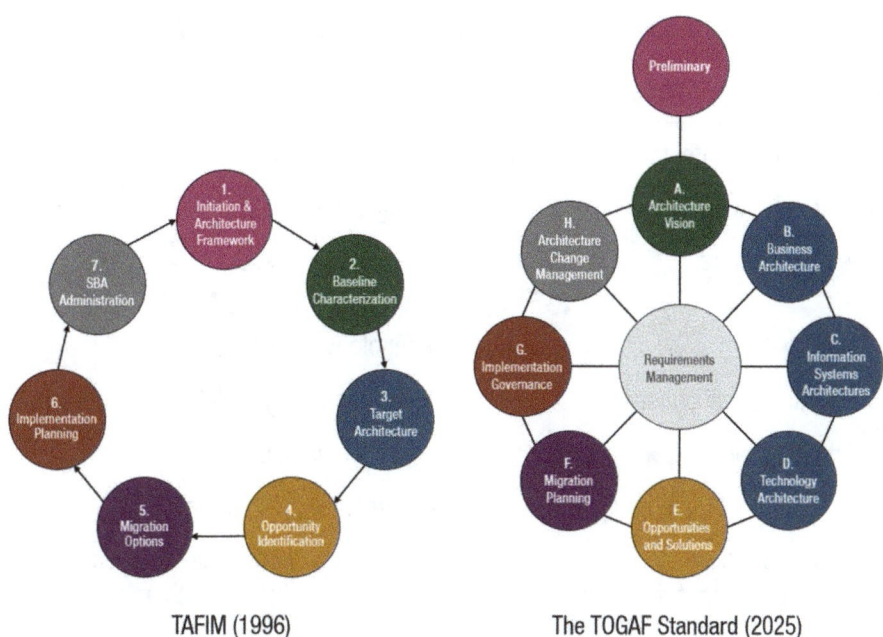

Figure 2-8. Phases of TAFIM and TOGAF ADM mapped

The phases in the TAFIM framework are logically similar to the phases in TOGAF's Architecture Development Method. Table 2-1 illustrates this mapping.

CHAPTER 2   PRELIMINARY NOTES

Table 2-1. Mapping of the TAFIM and TOGAF ADM Phases

TAFIM Framework Phase	TOGAF ADM Phase
Initiation and Architecture Framework	Preliminary Phase
Baseline Characterization	Architecture Vision
Target Architecture	Business Architecture
	Information Systems Architectures
	Technology Architecture
Opportunity Identification	Opportunities and Solutions
Migration Options	Migration Planning
Implementation Planning	Implementation Governance
SBA Administration	Architecture Change Management

It is worth noting that the third phase of the TAFIM framework was changed during the development of the TOGAF Standard. What is called *Target Architecture* in the TAFIM framework has been divided into three separate phases in the Architecture Development Method. These phases are known as Business Architecture, Information Systems Architectures, and Technology Architecture.

Over the years, the TOGAF Standard has always treated these three separate phases as truly distinct phases. However, with the advent of the tenth edition, you can see a further shift toward bringing the three phases more together.

The description of the phases in the framework is now the same for phases B, C, and D and focuses primarily on creating a Baseline and Target Architecture for that particular phase and everything that goes with it.

CHAPTER 2   PRELIMINARY NOTES

This seems to be a first step toward reversing the division that was previously thought to be necessary. Whether this will actually happen in the future remains to be seen.

The TAFIM phase called Implementation Planning is called Implementation Governance in the TOGAF Standard. From this perspective, the focus is on governance of the architecture work rather than pure planning. The part that focuses on planning is housed in the phase that precedes the governance phase, Migration Planning.

The phase in the TAFIM framework called SBA Administration – where SBA stands for Standards-Based Architecture – corresponds to Architecture Change Management on the TOGAF side. This phase manages changes to the architecture. It is designed to keep the architecture current and up to date.

Further Reading
The phases of the TAFIM framework are detailed in the document "Technical Architecture Framework for Information Management. Vol. 4. April 1996". They can be viewed at the URL below:

https://apps.dtic.mil/sti/pdfs/ADA321173.pdf

2.3. Summary

The emergence of architecture frameworks has been the catalyst for the creation and evolution of the TOGAF Standard as we know it today. The profession emerged in the 1960s and 1970s in response to the increasing complexity of IT environments and associated business processes.

CHAPTER 2   PRELIMINARY NOTES

One of the better-known frameworks is the Zachman Framework (actually an ontology). Published in 1987 by John Zachman, it was the first framework to focus on structuring IT infrastructure and business processes. In 1995, The Open Group introduced what is now called the TOGAF Standard, which provides a comprehensive approach to Enterprise Architecture.

The TOGAF Standard is derived from the TAFIM framework. The phases in the TAFIM framework are logically similar to the phases in the TOGAF Architecture Development Method, although they have evolved over the years.

In the mid- to late 2000s, Enterprise Architecture evolved to address the complexity of globalized and networked organizations. The focus shifted to a more holistic approach to Enterprise Architecture that included not only IT systems, but also business processes, people, and organizational structures. The evolution of Enterprise Architecture brought a new level of focus to the strategic importance of IT within organizations.

In the 2010s, Enterprise Architecture became an essential tool for guiding digital transformation initiatives, cloud adoption, and agile development practices. Throughout its history, Enterprise Architecture has evolved from a primarily IT-focused discipline to a strategic and business-centric practice.

Over the past few decades, the TOGAF Standard has matured from an enhanced version of the TAFIM framework to a full-fledged Enterprise Architecture framework, with the tenth edition released in 2022. A Technical Corrigendum was published in 2025 as a follow-up to the latest edition.

CHAPTER 3

The TOGAF Standard

This chapter is part of the PINK and GREEN reading tours. See Figure 1-1 in Chapter 1, Section 1.3, for alternative reading tours.

The TOGAF Standard, 10th Edition, is the first edition to have the word Standard *added to the title. Since The Open Group is an organization dedicated to creating and making available vendor-neutral standards, it is understandable that they chose this addition.*

Over the years, the framework has had several subtitles. Some additions have served the framework better than others, but this latest addition has managed to send a clear message: this is a standard, for Enterprise Architecture.

The TOGAF Standard consists of about six documents that form the basis of the framework. This set of documents is supplemented by additional guidance in the form of Series Guides. The complete set of documents, together with the TOGAF Reference Library, is part of the TOGAF Library.

CHAPTER 3   THE TOGAF STANDARD

3.1. What Is the TOGAF Standard

The TOGAF Standard is an open Enterprise Architecture framework. It is developed and maintained by an organization called The Open Group.

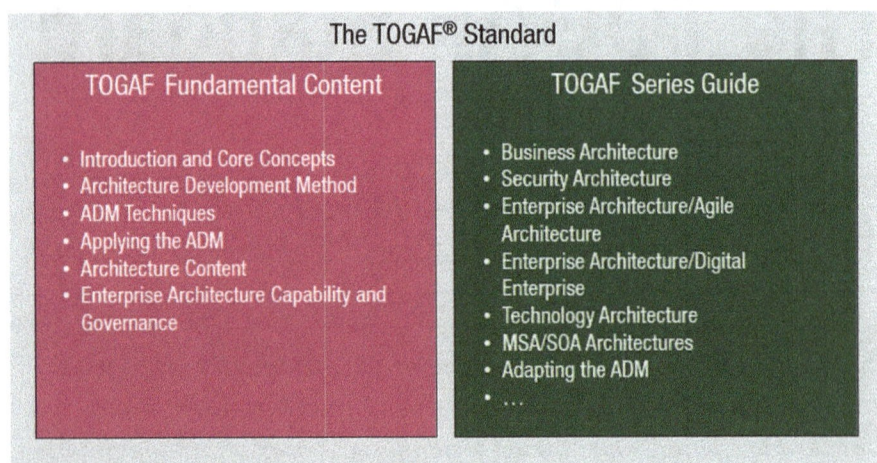

Figure 3-1. *The TOGAF Standard*

The TOGAF Standard describes the general approach to establishing an Enterprise Architecture. It is presented as a series of independent but closely related documents, as shown in Figure 3-1.

The TOGAF Standard is a standard of The Open Group. This organization works closely with customers and suppliers of technology products and services and with consortia and other standards organizations to capture, clarify, and integrate current and emerging requirements, establish standards and policies, and share best practices. Standards ensure openness, interoperability, and consensus [2].

The TOGAF Standard is a fundamental, product- and vendor-independent framework that can be applied to developing any type of architecture in almost any context. The TOGAF Library complements the foundational nature of the framework. The library offers a growing portfolio of practical guidance on applying the TOGAF Framework to specific contexts.

The TOGAF Standard's structure reflects that of an Architecture Capability within an organization. Figure 3-2 shows the relationship between the TOGAF Standard and the Architecture Capability.

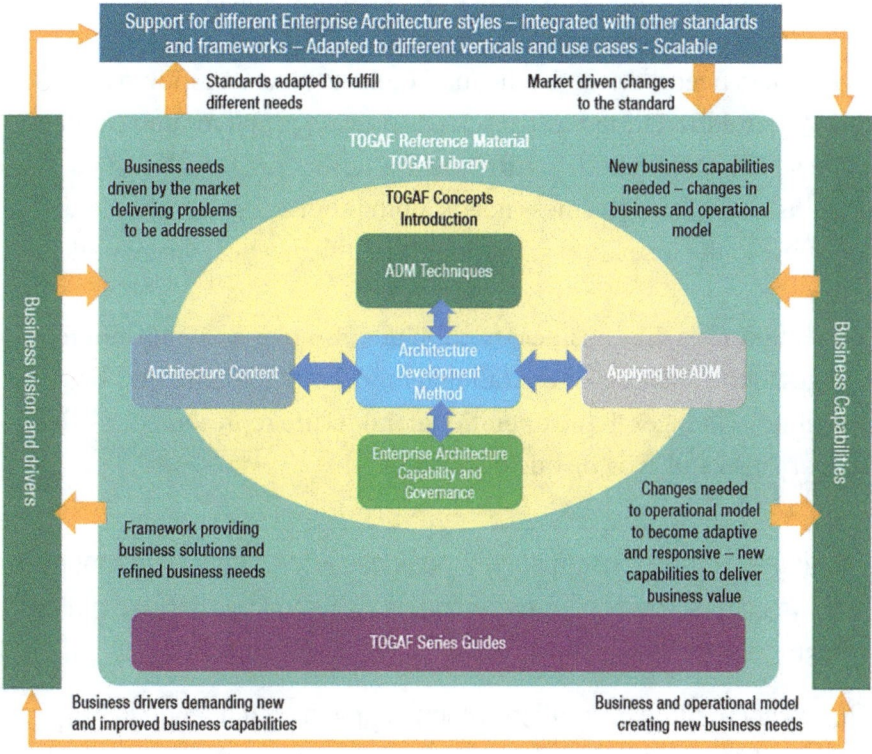

Figure 3-2. *Structure of the TOGAF Standard*

CHAPTER 3   THE TOGAF STANDARD

Looking at Figure 3-2, it is clear that the TOGAF Reference Material and the TOGAF Library are at the center. Inside are the TOGAF Concepts, which are shaped by the Core Content of the TOGAF Standard. At the center is the Architecture Development Method, the heart of the framework. It is described in more detail in Chapter 12. The Architecture Development Method is supported and complemented from all sides by the Architecture Content, techniques, and practical applications of the method. The Architecture Capability also plays an important role. This capability is described in more detail in Chapter 9.

Part of the TOGAF Library that provides additional guidance are the Series Guides (see Section 3.3).

Business needs, driven by the marketplace, serve as input to the TOGAF Standard. These business needs provide problems and challenges that need to be addressed. In turn, the framework provides business solutions and refined business needs. Both input and output streams contribute to the formation and updating of the business vision and drivers.

In addition to this, business capabilities also play an important role. Changes in the business or operational model often require new business capabilities. Changes in the operational model are represented in the form of capabilities and provide business value.

A fitting example of when such a business and/or operating model change occurs is when an organization decides to undertake a digital transformation.

During such a transformation, which is driven by a change in the organization's strategy and results in changes to the business and operating model, the need to create new business capabilities often arises.

Consider, for example, the transition from old-fashioned video stores to its modern version in the form of a streaming service such as Netflix (see Figure 3-3). This transformation from a company that offered physical products in a physical store to an organization that offers fully digital services in the form of streaming services has a significant impact on the business and operating model. This in turn leads to the creation of entirely new capabilities.

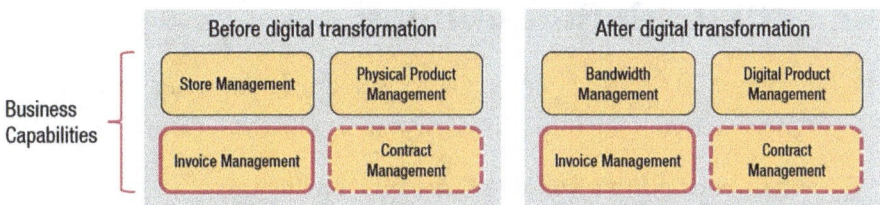

Figure 3-3. Simplified example of changed capabilities after digital transformation

Whereas before the digital transformation the video store had to deal with the capabilities Store Management and Physical Product Management, after the digital transformation, this has become Bandwidth Management and Digital Product Management.

The Contract Management capability (dashed line) appears to have remained the same, but in practice, before the transformation, it would have consisted mainly of contract agreements with DVD and Blu-ray Disc suppliers. After the transformation, Contract Management includes agreements with movie and series production studios.

Invoice Management (thick solid line) will be little different after digital transformation than before.

CHAPTER 3   THE TOGAF STANDARD

The entire TOGAF Standard supports different architecture styles and can be integrated with other standards and frameworks. The framework can be easily adapted to different verticals (industries) and a variety of use cases. The TOGAF Standard is also highly scalable.

Reduced to its essentials, the TOGAF Standard consists of the TOGAF Fundamental Content and a collection of TOGAF Series Guides that provide practical guidance on the application of the framework.

3.2. Fundamental Content

The Fundamental Content of the TOGAF Standard comprises six pivotal documents, each delving into distinct aspects of the framework:

- **Introduction and Core Concepts:** Laying the foundational understanding of the TOGAF Standard

- **Architecture Development Method (ADM):** Providing a step-by-step approach to developing Enterprise Architecture

- **ADM Techniques:** Offering tools and methodologies to enhance the Architecture Development Method process

- **Applying the ADM:** Guidance on practical implementation of the Architecture Development Method in various scenarios

- **Architecture Content:** Defining the artifacts and deliverables within the architecture

- **EA Capability and Governance:** Focusing on establishing and maintaining effective architectural practices and oversight

As said, each of the six documents covers a distinct part of the TOGAF Standard. However, they are all part of the core of the framework.

Introduction and Core Concepts: Describes, as the title suggests, the core concepts of the framework. The document consists primarily of concept terms with an accompanying explanation.

This includes the definition of Enterprise Architecture and the different architecture domains that the TOGAF Standard distinguishes. The Architecture Development Method is also briefly explained, but is discussed in much more detail in the documents "Architecture Development Method (ADM)", "ADM Techniques", and "Applying the ADM". The "Introduction and Core Concepts" document also consists of an explanation and use of deliverables, artifacts, and building blocks, as well as a brief mention of the Enterprise, Architecture and Solutions Continua, the Architecture Repository, and the Content Framework. The "Introduction and Core Concepts" document introduces a number of important topics but refers to other documents in the Fundamental Content for more detailed explanations.

Architecture Development Method (ADM): Describes the Architecture Development Method cycle and how to adapt the Architecture Development Method to architecture issues. The document also discusses issues such as the scope of an architecture problem and the integration of architecture.

The Architecture Development Method offers a structured approach to crafting and overseeing the lifecycle of an Enterprise Architecture. It is the heart of the TOGAF Standard. By integrating various components of the TOGAF Standard alongside external architectural assets, the Architecture Development Method is tailored to meet organizational needs. It's flexible, iterative, and designed to be adaptable, ensuring that architects can address specific business challenges effectively while aligning architecture with the broader objectives of the organization.

ADM Techniques: This document is a collection of techniques that are available for applying the Architecture Development Method. This includes the development of a set of general rules and guidelines (for Architecture Principles, see Chapter 14, Section 14.2.4) for the architecture being developed.

The document also describes *Stakeholder Management*, an important discipline that successful architecture practitioners can use to gain support for their projects. Another topic that is described is called *Architecture Alternatives and Trade-Offs*. This is a technique for identifying alternative target architectures and making trade-offs between the alternatives. It is about identifying more than one possible way forward and highlighting the pros and cons of each potential solution.

Applying the ADM: The Architecture Development Method is easily adaptable to the different scenarios an architect faces. The tailorability of the method is ideal for applying to different process styles (think of the iterative nature of the Architecture Development Method as described in Chapter 13.3) or for developing different specialized types of architectures (such as Security Architecture).

The document "Applying the ADM" provides guidelines for implementing these adaptations to make it applicable to different scenarios, challenges, and architectures. Chapter 13 goes into more detail about the various use cases for applying the Architecture Development Method.

Architecture Content: The Architecture Content Framework defines the *building blocks* and *architectural artifacts* used to describe the Enterprise Architecture.

Building blocks are a key element used to build the Enterprise Architecture. They represent the *core components*, such as business processes, information concepts, data entities, and application systems. Building blocks are crucial to shaping an organization's architecture.

CHAPTER 3 THE TOGAF STANDARD

Architectural artifacts are the deliverables produced during the phases of the Architecture Development Method, such as catalogs, matrices, diagrams, and maps that *describe* the architecture.

In addition to the artifacts, the Fundamental Content also includes the *Enterprise, Architecture,* and *Solutions Continua* (*classification systems* used to categorize architectural assets).

EA Capability and Governance: The Fundamental Content is further complemented by the *Architecture Capability Framework* and a *Governance Framework.*

The Architecture Capability Framework outlines the organizational structure, roles, and processes needed to establish and operate an effective Enterprise Architecture Capability within an organization. The Governance Framework is primarily focused on mapping the design of the governance of architecture work. The Governance Framework from the TOGAF Standard serves as a starting point and can be adapted to fit the organization.

A section that provides additional guidelines, techniques, and reference materials to support the effective use of the framework is called the TOGAF Library (see Section 3.4). It provides guidance on how to tailor the TOGAF Standard to an organization's specific needs and requirements.

Overall, the Fundamental Content of the TOGAF Standard serves as a comprehensive guide and toolbox for organizations to develop, manage, and evolve their Enterprise Architecture, promote alignment between IT and business strategies, and foster efficient and effective IT systems.

The reason the TOGAF Standard is divided into independent documents is to allow different areas of specialization to be considered in detail and potentially addressed in isolation. While the various documents within the TOGAF Framework function as a cohesive whole, they don't have to be adopted all at once. In fact, it's perfectly fine to cherry-pick the ones that suit the organization's needs. For instance, it would be totally fine to implement the Architecture Development Method but skip the

CHAPTER 3   THE TOGAF STANDARD

Architecture Capability components. That's the beauty of the TOGAF Standard – it's flexible by design, which is particularly useful in these scenarios:

- If the organization is *just starting* with the TOGAF Standard, it makes sense to gradually adopt its concepts. The organization could focus on the core areas first, like the Architecture Development Method, and explore the other sections later.
- If there *already is* some kind of architecture framework in place, it can blend what's already working with parts of the TOGAF Standard to add extra value.

This modular approach allows an organization to align the framework with their unique goals without feeling overwhelmed by having to adopt everything at once.

In addition to the core framework content covered by the six documents described above, the TOGAF Standard also provides guidance to address specific concerns and use cases through the TOGAF Series Guides.

3.3. Series Guides

The TOGAF Series Guides were developed in response to the need for better guidance on developing useful Enterprise Architecture. Stakeholders want useful Enterprise Architecture guidance to support their decisions and guide the implementation of necessary organizational changes.

The TOGAF Series Guides cover a range of topics, from general guidance on how to set up an Enterprise Architecture Capability, to domain-specific material for Business and Security Architecture, to using agile methods and agile software development. An approach to developing

Enterprise Architecture following the Architecture Development Method provides guidance on using the framework to develop, maintain, and use an Enterprise Architecture. It is a companion to the Fundamental Content and brings the concepts and generic constructs to life. Other guides offer insights into using the TOGAF Standard in the digital enterprise, emphasizing how to establish and enhance an Enterprise Architecture Capability that is aligned with the organization and what the Enterprise Architecture team is expected to support.

> Not all the TOGAF Series Guides will be relevant in every situation. However, Enterprise Architects who are planning the deployment of the TOGAF Standard should be aware of the guidance available [2].

The Series Guides are set to evolve more quickly than the Core Content, responding to emerging industry trends and market demands. This dynamic content is driven by continuous activities within The Open Group Architecture Forum, operating through a streamlined, incremental delivery process. This ensures that the evolving needs of the industry are met efficiently and consistently, reflecting the Forum's ongoing commitment to flexibility and adaptability in delivering valuable resources.

Examples of the areas covered by this guidance material are:

- Domain-specific guidance, such as integrating risk and security into an Enterprise Architecture.
- A foundation for understanding and using business models.
- An explanation of what business capabilities are and how to use them to improve business analysis and planning.

- Details on how business scenarios can develop resonant business requirements and how they support and enable the organization to achieve its business goals.

In addition, the Series Guides describe how to apply the Architecture Development Method in an agile delivery environment by breaking an architecture development project into small time-boxed increments and applying common agile techniques (see Chapter 18). Finally, it provides reference models, techniques for assessing and quantifying an organization's Enterprise Architecture maturity, and documents that provide guidance on using project management techniques to manage the development of the Enterprise Architecture.

Further Reading

The full TOGAF Standard is available online at no additional cost (a free Open Group account is required):

https://pubs.opengroup.org/togaf-standard/

The complete set of TOGAF Series Guides can be found at the following URL:

http://www.opengroup.org/library/guides/togaf/togaf-series-guides

The Open Group Architecture Forum can be accessed by following this link:

https://www.opengroup.org/architecture-forum

3.4. The TOGAF Library

Accompanying the TOGAF Standard is a broad portfolio of guidance material, known as the *TOGAF Library*, to support the practical application of the TOGAF approach. The TOGAF Library is a portfolio of additional guidance material, which supports the practical application of the TOGAF Standard. It contains guidelines, templates, patterns, and other forms of reference material to accelerate the creation of new architectures for the organization. The TOGAF Documentation Set (as can be seen in Figure 3-4) is part of the TOGAF Library.

The TOGAF Library, in turn, is part of The Open Group Library, which contains all kinds of Enterprise Architecture-related content.

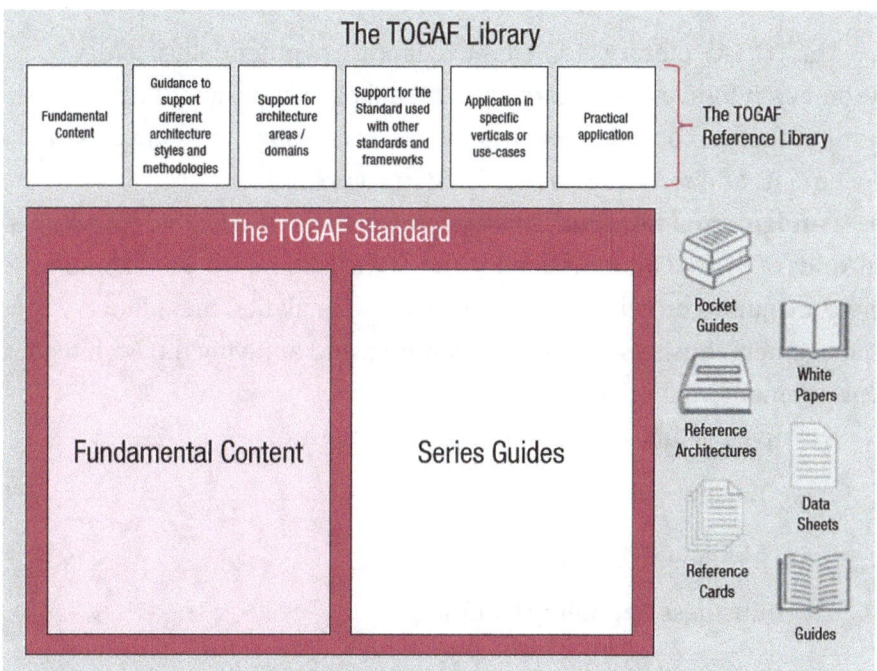

Figure 3-4. *The TOGAF Library*

CHAPTER 3   THE TOGAF STANDARD

The TOGAF Library supports one of the classes of information in the Architecture Repository, the Reference Library.

The Reference Library provides guidelines, templates, patterns, and other forms of reference material that can be leveraged in order to accelerate the creation of new architectures for the enterprise. Figure 3-5 illustrates the various reference materials.

Figure 3-5. The TOGAF Reference Library

The TOGAF Library is a Reference Library of potentially useful resources. It follows a categorization model based on capabilities and features that can be delivered into the market through different sets of documents and resources, as depicted in Figure 3-5.

Fundamental Content: This category of the TOGAF Library contains all kinds of Series Guides related to the TOGAF Standard. For example, there are guides on business intelligence and analytics, metadata management, business capability planning, and applying the Architecture Development Method using agile sprints.

Examples of this domain-specific guidance are:

- Business Intelligence & Analytics
- Metadata Management
- Business Capability Planning
- Applying the TOGAF® ADM using Agile Sprints

Guidance to Support Different Architecture Styles and Methodologies: The guidance section on different styles and methodologies offers information on architecture project management and establishing and evolving an Enterprise Architecture Capability. Guidance on how to use the TOGAF Framework to define and govern Microservices Architecture (MSA) is also available.

Support for different styles and methodologies is provided in part by:

- The TOGAF® Leader's Guide to Establishing and Evolving an EA Capability
- Microservices Architecture (MSA)

Support for Architecture Areas/Domains: This category contains valuable information on topics such as sustainable information systems, architecture maturity models, and business models. A few examples of support for architecture areas are:

- Environmentally Sustainable Information Systems
- Architecture Maturity Models
- Business Models

Support for the Standard Used with Other Standards and Frameworks: For example, guidance on using the Architecture Content Framework of the TOGAF Standard in combination with the ArchiMate standard can be found in this category of the TOGAF Library. Another interesting guide that can be found here is the document that provides information on the integration of the TOGAF Standard and SABSA.

Application in Specific Verticals or Use Cases: This is one of the lesser categories when it comes to supporting resources. One of the guides that stands out in this category is the "Cloud Ecosystem Reference Model".

Practical Application: One resource that may be useful is the "Template Deliverables" document set. This set of documents contains a number of useful templates for performing a Capability Assessment, drafting a Communications Plan, or creating an Architecture Roadmap.

All of the content can be downloaded from the TOGAF Library at no additional cost.

Further Reading

The TOGAF Library is a publicly accessible resource located at the following URL:

http://www.opengroup.org/togaf-library

The Open Group also offers a wide range of support materials in video format. These can be found at the following URL:

https://www.youtube.com/@theopengroup/videos

3.5. Key Advantages of the TOGAF Standard

In today's rapidly evolving business landscape, organizations face the constant challenge of aligning their operations with technological advancements to maintain competitiveness. Architecture frameworks, such as the TOGAF Standard, have emerged as essential tools in this endeavor.

The architecture framework described in this book provides a comprehensive approach to designing, planning, implementing, and governing Enterprise Architectures. Its tailorability and thorough coverage of architecture domains and segments make it an invaluable asset for organizations seeking to enhance efficiency, agility, and strategic alignment.

The key advantages of using the TOGAF Standard are as follows:

Comprehensive Coverage of Architectural Domains: The TOGAF Standard addresses four primary architecture domains: Business, Data, Application, and Technology. This holistic approach ensures that all facets of an organization's architecture are considered, promoting *coherence* and *integration* across various functions.

- **Business Architecture:** Defines the business strategy, governance, organization, and key business capabilities. By articulating these elements, organizations can ensure that their IT systems support and enhance business objectives.

- **Data Architecture:** Describes the structure of an organization's logical and physical data assets and data management resources. A well-defined Data Architecture facilitates effective data governance and utilization, enabling informed decision-making.

- **Application Architecture:** Provides a blueprint for individual systems to be deployed, their interactions, and their relationships to core business capabilities. This ensures that applications are aligned with business needs and can adapt to changing requirements.

- **Technology Architecture:** Details the hardware, software, and network infrastructure required to support important applications. A robust Technology Architecture underpins the reliable and efficient operation of IT services.

By encompassing these domains, the TOGAF Standard facilitates a unified approach to Enterprise Architecture, ensuring that all components work synergistically to achieve organizational goals and objectives.

Adaptability and Tailoring: One of the standout features of the TOGAF Standard is its emphasis on tailorability. The framework encourages organizations to tailor its Architecture Development Method to fit specific needs and contexts. This flexibility allows organizations of varying sizes and industries to apply the TOGAF methodology effectively, ensuring relevance and practicality in diverse scenarios.

CHAPTER 3   THE TOGAF STANDARD

> The TOGAF Framework recommends that the ADM be adapted to meet the needs of the enterprise and to support different architecture styles [2].

The Architecture Development Method, central to the TOGAF Framework, is designed to be iterative (see Chapter 13, Section 13.3) and customizable (see Chapter 17). Organizations can adjust the Architecture Development Method to accommodate specific project requirements, resource constraints, and strategic objectives. This tailored approach ensures that the architectural process remains aligned with business priorities and can respond to emerging challenges and opportunities.

Enhanced Business and IT Alignment: Achieving alignment between business objectives and IT capabilities is a perennial challenge for many organizations. The TOGAF Standard provides structured steps to bridge this gap, ensuring that IT initiatives support and drive business strategies.

For example, by clearly defining business processes and mapping them to other entities, organizations can identify redundancies, gaps, and opportunities for improvement. This alignment enhances operational efficiency and ensures that technological investments deliver actual and tangible business value.

Moreover, the TOGAF Standard's emphasis on stakeholder engagement (see Chapter 14, Section 14.3) ensures that both business and IT perspectives are considered throughout the architectural process. This collaborative approach fosters mutual understanding and commitment to shared goals, reducing the risk of misalignment and project failure.

Facilitation of Digital Transformation: In the context of digital transformation (see Chapter 18), the TOGAF Standard serves as a critical enabler. A study by O'Higgins [6] highlights that effective Business Architecture practices significantly improve business alignment, efficiency, service delivery, and strategic outcomes, leading to successful digital transformation.

The TOGAF Standard's comprehensive approach allows organizations to assess their current state, envision a desired future state, and develop a roadmap to transition between the two. This structured methodology ensures that digital transformation initiatives are strategically planned and executed, minimizing disruption and maximizing value realization.

Support for Process Innovation: Dynamic Enterprise Architecture Capabilities, as facilitated by the TOGAF Standard, positively influence an organization's process innovation and business and IT alignment.

By providing a clear understanding of existing processes and systems, the TOGAF Framework enables organizations to identify areas ripe for innovation. This insight supports the development of new processes and the optimization of existing ones, driving efficiency and competitive advantage.

Common Language and Communication: The TOGAF Standard offers a common vocabulary and set of concepts that facilitate effective communication among stakeholders. This shared language reduces misunderstandings and ensures that all parties have a clear understanding of architectural objectives, processes, and outcomes.

Effective communication is crucial for the successful implementation of Enterprise Architecture initiatives. The TOGAF Framework's standardized terminology and documentation practices ensure that information is conveyed clearly and consistently, supporting collaboration and informed (strategic) decision-making.

Scalability and Resource Optimization: The modular nature of the TOGAF Standard allows organizations to apply its principles at various scales, from small projects to organization-wide initiatives. This scalability ensures that resources are utilized efficiently and architectural efforts are commensurate with organizational needs and capacities.

By providing a structured approach to architecture development, the TOGAF Standard helps organizations avoid the pitfalls of ad hoc planning and implementation. This structured approach ensures that resources are allocated effectively, risks are managed proactively, and outcomes are aligned with strategic objectives.

CHAPTER 3 THE TOGAF STANDARD

The TOGAF Standard stands as a robust and versatile framework for Enterprise Architecture, offering comprehensive coverage of architectural domains, tailorability, and a structured methodology for aligning business and IT. It enables this alignment with the other architecture domains and segments. Its emphasis on stakeholder engagement, process innovation, and effective communication further enhances its value to organizations.

By adopting the TOGAF Standard, organizations can navigate the complexities of modern business environments, drive digital transformation, and achieve strategic objectives with greater efficiency and effectiveness. Its proven methodologies and adaptable nature make the TOGAF Standard an invaluable tool for organizations seeking to harness the full potential of their architectural capabilities.

3.6. Summary

The TOGAF Standard describes the generally applicable approach to establishing an Enterprise Architecture. It is an open framework for Enterprise Architecture and a foundational and product- and vendor-independent framework, meaning that it is applicable to the development of any type of architecture in virtually any context.

The framework consists of the *Fundamental Content* and a collection of *Series Guides*. The Fundamental Content consists of six documents, each of which covers a different part of the TOGAF standard, but all of which belong to the core of the framework. The Series Guides provide practical guidance in the application of the TOGAF Standard. The structure of the TOGAF Standard reflects the structure and content of an Architecture Capability within an organization.

The Fundamental Content comprises six pivotal documents, each delving into distinct facets of the framework:

- **Introduction and Core Concepts:** Laying the foundational understanding of the TOGAF Standard.

- **Architecture Development Method (ADM):** Providing a step-by-step approach to developing Enterprise Architecture.

- **ADM Techniques:** Offering tools and methodologies to enhance the Architecture Development Method process.

- **Applying the ADM:** Guidance on practical implementation of the Architecture Development Method in various scenarios.

- **Architecture Content:** Defining the artifacts and deliverables within the architecture.

- **EA Capability and Governance:** Focusing on establishing and maintaining effective architectural practices and oversight.

This segmentation into independent documents facilitates in-depth exploration of specialized topics, allowing practitioners to address specific areas of interest with precision.

Beyond its foundational texts, the TOGAF Standard extends its utility through Series Guides, which offer tailored advice on unique concerns and use cases. These guides serve as practical companions, aiding organizations in the nuanced application of the framework.

Complementing these resources is the TOGAF Library, an integral part of The Open Group Library. This repository is a treasure trove of materials focused on Enterprise Architecture, enriching the practical application of the TOGAF Standard. It supports organizations in customizing the framework to align with their specific needs, ensuring that the TOGAF Standard remains a dynamic tool in the architect's arsenal.

CHAPTER 3   THE TOGAF STANDARD

The TOGAF Standard is not merely a framework but a living, evolving guide that empowers organizations to craft robust and flexible architectures, driving strategic objectives and fostering innovation.

Using an architecture framework allows an organization to take advantage of existing standards and methodologies, thus avoiding the need to reinvent the proverbial wheel. Architecture frameworks - the TOGAF Standard included - must be tailored to the specific needs of an organization. Because every organization is unique, tailoring a framework is more of a necessity than an option.

By selecting those parts of an architecture framework that add value, an organization can take advantage of the tools that such a framework provides. Using an architecture framework – and in this specific case, the TOGAF Standard – gives an organization the ability to apply the approach and methodology to its own specific situation. In terms of business value, this saves time, money, and resources.

CHAPTER 4

Architecture Definition

 This chapter is part of the PINK and GREEN reading tours. See Figure 1-1 in Chapter 1, Section 1.3, for alternative reading tours.

Definitions are of the utmost importance in creating clarity about what is being talked about or what is being dealt with. A definition provides an unambiguous description of the subject, leaving no room for interpretation. However, when this freedom of interpretation is provided, it can lead to different perceptions of the subject, with all the consequences that entails.

Looking at the field of Enterprise Architecture, it is important to have an unambiguous definition. In this chapter, the many definitions of Enterprise Architecture are presented. It is also pointed out that some essential parts are missing.

CHAPTER 4   ARCHITECTURE DEFINITION

4.1. What Is Enterprise Architecture

One of the topics that has been debated over the years is what exactly is meant by architecture (or Enterprise Architecture, for that matter). In the current version of the TOGAF Standard, the following definition is given:

> The fundamental concepts or properties of an entity in its environment and governing principles for the realization and evolution of this entity and its related lifecycle processes [7].

With the publication of the Technical Corrigendum 1 (2025), the TOGAF Standard currently adopts the ISO/IEC/IEEE 42010:2022 [7] standard for defining Enterprise Architecture. The given definition still leaves room for interpretation. To fill this gap, the framework has added additional text to the above, a sort of second definition:

> The structure of components, their interrelationships, and the principles and guidelines governing their design and evolution over time [2].

Both definitions - or rather the combination of the two definitions - indicate that architecture defines the characteristics of a system and that the system is surrounded by frameworks that oversee its development. According to the TOGAF Standard, the word *system* can mean several things and is not directly related to an IT system. The word *system* can just as easily refer to an organization (an enterprise) or be even larger in nature. On the other hand, it can just as easily refer to a project. So there is plenty of room for interpretation.

CHAPTER 4 ARCHITECTURE DEFINITION

Chris Greenslade, who was Chairman of The Open Group in 2003, gave a presentation on TOGAF 7, including a look ahead to version 8. There were some interesting things mentioned in the presentation that can be considered progressive, especially considering the time they were in at that point. More about this is described in Chapter 2, Section 2.2.

The presentation included a definition of what was understood to be architecture in the year 2003:

"Conceptually an (IT) Architecture is the fundamental organization of a system, embodied in its components, their relationships to each other and the environment, and the principles governing its design and evolution.

Practically it is represented in Architectural Descriptions from the viewpoints of the Stakeholders" [8].

What is surprising about today's definition of Enterprise Architecture in the TOGAF Standard is that it is missing an important component: its relation to the achievement of business objectives. Enterprise Architecture must ensure that these objectives and their intended results are achieved, and it must shape and organize itself to make this possible. In doing so, Enterprise Architecture must ensure consistency and coherence between concepts (from organizational structures to information systems and from capabilities to applications).

CHAPTER 4   ARCHITECTURE DEFINITION

> Enterprise Architecture is perhaps best described as a framework for understanding and managing the overall structure and strategy of an organization.
>
> It is about creating a holistic view of the organization's activities, including its business processes, information systems, and technology infrastructure, and maintaining the coherence between them.
>
> The purpose of Enterprise Architecture is to align these various elements with the goals and objectives of the organization and to ensure that the elements work together effectively and efficiently.
>
> Enterprise Architecture focuses on identifying and resolving inconsistencies in business operations and enables planning for future growth and development.

Definitions are very important because they provide clarity about what is being talked about or what is being dealt with. A definition provides an unambiguous description of the subject, thus avoiding confusion.

Further Reading
Chris Greenslade's 2003 presentation can be found at this URL:

https://www.opengroup.org/architecture/0307bos/presents/greenslade_togaf.pdf

4.2. Confusion Over Definitions

Enterprise Architecture is still often confused with IT Architecture. As was made clear in Chapter 2, Section 2.2, Enterprise Architecture has its origins in the technology corner. Perhaps part of the problem is that Enterprise Architecture is still lumped together with IT Architecture. In reality, the two are as different as night and day.

Enterprise Architecture is first and foremost about enabling an organization to achieve its goals and objectives, thereby realizing the organization's strategy. It is used to address stakeholder concerns and answer any questions they may have regarding the organization's portfolio or specific projects.

IT Architecture, on the other hand, is far more technical in nature compared with Enterprise Architecture. Therefore, the two are miles apart from being the same. There is absolutely *nothing technical* to Enterprise Architecture. The diagrams and matrices used in architecture efforts are often being referred to as being technical in nature. However, when used in an Enterprise Architecture practice, they answer stakeholders' questions and address their concerns. They do not illustrate technical solutions.

Enterprise Architecture oversees all the architecture domains (of which Technology Architecture is one) and plays a coordinating role. It leaves the execution of Technology Architecture to the IT Architects.

As such, Enterprise Architecture is considered a strategic business management tool, not a technical instrument [9].

The definition of Enterprise Architecture used by the TOGAF Standard is only one of many that exist. No single definition can be considered the ultimate one. Perhaps that is why even the TOGAF Standard uses more than one definition of Enterprise Architecture.

CHAPTER 4   ARCHITECTURE DEFINITION

4.3. Multiple Definitions

As mentioned earlier, the definition shown at the beginning of this chapter is the one included in the Core Concepts Guide of the TOGAF Standard. The definition is based on the ISO/IEC/IEEE 42010:2022 standard, which is maintained and updated by ISO.

Chapter 3 attempted to paint a picture of the composition of the TOGAF Standard. The section called Fundamental Content was explained, as well as the additional Series Guides. One of these Series Guides, the Leader's Guide to be precise, describes a second definition of Enterprise Architecture:

> The process of translating business vision and strategy into effective enterprise change by creating, communicating, and improving the key principles and models that describe the enterprise's future state and enable its evolution [3].

This definition was originally provided by Gartner [10] and was created in 2008, *years* before the 2022 ISO definition. Given the speed with which the profession evolves (and has evolved) over the years, it's safe to say that using a definition drawn up in 2008 in an architecture framework updated in 2025 might raise a few eyebrows. Not to mention the fact that there is a real chance that this definition is outdated and superseded.

Figure 4-1 shows the collection of all Enterprise Architecture definitions used within the TOGAF Standard or referenced directly or indirectly through the use of other bodies of knowledge.

CHAPTER 4   ARCHITECTURE DEFINITION

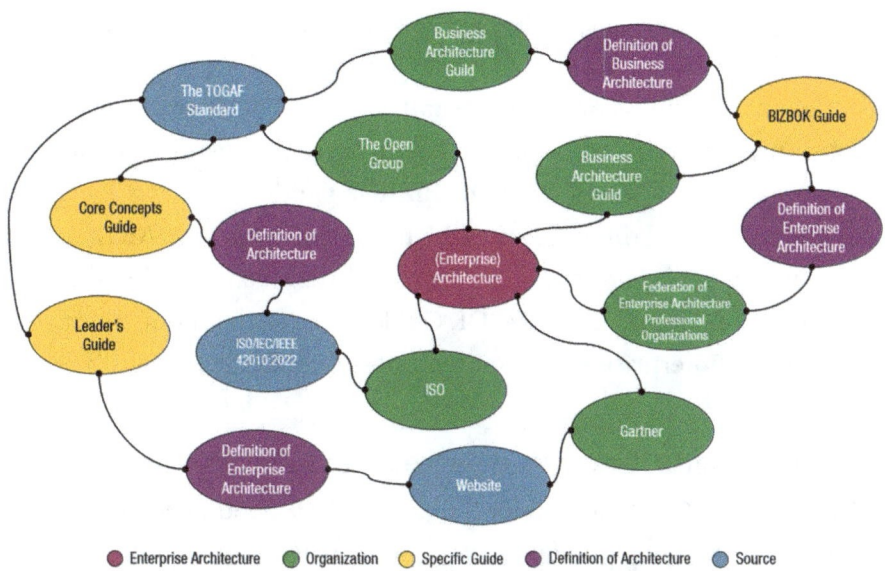

Figure 4-1. *The many definitions of Enterprise Architecture*

Figure 4-1 can be read as follows. At the center is Enterprise Architecture as the primary object to be defined. Immediately around it are The Open Group, the Business Architecture Guild, the Federation of Enterprise Architecture Professional Organizations (FEAPO), Gartner, and ISO. All are organizations that either provide their own definition of Enterprise Architecture or played a significant role in formulating the definition as ultimately adopted by the TOGAF Standard.

The various sources from which the individual definitions can be found are indicated in blue. For example, the definition of the profession as described in the TOGAF Standard can be found in the tenth edition of the framework. Gartner, for example, has published the definition of Enterprise Architecture on its website.

The yellow ovals indicate the specific references to the sources in which the definitions are articulated. Finally, the purple ovals indicate a definition of either Enterprise Architecture, Business Architecture, or Architecture.

CHAPTER 4   ARCHITECTURE DEFINITION

One of the bodies of knowledge referenced in the TOGAF Standard is the BIZBOK Guide. This body of knowledge deals with the concepts within the domain of Business Architecture and goes into considerable depth. It is not surprising, therefore, that the TOGAF Standard makes grateful use of the material already contained in the BIZBOK Guide.

A notable part of this guide is the definition of Enterprise Architecture. This results in a third definition of Enterprise Architecture within the same TOGAF Standard. The BIZBOK Guide bases its definition on the Federation of Enterprise Architecture Professional Organizations (FEAPO):

Enterprise Architecture represents the holistic planning, analysis, design, and implementation for the development and execution of strategy by applying principles and practices to guide organizations through the integration and interoperation of all other architecture domains.

What is good to see reflected in the above definition is the relationship to organizational strategy (*for the development and execution of strategy*) and thus to business goals and objectives. This link is missing from the TOGAF Standard's own definition.

Definitions are incredibly important and can make all the difference in understanding what things are about (or what is out of scope or specifically not about).

Ask someone to describe an apple. Nine times out of ten, the person will come up with something like *"a round piece of fruit with a stem (and a small leaf) that contains vitamins"*. Some will also add a color to the description.

However, this description could just as easily refer to a cherry or a grape.

Granted, cherries and grapes are also fruits, but they are not apples. To avoid *comparing apples and oranges*, definitions are used. They really matter [9].

The fact that there are multiple definitions of the same topic within a single standard, such as the TOGAF Standard, is an indication of how difficult a topic this is. The field of Enterprise Architecture has proven over the years to be hard to capture in a single, conclusive definition. In addition, the discipline manifests itself in different ways and at different levels, making it difficult to define what it is.

The conclusion is that there is no single, definitive definition of Enterprise Architecture. Therefore, it is particularly important to define the way Enterprise Architecture is practiced in an organization or how it is represented, so that there is no ambiguity within the organization.

Further Reading

An interesting article on the variety of definitions can be found at the URL below:

https://eawheel.com/blog/2023/04/the-many-definitions-of-enterprise-architecture/

4.4. Summary

Enterprise Architecture is a difficult topic to define. This has been proven over the years. The fact that there are multiple definitions for the same thing is ample evidence of this. Within the TOGAF Standard alone, three definitions are used:

CHAPTER 4 ARCHITECTURE DEFINITION

- A definition based on the ISO/IEC/IEEE 42010:2022 standard
- A second definition from Gartner
- A third definition, taken indirectly from the BIZBOK Guide

For the first definition, the TOGAF Standard adopts the ISO/IEC/IEEE 42010:2022 standard for defining Enterprise Architecture. The framework adds a second definition to the existing one, in order to make it more complete and fitting. What is interesting about the TOGAF Standard's definition of Enterprise Architecture is that a key component is missing. It is about the realization of business goals.

For the second definition - described in the Leader's Guide - the TOGAF Standard refers to Gartner. This organization created a definition of Enterprise Architecture in 2008. This is the one adopted in the TOGAF Standard, which was updated in 2025 with the Technical Corrigendum 1. The risk of adopting a definition that is several years old is that it may now appear outdated. There is a real chance that the definition being referenced has since been updated and revised.

The third definition of Enterprise Architecture is indirectly adopted by the TOGAF Standard from the BIZBOK Guide. Since the TOGAF Standard refers to the BIZBOK Guide for the design and application of the Business Architecture domain, the definition of Enterprise Architecture used there is also adopted.

Three definitions in one standard.

The profession of Enterprise Architecture is difficult to capture in a single definition. Therefore, it is more important to define the way Enterprise Architecture is practiced in an organization so that there is no ambiguity within the organization.

CHAPTER 5

Enterprise Metamodel

 This chapter is part of the GREEN, BLUE, and GOLD reading tours. See Figure 1-1 in Chapter 1, Section 1.3, for alternative reading tours.

Every entity used in architecture work and described in the (Extended) Content Framework is rooted in the Enterprise Metamodel. This makes the Enterprise Metamodel a fundamental part of the TOGAF Standard.

Since no two organizations are alike, the Enterprise Metamodel serves as a Foundation Enterprise Metamodel. This means that it must be tailored to the needs of the organization. The metamodel described in the TOGAF Framework is best viewed and used as a starting point.

The Content Framework provides a categorization mechanism that can be used to structure a representation of the Enterprise Metamodel, consisting of five interrelated sections. It is not to be confused with an architecture diagram that shows domains or layers of the architecture.

5.1. Sections of the Enterprise Metamodel

One of the core components of the TOGAF Standard is the Enterprise Metamodel. This metamodel contains the most common entities and their relationships as they are likely to occur in practice in most organizations. The Enterprise Metamodel *provides the context* for the specific artifacts

CHAPTER 5 ENTERPRISE METAMODEL

referenced in the descriptions of the Architecture Development Method phases (see Chapter 12). The Content Framework described in Chapter 6 *defines the types of architecture work products* and their relationships, based on the *entities* from the Enterprise Metamodel. Both the Enterprise Metamodel and the Content Framework are kept in sync; the Enterprise Metamodel provides the entities used in the Content Framework, and the Content Framework evolves the Enterprise Metamodel.

The Enterprise Metamodel comprises five interconnected *sections*, which correspond to the four architecture domains (Business, Data, Application, and Technology) and one generic section. As the name suggests, the Enterprise Metamodel is a metamodel, not an architecture diagram. Therefore, the term *sections* is used.

The five interrelated sections of the Enterprise Metamodel are:

- General Entities
- Business Architecture
- Data Architecture
- Application Architecture
- Technology Architecture

To improve readability, the relations between entities have been omitted from Figures 5-1, 5-2, 5-5, 5-14, 5-17, and 5-19.

CHAPTER 5 ENTERPRISE METAMODEL

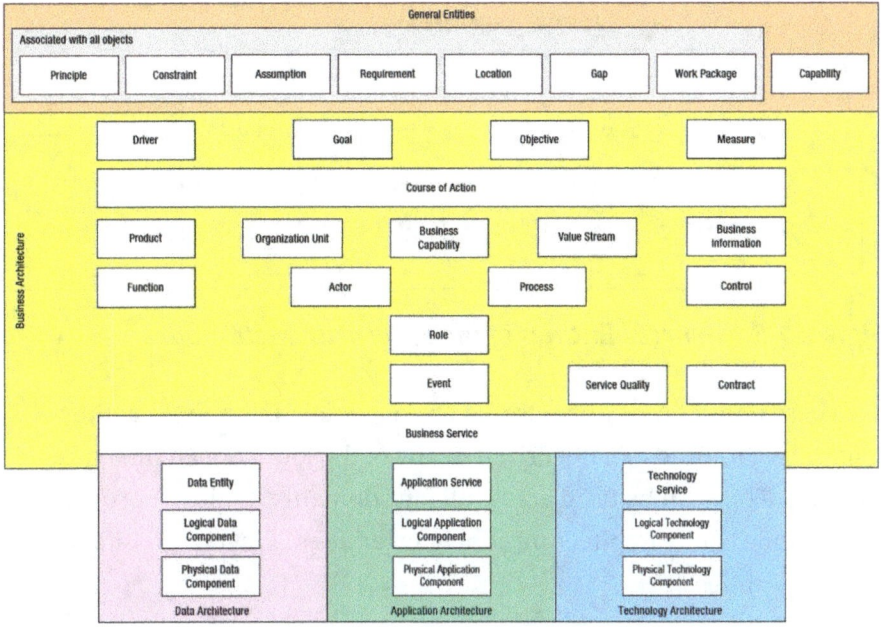

Figure 5-1. *The Foundation Enterprise Metamodel*

Each section consists of a set of model entities [11]. These entities are architecture concepts, each with its own definition and explanation. Understanding what the various entities stand for, what they mean, and how they can and should be used is critical to understanding the architecture products described in the TOGAF Framework. Having a clear definition ensures that there is no misunderstanding about how to use the concepts. For the above reasons, the entities are described in more detail in the following paragraphs.

5.1.1. General Entities

The upper section of the Enterprise Metamodel consists of what is called *General Entities* in the TOGAF Standard. It contains eight entities.

CHAPTER 5 ENTERPRISE METAMODEL

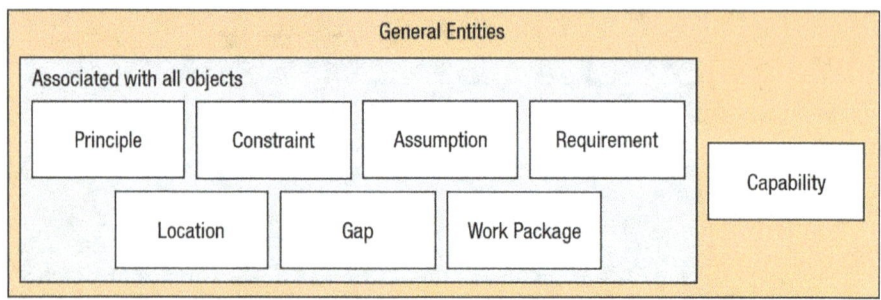

Figure 5-2. General Entities of the Enterprise Metamodel

These entities are concepts that are so generic in nature that they cannot be captured or categorized in any of the four core architecture domains. They can be used across all four domains. Table 5-1 provides an overview of the architecture concepts (entities) that are classified as General Entities.

Table 5-1. General Entities

Entity	Description
Principle	A qualitative statement of intent that should be met by the architecture. Aside from the statement, it has at least a descriptive name, supporting rationale, and one or more implications.
Constraint	An external factor that prevents an organization from pursuing particular approaches to meet its goals.
Requirement	A quantitative statement of business need that must be met by a particular architecture or work package.
Assumption	A statement of probable fact that has not been fully validated at this stage, due to external constraints.
Location	A place where activities occur. Locations can be composed and decomposed.

(continued)

Table 5-1. (*continued*)

Entity	Description
Gap	A statement of difference between two states. Used in the context of gap analysis, where the difference between the Baseline and Target Architecture is identified.
Work package	A set of actions identified to achieve one or more objectives for the business. A work package can be a part of a project, a complete project, or a program.
Capability	An ability that an organization, person, or system possesses.

As mentioned earlier, the architecture entities from the upper section of the Enterprise Metamodel can be used or related in the context of any architecture domain. For example, the *principle* entity can be related to an element from the Business Architecture domain or from the Technology Architecture domain. The *location* entity can be related to either a (logical or physical) application component or a role or actor. Due to their ability to be related to all architecture domains and their underlying elements, the TOGAF Standard refers to these entities as General Entities.

In this book, I will explain certain topics, concepts, and methods by providing examples. These examples are based on a fictional healthcare organization for which an architecture has been developed.

All examples in this and subsequent chapters relate to this organization and its situation. This approach ensures consistency throughout the book.

Principle: A qualitative statement of intent that should be met by the architecture. It has at least a supporting rationale and a measure of importance.

Principles are used to provide a high-level framework for architecture initiatives. There are Architecture Principles and Enterprise Principles (see Chapter 14, Section 14.2). *Principles* are similar to the constitution of a country. They have a high level of abstraction and serve as a kind of guiding light that can be used to give further direction and interpretation to their application and use. Principles are highly influential and strategic in nature and carefully designed to fit the organization and its business model. Like a constitution, these principles change little or nothing and have a strong binding character. The high-level or basic principles are based on business goals and objectives [12]. An example of the entity *principle* is provided in Figure 5-3.

Constraint: A factor – mostly external – that prevents an organization from pursuing particular approaches to meet its goals and objectives.

Stakeholders often bring constraints along with their concerns. For example, there may be a desire or necessity within the organization to create a new function (business role). However, if contractual arrangements have been made with an external party that provides staff to the organization, and that party uses standard job profiles, then the new role may not fit within the applicable standard. This restriction – *constraint* – makes it impossible to create the desired position.

Requirement: A *requirement* indicates the changes needed to a new or existing situation, environment, process, or system within the context of an initiative (work package). Requirements further specify the objective to be achieved and therefore represent a statement of a business need that must be met by a particular architecture or work package. They represent the essential attributes or behaviors that an enterprise system, architecture, or solution must have to achieve the organization's goals, objectives, and business outcomes. Requirements are used to guide the design, development, and validation of the architecture.

The relationship between a principle, a requirement, and a constraint is shown in Figure 5-3.

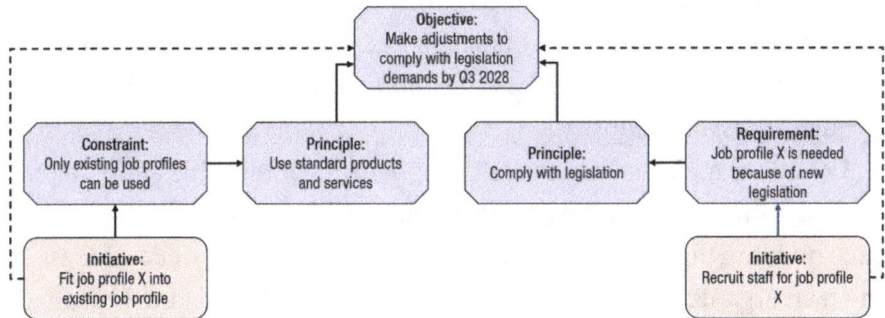

Figure 5-3. *Relationship between a principle, requirement, and constraint*

Figure 5-3 shows two principles. The principle on the left has a relationship with a constraint, and the principle on the right has a relationship with a requirement. The constraint has a relationship with an initiative (work package), and the requirement has a relationship with another initiative. Both initiatives affect the achievement of the objective, and both principles are also related to it. The situation described in Figure 5-3 demonstrates that the objective has been given a framework in the form of two principles. The principle on the right can be further detailed by the requirement and implemented by the associated initiative. The principle on the left is limited in its application by the constraint. However, an initiative is linked to the constraint in the hope of removing it.

Assumption: A statement of probable fact that has not been fully validated at this stage, due to external constraints.

An *assumption* is a foundational element used in architectural practices to define and communicate uncertainties or gaps in information that could influence decisions during the architecture development process. Assumptions are necessary when there is incomplete knowledge or uncertainty about factors impacting the architecture, but decisions still need to be made in order to proceed.

For instance, it may be assumed that an existing application will support a certain set of functional requirements, although those requirements may not yet have been individually validated. Assumptions are commonly used in project management. The TOGAF Series Guide "Architecture Project Management" elaborates on the added value of formulating assumptions.

Location: A place where activity happens. Literally. This entity refers to actual (physical) locations, for example, the geographic location of the office building the organization occupies or the countries or continents in which an organization operates, to name a few. The entity *location* can also be used in relation to information systems, applications, and technology components. A common diagram, often used by IT Architects, is that of the distribution of systems (think of server systems) across different data centers. In this diagram, the entity *location* is used to denote the different data centers.

Gap: This entity represents the difference between two states of architecture. It is most commonly used in the context of performing a gap analysis, where the difference between the Baseline and Target Architecture is identified.

A *gap* refers to a specific difference or shortfall between two states of an Enterprise Architecture: the *Baseline Architecture* (current state) and the *Target Architecture* (desired future state[1]). Identifying and analyzing these gaps helps organizations understand what is missing, incomplete, or needs to change to move from the baseline to the target. A gap details the results of a gap analysis in phases B, C, and D of the Architecture Development Methodology and is an important input for subsequent implementation and migration planning.

---

[1] The *desired future state* should not be considered definitive in any way. Rather, it refers to a state at a specific point in time.

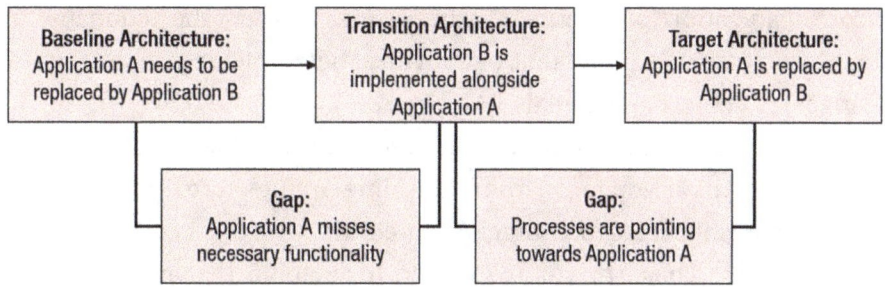

Figure 5-4. *Gaps between Baseline, Transition, and Target Architectures*

Continuing with the example shown in Figure 5-3, an application may be required to support certain functionality in order to execute the initiatives illustrated in the figure. This functionality is currently missing from the existing application (Application A). In order to achieve the objective shown in Figure 5-3, it has been decided to replace Application A with Application B. However, this will not happen overnight, so a phased approach has been chosen. First, Application B is implemented alongside Application A, and then all data and interfaces are transferred to Application B. Finally, Application A is decommissioned. This scenario is illustrated in Figure 5-4 using the *gap* entity.

Work Package: A set of actions identified to achieve one or more objectives for the business. A *work package* can be a part of a project, a complete project, or a program. It is a key concept used to manage and structure the implementation of an architecture. It represents a defined set of activities designed to achieve specific business objectives and outcomes as part of the overall organizational transformation.

Another word for work package is *initiative*. An initiative is a concept used to represent a planned action or project designed to achieve specific goals or objectives within the organization. Initiatives are concrete activities that are undertaken to implement changes, improve processes, or achieve desired outcomes as part of the overall strategy of the organization. They are directly linked to specific goals or objectives of the organization (see Figure 5-3). Initiatives represent the actionable steps taken (projects, programs, or activities that implement architectural changes) to achieve these goals.

Work packages are also used to represent the activities required to close the gaps as you move from a Baseline Architecture through a Transition Architecture to a Target Architecture.

Capability: The term *capability* refers to an ability that an organization, system, or individual must possess. It represents a key construct in the TOGAF Framework, central to *capability-based planning*, which is an approach to Enterprise Architecture focusing on what an organization is able to do, rather than simply what it has.

A capability is an entity that focuses on what an organization (or a person, or even an object) is capable of doing. For example, consider an airplane. It has the ability to transport cargo or people from point A to point B. Of course, this capability is not unique to an airplane, because a car has the same capability. What makes an airplane different from other forms of transportation is that it has the ability *to fly*.

Another example of a capability, in this case of a motorcycle, is that it has the ability to *maneuver between cars* (think of traffic jams). Even though the motorcycle does not have a button that can be pressed to activate this ability, it still *has* this ability. Unlike an airplane that can actually fly on autopilot, the motorcycle would need its rider to perform the capability.

Capabilities define what an organization, person, or object should be able to do. The ability to actually execute the capabilities depends on having people, processes, applications, and/or technology available to perform the tasks necessary to execute the abilities. In architectural terms, a capability defines what an organization can (or should) do. It would then require actors, organizational units, processes, applications, and even technology to execute it. A capability therefore defines, at a conceptual level, what abilities an organization must have in order to do what it does (or wants to do).

CHAPTER 5   ENTERPRISE METAMODEL

The capability entity stands out from the rest of the General Entities because in addition to this entity, the Enterprise Metamodel also includes the *business capability* entity (see Section 5.1.2). However, these two types of entities are different from each other. The business capability entity has the following definition:

A business capability is a particular ability or capacity that a business may possess or exchange to achieve a specific purpose or outcome [13].

A business capability thus differs from a capability in that the former has an *actionable component* in its definition. The bit about *to achieve a specific purpose or outcome* indicates that it must indeed have a specific purpose. The regular capability does not have this addition in its definition. This makes business capabilities more meaningful compared with regular capabilities. After all, simply having an *ability* says nothing about its usefulness and necessity. By adding a clear purpose (*to achieve a specific purpose or outcome*) to a regular capability, it suddenly becomes valuable.

Further Reading
The TOGAF Series Guide "Architecture Project Management", document number G188, provides additional information on applying the TOGAF Standard to project management.

73

CHAPTER 5   ENTERPRISE METAMODEL

5.1.2. Business Architecture

The most comprehensive section of the Enterprise Metamodel consists of Business Architecture Entities. Nineteen in total.

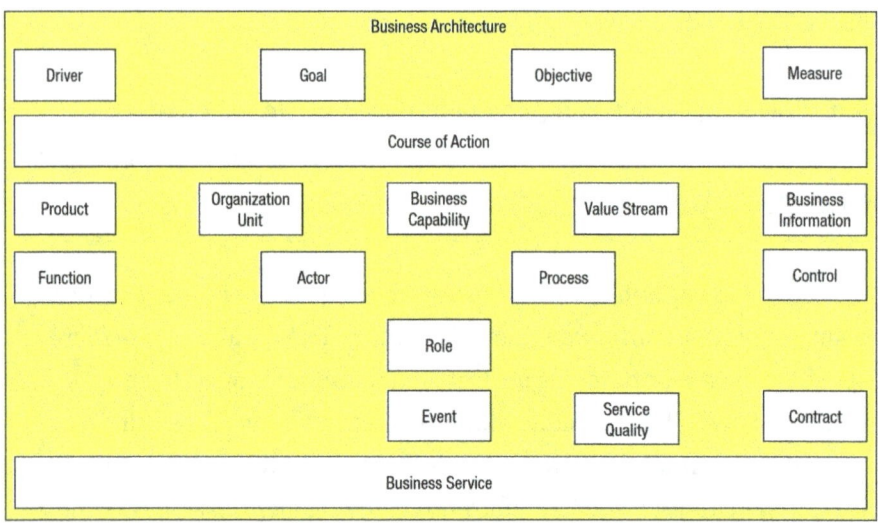

Figure 5-5. *Business Architecture Entities of the Enterprise Metamodel*

Table 5-2 lists the architecture concepts (entities) that can be classified as Business Architecture Entities.

Table 5-2. *Business Architecture Entities*

Entity	Description
Driver	An external or internal condition that motivates the organization to define its goals.
Goal	A high-level statement of intent or direction for an organization. Typically used to measure success of an organization.

(*continued*)

Table 5-2. (*continued*)

Entity	Description
Objective	An organizational aim that is declared in a Specific, Measurable, Achievable, Realistic, and Timebound (SMART) way.
Measure	An indicator or factor that can be tracked, usually on an ongoing basis, to determine success or alignment with objectives and goals.
Course of action	Direction and focus provided by strategic goals and objectives, often to deliver the value proposition characterized in the business model.
Business capability	A particular ability that a business may possess or exchange to achieve a particular purpose.
Value stream	A representation of an end-to-end collection of activities that create an overall result for a customer, stakeholder, or end user.
Function	A set of business behaviors based on a chosen set of criteria. Functions are usually close-coupled to/with organizational units.
Process	A process represents a sequence of activities that together achieve a specified outcome, can be decomposed into sub-processes, and can show operation of a business capability or service.
Business service	Supports the business by encapsulating a unique element of business behavior; a service offered external to the organization may be supported by business services.
Organization unit	A self-contained unit of resources with goals, objectives, and measures. Organization units may include external parties and business partner organizations.

(*continued*)

Table 5-2. (*continued*)

Entity	Description
Actor	A person, organization, or system that has a role that initiates or interacts with activities. Actors may be internal or external to an organization.
Role	The usual or expected behavior of an actor or the part somebody or something plays in a particular process or event. An actor may have a number of roles.
Product	An outcome generated by the business to be offered to customers. Products include materials and/or services.
Contract	An agreement between a consumer and a provider that establishes functional and non-functional parameters for interaction. This applies to all types of service interactions within the metamodel.
Business information	Represents a concept and its semantics used within the business.
Control	A decision-making step with accompanying decision logic used to determine execution approach for a process or to ensure that a process complies with governance criteria.
Event	An organizational state change that triggers processing events. It may originate from inside or outside the organization and may be resolved inside or outside the organization.
Service quality	A configuration of non-functional requirements or attributes that may be assigned to a business, application, or technology service.

CHAPTER 5 ENTERPRISE METAMODEL

The entities listed in Table 5-2 are further specified in the following paragraphs:

Driver: The condition that motivates the organization to define its goals. This condition can be internal or external.

A *driver* refers to any influence, factor, or condition that pushes or drives the organization to make decisions or take actions that affect its business activities and strategic direction. Drivers are often the reasons behind changes or new initiatives within an organization. They could arise from external environments such as *market share, legislation*, or *technological advancements* or from internal conditions like *leadership changes* or *operational inefficiencies*.

Drivers can also be stakeholder related. In this case, they are called *concerns*. In the context of Enterprise Architecture, identifying and analyzing drivers helps architects understand the underlying reasons for change and shape architecture decisions accordingly. Figure 5-6 demonstrates the use of a driver in relation to goals and objectives.

Goal: A high-level statement of intent or direction for an organization, typically used to measure success of an organization.

A *goal* can be anything a stakeholder wants, such as a state of affairs or a value produced. Common examples of goals are *increase profits, reduce help desk wait times,* or *implement new regulatory requirements*. Goals are usually expressed in qualitative terms, using words such as *increase, improve,* or *easier*. Figure 5-6 shows an example of a goal in the context of the fictional company mentioned in Section 5.1.1.

Objective: An organizational aim that is declared in a SMART[2] way.

Objectives further detail goals, as they express the measurable aims of the organization with regard to the goals they are related to. For instance, the goal *implement new regulatory requirements* could be related to the objective *make adjustments to comply with legislation demands by Q3 2028 (see* Figure 5-6*)*.

---

[2] Specific, Measurable, Achievable, Realistic, and Timebound (SMART).

CHAPTER 5   ENTERPRISE METAMODEL

Measure: An indicator or factor that can be tracked, usually on an ongoing basis, to determine success or alignment with goals and objectives.

A *measure* is an entity related to managing and assessing Enterprise Architecture performance. It is used to evaluate how effectively an organization is achieving its goals and objectives through architecture initiatives and projects. A measure is a quantitative indicator used to track progress toward achieving a specific goal or objective. It provides evidence-based insights into the performance and success of an architectural element, initiative, or strategy.

There are three types of measures:

- **Key Performance Indicators (KPIs):** Specific measures tied to business outcomes and critical success factors (CSFs). KPIs are used to monitor progress against strategic goals and are typically outcome-focused.

- **Metrics:** Broader measures that can include both quantitative and qualitative data. These are not always directly tied to outcomes but can measure process efficiency, resource usage, and other operational aspects.

- **Target Values:** Measures often have associated target values that indicate the desired performance level. These target values provide a reference point for evaluating current performance against expectations.

Using Figure 5-6 as an example, a *measurable* indicator could be the *Q3 Legal Department Report*. This report should indicate whether or not the requirements of the new legislation defined in the objective have actually been met, based on certain metrics or target values. Figure 5-6 illustrates the relationship of the measure to entities such as goal and objective.

CHAPTER 5   ENTERPRISE METAMODEL

Course of Action: Direction and focus provided by strategic goals and objectives, often to deliver the value proposition characterized in the business model. A course of action represents what an organization has decided to do.

A *course of action* is an entity that represents an approach or plan to achieve specific objectives or solve a problem. It defines a coordinated set of activities, decisions, and policies that an organization can follow to achieve desired outcomes.

Courses of action can be divided into *strategies* and *tactics*. It is not possible to make a sharp distinction between the two, but *strategies* are typically long-term and fairly broad in scope, while *tactics* tend to be shorter-term and narrower in scope. An example of a (tactical) course of action is shown in Figure 5-6.

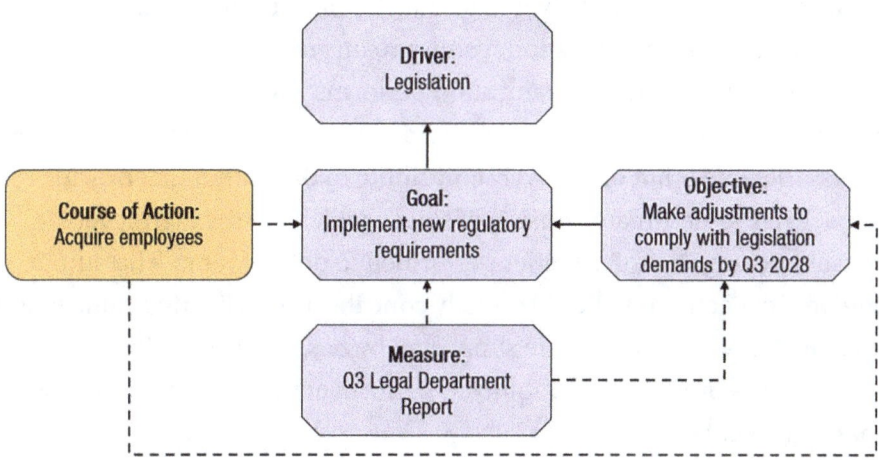

Figure 5-6. *Relationship between a driver, goal, objective, measure, and course of action*

Figure 5-6 shows that the aforementioned fictional organization (see Section 5.1.1), faced with soon-to-be-enacted legislation, identified legislation as a key driver. In order to comply with the upcoming legislation, the organization has set a goal to implement measures. These measures must be implemented by the third quarter of the year. One measure to monitor implementation is the preparation of a report by the legal department. This report will indicate whether the requirements have been met. Once the measures have been implemented, a course of action will be set in order to acquire the staff needed to implement them.

Business Capability: A particular ability that a business may possess or exchange to achieve a particular purpose.

The *business capability*, in relation to the entity *capability*, is explained in Section 5.1.1.

Value Stream: A representation of an end-to-end collection of activities that create an overall result for a customer, stakeholder, or end user.

A *value stream* is a core concept used in Enterprise Architecture to focus on how value is delivered to stakeholders, both internal and external, across an organization. Value streams represent the flow of activities or steps that an organization performs to create and deliver products or services that meet stakeholder needs. These activities are not necessarily linear but collectively contribute to the output that provides value. Each value stream consists of *stages*, which are the high-level phases or categories of activities performed to deliver value. The output or outcome of each stage should directly contribute to delivering value to the stakeholder. Value streams are supported by *business capabilities*, which represent the organization's ability to perform activities that support or enable the value stream.

Value streams and *business processes* may seem similar, but they are defined at different levels of abstraction and serve different purposes. A business process describes the (time-ordered) sequence of behaviors required to produce a result for an individual case, and it may describe alternative paths and decision points. In contrast, a value stream focuses on the overall value-creating behavior from the perspective of the importance, worth, or usefulness of what is produced and is not a

CHAPTER 5　ENTERPRISE METAMODEL

description of time-ordered tasks for individual cases. Value streams reflect an organization's *business model* and value proposition, while business processes reflect its *operating model*.

Let's consider the acquisition of new talent (as already indicated by the course of action in Figure 5-6) as an example of the relationship between value streams and business processes. Figure 5-7 shows a simplified example of the stages of the value stream *Acquire Talent*.

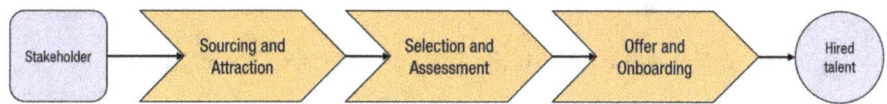

Figure 5-7.　*Value stream stages of Acquire Talent*

Each of the value stream stages has a relationship to the underlying process steps being performed:

- **Sourcing and Attraction:** Identify and attract potential candidates who fit the organization's needs. During this stage, the process steps that are being followed are creating and posting job descriptions, reaching out to potential candidates, and building a talent pipeline. This stage also includes employer branding efforts to make the organization attractive to prospective candidates.

- **Selection and Assessment:** Evaluate and select the most suitable candidates from the pool of applicants. The steps executed by the underlying process are screening resumes, conducting interviews and administering assessments, and narrowing down the candidate pool to those who best meet the job requirements and organizational fit.

- **Offer and Onboarding:** Finalize the hiring process and smoothly integrate the new hire into the organization. The business process steps that support this value stream stage are extending job offers, negotiating terms, and onboarding the new hire. This stage ensures that the candidate transitions effectively into their new role and aligns with the organization's culture and expectations.

Figure 5-8 shows the addition of process steps to the value stream stages.

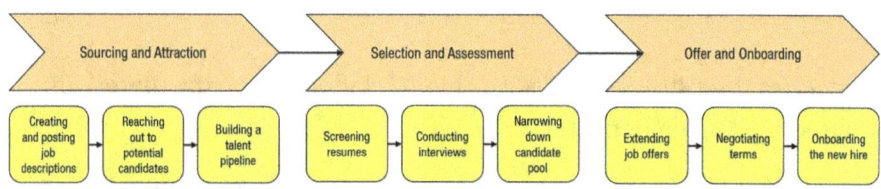

Figure 5-8. *Value stream supplemented by business processes*

The value created by moving through the value stream is *hired talent* (something that is not tangible). A business process describes the steps taken at each stage of the value stream. More specifically, they describe the activities that take place before the acquisition is actually completed.

The end product of the described process steps is a signed employee agreement with the new hire. This is a product that, when printed, can actually be grasped. As such, it becomes a tangible product. Value streams never deliver a tangible end product. They deliver *value*.

A key principle of value streams is that value is always defined from the perspective of the stakeholder – the customer, end user, or recipient of the product, service, or deliverable produced by the work. The value obtained is in the eye of the beholder [14].

An acquired talent (which is the value produced by the value stream) is not tangible (the person is, of course, but the fact that they were hired is not).

Function: A set of business behaviors based on a chosen set of criteria. Functions are usually close-coupled to or with organizational units.

The entity *function* refers to a fundamental building block within the architecture framework. Specifically, it is a component that represents a distinct, cohesive set of activities or services that contribute to achieving a particular outcome or purpose within the organization. A function is defined by its purpose and the responsibilities it holds. It is meant to achieve specific goals or deliver certain outcomes that align with the organization's strategic objectives. Functions are composed of various components such as processes, procedures, and interactions. These components work together to perform the function's role within the broader architecture.

Put more simply, a function is a logical *grouping of activities* that together support the overall operation of a business. Business functions often reflect the organizational structure. However, a function is not tied directly to departments but to the activities themselves.

Process: A process represents a sequence of activities that together achieve a specified outcome, can be decomposed into sub-processes, and can show operation of a business capability or service.

A *process* refers to a set of related, sequential activities or tasks designed to achieve a specific goal or output within the architecture development framework. The TOGAF Standard uses processes to structure and guide the execution of its methodologies and practices, ensuring that activities are carried out systematically and effectively.

Figure 5-8 shows three series of activities associated with three value stream stages. Each set of activities represents the process stages of the corresponding parent processes. The first set of activities is part of the parent process called Acquiring Talent. The second set of activities is part of the Evaluating Candidates process. The third and final set of activities is part of the Employee Hiring process. Figure 5-9 shows the series of activities, each with its corresponding parent process.

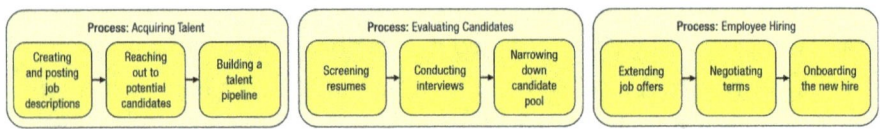

Figure 5-9. *Processes containing series of activities*

The entities *function* and *process* are still sometimes confused. However, the distinction between the two is easy to make. The entity function is broader and more abstract than the entity process. Figure 5-10 illustrates this. Processes define *how things get done step by step*, while functions define *what activities are necessary* to support the organization's goals.

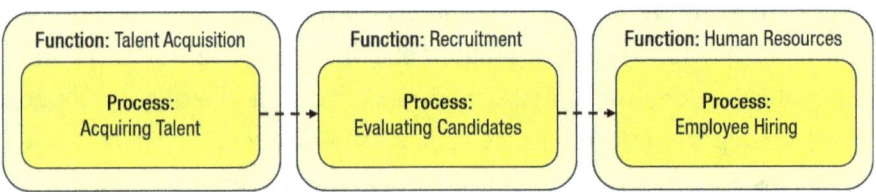

Figure 5-10. *Functions containing corresponding processes*

Organization Unit: A self-contained unit of resources with goals, objectives, and measures. *Organization units* may include external parties and business partner organizations.

Beyond the formal definition, an organization unit can quite easily be compared to a department within an organization. From this perspective, the similarity between an organization unit and a business actor seems quite logical. The TOGAF Standard follows this logic. However, there is a second meaning for this entity. When considering goals, objectives, and measures, the element that comes closest to the way the framework

CHAPTER 5   ENTERPRISE METAMODEL

envisions the organization unit is the *stakeholder*. A stakeholder is a role of an individual, team, or organization (or classes thereof) that represents their interests in the outcome of the architecture. Therefore, from a TOGAF perspective, an organization unit is best represented as a stakeholder when considering the relationships to strategic elements such as goals, objectives, and measures.

Actor: A person, organization, or system that has a role that initiates or interacts with activities. Actors may be internal or external to an organization.

The concept of the *actor* describes an entity that interacts with or is affected by an Enterprise Architecture. This interaction can be direct or indirect, and actors are typically external to the architecture itself.

An important distinction that the TOGAF Standard makes here between the entity *actor* and the *organization unit* is that the entity actor can represent an entire organization, but not a department within it. This representation is exclusively for the entity organization unit. Actors therefore usually represent individuals within an organization or external parties. Figure 5-11 illustrates the use of and differences between actors and roles within an organization unit.

Role: The usual or expected behavior of an actor or the part somebody or something plays in a particular process or event. An actor may have a number of roles.

A *role* is defined as a key concept that describes a set of responsibilities, behaviors, and expectations associated with a position or function within an organization or enterprise. The role is typically linked to the tasks and duties that a person or team performs in relation to specific activities or processes. Roles can be assigned to individuals or teams based on their expertise, authority, and function in the organization. A role outlines specific activities, decisions, and accountabilities, which can span across different domains such as business, data, application, and technology.

85

CHAPTER 5   ENTERPRISE METAMODEL

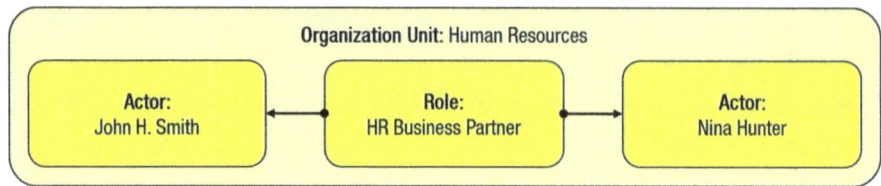

Figure 5-11. *Actors within an organization unit with an assigned role*

Figure 5-11 shows an organizational unit called Human Resources (HR). The unit consists of two people (John H. Smith and Nina Hunter) who both have the role of HR Business Partner.

Product: An outcome generated by the business to be offered to customers. Products include materials and/or services (accompanied by a contract) and are often the end result of a customer-centric process.

A *product* generally refers to an outcome of a process and should always be of value to a customer. Consider, for example, the process of hiring a new employee. At the end of the process, the *employment agreement* is the *product* offered to the *customer*, in this case the employee. The *terms of employment* are set out in the accompanying *contract*.

In some cases, a product can also be compared to a *business object*, which is a concept that is used within a particular business domain. Figure 5-12 illustrates the use of a product, business objects, and a contract.

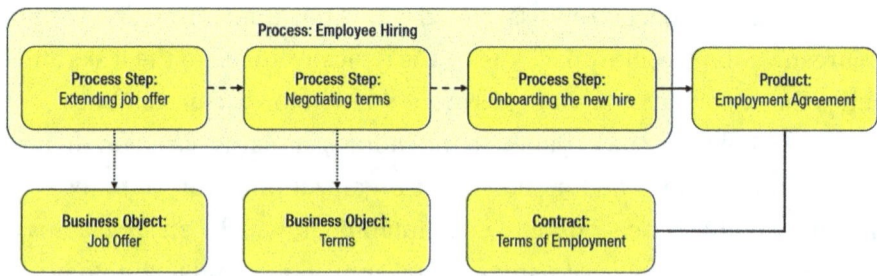

Figure 5-12. *A process creating an end product accompanied by a contract*

Contract: An agreement between a consumer and a provider that establishes functional and non-functional parameters for interaction. This applies to all types of service interactions within the metamodel.

This is a formal agreement that outlines the expectations, responsibilities, and obligations between parties involved in an architectural endeavor. The concept of a *contract* is crucial for defining clear relationships and ensuring alignment between different stakeholders within an architecture.

Contracts in the light of the TOGAF Standard are used to formalize agreements (often related to architecture development and implementation). They help ensure that all parties involved have a mutual understanding of their roles, responsibilities, and deliverables. An example of the entity contract is provided in Figure 5-12.

Business Information: Represents a concept and its semantics used within the business.

The entity *business information* is part of the broader framework for Enterprise Architecture and plays a crucial role in aligning business processes with IT infrastructure. Business information within the context of the TOGAF Standard is concerned with the data and information necessary for business operations and decision-making. It helps to create a clear understanding of what information is critical and of value to the organization, how it should be managed, and how it contributes to the overall business strategy.

Although the definition that the TOGAF Standard gives to this entity is very brief, there is more to be said about it. The beauty of this entity is that it provides a direct link to the information concept from the Business Architecture Body of Knowledge (BIZBOK) Guide. Where the Information Architecture domain disappeared from the TOGAF Standard after version 8 (2002), there now seems to be a movement that could indicate the return of the Information Architecture domain. Perhaps future versions of the TOGAF Standard will confirm this observation.

CHAPTER 5 ENTERPRISE METAMODEL

The BIZBOK Guide defines an information concept as follows:

An information concept is comprehensive in terms of representing objects across a business ecosystem, but may or may not be defined in a data model or database.

The information concept realizes or makes explicit a business object, with the object serving as the basis for the information concept name [5].

In the BIZBOK Guide, the term *information concept* refers to a high-level abstraction that represents a fundamental piece of information that a business uses, manages, and interacts with. Information concepts are not specific data points but rather a representation of the information a business uses to support its processes, products, services, and strategies. Information concepts provide clarity on how information is used across the business.

An information concept is *a key piece of business information*, often derived from business objects. It's not the raw data but a *conceptual model of data* that is meaningful to the business, such as a *strategy* and *vision* or *human resource* and *competency*. These concepts are the foundation for structuring and understanding the information needs of the business, independent of the specific systems or technologies in use.

It's an abstract representation, which means it is not concerned with technical implementation (e.g., databases) but with how the business views and categorizes the information.

Information concepts are relevant in the context of the business and its architecture, helping to ensure alignment between business processes, strategies, and the information needed to support them. They often span multiple domains or functions within an organization, linking processes, capabilities, and value streams.

CHAPTER 5   ENTERPRISE METAMODEL

According to the TOGAF Standard, a data entity (see Section 5.1.3) corresponds to the information concept from the BIZBOK Guide.

The BIZBOK Guide describes Business Architecture (including the information concept, but also information mapping) much more extensively than is done here. For a complete description, it is recommended to consult the BIZBOK Guide.

Control: A decision-making step with accompanying decision logic used to determine an execution approach for a process or to ensure that a process complies with governance criteria.

The entity *control* refers to the mechanisms and processes used to ensure that an organization's architecture aligns with its strategic goals and operates effectively. It encompasses the methods and practices employed to manage, monitor, and guide the implementation of the architecture.

Within the TOGAF Standard, the entity *control* is typically used for the following situations:

- **Governance:** Establishing structures and processes to guide and oversee the development and implementation of the Enterprise Architecture. This includes defining roles, responsibilities, and decision-making processes.

- **Compliance:** Ensuring that the architecture and its implementation adhere to relevant regulations, standards, and policies. This involves regular checks and audits to verify compliance.

- **Monitoring and Reporting:** Implementing mechanisms to track the performance and progress of architectural initiatives. This includes setting up metrics, monitoring systems, and reporting structures to provide visibility into the effectiveness of the architecture.

- **Risk Management:** Identifying, assessing, and mitigating risks associated with the architecture. This involves understanding potential threats and vulnerabilities and implementing strategies to address them.

- **Change Management:** Managing changes to the architecture in a controlled and systematic way. This includes assessing the impact of changes, coordinating updates, and ensuring that changes align with the overall architectural vision and strategy.

Event: An organizational state change that triggers processing events. It may originate from inside or outside the organization and may be resolved inside or outside the organization.

An *event* is something that has a notable impact or significance, whether it's a change in market conditions, a new regulatory requirement, or something as small as hiring new employees. Events often represent a shift from one state to another, such as candidates being selected after the initial evaluation. Events can be triggered by external factors (new laws, competitive actions) or internal factors (process changes, project milestones). Organizations need to be able to respond to events effectively. This involves understanding the potential impact on the architecture and making necessary adjustments to align with the new conditions.

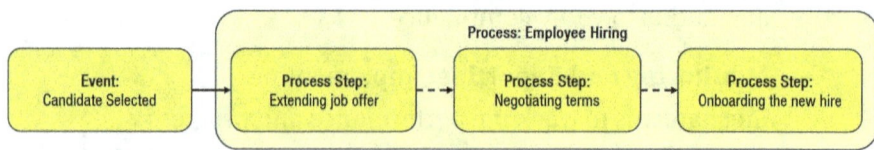

Figure 5-13. Event triggering a process

Figure 5-13 shows the process Employee Hiring being triggered by the event of a candidate that is selected.

Service Quality: A configuration of non-functional requirements or attributes that may be assigned to a business, application, or technology service.

The entity *service quality* refers to the measure of how well a service meets or exceeds the expectations of its stakeholders. It encompasses various attributes such as reliability, responsiveness, and user satisfaction. Within the TOGAF Standard, this concept is tied to the broader framework of *IT service management*, which aims to ensure that IT services deliver value to the business and meet defined service levels. The TOGAF Standard emphasizes that service quality should be considered when developing and refining architectural views, particularly in the Technology Architecture domain, with regard to Solution Architecture. The quality of services provided by IT systems and applications is crucial to the overall success of the Enterprise Architecture.

Business Service: Supports the business by encapsulating a unique element of business behavior; a service offered external to the organization may be supported by business services.

A *business service* entity represents a logical, self-contained unit of business functionality that is valuable to the organization or its customers. Business services encapsulate a set of related functions or activities that together support a particular business process or goal. A business service can be provided either to external stakeholders, such as customers or partners, or to internal stakeholders, such as other business units. The consumers of the service typically do not need to know the internal details of how the service is delivered.

The business service appears in the Enterprise Metamodel as the link or layer between the Business Architecture and the Data, Application, and Technology Architectures. This entity is the interface between the more operational architectures on the one hand and the strategic and tactical architectures on the other.

Further Reading

To learn more about the differences and similarities between value streams and business processes, read the "Business Streams And Value Processes" blog:

https://eawheel.com/blog/2023/05/business-streams-and-value-processes/

An interesting video on the topic of Business Architecture and value streams can be at the following URL:

https://m.youtube.com/watch?v=W3yHCffXDM0&t=314s

See the BIZBOK Guide for more information on Business Architecture (including information concepts and information mapping):

https://www.businessarchitectureguild.org/page/BIZBOK

Additional information about information mapping can also be found in the TOGAF Series Guide "Information Mapping", document number G190.

5.1.3. Data Architecture

The previous section ended with the business service entity as the connecting concept between the Business Architecture on the one hand and the more substantive and operational architectures such as Data, Application, and Technology on the other. This section describes the entities used within the Data Architecture area of the Enterprise Metamodel.

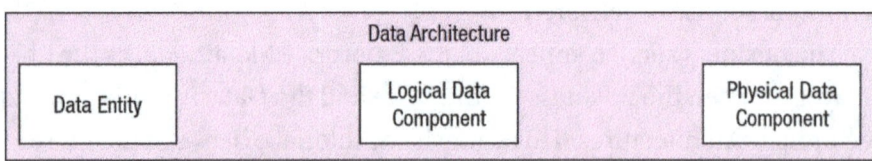

Figure 5-14. Data Architecture Entities of the Enterprise Metamodel

CHAPTER 5　ENTERPRISE METAMODEL

The first of the three architectures located at the bottom of the Content Framework is called Data Architecture. Like Application Architecture and Technology Architecture, it contains three entities (see Table 5-3).

Table 5-3. *Data Architecture Entities*

Entity	Description
Data entity	Represents data that is recognized by the business as a distinct concept.
Logical data component	A data structure composed of logically related data entities.
Physical data component	A data structure that realizes related logical data components represented in the format or schema required by a particular technology.

The Data Architecture section of the Enterprise Metamodel is directly related to the business service entity from the Business Architecture section.

Data Entity: Represents data that is recognized by the business as a distinct concept.

Data entities are usually represented in models such as entity-relationship diagrams (ERDs) or class diagrams. These representations include relationships between different entities, constraints, and attributes that further define the nature of the data stored. Data entities are closely tied to the Business Architecture domain. They help to define Business Information Models that guide how data is structured, stored, and used within an organization, ensuring alignment with the business goals.

The data entity has three relationships with entities in the Business Architecture section. First, a data entity is accessed and updated by a business service. Second, it is consumed by an actor, and finally, the data entity realizes a business information entity (see Section 5.1.2).

This realization relationship allows a data entity to exchange data with the business information concept so that it can be transformed into information of value to the organization. Finally, a business information entity helps to create a clear understanding of what information is critical and of value to the organization, how it should be managed, and how it contributes to the overall business strategy.

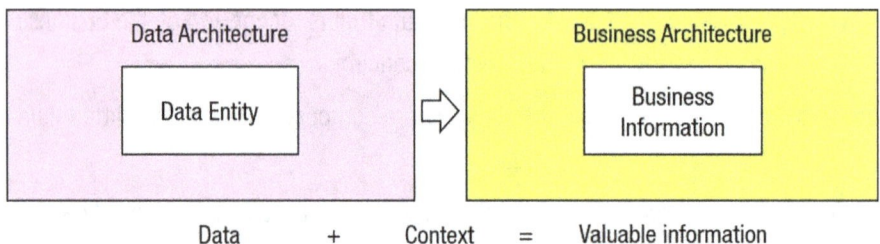

Figure 5-15. Relationship between the data entity and business information

By associating a data entity with a business information entity, it is possible to turn data into valuable information.

For example, consider the numbers 22, 23, 22, 25, and 21. At first glance, these numbers do not mean much. It is only when context is added, such as indicating that this is the expected temperature for the next five days, that the data set gains value and becomes information. This is illustrated in Figure 5-15.

Logical Data Component: A data structure composed of logically related data entities.

The purpose of defining *logical data components* is to represent the abstract, technology-agnostic structures of the data needed by the business, independent of any specific technology or implementation. As the name of the entity already suggests, they represent the *logical structure of data* without regard to how the data is physically stored or implemented. The focus is on *what the data is* and how it relates to business operations rather than how or where it is stored. Logical data component entities are *abstract* representations, meaning they do not concern themselves with databases, file systems, or other technologies.

Physical Data Component: A data structure that realizes related logical data components represented in the format or schema required by a particular technology.

A *physical data component* represents the *tangible* implementations of data systems, including the infrastructure (hardware and software) needed to store, access, and manipulate data. It could refer to databases, file servers, data warehouses, cloud storage solutions, and similar technologies.

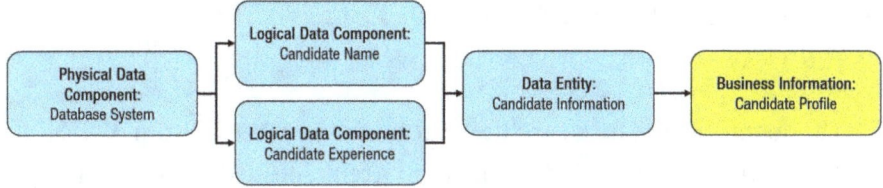

Figure 5-16. *Physical and logical data components in relation to a data entity and business information*

Physical data components are implementations of logical data components. While a logical data component focuses on abstract data structures and conceptual representations of data (such as database schema or data models), the physical component is the actual infrastructure used to implement these structures. For example, a logical component might describe the type of data being stored (e.g., candidate information), while the physical component specifies which database system stores that data. Figure 5-16 illustrates the relationship between physical and logical data components on the one hand and data entities and business information on the other.

5.1.4. Application Architecture

The business service entity serves as the connecting concept between the Business Architecture on the one hand and architectures such as Data, Application, and Technology on the other. This section describes the entities used within the Application Architecture area of the Enterprise Metamodel.

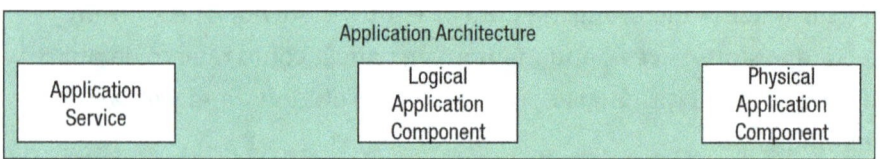

Figure 5-17. Application Architecture Entities of the Enterprise Metamodel

The second of the three architectures located at the bottom of the Content Framework is called Application Architecture. It contains three entities (see Table 5-4).

Table 5-4. Application Architecture Entities

Entity	Description
Application service	The automated elements of a business service. An application service may deliver or support part or all of one or more business services.
Logical application component	An encapsulation of application functionality that is definable by services offered and data maintained, independently of implementation and technology.
Physical application component	A realization of logical application functionality using components of functionality in applications that may be hired, procured, or built.

Application Service: The automated elements of a business service. An application service may deliver or support part or all of one or more business services.

An *application service* entity delivers specific functionality that an organization needs. It typically supports the business processes and operations by offering automated or semi-automated functions.

Logical Application Component: An encapsulation of application functionality that is definable by services offered and data maintained, independently of implementation and technology.

The *logical application component* entity refers to an abstract representation of software that performs specific *functions* or *services* in an architecture. This component encapsulates the behavior and responsibilities of a system without getting into the physical implementation details, which allows architects to focus on *what* the system should do rather than *how* it is realized.

The logical application component entity represents the *abstract* or *conceptual* view of an application. It describes its functionality, interfaces, and behaviors without reference to technology specifics.

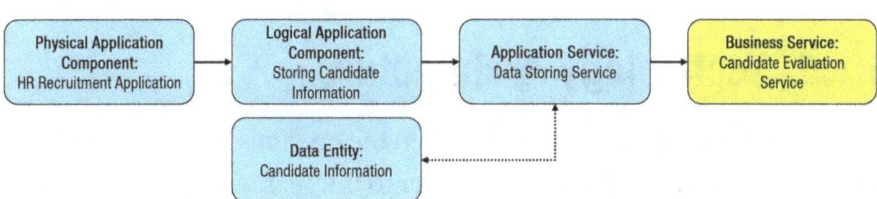

Figure 5-18. *Relationship between physical and logical application components and application and business services*

The use of a logical application component and an application service is depicted in Figure 5-18. A logical application component entity may represent the system that handles HR recruitment *functionalities* like storing candidate information and tracking interviews. This logical component doesn't specify if the system will be implemented using HR

Recruitment Application A, B, or a custom-built application or platform – that decision happens in the physical design phase. A logical application component focuses on the functionality it provides.

The application service entity is implemented by the logical application component and uses a business service to provide functionality to the organization. The application service may also use one or more data entities. Figure 5-18 illustrates this example scenario.

Physical Application Component: A realization of logical application functionality using components of functionality in applications that may be hired, procured, or built.

The *physical application component* entity is an actual, concrete implementation of an application system or software that runs on a specific technology platform. It refers to the *tangible* aspects of an application, including the installed software, the configurations, and the underlying hardware resources required to operate the application. The physical application component entity describes the implementation-specific details, such as the software packages, configurations, and deployment environments (see Figure 5-18).

5.1.5. Technology Architecture

The business service entity mentioned in the previous two sections also connects the Business Architecture layer to the Technology Architecture layer. This section describes the entities used within the Technology Architecture area of the Enterprise Metamodel.

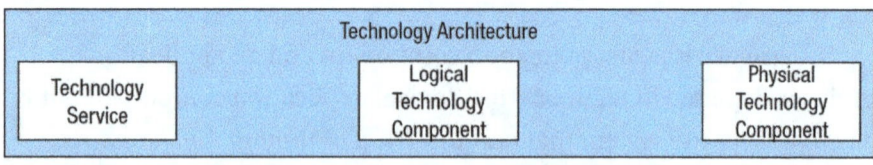

Figure 5-19. Technology Architecture Entities of the Enterprise Metamodel

The third of the three architectures located at the bottom of the Content Framework is called Technology Architecture. Like Data Architecture and Application Architecture, it contains only three entities (see Table 5-5).

Table 5-5. *Technology Architecture Entities*

Entity	Description
Technology service	A technical capability required to provide enabling infrastructure that supports the delivery of applications.
Logical technology component	An implementation-independent encapsulation of technology services.
Physical technology component	A realization of logical technology functionality using a particular technology product that may be deployed.

Technology Service: A technical capability required to provide enabling infrastructure that supports the delivery of applications.

Much like the application service entity mentioned in Section 5.1.4, and as the name suggests, the *technology service* provides a service that other entities can use. It represents the technical capabilities provided by the underlying technology infrastructure that support the higher-level business, application, and data services in the organization.

Consider a cloud storage service within an Enterprise Architecture. This cloud-based service provides scalable, secure, and on-demand storage resources to various applications and users across the organization. Think, for example, about storing files, handling data backups, and ensuring data accessibility from different geographical locations. Figure 5-20 demonstrates the way a technology service entity delivers its services to other entities.

Logical Technology Component: An implementation-independent encapsulation of technology services.

CHAPTER 5   ENTERPRISE METAMODEL

The *logical technology component* entity is an abstraction representing a technology-related function or capability that does not include any specific physical implementation details. It focuses on the functionality required rather than the actual physical hardware, software, or network that performs that function.

An example of a logical technology component could be a Database Management System (DBMS). At this level, the logical technology component does not specify whether the DBMS will be MySQL, Oracle, SQL Server, or another product. It simply indicates the need for a database system that can manage, store, and retrieve data in a structured manner.

Consider Figure 5-20. The aforementioned fictitious company's architecture (see Section 5.1.1) needs to manage large amounts of candidate data. The architecture specifies a logical technology component entity called Database Management System (DBMS) as part of the Technology Architecture. The DBMS manages structured candidate data, supports SQL queries, ensures data integrity, provides backup capabilities, etc. It does not, however, specify the specific database product, vendor, or version. Should the organization select MySQL as the database system (the actual solution), the logical component DBMS would then be realized by a specific physical technology component.

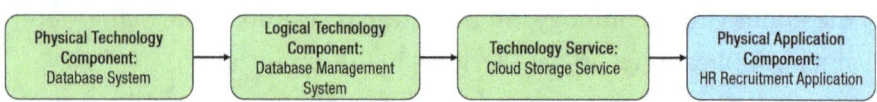

Figure 5-20. *Relationship between physical and logical technology components and technology and application services*

Physical Technology Component: A realization of logical technology functionality using a particular technology product that may be deployed.

The entity *physical technology component* refers to the actual, tangible parts of an IT system, such as hardware, software, and networking devices, that are deployed in an organization's Technology Architecture. These components physically exist and operate in the real world, enabling the functioning of the overall system. Figure 5-20 illustrates the coherence and relations between a technology service, a logical technology component, and a physical technology component.

Further Reading

The "TOGAF Content Framework and Enterprise Metamodel" chapter of the TOGAF Standard provides additional information on the form and application of the Content Framework and Enterprise Metamodel.

For more information on the TOGAF Standard's view of business capabilities, see the "Business Capabilities, Version 2" Series Guide, document number G211.

5.2. Tailoring the Enterprise Metamodel

The Enterprise Metamodel, the Content Framework, and the Architecture Development Method are inextricably linked. The Architecture Development Method is often considered the heart of the TOGAF Standard and plays a very important role in the triangle just described. The Architecture Development Method allows architects to create an *organization-specific* version of the Enterprise Metamodel. The necessary steps for this take place during the Preliminary Phase (refer to Chapter 12, Section 12.3, for additional information).

CHAPTER 5   ENTERPRISE METAMODEL

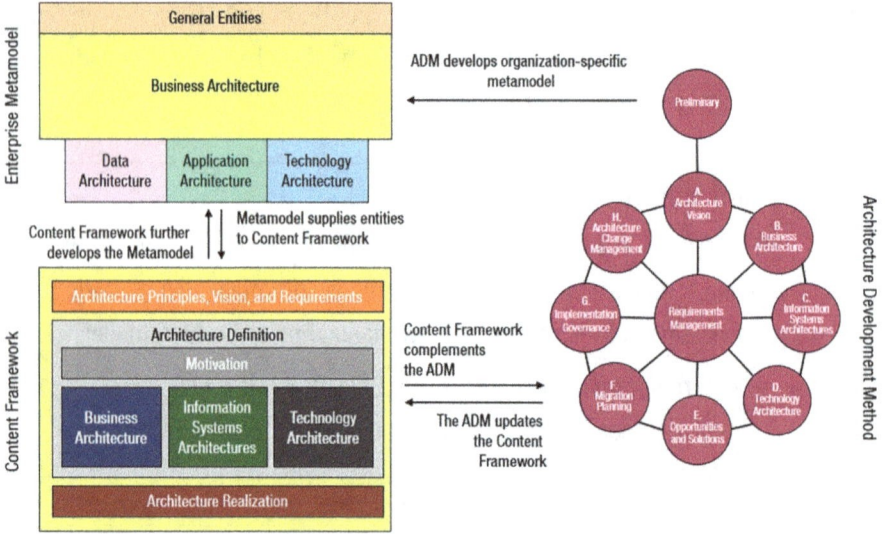

Figure 5-21. Relation between ADM, Enterprise Metamodel, and Content Framework

---

Please note: The arrows in Figure 5-21 do not represent a direct connection between a model on the one hand and a specific phase of the Architecture Development Method on the other. All arrows point *from one model to another*, not to specific parts of a model.

Once the Enterprise Metamodel - of which the TOGAF Standard describes a basic model that must be adapted to the situation, wishes, and requirements of the organization - is formed, this model serves as input for the Content Framework. The organization-specific Enterprise Metamodel contains and describes all the entities that exist and are needed within the organization. These entities are made available to the Content Framework.

CHAPTER 5 ENTERPRISE METAMODEL

The Content Framework describes and structures architecture within an organization. It consists of a collection of models that represent the architecture, making use of the entities from the Enterprise Metamodel. The Content Framework uses the described entities to create architecture deliverables and artifacts. It also enables the evolution of the Enterprise Metamodel.

The Enterprise Metamodel serves as a handle or starting point, and it is strongly recommended that the metamodel be adapted to one's organization. This means that the Enterprise Metamodel allows - even suggests - adding new entities or removing entities that are not relevant to the organization.

Note that the TOGAF Standard does not aim to constrain an enterprise's selection of artifacts [2].

The TOGAF Standard's view of the Enterprise Metamodel is that it can be used as a basis for developing an organization-specific metamodel. This development takes place during the Preliminary Phase when the Enterprise Architecture Capability is being established.

When developing an organization-specific metamodel, architects may choose not to include entities and relationships from the Enterprise Metamodel that are not relevant and/or add additional entities and relationships [2].

The tailorability of the TOGAF Standard is highlighted once again by allowing this flexibility in the application of the Enterprise Metamodel.

CHAPTER 5   ENTERPRISE METAMODEL

> Although the TOGAF Standard does not explicitly state it, the organization unit entity in the Enterprise Metamodel is used to refer to stakeholders.
>
> The TOGAF Standard indicates that an organization unit is *"a self-contained unit of resources with goals, objectives, and measures"*. This is consistent with the fact that stakeholders are typically associated with goals and objectives.
>
> However, I would strongly suggest adding the entity stakeholder to the Enterprise Metamodel. This is an essential entity that is used quite often when creating architecture deliverables and artifacts.
>
> ArchiMate, on the other hand, does recognize the stakeholder element and recommends using this element when the context is about goals and objectives and therefore strategic in nature:
>
> "We propose to use the ArchiMate definition of Stakeholder when using the TOGAF definition of Organization Unit with relations to Measure, Goal, or Objective" [15].
>
> The addition of the stakeholder entity to the Enterprise Metamodel contributes to the integration between the TOGAF Standard and the ArchiMate modeling language, also a standard managed by The Open Group.

The first iteration of the Architecture Development Method focuses specifically on creating the Architecture Capability. Part of the capability creation is the development of an initial version of the Enterprise Metamodel. This involves creating an organization-specific variant of the Enterprise Metamodel as presented in the TOGAF Standard. This step is important and deserves to be emphasized again, as many architects tend to overlook it.

It makes perfect sense to add entities such as stakeholder, business collaboration, business interaction, business interface, resource, risk, and value to the organization-specific metamodel. For example, the stakeholder entity is used in the Stakeholder Catalog, and most of the other listed entities appear in the Business Model Diagram (see Chapter 12, Section 12.4.2).

The Enterprise Metamodel is populated with the entities that exist in the organization. If an organization has exceptional entities, they can (and need to) be added to the metamodel. Once the Enterprise Metamodel is customized, it is used as input to the Content Framework. This framework describes all the architecture deliverables (consisting of artifacts and building blocks) that are meaningful to produce during the architecture work. The existing Content Framework in the TOGAF Standard shows a basic model. And again, this base model needs to be tailored according to which entities are described in the organization-specific Enterprise Metamodel.

When the Content Framework is populated with architecture deliverables, they can in turn be made available to the phases from the Architecture Development Method. As you move through the phases, the architecture deliverables are created according to how they are included in the Content Framework. Of course, it is and will remain possible to make adjustments to both the Enterprise Metamodel and the Content Framework as necessary. Architecture constantly evolves, and the concepts used must evolve with it.

CHAPTER 5   ENTERPRISE METAMODEL

 Understand the Enterprise Metamodel: Get familiar with the Foundation Enterprise Metamodel provided in the TOGAF Standard. This model outlines standard entities and relationships used in Enterprise Architecture.

Customize the Metamodel: Identify and include any exceptional entities that are unique to the organization. Adding these entities ensures the metamodel accurately represents the organization's architecture.

Integrate the Customized Metamodel: Use the tailored Enterprise Metamodel as input for the Content Framework. This framework defines all architecture deliverables, including artifacts and building blocks, that are relevant during architecture development. Adding or removing architecture deliverables to and from the Content Framework will customize it to the organization's needs.

Align Deliverables: When progressing through the phases of the Architecture Development Method, create architecture deliverables in accordance with the customized Content Framework. This alignment ensures consistency and relevance in the architecture work.

Continuously Evolve: Recognize that architecture is an evolving discipline. Regularly review and update both the Enterprise Metamodel and the Content Framework to reflect changes in the organization and its environment. This ongoing refinement maintains the effectiveness and accuracy of the Enterprise Architecture.

Tailoring the Enterprise Metamodel is an absolute necessity. The foundation model provided in the TOGAF standard is only a starting point. Be sure to spend sufficient time creating an organization-specific model before proceeding with the development of the Content Framework.

CHAPTER 5   ENTERPRISE METAMODEL

Further Reading

For more information on customizing the Enterprise Metamodel, see "The TOGAF® Leader's Guide to Establishing and Evolving an EA Capability", Chapter 8, Customization of Architecture Contents and Metamodel, document number G184.

More information on using ArchiMate to support the TOGAF Standard can be found in the document "How to Use the ArchiMate® Modeling Language to Support the TOGAF® Standard", document number G21E.

5.3. Summary

One of the core components of the TOGAF Standard is the Enterprise Metamodel. This metamodel contains the entities and their relationships as they are likely to occur in practice in most organizations. The Enterprise Metamodel provides the context for the specific artifacts referenced in the descriptions of the Architecture Development Method phases.

The entities from the Enterprise Metamodel form the basis for the Content Framework. Both the Enterprise Metamodel and the Content Framework are kept in sync; the Enterprise Metamodel provides the entities used in the Content Framework, and the Content Framework evolves the Enterprise Metamodel.

The Enterprise Metamodel comprises five interconnected sections:

- General Entities
- Business Architecture
- Data Architecture
- Application Architecture
- Technology Architecture

Each section consists of a set of model entities. These entities are architecture concepts, each with its own definition and explanation.

The *General Entities* are concepts that are so generic in nature that they cannot be captured or categorized in any of the four architecture domains. They can be used across all four domains. The General Entities consist of eight entities.

All 19 architecture concepts that are in the *Business Architecture* environment can be classified as entities in this context. However, the way the TOGAF Standard views certain entities such as driver, goal, objective, and measure is different from the way the ArchiMate modeling language views them. The ArchiMate standard assigns these entities to the *motivation elements*. The TOGAF Standard places all these entities under one heading: Business Architecture. The reason for this is that the TOGAF Standard does not have motivational or strategic elements, but uses the architecture domains (Business, Data, Application, and Technology).

The *Data Architecture* section of the Enterprise Metamodel is directly related to the business service entity from the Business Architecture section. It contains three entities. The same goes for the *Application Architecture* and *Technology Architecture* sections. They contain three entities as well.

All 36 entities have been described in detail in this chapter, along with examples of how to use them in practice. The entities in the Enterprise Metamodel are part of what is called the Foundation Enterprise Metamodel, a standard set of entities to start with. One of the key aspects of using the framework is to tailor it to the needs of the organization. This includes the basic set of entities that are part of the Enterprise Metamodel. Entities can be added to or removed from the base set, depending on the needs of the organization.

The Enterprise Metamodel, the Content Framework, and the Architecture Development Method are inextricably linked. The Architecture Development Method allows architects to create an organization-specific version of the Enterprise Metamodel. The necessary steps for this take place during the Preliminary Phase. Once the Enterprise Metamodel is formed, it serves as input for the Content Framework. The organization-specific Enterprise Metamodel contains and describes all the entities that exist and are needed within the organization. These entities are made available to the Content Framework.

The steps for creating a tailored Enterprise Metamodel are as follows:

- **Understand the Enterprise Metamodel:** Get familiar with the Foundation Enterprise Metamodel provided in the TOGAF Standard. This model outlines standard entities and relationships used in Enterprise Architecture.

- **Customize the Metamodel:** Identify and include any exceptional entities that are unique to the organization. Adding these entities ensures the metamodel accurately represents the organization's architecture.

- **Integrate the Customized Metamodel:** Use the tailored Enterprise Metamodel as input for the Content Framework. This framework defines all architecture deliverables, including artifacts and building blocks, that are relevant during architecture development. Adding or removing architecture deliverables to and from the Content Framework will customize it to the organization's needs.

- **Align Deliverables:** When progressing through the phases of the Architecture Development Method, create architecture deliverables in accordance with the customized Content Framework. This alignment ensures consistency and relevance in the architecture work.

CHAPTER 5 ENTERPRISE METAMODEL

- **Continuously Evolve:** Recognize that architecture is an evolving discipline. Regularly review and update both the Enterprise Metamodel and the Content Framework to reflect changes in the organization and its environment. This ongoing refinement maintains the effectiveness and accuracy of the Enterprise Architecture.

The Enterprise Metamodel is the foundation of the Enterprise Architecture. It takes time and care to create a customized and organization-specific version of the metamodel. The model serves as input to both the Content Framework and the Architecture Development Method.

CHAPTER 6

Content Framework

  This chapter is part of the GREEN and GOLD reading tours. See Figure 1-1 in Chapter 1, Section 1.3, for alternative reading tours.

The Content Framework is best thought of as a categorization framework. The primary purpose of the Content Framework is to structure the work products and models that make up the architecture and to describe the architecture products used in an organization. It promotes consistency and coherence in the deliverables produced by following the phases of the Architecture Development Method. The Content Framework – in its extended form – also provides a comprehensive checklist of architecture deliverables that can be produced and reduces the risk of gaps in the final architecture deliverables. In other words, the Content Framework helps an organization use standard architecture concepts, terms, and deliverables.

6.1. A Categorization Framework

The Architecture Development Method provides a way to develop and manage architectures within an organization. It consists of eight main phases and is preceded by a Preliminary Phase. Central to the Architecture Development Method is Requirements Management – not a separate phase, but a mechanism for continuously testing the architecture against it. Each phase of the Architecture Development Method – with the exception of Requirements Management – describes a set of architecture work products.

Because not every organization is the same, it is strongly recommended that the TOGAF Standard be tailored to the requirements and needs of the organization so that the framework is a good fit for the type of organization and its way of working. The TOGAF Standard should be tailored to represent the organization in which it is used.

One of the first steps in establishing the architecture function is to determine the use of the Architecture Capability. Chapter 9 provides further details on this process.

An essential task when establishing the organization-specific Architecture Capability during the Preliminary Phase of the Architecture Development Method is to define the following:

- **An Understanding of the Types of Entities:** It is necessary to understand the types of entities used in the organization and the relationships between them. This is critical to the creation of the architecture description. The *Enterprise Metamodel* represents this information in the form of a formal model.

- **A Categorization Framework:** In order to structure the architecture descriptions, the work products used to express an architecture, and the collection of models that describe the architecture, a categorization framework needs to be used – in other words, the *Content Framework*.

- **The Specific Artifacts:** After both the metamodel with the entities to be used and the categorization framework have been created, the artifacts to be developed can be determined.

The Content Framework and the Enterprise Metamodel are closely related. Where the Enterprise Metamodel is a detailed representation of the entities that exist in the organization, the Content Framework is a kind of high-level summarization. A more fitting term is a *categorization framework*.

The content of the Content Framework consists of a collection of models that represent the architecture. The categorization mechanism of the Content Framework may be used to structure a representation of the Enterprise Metamodel, which, as was mentioned earlier, is a detailed representation of the entities used in an organization. The Content Framework is intended to [11]:

- **Describe a detailed model** of the architecture products used in an organization. Since not all organizations are alike, the models of the architecture products may differ per organization. This is why it is so important to tailor the architecture framework to the needs and requirements of the organization in which it will be used.

- **Promote consistency and coherence** in the deliverables produced by following the phases of the Architecture Development Method. Using the Content Framework ensures that every time a particular deliverable is needed, it is developed in the same way. The entities behind the Content Framework's categorization mechanism ensure that the same entities are always used and that those entities always mean the same thing.

- **Provide a comprehensive checklist** of architecture deliverables that can be produced. The Content Framework allows a checklist to be created for each category of entities, specifying which architecture deliverables can (or should) be produced.

- **Reduce the risk of gaps** in the final architecture deliverables. Because it has been determined in advance which models use which entities, the chance of gaps is minimal, if not non-existent. The condition for this to work is that careful thought is given to filling out the Content Framework before it is implemented.

- **Help an organization use standards** such as architecture concepts, terms, and deliverables. The predefined entities within the Content Framework categories ensure that there is no ambiguity in the use of the entities. Each of the entities is given a definition, the terms to be used, and the deployment of the entity within the architecture deliverables.

The Content Framework is a well-thought-out categorization mechanism that seamlessly connects to the phases of the Architecture Development Method. Partly because of this connection, the Content Framework is a very valuable addition to the TOGAF Standard and therefore to any organization.

6.2. Advantages of the Content Framework

One of the advantages of the Content Framework is that it is central to defining and organizing architectural work products throughout the architecture development process. It is designed to provide a more modern, flexible, and outcome-driven approach to Enterprise Architecture.

CHAPTER 6 CONTENT FRAMEWORK

The Content Framework defines a categorization framework to be used to structure the architecture description, the work product used to express an architecture, and the collection of models that describe the architecture [11].

The Content Framework aims to be more adaptable to various organizational contexts while ensuring that all relevant architecture elements are covered comprehensively. It contains the following key aspects:

- **Modular Structure:** The TOGAF Standard is more modular than previous versions, allowing organizations to adopt only the parts of the framework that are relevant to their needs. This modularity also extends to the Content Framework.

- **Core Content Model:** The Content Framework provides a comprehensive content model (Figure 6-1) that defines the types of building blocks and artifacts that an architecture effort should produce. It includes core concepts such as actors, processes, applications, data entities, and technology components. This model is designed to ensure consistency across various architectural views (Business, Data, Application, and Technology).

- **Extended Content Model:** The framework also allows for extended content models that provide additional depth in specific areas, such as security, governance, or data management. This makes it easier for architects to tailor the framework to their organizational context.

115

- **Architecture Artifacts:** The Content Framework defines key artifacts like catalogs, matrices, and diagrams that help document different aspects of the architecture, for example, a Business Process Catalog or an Application Portfolio Matrix. These artifacts are designed to align with the phases of the Architecture Development Method.

- **Views and Viewpoints:** The framework emphasizes the creation of views and viewpoints, which are ways of looking at the architecture to address the concerns of different stakeholders. Viewpoints help define the conventions for constructing and using the views, ensuring consistency and relevance.

- **Alignment with Agile and Digital Transformation:** The TOGAF Standard has placed a stronger emphasis on supporting agile methodologies and digital transformation efforts. The Content Framework is designed to be used in conjunction with modern practices like DevOps and cloud architecture.

- **Guidance for Tailoring:** The TOGAF Standard includes extensive guidance on how to tailor the metamodel and artifacts to fit specific organizational needs. This makes it easier to apply TOGAF in diverse environments. Tailoring the TOGAF Standard is done by selecting, adapting, and using those parts of the framework that are valuable to the situation or organization. It establishes a set of tools and a common vocabulary.

- **Outcome-Focused:** A major shift in the TOGAF Standard when compared with earlier versions is its focus on outcomes rather than just processes. The Content Framework reflects this by emphasizing the creation of valuable architecture artifacts that deliver business outcomes, rather than just following a rigid structure.

The Content Framework differs from previous versions in *flexibility*, *integration*, and its emphasis on *outcomes*. The modular and adaptable nature of the TOGAF Standard makes the Content Framework more flexible than in previous versions. It better integrates with other frameworks and methodologies such as the agile methodology. There is a stronger focus on creating actionable outputs that directly contribute to achieving business objectives.

6.3. Mapping to the Architecture Development Method

Looking at the Content Framework from a high-level perspective, it is not that difficult to see the implementation of the Architecture Development Method. Figure 6-1 illustrates the phases of the Architecture Development Method.

CHAPTER 6   CONTENT FRAMEWORK

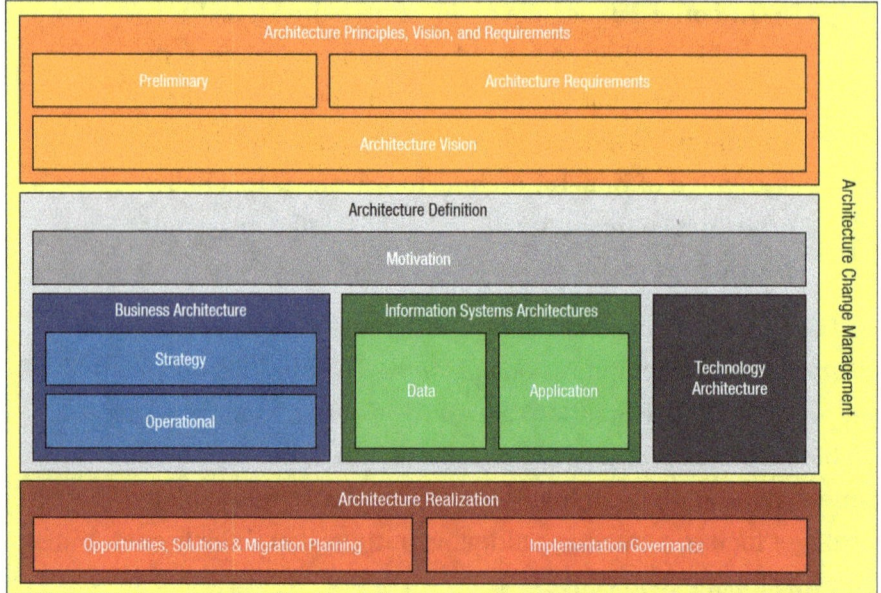

Figure 6-1. *The Content Framework by ADM phase*

The Content Framework consists of the following sections:

Architecture Principles, Vision, Motivation, and Requirements: The models in this section are intended to capture the surrounding context of formal architecture models, including general Architecture Principles, strategic context that forms input for architecture modeling, and requirements generated from the architecture.

The relevant aspects of the business context that have given rise to the Request for Architecture Work are typically recorded in the Preliminary and Architecture Vision phases.

Business Architecture: This part of the Content Framework captures architecture models of the business, looking specifically at factors that motivate the organization, its structure, and its capabilities. Phase B (Business Architecture) of the Architecture Development Method aligns with this part of the Content Framework.

Information Systems Architectures: The models described in this part of the Content Framework capture architecture models of applications and data. The Information Systems Architectures phase is used here.

Technology Architecture: The Content Framework describes models that capture technology assets, which are used to implement and realize information system solutions. This is realized by applying Phase D (Technology Architecture) of the Architecture Development Method.

Architecture Realization: The models in this section capture change roadmaps that show the transition between architecture states. They are used to guide and drive the implementation of the architecture. Phase F (Migration Planning) and Phase G (Implementation Governance) align with this part of the Content Framework.

Architecture Change Management: Internal and external events that affect the Enterprise Architecture are captured in this section of the Content Framework. They generate action requirements. Architecture Change Management is provided by Phase H of the Architecture Development Method.

The Content Framework should be used to *complement* the Architecture Development Method, as it describes what the architecture should look like when it is complete.

CHAPTER 6 CONTENT FRAMEWORK

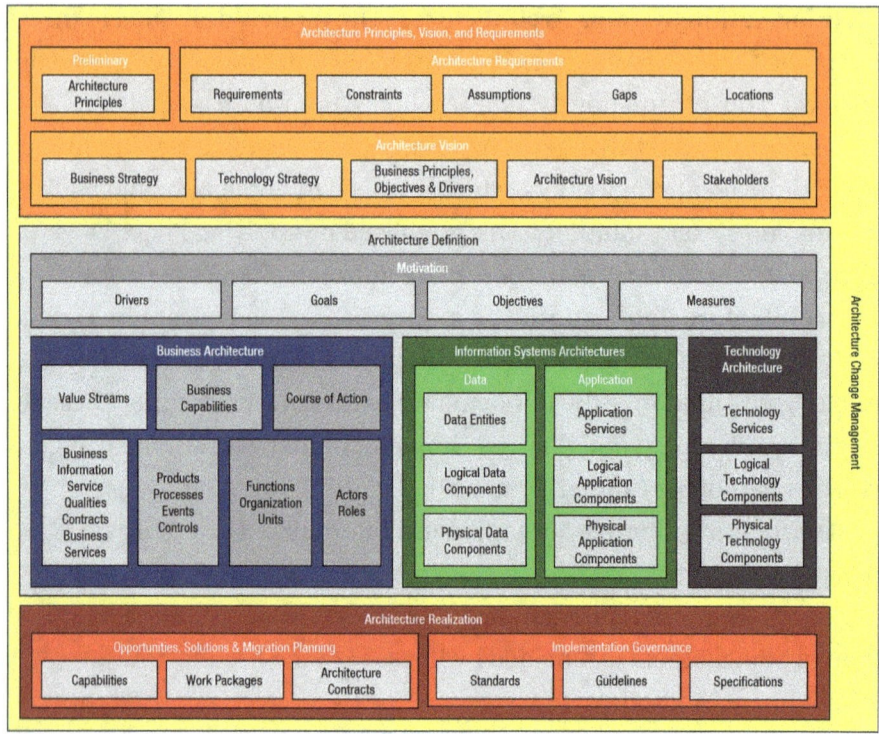

Figure 6-2. *The Extended Content Framework*

The *Extended* Content Framework, as the model in Figure 6-2 is called, contains the entities to be used or applied, in addition to the relationship to the phases of the Architecture Development Method mentioned earlier. The entities used and mapped in the Extended Content Framework are derived from the Enterprise Metamodel, which is described in detail in Chapter 5.

For example, consider the top-left section called Preliminary, which is part of the top layer called Architecture Principles, Vision, and Requirements. The Extended Content Framework shows that the development of Architecture Principles is what unfolds during this phase of the Architecture Development Method. Frankly, there is much more that happens during the Preliminary Phase, but the Extended Content Framework focuses on the development of Architecture Principles.

CHAPTER 6   CONTENT FRAMEWORK

To the right of the Preliminary section is the Architecture Requirements section. This area contains the entities *requirements, constraints, assumptions, gaps,* and *locations* – most of the General Entities.

Just below the Architecture Principles, Vision, and Requirements section is the Architecture Definition. This section contains by far the most entities. It is divided into the sections Motivation, Business Architecture, Information Systems Architectures, and Technology Architecture. These last three areas naturally relate to phases B, C, and D of the Architecture Development Method.

The section called Motivation differs slightly from the previous one-to-one references to the Architecture Development Method phases. Motivation contains the *driver, goal, objective,* and *measure* entities, all of which belong to the Business Architecture section according to the Enterprise Metamodel. One would expect these entities to have been made part of the Business Architecture section in the Extended Content Framework, but for some unknown reason the TOGAF Standard has not chosen to do so.

This is one of the inconsistencies in the TOGAF Standard.

Originally, the Business Architecture section had a relatively well-defined set of entities.

However, it may well be that the desire to emphasize the coherence of architectural entities, elements, and concepts in the development of the TOGAF Standard led to the addition of the driver, goal, objective, and measure entities. Add to this the fact that many different people have worked on the Standard, and it is less surprising that such inconsistencies sometimes creep in unintentionally.

CHAPTER 6   CONTENT FRAMEWORK

The Business Architecture section consists of 19 entities, all of which are described and explained in detail in Section 5.1.2 of Chapter 5. The same goes for the Information Systems Architectures (Data and Application Architecture) and Technology Architecture sections. The entities found in the Data, Application, and Technology Architectures are covered in Chapter 5, Sections 5.1.3, 5.1.4, and 5.1.5, respectively.

The final section of the Extended Content Framework is called Architecture Realization and consists of two parts, Opportunities, Solutions, and Migration Planning on the one hand and Implementation Governance on the other. The Opportunities, Solutions, and Migration Planning section has a one-to-one relationship with phases E and F of the Architecture Development Method. The entities *capabilities* and *work packages* mentioned in this section are part of the General Entities in the Enterprise Metamodel, as are the previously mentioned requirements, constraints, assumptions, gaps, and locations. The name of the Implementation Governance section indicates that it is associated with the phase of the Architecture Development Method of the same name. As an exception to the rule, the entities it contains are not found in the Enterprise Metamodel. The entities *standards*, *guidelines*, and *specifications* are taken from the Architecture Repository (see Chapter 7).

Architecture Change Management, as a section related to Phase H of the Architecture Development Method, is located in the Extended Content Framework on the right-hand side and runs from top to bottom.

The Content Framework and the Extended Content Framework are directly related to the phases of the Architecture Development Method. The TOGAF Standard does this very cleverly by directly showing that the schematic representation of an architectural approach can be seamlessly related to the well-known approach for its implementation, the Architecture Development Method.

6.4. Tailoring the Content Framework

As mentioned in Chapter 5, Section 5.2, the Architecture Development Method can update the Content Framework and the Enterprise Metamodel, but only under specific circumstances. While the Content Framework is a set of reference materials, the iterative nature of the Architecture Development Method allows the architecture content to be updated and refined as the architecture project progresses, specifically during the following phases:

Phase A (Architecture Vision): The initial architecture content is established and aligned with the business objectives. This phase may identify the need for additional content or specific viewpoints that weren't originally part of the Content Framework.

Phases B, C, and D (Business, Information Systems, and Technology Architectures): As detailed architectures are developed in these phases, the specific architecture artifacts created may add to or modify existing components of the Content Framework.

Phase H (Architecture Change Management): This phase is explicitly designed to manage changes in the architecture, which may necessitate updates to the Content Framework to reflect new or modified architecture standards, models, or deliverables.

Throughout the Architecture Development Cycle: New insights, stakeholder feedback, or the identification of new architectural requirements may drive changes or additions to the Content Framework, but to the Enterprise Metamodel as well.

CHAPTER 6   CONTENT FRAMEWORK

For example, during Phase F of the Architecture Development Method, it is recommended that a roadmap be created that includes all the activities required to implement the architecture work. Activities in the context of the TOGAF Standard are referred to as *initiatives* or *work packages*.

The Content Framework and the Enterprise Metamodel both include a reference to the work package entity. However, the TOGAF Standard doesn't seem to use this entity in any way other than in the Project Context Diagram artifact (see Chapter 12, Section 12.8.2, Figure 12-41).

It is therefore recommended that an additional architecture artifact be included in the Content Framework. Specifically, this would be a Work Package Portfolio Map (see Chapter 14, Section 14.8). It would also be a good idea to include the Work Package Portfolio Map artifact in the model as shown in Figure 7-13 in Chapter 7, Section 7.5.1.

During Phase E (Opportunities and Solutions) and Phase F (Migration Planning) of the Architecture Development Method, the Work Package Portfolio Map comes in handy. This artifact contains all the initiatives (work packages) needed to perform the architecture work. Additionally, this artifact illustrates the relationships between initiatives and organizational goals and objectives. Therefore, the Work Package Portfolio Map is an essential artifact for establishing the relationship between the organization's business goals and their planning using the Architecture Roadmap.

While the Enterprise Metamodel and the Content Framework serve as a guide, the iterative execution of the Architecture Development Method ensures that they evolve and are refined as part of the overall architecture governance and development process.

Further Reading

For more information on customizing the Content Framework, see "The TOGAF® Leader's Guide to Establishing and Evolving an EA Capability", Chapter 8, Customization of Architecture Contents and Metamodel, document number G184.

6.5. Summary

This chapter described the Content Framework, an important part of the TOGAF Standard. The Content Framework is a categorization framework that is used to make sense of descriptions of architectures. It is also used to place and clarify the different types of entities in the context of the Architecture Development Method. The Content Framework is a one-to-one translation of the phases of the Architecture Development Method.

In addition to the compact form of the framework, there is also a more extensive form called the Extended Content Framework. In this variant, all entities used in the architecture work are mapped to the phases of the Architecture Development Method. The entities used and mapped in the Extended Content Framework are derived from the Enterprise Metamodel.

CHAPTER 6   CONTENT FRAMEWORK

The Content Framework aims to be more adaptable to various organizational contexts while ensuring that all relevant architecture elements are covered comprehensively. It is more modular than previous versions, making it more adaptable. It also defines key artifacts like catalogs, matrices, and diagrams that help document different aspects of the architecture. The framework emphasizes the creation of views and viewpoints and places a stronger emphasis on supporting agile methodologies and digital transformation efforts.

A major shift in the TOGAF Standard when compared with earlier versions is its focus on outcomes rather than just processes. The Content Framework reflects this by emphasizing the creation of valuable architecture artifacts that deliver business outcomes, rather than just following a rigid structure.

The Content Framework differs from previous versions in *flexibility*, *integration*, and its emphasis on *outcomes*. The modular and adaptable nature of the TOGAF Standard makes the Content Framework more flexible than in previous versions. It better integrates with other frameworks and methodologies such as the agile methodology. There is a stronger focus on creating actionable outputs that directly contribute to achieving business objectives.

CHAPTER 7

Architecture Repository

 This chapter is part of the GREEN reading tour. See Figure 1-1 in Chapter 1, Section 1.3, for alternative reading tours.

This chapter explores the concept of the Architecture Repository. The TOGAF Standard describes the repository as an intangible concept in which all aspects of the architecture are collected and stored. The chapter delves into the various aspects of the Architecture Repository and explains each one in detail. The relationship and connection between components are discussed at length, as is a method for making the repository tangible through appropriate architectural tools. One of the three continua described in this chapter can be used to access stored assets at any time. Key components that are stored in the repository, such as principles, building blocks, viewpoints and views, and the three continua are explained in more detail. How they can be created and used in the context of an architecture problem or challenge is covered in this chapter.

CHAPTER 7   ARCHITECTURE REPOSITORY

7.1. The Central Hub

Many organizations and their architects use architecture tools. These tools range from spreadsheets and slide decks to robust applications that provide nearly all the functionality of an Architecture Repository. Therefore, they are often considered to be equivalent to an Architecture Repository. More often than not, these tools contain numerous diagrams and other visualizations of solutions developed over time.

Yet, according to the TOGAF Standard, the Architecture Repository is an intangible concept. It is *not a tool* and should not be directly compared to one. Furthermore, an Architecture Repository should only contain reusable building blocks, an organization-specific metamodel, and additional architecture governance information. A well-designed repository *does not* contain diagrams of specific solutions.

The Architecture Repository acts as the backbone or central hub for the Enterprise Architecture practice. It's where all critical *architectural assets* – like frameworks, templates, and reference materials – come together. By offering a structured environment, it helps ensure that architectural development is smooth, efficient, and aligned with best practices. Whether this requires handling completed artifacts or works in progress, the Architecture Repository provides a clear organization for everything, streamlining management and making it easier to evolve the Enterprise Architecture as the needs of the organization grow.

The Architecture Repository (Figure 7-1) consists of a number of key components. Each of these components serves its own use. It also acts as an input to the perspective provided by the Enterprise, Architecture, and Solutions Continua.

CHAPTER 7   ARCHITECTURE REPOSITORY

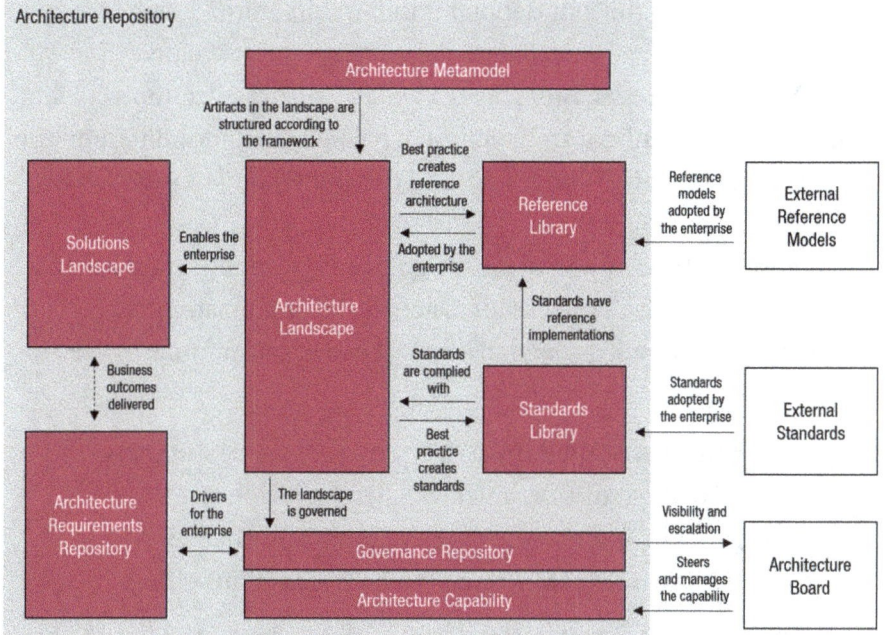

Figure 7-1. *The Architecture Repository*

7.2. Key Components of the Architecture Repository

Reading the visualization of the Architecture Repository from top to bottom, the first item encountered is the *Architecture Metamodel*.

Architecture Metamodel: Defines the structure for how architecture artifacts are organized and how different types of information (like business, data, application, technology, and security) relate to each other.

The Architecture Metamodel ensures consistency across architecture projects and helps architects to understand and relate different components of the architecture. It provides a common language, helping teams align across the organization. The Enterprise Metamodel (see

Chapter 5) serves as the foundation for the Architecture Metamodel and provides guidance for the *Architecture Landscape* to be created.

Architecture Landscape: This is a view of the organization's current, planned, and aspirational architecture. It typically includes different levels of architectures (strategic, segment, and capability) and timelines (past, present, and future).

Because the landscape can grow significantly over time and because of the different needs of stakeholders across the organization, the Architecture Landscape is divided into three levels of granularity (see Figure 7-2):

- **Strategic Architectures:** Provide a long-term summary view of the entire organization. They provide an organizing framework for operational and change activities and enable executive-level direction setting.

- **Segment Architectures:** Provide more detailed operating models for areas within an organization. Segment Architectures can be used at the program or portfolio level to organize and operationally align more detailed change activities. Another term often used for Segment Architectures is Domain Architectures. Figure 10-7 in Section 10.3.2 of Chapter 10 elaborates on Segment Architecture.

- **Capability Architectures:** Show in more detail how the organization can support a particular unit of capability. Capability Architectures are used to provide an overview of current capability, target capability, and capability increments and allow individual work packages and projects to be grouped into managed portfolios and programs.

CHAPTER 7 ARCHITECTURE REPOSITORY

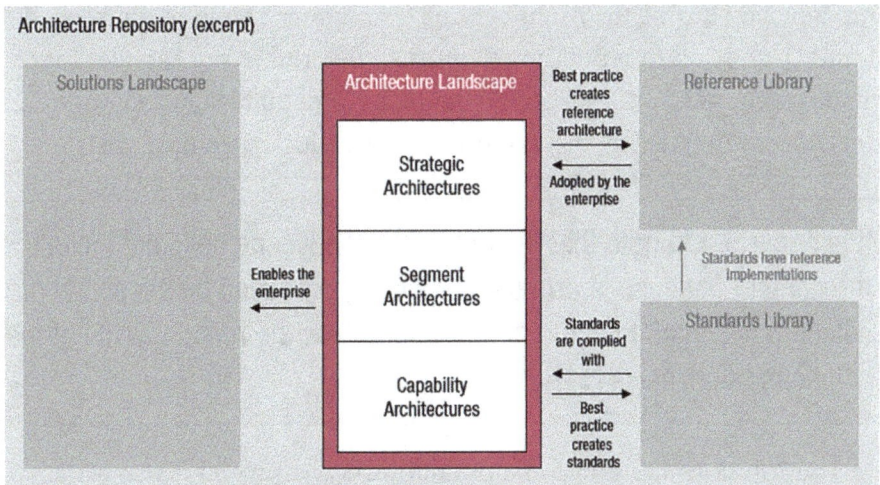

Figure 7-2. *Excerpt of the Architecture Repository, showing the three levels of architecture granularity*

Strategic, Segment, and Capability Architectures play an important role in developing architecture using an agile approach. Chapter 18 details the use of agile architecture.

The Architecture Repository also provides a container for the Architecture Building Blocks associated with the Architecture Landscape. These reusable building blocks are created during the architecture work as a result of following the Architecture Development Method.

Furthermore, the Architecture Landscape enables the organization to do what it does by providing solutions through the *Solutions Landscape* in the form of reusable Solution Building Blocks. These building blocks are stored in the Solutions Landscape of the Architecture Repository.

Solutions Landscape: This contains models and documents related to individual Solution Architectures within the organization. It is typically more detailed than the broader Architecture Landscape allows for the creation of specific systems solutions or business solutions by using Solution Building Blocks.

The Solution Building Blocks stored in the Solutions Landscape support the Architecture Building Blocks that were specified, developed,

131

and deployed during the Architecture Development Method phases A, B, C, and D. The Solution Building Blocks can be products or services that can be categorized as Strategic, Segment, or Capability Solution Building Blocks according to the Architecture Building Block specifications.

It is important to note that the Solutions Landscape does not include the content (such as information and data) produced by the solutions selected within and by an organization; this is the responsibility of the solutions themselves [11].

The Solutions Landscape helps architects and project teams work on individual projects while keeping them aligned with the broader Enterprise Architecture. It is essential for solution delivery and integration efforts. Just below the Solutions Landscape is the *Architecture Requirements Repository*.

Architecture Requirements Repository: This repository captures, manages, and tracks architecture requirements, including both business and technical requirements.

The Architecture Requirements Repository helps ensure that architecture projects are aligned with the actual needs of the business. It also allows for the consistent management of requirements across different phases and projects, which is critical for ensuring that architectural outcomes meet organizational goals. All requirements created or updated during the course of the Architecture Development Method are captured and managed in this section of the Architecture Repository.

The architecture requirements associated with the Strategic, Segment, and Capability Architectures are stored in the Architecture Requirements Repository:

- **Strategic Architecture Requirements:** Offer a forward-looking perspective on the needs of an organization, outlining the necessary adjustments to support its strategic direction. These requirements are pivotal in steering long-term transformations, ensuring that the organization aligns with its intended goals. Typically crafted at the executive level, they identify both the high-level direction and the specific shifts required to actualize it. Through this approach, Strategic Architecture Requirements serve as a roadmap for the organization's overarching vision and operational evolution.

- **Segment Architecture Requirements:** Offer precise, content-focused guidance for operational execution. Aligned with the organization's operating model, they break down broader strategic requirements into detailed, actionable steps. Typically crafted at the portfolio or project level, these requirements provide a more granular view, ensuring that necessary changes in architecture work are addressed comprehensively. In essence, Segment Architecture Requirements function as the tactical enablers that translate high-level goals into concrete, manageable actions within the architecture process. This level of detail ensures effective implementation and alignment with the organization's objectives.

- **Capability Architecture Requirements:** Focus on frameworks tied to specific capabilities or groups of capabilities. These requirements are crucial for defining the needs of individual initiatives, projects, or work packages within a broader project portfolio. The goal is to ensure that each piece aligns with the larger strategy, supporting the overall architecture while addressing the distinct objectives of each initiative.

CHAPTER 7 ARCHITECTURE REPOSITORY

The Architecture Requirements Repository is used during the execution of all phases of the Architecture Development Method. All requirements created or updated during the execution of the architecture process find their place in the Architecture Requirements Repository, where they are captured and managed.

To the right of the Architecture Landscape (for reference, see Figure 7-1) are two libraries situated: the *Reference Library* and the *Standards Library*. The first, the Reference Library, contains all kinds of reference material that can be used to develop architecture. This includes various bodies of knowledge, standards templates, and best practices.

Reference Library: This contains best practices, guidelines, templates, and patterns that can be reused in different architectural projects.

The Reference Library is a ready-to-use resource that accelerates the architecture development process. It includes reference architectures and models, templates, and the Viewpoint Library. The latter is a collection of views and viewpoints.

Architects can reference existing solutions, saving time and ensuring that the organization follows proven and standardized methods. This keeps solutions aligned and reusable across projects.

To make sense of the different classifications of reference materials, the TOGAF Standard refers to the Architecture Continuum (see Section 7.3.2). This continuum illustrates how reference architectures can be organized across a range – from Foundation and Common Systems Architectures to Industry and Organization-Specific Architectures [11].

The second library, as the name implies, contains standards in the form of specifications that the architecture must conform to.

Standards Library: A collection of standards and policies that guide the development and governance of the architecture. This includes IT standards, technology protocols, business process standards, and industry guidelines.

CHAPTER 7   ARCHITECTURE REPOSITORY

There are several types of standards, such as legal and regulatory mandates and industry and organizational standards. Both libraries (the Reference Library and the Standards Library) are fed by external reference models and standards. An example of an external reference model is the Business Architecture Guild Body of Knowledge, the BIZBOK Guide [5]. This guide serves as a reference for the practice and application of Business Architecture. Interestingly, the TOGAF Standard uses this body of knowledge as a reference. As for an external standard, the TOGAF Standard itself comes to mind. This open standard for architectural frameworks is called an external standard.

By providing architects and developers with a predefined set of standards, the Standards Library ensures compliance and consistency. It helps avoid redundancies and conflicts between different architecture components and projects.

The Architecture Landscape is governed by the Governance Repository and the Architecture Capability.

Governance Repository: A repository of decisions, rules, governance guidelines, and principles applied to architecture projects. It includes records of approvals, waivers, and compliance assessments.

The Architecture Landscape and Governance Repository provide decision-makers with the data and context needed for strategic planning, Risk Management, and operational alignment. Together with the Standards Library, the Governance Repository ensures that projects adhere to internal and external requirements, preventing non-compliance and ensuring regulatory alignment.

The Governance Repository thus provides accountability and traceability in architectural decisions, ensuring that all actions and developments adhere to the governance processes of the organization. This is critical for Risk Management and regulatory compliance.

Architecture Capability: This refers to the skills, processes, organizational structures, and roles that support the development and management of Enterprise Architecture.

The Architecture Capability helps in managing resources efficiently, ensuring that the organization has the right capabilities to deliver on architectural initiatives. It also supports continuous improvement in the Enterprise Architecture practice.

Both the Governance Repository and the Architecture Capability gain visibility through the Architecture Board. The Architecture Board is also used as an escalation mechanism. Conversely, from the perspective of the Architecture Board, it provides direction and management of the Architecture Capability.

The Architecture Repository is a powerful tool for managing architectural assets and supporting the governance, compliance, and collaboration necessary to deliver effective Enterprise Architecture. Its practical uses range from accelerating project delivery to ensuring consistency across all architecture initiatives within an organization.

Further Reading

Additional information about the Architecture Repository can be found in *The TOGAF Standard, 10th Edition*, document number C220.

7.3. The Three Continua

As mentioned in the previous section, the TOGAF Standard refers to what is called a *continuum* to make sense of the different classifications of reference materials. It introduces the concept to enable a modular and adaptable approach to shaping an organization's architecture. A continuum allows architects to scale and customize development processes, adjusting the level of detail, complexity, and governance to fit specific needs. It's a versatile tool designed to accommodate the shifting requirements of various architectural landscapes, ensuring that organizations can evolve and structure their architectures effectively, regardless of the challenges they face. This aligns with the TOGAF Standard's overarching flexibility and adaptability as a framework.

CHAPTER 7　ARCHITECTURE REPOSITORY

A continuum in the context of the TOGAF Standard can be seen *as a lens through which to view the Architecture Repository*. A continuum is a *conceptual framework* that helps guide the creation and evolution of architectures and solutions.

The framework distinguishes three continua: the *Enterprise Continuum, Architecture Continuum,* and *Solutions Continuum* (see Figure 7-3). They allow the categorization of architecture assets stored and made available in the Architecture Repository. When applied from left to right, these three continua provide an increasingly detailed perspective (view) of the contents of the Architecture Repository.

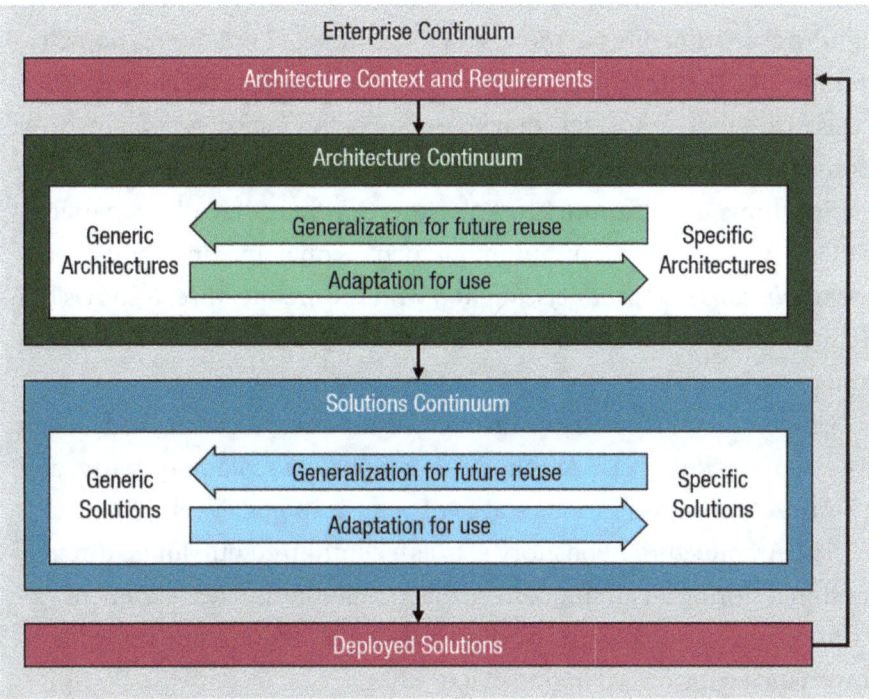

Figure 7-3. *The Enterprise Continuum*

CHAPTER 7 ARCHITECTURE REPOSITORY

Enterprise Continuum: The Enterprise Continuum acts as an *umbrella* term that covers both the Architecture Continuum and the Solutions Continuum, bridging the gap between conceptual architectures (part of the Architecture Continuum) and practical solutions (part of the Solutions Continuum). The Enterprise Continuum provides a structure for understanding how assets are developed and reused across different levels, from the highly abstract to the practical. It helps an organization manage assets related to its architecture and solutions and allows the reuse of those architectural assets (such as building blocks).

Architecture Continuum: This continuum guides the creation and evolution of *architectures* by offering a range of options, from generalized, high-level architectures to more specific, concrete implementations. The Architecture Continuum allows architects to organize and classify different levels of architectural assets that an organization may need as it matures. It spans from very generic and abstract to specific and detailed.

Solutions Continuum: The Solutions Continuum works in parallel with the Architecture Continuum but focuses on delivering actual, real-world *solutions and products*, aligned with the architecture. It shows how architectures are transformed into actual, deployable solutions, ranging from generic solutions to specific implementations.

Combined together, these three continua ensure a flexible approach to developing Enterprise Architecture, supporting scalability, reuse, and customization across various stages of organizational development.

The Architecture Repository acts as a centralized hub for all three continua, storing all architecture artifacts, standards, and references. It's the go-to resource for retrieving, managing versions, and applying governance controls, all in one place.

The setup demonstrated in Figure 7-4 streamlines the architecture effort, making it easier to maintain oversight. By organizing architectural content in a clear and structured manner, the Architecture Repository ensures that different teams and stakeholders stay on the same page.

CHAPTER 7 ARCHITECTURE REPOSITORY

Figure 7-4 highlights the locations of the Architecture Repository where the Architecture and Solutions Continua perspectives are concentrated.

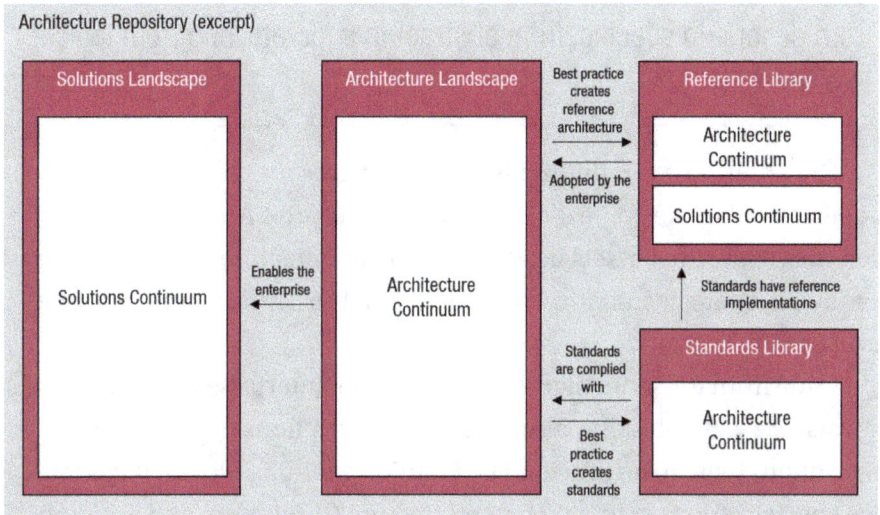

Figure 7-4. *Excerpt of the Architecture Repository, showing the views of the Architecture and Solutions Continua*

7.3.1. The Enterprise Continuum

As with all three continua, the Enterprise Continuum is a conceptual framework and provides a lens through which to view the entire collection of architecture assets stored in the Architecture Repository. This can include everything from architecture descriptions and models to building blocks, patterns, and viewpoints. These assets are not limited to the organization, but also include best practices and standards from the broader industry.

CHAPTER 7   ARCHITECTURE REPOSITORY

> The Enterprise Continuum provides a view of the Architecture Repository that shows the evolution of these related architectures from generic to specific, from abstract to concrete, and from logical to physical [11].

The Enterprise Continuum is the *outermost continuum* and classifies assets related to the context of the overall Enterprise Architecture. It classifies contextual assets used to develop architectures, such as policies, standards, strategic initiatives, organizational structures, and organization-level capabilities.

Aside from the more high-level assets, the Enterprise Continuum can also classify solutions, as opposed to descriptions or specifications of solutions, which are provided by Solution Building Blocks. It contains two specializations, namely, the Architecture Continuum and Solutions Continuum.

The Enterprise Continuum provides a way to consider resources on a scale ranging from the most general (foundation) to most specific (organization-specific). This scale is visualized in Figure 7-5.

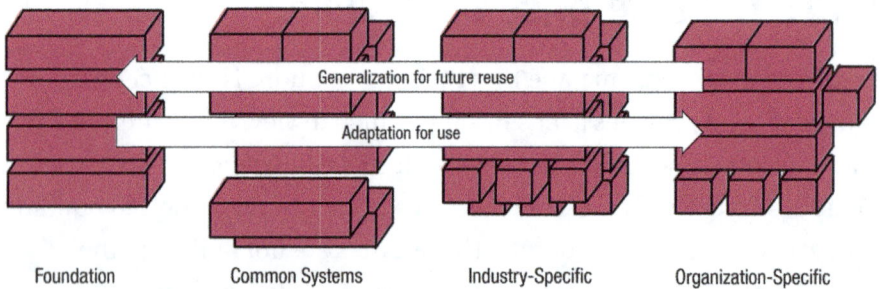

Figure 7-5. *The four classification models of the Enterprise Continuum*

CHAPTER 7   ARCHITECTURE REPOSITORY

> Remember the Enterprise Metamodel described in Chapter 5?
>
> This metamodel is a *Foundation* Enterprise Metamodel. This emphasizes yet again that the model should be used as a starting point and especially not as a single truth or a complete model that will suffice in every situation and for every organization.
>
> The model expects (or even requires) modification to make it organization-specific.

7.3.2. The Architecture Continuum

The structuring and categorization mechanism used for storing *architecture artifacts*, *standards*, and *reference architectures* in the Architecture Landscape, Reference Library, and Standards Library of the Architecture Repository is called the *Architecture Continuum*. This continuum guides the creation and evolution of architectures. It offers a range of options, from generalized, high-level architectures to more specific, concrete implementations.

The Architecture Continuum consists of four different architecture types that range from *foundational architectures* (generic) to *organization-specific architectures* (tailored for a particular organization). These architectures are used to define and deliver specific business capabilities, and the Architecture Continuum helps architects choose the appropriate architecture type based on the organization's needs.

The Architecture Continuum (Figure 7-6) illustrates how architectures are developed and evolved across the continuum ranging from Foundation Architectures through Common Systems Architectures and from Industry Architectures to an organization's own Organization-Specific Architecture.

CHAPTER 7 ARCHITECTURE REPOSITORY

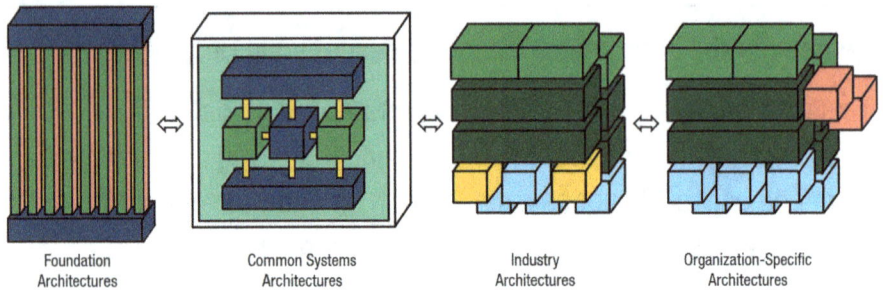

Figure 7-6. The four architecture types of the Architecture Continuum

---

Figure 7-6 may be a little difficult to interpret. In particular, because parts of the Foundation Architectures and Common Systems Architectures models (the object shapes and sizes) are not clearly reflected in the models representing Industry Architectures and Organization-Specific Architectures.

It would be more evocative if, for example, parts of the Foundation and Common Systems Architectures are used in the model of, say, the Organization-Specific Architectures.

This has been done with the Industry Architectures model; parts of it can be found in the Organization-Specific Architectures model.

Foundation Architectures: A Foundation Architecture is a fundamental architecture upon which other, more specific architectures can be based. It consists of generic components, interrelationships, principles, and guidelines that provide a foundation on which more specific architectures can be built.

Foundation Architectures are the most generalized and abstract of the four architecture types. They provide a high-level reference model for universal services and technologies that can be used across various industries. These architectures focus on fundamental building blocks such as network protocols, operating systems, security frameworks, and service-oriented architectures. Foundational components are crucial for ensuring that systems can operate efficiently and securely, but they are not specific to any specific industry or organization.

Common Systems Architectures: Common Systems Architectures build on top of the Foundation Architectures but are still generic across multiple industries. These architectures focus on more specific, common solutions that are applicable across different domains. They may include system-wide applications like content management systems, CRM, or data analytics platforms. Each is *incomplete* in terms of *overall system functionality*, but *complete* in terms of *a particular problem domain* (security, manageability, networking, operations, etc.).

For example, in healthcare, Common Systems Architectures might include general patient management systems, scheduling systems, or financial systems for billing and invoicing. These systems are necessary for running any healthcare organization, but they aren't necessarily unique to healthcare. They could be adapted from systems used in retail, education, or other industries. However, they must meet healthcare-specific regulatory requirements.

Industry Architectures: Industry Architectures are tailored to the needs of a specific industry. They provide reference models, guidelines, and standards that address the unique processes, requirements, and regulations of that particular industry. These architectures define best practices and common frameworks for industry-specific capabilities.

To illustrate, Industry Architectures would address the *unique workflows*, *standards*, and *regulations* relevant to the industry. To continue the example for the healthcare sector, this might include electronic health record (EHR) systems, telemedicine platforms, and health information exchange (HIE) standards[1]. Industry-specific regulations like HL7 for data exchange[2] and ICD-11 coding for medical records[3] are key aspects of this architecture type. These architectures ensure that healthcare organizations can efficiently deliver patient care while meeting industry regulations.

Organization-Specific Architectures: Organization-Specific Architectures are the most tailored and customized of the four architecture types. These architectures are *designed specifically* for the unique needs of a *particular organization*, based on its goals, strategies, and environment. They reflect the specific workflows, business processes, and organizational culture.

For a specific hospital or healthcare provider, the Organization-Specific Architecture might focus on implementing a customized EHR system that integrates seamlessly with the organization's unique workflows and patient management processes. It could involve bespoke features, such as a custom patient portal, integration with local pharmacy systems, and specific tools for managing population health. This type of architecture would also include specific configurations for mobile health applications, support for telemedicine, and integration with external systems, such as a local laboratory.

---

[1] A health information exchange (HIE) is a regional collaboration among independent healthcare organizations for sharing clinical information. Source: https://www.gartner.com/en/information-technology/glossary/hie

[2] Health Level Seven, abbreviated to HL7, is a range of global standards for the transfer of clinical and administrative health data between applications with the aim to improve patient outcomes and health system performance. Source: https://www.hl7.org/

[3] *International Classification of Diseases 11th Revision*. Source: https://www.who.int/standards/classifications/classification-of-diseases

CHAPTER 7   ARCHITECTURE REPOSITORY

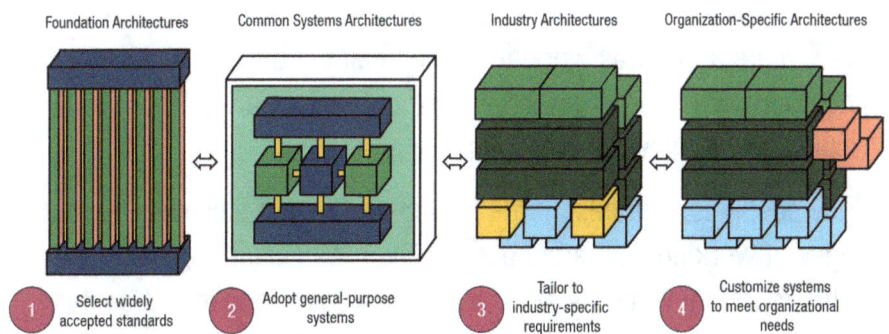

Figure 7-7. Four steps for applying the four architecture types of the Architecture Continuum

As illustrated by Figure 7-7, the four types of architectures from the Architecture Continuum are typically applied in the following ways.

1. **Start with Foundation Architectures** to ensure that the components such as infrastructure and security are built on widely accepted standards.

2. **Move to Common Systems Architectures** to adopt general-purpose systems like billing or scheduling software.

3. **Use Industry Architectures** to tailor the systems according to industry-specific requirements. For healthcare, this could be patient records and regulatory compliance.

4. **Develop Organization-Specific Architectures** to customize those systems based on the specific needs of the organization.

CHAPTER 7   ARCHITECTURE REPOSITORY

A few years ago, I was working in the healthcare industry. I had the opportunity to work on the implementation of an Enterprise Architecture for several hospitals.

Many hospitals – at least in the Netherlands – are still in the early stages of working with architecture. The methodical approach and architectural thinking are still rather underdeveloped in this sector.

The challenge I faced at that time was to map the hospital across the axis of the things people do there. To put it in architectural terms, to map the business functions and the corresponding care processes.

Now, each hospital – depending on the specializations within the organization – fills out the processes and business functions in a slightly different way than the next hospital. This led me to look for a generic view of business functions and healthcare processes.
I wanted something that I could use as a base and just add to or customize for the particular hospital I was working in.

That is how I ended up with ZiRA. This stands for the Dutch Hospital Reference Architecture.

ZiRA offered me a set of reference architectures that I could use as a foundation. For example, I was grateful to be able to use the existing business function model and the care process model. I was able to adapt both models to provide an accurate representation of the hospital I was working in.

Reusing an existing reference architecture, in this case an Industry Architecture, helped me tremendously, not only in terms of design and interpretation, but especially in terms of time.

CHAPTER 7   ARCHITECTURE REPOSITORY

7.3.3. The Solutions Continuum

The Solutions Continuum is aimed specifically at artifacts related to solutions. Therefore, this continuum categorizes the components and artifacts stored in the Reference Library or Solutions Landscape with increasing detail and specialization (see Figure 7-4). It provides a consistent way to describe and understand the implementation of the assets defined in the Architecture Continuum.

The Solutions Continuum represents the detailed specification and construction of the architectures at the corresponding levels of the Architecture Continuum. At each level, the Solutions Continuum is a population of the architecture with reference building blocks – either purchased products or built components – that represent a solution to the organization's business need expressed at that level. A populated repository based on the Solutions Continuum can be regarded as a solutions inventory. Having such an inventory can add significant value to the task of managing and implementing improvements in the organization.

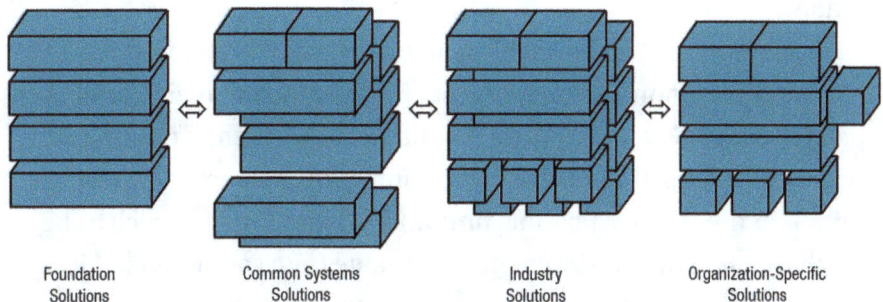

Foundation Solutions ⇔ Common Systems Solutions ⇔ Industry Solutions ⇔ Organization-Specific Solutions

Figure 7-8. *The four solution types of the Solutions Continuum*

Moving from *left to right* on the Solutions Continuum (Figure 7-8) focuses on delivering solution value. For example, Foundation Solutions deliver value by creating Common Systems Solutions. These solutions are then used to create Industry Solutions, and Industry Solutions are used to create Organization-Specific Solutions.

Moving from *right to left* on the Solutions Continuum focuses on addressing business needs.

These two perspectives are important to an organization that is trying to focus on its needs while maximizing the use of available resources through leverage. This is illustrated by Figure 7-9.

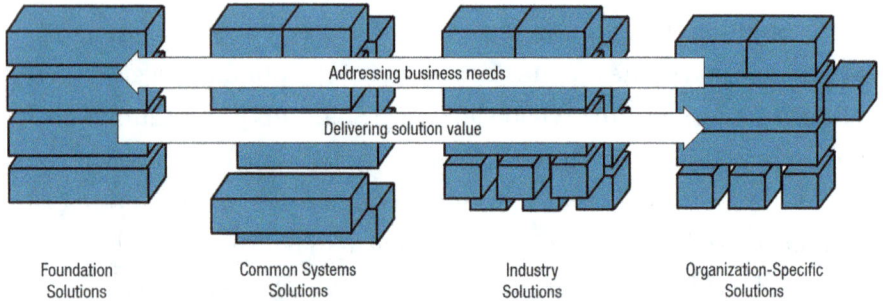

Figure 7-9. Two perspectives of the Solution Continuum

The Solutions Continuum contains the following four types of solutions:

Foundation Solutions: Highly generic concepts, tools, products, services, and solution components that are the fundamental providers of capabilities. Services include professional services, such as training and consulting services, that ensure the maximum investment value from solutions in the shortest possible time and support services, such as help desk, that ensure the maximum possible value from solutions (services that ensure timely updates and upgrades to the products and systems).

Common Systems Solutions: These are implementations of a Common Systems Architecture comprised of a set of products and services, which may be certified or branded. They represent the highest common denominator for one or more solutions in the industry segments that the Common Systems Solution supports.

Common Systems Solutions represent collections of common requirements and capabilities, rather than those specific to a particular customer or industry. They provide organizations with operating environments specific to operational and informational needs, such as high-availability transaction processing and scalable data warehousing systems. Examples of Common Systems Solutions include an enterprise management system product or a security system product.

Industry Solutions: Implementations of an Industry Architecture, which provides reusable packages of common components and services specific to an industry.

Fundamental components are provided by Common Systems Solutions and/or Foundation Solutions and are augmented with industry-specific components. Examples include a physical database schema or an industry-specific point-of-service device. Industry Solutions are – as the name implies – industry-specific and aggregate procurements that are ready to be tailored to an individual organization's requirements.

Organization-Specific Solutions: These solutions are the embodiment of implementations of the Organization-Specific Architecture that provides the required business capabilities. Because solutions are designed for specific business operations, they contain the highest amount of unique content in order to accommodate the varying people and processes of specific organizations.

Building Organization-Specific Solutions on Industry Solutions, Common Systems Solutions, and Foundation Solutions is the primary purpose of connecting the Architecture Continuum to the Solutions Continuum.

Further Reading

More information about the Enterprise, Architecture, and Solutions Continua can be found in *The TOGAF Standard, 10th Edition*, document number C220.

7.4. Catalogs, Diagrams, Maps, and Matrices

Many of the architecture assets or deliverables stored and categorized in the Architecture Repository consist of catalogs, diagrams, maps, and matrices.

Architecture deliverables are products created in the course of performing architecture work. They consist of one or more architecture artifacts, which are standardized architecture products for capturing and visualizing information related to the architecture work being performed.

Architecture artifacts, in turn, consist of one or more building blocks. These building blocks are documented using catalogs, diagrams, maps, and matrices and stored and categorized in the Architecture Repository.

The relationship of catalogs, diagrams, maps, and matrices to building blocks and architecture artifacts is shown in Figure 7-10. The figure shows that an architecture deliverable consists of one or more artifacts and that each artifact consists of one or more building blocks. A building block typically consists of one or more architecture entities. As mentioned earlier, the Architecture Repository provides storage for the building blocks.

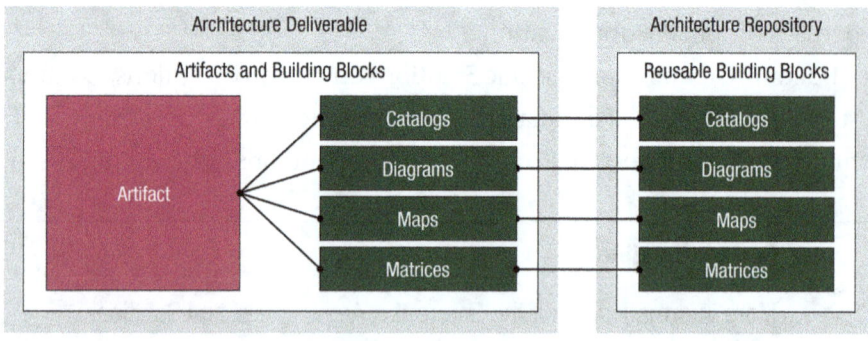

Figure 7-10. Relationship between architecture deliverables, artifacts, and building blocks

CHAPTER 7   ARCHITECTURE REPOSITORY

Sections 7.4.1 through 7.4.3 clarify exactly what is meant by catalogs, diagrams, maps, and matrices. Examples are used to illustrate the differences and similarities.

7.4.1. Catalogs

Catalogs are *lists of building blocks* of a specific entity, type, or related types that are used for governance or reference purposes. A catalog addresses a specific entity (e.g., processes, applications, technology components) and supplements the information about the concept with additional data. In a catalog, entities are not linked to other entities. An example of a catalog is shown in Table 7-1.

Table 7-1. Example of a Catalog

Entity Name	Value A	Value B
Entity A	...	...
Entity B	...	...
Entity C	...	...

A commonly used catalog is the Stakeholder Catalog (see Chapter 12, Section 12.4, Table 12-8, and Chapter 14, Section 14.3.2, Table 14-7). This catalog lists all stakeholders relevant to the architecture effort, along with their key concerns, classification, and interest (in the form of information provided).

7.4.2. Diagrams and Maps

Diagrams and maps are *representations of architectural content* in either *graphical format* (diagrams) or *textual format* (maps). Diagrams and maps can also be used as a technique for graphically populating architectural content or for verifying the completeness of collected information (think,

CHAPTER 7   ARCHITECTURE REPOSITORY

e.g., of an organizational diagram in which locations and actors are added). A diagram or map visualizes information captured in a catalog or matrix.

Table 7-2. *Example of a Map*

Entity A (Actor)	Entity B (Process)	Entity C (Business Service)
Actor A	Process A	Business Service A
Actor B	Process B	Business Service B
Actor C	Process C	Business Service C

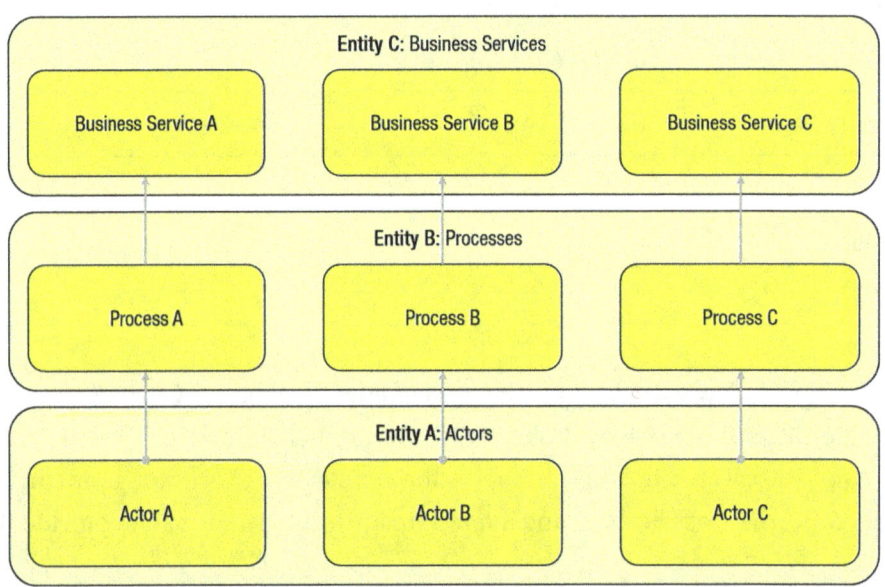

Figure 7-11. *Example of a Diagram*

Figure 7-11 shows the same information as Table 7-2, but in a graphical form. Figure 7-11 is therefore a diagram, as opposed to Table 7-2, which is called a map.

One of the more known and often used diagrams is called the Process Flow Diagram. Such a diagram visualizes the steps necessary to perform a certain process in a graphical manner.

A well-known example of a map is the Business Capability Map or Organization Map (see Chapter 12, Section 12.4). The Business Capability Map lists the capabilities that an organization needs to fulfill its purposes. The capabilities may be supplemented by other entities such as processes and applications.

The Organization Map shows the relationships between the primary entities that make up the organization, its partners, and stakeholders.

7.4.3. Matrices

A matrix is a grid that represents the relationships between two or more model entities. The relationship between the entities – typically spread across two or more catalogs – is brought together in a matrix. The method of bringing entities together is called *cross-mapping*.

Table 7-3. Example of a Matrix

	Entity B		
	Entity B Value A	Entity B Value B	Entity B Value C
Entity A ▼			
Entity A Value A	X		
Entity A Value B		X	
Entity A Value C			X

Table 7-3 shows two entities mapped together to create a cross-mapping. In the example, Entity A is cross-mapped to Entity B, to demonstrate their relationship. Entity A Value A has a direct relation with Entity B Value A, but not with Entity B Values B and C.

Knowing what to visualize – and what not to – is important for communicating with organizational stakeholders. As a general rule, less is more. Therefore, it is recommended to limit the amount of information used and displayed in a catalog, diagram, map, or matrix. It is better to create more views than to overload existing ones with information at the expense of the readability (and thus usability) of the architecture artifacts.

7.5. Architecture Deliverables

Architecture deliverables are an important part of the architecture being developed for an organization. They support the conversations that take place around architecture issues and provide insight into what certain situations or scenarios look like. Architecture deliverables are an essential link in the use of architecture artifacts to get the message across.

As mentioned earlier in Section 7.4, architecture deliverables are composed of one or more artifacts. Artifacts, in turn, consist of one or more building blocks. Building blocks include catalogs, diagrams, maps, and matrices.

The example in Figure 7-12 shows an architecture deliverable called the Architecture Definition Document. This deliverable consists of three architecture artifacts: an Organization Map, a Process Flow Diagram, and an Application Landscape. Each of the architecture artifacts consists of one or more building blocks. Architecture artifacts describe the (contents of) building blocks.

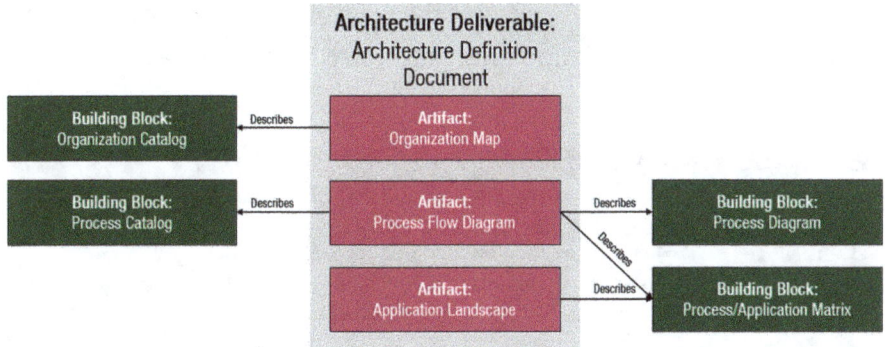

Figure 7-12. Example of an architecture deliverable and its contents

7.5.1. Architecture Artifacts

As one moves through the phases of the Architecture Development Method – whether through one of the iterative forms or not – a number of architecture artifacts may be produced.

It's important to note that creating all the artifacts listed in the TOGAF Standard is by no means a necessity. The framework only indicates the possibilities, but in no way requires that all artifacts be created. Again, use those artifacts that make sense and are useful to the organization or architecture work.

Once an organization has adapted the Enterprise Metamodel to its specific situation or requirements, an overview of the necessary architecture artifacts is formed based on this model.

Figure 7-13 shows an overview of the possible architecture artifacts to be created based on the Foundation Enterprise Metamodel as provided in the TOGAF Standard[4].

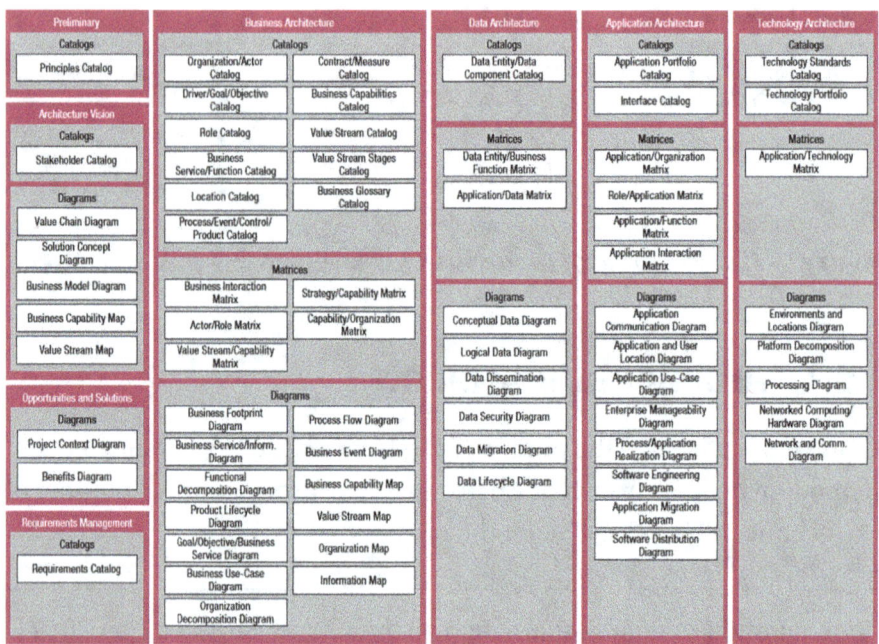

Figure 7-13. *Artifacts associated with the Foundation Enterprise Metamodel*

Architecture artifacts, as shown in Figure 7-13, are created during the Preliminary and Requirements Management phases and phases A–E of the Architecture Development Method (see Chapter 12, Sections 12.3 through 12.8 and Section 12.12).

---

[4] Figure 7-13 can be downloaded in a large format from https://eawheel.com/books/media/

CHAPTER 7 ARCHITECTURE REPOSITORY

As with all other components of the TOGAF Standard, Figure 7-13 and Table 7-4 show the possible artifacts that may be produced during the corresponding phases of the Architecture Development Method. Given the tailorability of the framework, it is certainly not necessary to produce all artifacts consistently.

It is very important to determine which artifacts actually add value to the architecture effort at hand and then create those artifacts. All other artifacts can then be ignored.

The recommended artifacts for production in each Architecture Development Method phase are as follows.

Table 7-4. *Artifacts and their entities created per ADM Phase*

ADM Phase	Artifacts	Used Entities
Preliminary Phase	Principles Catalog	Principle
Phase A: Architecture Vision	Stakeholder Catalog	– (or stakeholder, once added to the metamodel)
	Value Chain Diagram	Value stream
	Solution Concept Diagram	Objective, requirement, and constraint
	Business Model Diagram	Actor, role, stakeholder (see above), contract, function, process, business capability, business service, product
	Business Capability Map	Business capability
	Value Stream Map	Value stream

(*continued*)

Table 7-4. (*continued*)

ADM Phase	Artifacts	Used Entities
Phase B: Business Architecture	Organization/Actor Catalog	Organization unit, actor, location (when Location Catalog is not maintained)
	Driver/Goal/Objective Catalog	Organization unit, driver, goal, objective, measure (may optionally be included)
	Role Catalog	Role
	Business Service/Function Catalog	Organization unit, function, business service, application service (may optionally be included here)
	Location Catalog	Location
	Process/Event/Control/Product Catalog	Process, event, control, product
	Contract/Measure Catalog	Business service, application service (optionally), contract, measure
	Business Capabilities Catalog	Business capability
	Value Stream Catalog	Value stream
	Value Stream Stages Catalog	Business capability, value stream
	Business Glossary Catalog	–

(*continued*)

Table 7-4. (*continued*)

ADM Phase	Artifacts	Used Entities
	Business Interaction Matrix	Organization unit, function, business service
	Actor/Role Matrix	Actor, role
	Value Stream/Capability Matrix	Business capability, value stream
	Strategy/Capability Matrix	Business capability, value stream, course of action
	Capability/Organization Matrix	Business capability, value stream, organization unit
	Business Footprint Diagram	Organization unit, function, business service
	Business Service/Information Diagram	Business service, business information
	Functional Decomposition Diagram	Capability
	Product Lifecycle Diagram	Product
	Goal/Objective/Business Service Diagram	Goal, objective, business service
	Business Use-Case Diagram	Actor, role, process, function (later on, it can include data, application, and technology details)
	Organization Decomposition Diagram	Actor, role, location

(*continued*)

Table 7-4. (*continued*)

ADM Phase	Artifacts	Used Entities
	Process Flow Diagram	Process, all mappings related to the process entity
	Business Event Diagram	Event, process
	Business Capability Map	Business capability, entities that flesh out the business capability
	Value Stream Map	Business capability, value stream
	Organization Map	Organization unit, actor, role, stakeholder (if available)
	Information Map	Business information
Phase C: Data Architecture	Data Entity/Data Component Catalog	Data entity, logical/physical data component
	Data Entity/Business Function Matrix	Data entity, function (optionally, data entity relationship to owning organization unit)
	Application/Data Matrix	Logical application component, data entity
	Conceptual Data Diagram	Data entity
	Data Dissemination Diagram	Data entity, business service, logical/physical application component
	Data Security Diagram	Role, organization unit, logical/physical application component
	Data Migration Diagram	Data entity, location
	Data Lifecycle Diagram	Data entity, process

(*continued*)

Table 7-4. (*continued*)

ADM Phase	Artifacts	Used Entities
Phase C: Application Architecture	Application Portfolio Catalog	Application service, logical/physical application component
	Interface Catalog	Logical/physical application component
	Application/Organization Matrix	Logical application component, organization unit, actor, business service
	Role/Application Matrix	Logical application component, role
	Application/Function Matrix	Logical application component, function, business service
	Application Interaction Matrix	Application service, logical/physical application component
	Application Communication Diagram	Logical/physical application component, data entity
	Application and User Location Diagram	Logical/physical application component, location
	Application Use-Case Diagram	Actor, role, logical application component
	Enterprise Manageability Diagram	Logical/physical application component, logical technology component
	Process/Application Realization Diagram	Logical application component, process

(*continued*)

Table 7-4. (*continued*)

ADM Phase	Artifacts	Used Entities
	Software Engineering Diagram	Physical application component (composed of packages, modules, services, and operations)
	Application Migration Diagram	Logical/physical application component, staging areas (plateaus), logical technology component
	Software Distribution Diagram	Physical application component, physical technology component, location
Phase D: Technology Architecture	Technology Standards Catalog	Technology service, logical/physical technology component
	Technology Portfolio Catalog	Technology service, logical/physical technology component
	Application/Technology Matrix	Logical/physical application component, application service, logical/physical technology component
	Environments and Locations Diagram	Logical/physical technology component, logical/physical application component, location, actor
	Platform Decomposition Diagram	Logical/physical technology component

(*continued*)

Table 7-4. (*continued*)

ADM Phase	Artifacts	Used Entities
	Processing Diagram	Logical/physical technology component, logical/physical application component, actor, location
	Networked Computing/ Hardware Diagram	Logical/physical application component, logical/physical technology component, function
	Network and Communications Diagram	Logical/physical technology component
Phase E: Opportunities and Solutions	Project Context Diagram	Work package, organization unit, function, business service, process, logical application component, data entity, logical technology component
	Benefits Diagram	Stakeholder, once added to the metamodel, business capability, business service, product

The various artifacts mentioned during the implementation of the phases of the Architecture Development Method listed in Table 7-4 are detailed in Chapter 12.

7.5.2. Building Blocks

The previous section covered the possible artifacts that can be created during phases A–E of the Architecture Development Method. It is important to recognize that each of these artifacts consists of one or more building blocks.

CHAPTER 7   ARCHITECTURE REPOSITORY

Building blocks are the backbone of the architecture and serve as essential, reusable components in architectural solutions. These blocks form a modular foundation, designed to be generic and adaptable, ensuring they can seamlessly integrate with other building blocks. Their inherent reusability enables architects to combine them in various configurations, promoting efficiency and scalability in developing architectural solutions. By focusing on reusability and modularity, building blocks not only streamline architecture development but also provide flexibility, allowing architects to tailor solutions that meet evolving needs while maintaining consistency.

The comparison between building blocks and LEGO® is easy to make and also helps to understand how building blocks work.

The Danish toy company's bricks come in a variety of colors, shapes, and sizes. What they all have in common is that they are made so generically that they can easily be combined with other bricks to create new (sometimes larger) sets.

Building blocks in the context of the TOGAF Standard work much like LEGO bricks. They, too, can be stacked to create new building blocks.

For example, consider the Solution Building Blocks in Figure 7-14. The figure shows that stacking three Solution Building Blocks creates a larger one, an Architecture Building Block.

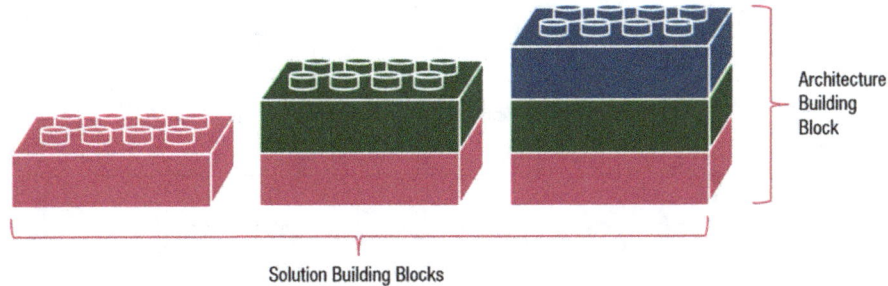

Figure 7-14. *Stackable LEGO bricks*

Flexible as they are, building blocks can be defined at varying levels of detail, depending on the stage of architecture development. In the early stages, a building block might just have a name or a brief description. As development progresses, it can be broken down into supporting components, complete with detailed specifications. Building blocks serve as key references for both *architecture* and *solution* design. This modular approach ensures adaptability across different phases and project requirements.

A building block represents a potentially reusable component that can be combined with other building blocks to deliver architectures and solutions [2].

Building blocks are packages of functionality defined to meet the business needs across an organization. They correspond to the entities used and defined in the Enterprise Metamodel (see Chapter 5), such as *organization unit, business service, application,* or *data entity*.

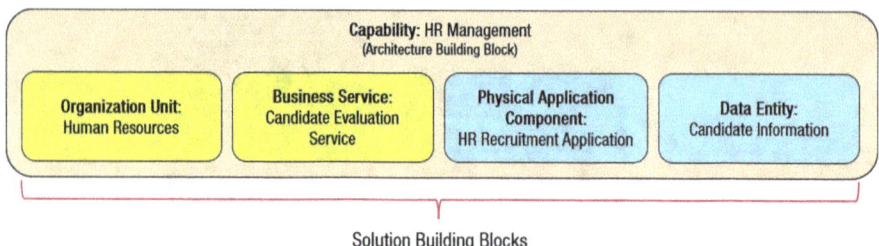

Figure 7-15. Architecture Building Block composed of multiple Solution Building Blocks

A building block is a well-defined unit with clear boundaries, designed to be useful and preferably reusable. Whether tangible or intangible, building blocks are meant to work seamlessly with other building blocks, enhancing their flexibility and adaptability in various architectural scenarios.

The TOGAF Enterprise Metamodel identifies the types of entity that need to be considered when generating a logical model, specifying a change (Architecture Building Blocks) and creating their associated physical and operational implementations (Solution Building Blocks) [16].

Because of the interaction with other components, it is imperative that the definition and scope of these building blocks should not be tightly bound to their implementation. In simpler terms, the building block's design should remain flexible enough to accommodate various implementations without altering its core structure. Different organizations will approach this assembly process uniquely, depending on their architecture.

When considering the level of detail for specifying a building block, it's crucial to align with the architecture's objectives. Sometimes, less detail can deliver more impact. For instance, when showcasing an organization's capabilities using a Business Capabilities Catalog or a Business Capability Map, a single, clear diagram often holds more value than a lengthy, complex 100-page document. It's about clarity and precision – tailoring the information to what truly matters, ensuring that the message is concise and accessible, without overwhelming the audience with unnecessary depth.

7.5.2.1. Building Blocks in Architecture Design

The various building blocks in an architecture specify the scope and approach that will be used to address a specific business problem.

An architecture is a set of building blocks depicted in an architectural model and a specification of how those building blocks are connected to meet the overall requirements of the business [11].

When designing specific architectures, certain core principles guide the use of building blocks:

- Only include building blocks that directly address the business problem at hand.
- Building blocks often interact in complex ways.
- A single building block may support multiple others or provide partial support to one.

These principles ensure that the architecture remains relevant and appropriately structured to meet the unique challenges of the business.

In the early phases of developing architecture, building blocks tend to be more generalized. This broad approach helps ensure smooth integration across various components. But as projects move forward and the details of implementation come into play, these building blocks need to become more refined and precise. A clearer view of these elements helps guide strategic decisions, highlights crucial choices, and evaluates the long-term impact of shared resources and reusability. By focusing on these critical aspects, the architecture becomes more aligned with both the strategic goals and operational demands of the organization. Each building block should adhere to the appropriate standards, Enterprise Principles, and any specific organizational guidelines.

7.5.2.2. Building Block Specification Process

The evolutionary and iterative process of building block definition takes place gradually as the Architecture Development Method is followed, mainly in phases A, B, C, and D. As the definition proceeds, detailed information about the functionality that is required, the constraints imposed on the architecture, and the availability of products may affect the choice and the content of building blocks.

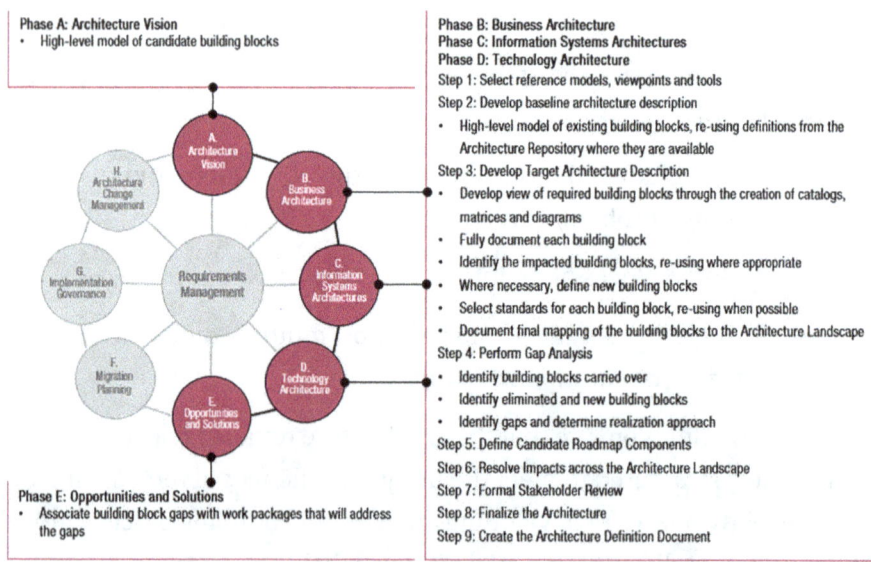

Figure 7-16. ADM Phases and steps at which building blocks are evolved and specified

During Phase A (Architecture Vision), a high-level model of the candidate building blocks is created. The purpose of such a model is to identify the proposed change(s) required from an architectural perspective. This could include working out the required capabilities of an organization to implement the proposed change. An example of a high-level model typically created during Phase A (and updated and further specified in Phase B) is a Business Capability Map.

The execution of phases B, C, and D clarifies what building blocks are needed to further interpret the desired change(s). Phases B, C, and D specify the entities that flesh out the Business Capability Map created in Phase A.

If the building blocks do not exist, they are created using catalogs, matrices, and diagrams. If they do exist, it is preferable to reuse them from the Architecture Repository. During phases A, B, and C, not only are new building blocks created or existing building blocks reused, but building blocks that are no longer needed are removed from the repository. This last activity ensures that the Architecture Repository is kept clean and up to date.

Building blocks are created during Phase E (Opportunities and Solutions) using work packages (initiatives). These initiatives fill in the gaps for the missing building blocks.

The key phases and steps of the Architecture Development Method at which building blocks are evolved and specified are summarized in Figure 7-16. The major work in these steps consists of identifying the Architecture Building Blocks required to meet the business goals and objectives. The selected set of Architecture Building Blocks is then refined in an iterative process to arrive at a set of Solution Building Blocks, which can either be bought off-the-shelf or custom developed.

7.5.2.3. Types of Building Blocks

Building blocks in the TOGAF Standard come in two flavors: *Architecture Building Blocks* and *Solution Building Blocks*. The distinction between them is simple but crucial. Architecture Building Blocks represent the *what* and *why*, providing high-level, technology-agnostic capabilities. Meanwhile, Solution Building Blocks dive into the *how*, focusing on practical implementation using specific technologies and products.

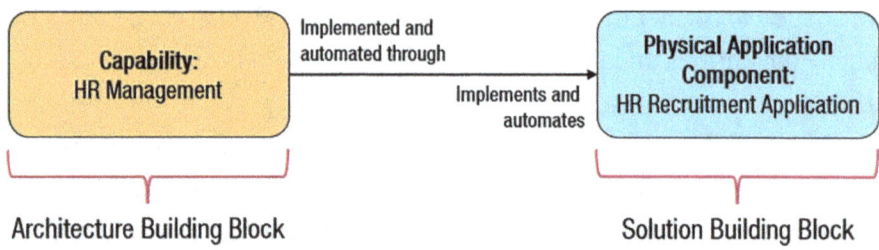

Figure 7-17. Architecture Building Block capability implemented and automated through a Solution Building Block

---

Note that Figure 7-17 shows an Architecture Building Block being implemented and automated by a Solution Building Block.

Normally, Solution Building Blocks only implement Architecture Building Blocks. However, in the case where the Architecture Building Block is a capability entity and the Solution Building Block is an entity from either the data, application, or technology domains, the Solution Building Block can also automate the Architecture Building Block [5].

By abstracting functionality and capabilities in Architecture Building Blocks and then implementing them through Solution Building Blocks (see Figure 7-17), the TOGAF Standard ensures the architecture remains flexible and scalable, while also grounded in practical solutions.

CHAPTER 7   ARCHITECTURE REPOSITORY

7.5.2.4. Architecture Building Blocks

When architecture work begins and the Architecture Development Methodology is applied during an initial iteration, Architecture Building Blocks are typically created during this round. These building blocks capture architecture requirements such as business, data, application, and technology requirements. The information stored in the building blocks from this activity can be used to direct and guide the development of Solution Building Blocks in a later phase of architecture development.

Architecture Building Blocks represent high-level conceptual components that define what needs to be delivered without specifying the technology or implementation details. They are concerned with requirements, capabilities, and strategic objectives, focusing on functionality, services, and constraints that are independent of specific technologies. Architecture Building Blocks typically help in shaping the architectural vision and form the blueprint for future solutions.

Architecture Building Blocks are architecture documentation and models from the enterprise's Architecture Repository [11].

Specifications for Architecture Building Blocks include fundamental functionality and attributes, such as semantic and unambiguous attributes, security capabilities, and manageability. They also ensure interoperability and relationships with other building blocks. A final important detail of building block specifications is that they map to business or organizational entities.

Architecture Building Blocks typically describe a required capability and shape the specification of Solution Building Blocks. For example, an organization may require a Human Resources Management capability that is supported by a number of Solution Building Blocks, such as an organization unit, a business service, an application, and a data entity (see Figure 7-15).

171

CHAPTER 7 ARCHITECTURE REPOSITORY

The Architecture Building Block in this example outlines the need to manage HR-related activities, including, to name one, the acquisition of new employees. The capability, in this example, does not specify a particular product, process, or service, so it remains high-level.

Although not explicitly named in the TOGAF Standard, Architecture Building Blocks are typically stored in the Architecture Repository. Since this repository serves as a central location for all architectural assets, it includes both current and reusable assets.

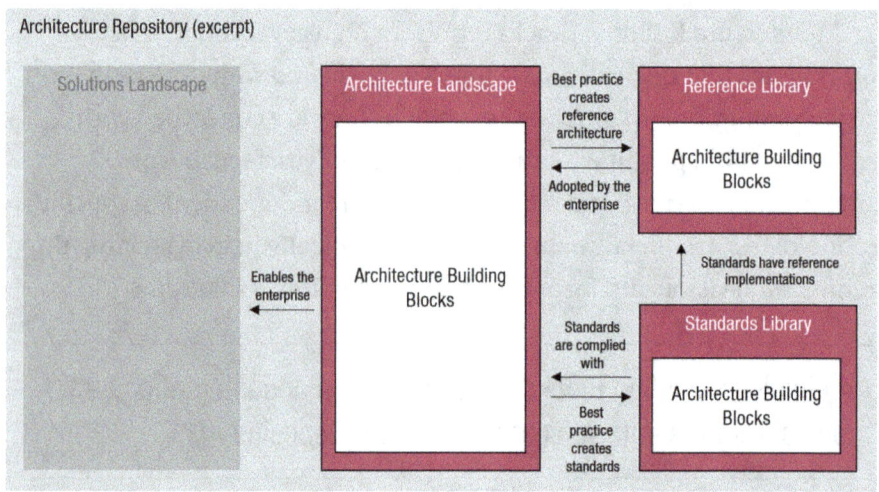

Figure 7-18. *Excerpt of the Architecture Repository, showing Architecture Building Blocks storage locations*

Within the Architecture Repository, Architecture Building Blocks are stored in the following sections (see Figure 7-18):

- **Architecture Landscape:** This stores various architectural models at different levels of abstraction, including the Architecture Building Blocks that define architecture elements at different stages of architecture development.

- **Reference Library:** This section contains reusable architecture assets like architecture patterns, principles, standards, and Architecture Building Blocks that can be referenced when developing architectures.
- **Standards Library:** Architecture Building Blocks can be stored here when they align with industry standards, helping to ensure consistency with established guidelines.

These sections within the Architecture Repository allow organizations to store, manage, and govern Architecture Building Blocks effectively as part of their architecture practice.

7.5.2.5. Solution Building Blocks

In contrast to the high-level character of Architecture Building Blocks, Solution Building Blocks are actually much more detailed in nature. In fact, these building blocks describe the entities that make up the parent Architecture Building Block. Solution Building Blocks define what products and components will implement the required functionality; they determine the actual implementation of (parts of) the architecture. Solution Building Blocks fulfil business requirements and are product- or vendor-aware, something that Architecture Building Blocks are not.

Solution Building Blocks represent actual implementation choices to realize a required capability [2].

For example, a business service such as the Candidate Evaluation Service mentioned in Figure 7-15 is a Solution Building Block, since it can be described through complementary artifacts and then put to use to realize solutions for the organization. Solution Building Blocks may be either procured or developed.

Specifications for Solution Building Blocks include specific functionality and attributes. These building blocks map to the IT topology and operational policies. The attributes of Solution Building Blocks are aimed at, for example, security, manageability, localizability, and scalability, but also at performance and configurability.

One final specification of Solution Building Blocks is their relationship to Architecture Building Blocks and other Solution Building Blocks.

Solution Building Blocks represent physical or concrete components that implement the functionalities or capabilities defined by the Architecture Building Blocks. They include specific technologies, products, and solutions that are deployed to realize the architecture. Solution Building Blocks are directly linked to real-world implementation and include the detailed design of systems, platforms, and technologies that meet the requirements laid out by Architecture Building Blocks.

The aforementioned HR Management capability (Architecture Building Block) did not provide specific details as to which application could support the capability. This is where the Solution Building Block comes into play. This building block defines the actual application (by using the logical or physical application component entity) that automates the HR Management capability. The specific product provides all the required functionalities.

Just as the TOGAF Standard does not mention where the Architecture Building Blocks are stored, it also omits a description of the location of the Solution Building Blocks.

CHAPTER 7   ARCHITECTURE REPOSITORY

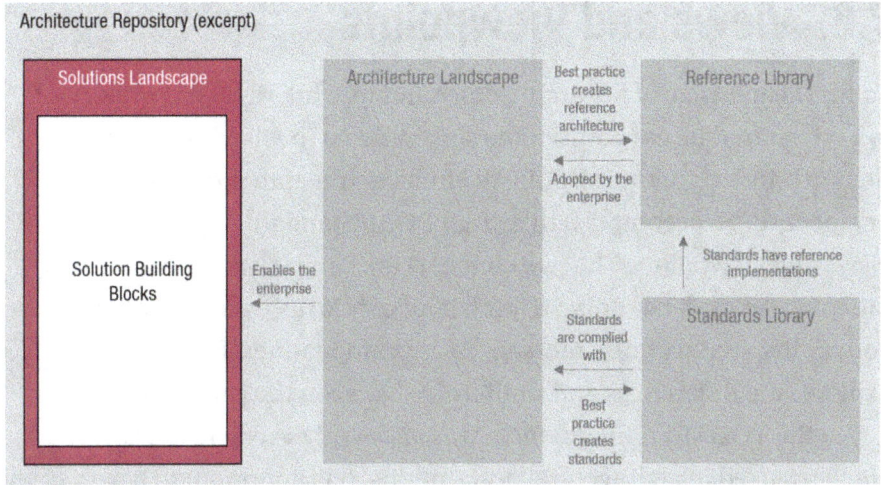

Figure 7-19. *Excerpt of the Architecture Repository, showing the Solution Building Blocks storage location*

Solution Building Blocks are stored in the Architecture Repository's following section (see Figure 7-19):

- **Solutions Landscape:** This stores the Solution Building Blocks at different levels of abstraction. These can be shaped as catalogs, diagrams and maps, or matrices.

 Further Reading

Additional information on building blocks can be found in *The TOGAF Standard, 10th Edition*, document number C220, and the TOGAF Series Guide "An Approach to Selecting Building Blocks", document number G248.

7.6. Views and Viewpoints

To get the message of the architecture and its work right, it is important to understand the concerns of the stakeholders involved. This is a prerequisite for communicating the final architecture from the right perspective. To accomplish this, the TOGAF Standard refers to the use of views and viewpoints. The two concepts are similar, yet fundamentally different. An understanding of both concepts is necessary in order to convey the architecture's message in a sound manner. This section explores the differences and similarities between the concepts.

In the TOGAF Standard, the terms *view* and *viewpoint* are key concepts used in architectural descriptions. It is important to have a clear understanding of what they are and how they differ from each other.

View: A representation of a system or architecture from the *perspective of a specific stakeholder*. It is essentially a *visualization* or *presentation* of the architecture that addresses particular concerns or interests of one or more stakeholders.

The purpose of a view is to help communicate architectural decisions or system properties to stakeholders in a way that is meaningful to them. For example, a security view might describe the security measures in place for a system, while a process view could show how business processes flow through the architecture.

Viewpoint: Defines the *perspective* and *rules* for creating a view. It is essentially a *template* or a *set of guidelines* for constructing views to address specific concerns. Viewpoints outline *how* to structure views and which stakeholders they are aimed at.

It ensures that views are created in a consistent manner and address relevant concerns of stakeholders. A performance viewpoint would specify how to create views that deal with performance-related aspects of the system, like response times or throughput.

7.6.1. Relationship Between Views and Viewpoints

A viewpoint is like a *blueprint* for constructing views. Multiple views can be created using a single viewpoint, but each view is tailored to specific stakeholder concerns.

In summary, a *view* is what you actually present or show to stakeholders, while a *viewpoint* is the set of rules or guidelines that dictate how you create that view.

Figure 7-20. *Two viewpoints and one view*

Views and viewpoints are part of the Reference Library of the Architecture Repository. This part of the repository contains organizational reference material, which includes views and viewpoints. The differences and similarities between the two concepts are best illustrated by a simple example (as depicted in Figure 7-20).

Imagine the following situation: There is a small village located in a valley. This valley is surrounded by two mountain peaks, one to the left and one to the right of the village. On each mountaintop there is a pair of binoculars used to look at the village. Each pair of binoculars is used by a different person, one on each mountaintop. Both people are looking through the binoculars at the village in the valley. The person on the left mountaintop may be focusing on the church tower in the village, while the person on the right mountaintop is focusing on the green park surrounding the village.

Both people are looking at the same village, but may see it slightly differently. Perhaps this is due to the position of the binoculars, so that the person who focuses on the church tower cannot see the green square at all. Another possibility is that this person is not interested in the green square at all and therefore focuses on the church tower. Two people looking at the same village from two different vantage points.

In this example, the individuals are the stakeholders within an organization, and the binoculars used on the two mountaintops are the viewpoints from which the village is viewed. In this example, the village represents the view.

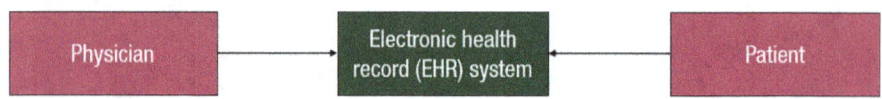

Figure 7-21. *Two viewpoints and one view in a real-world scenario*

Let's consider a real-world scenario as shown in Figure 7-21. A physician and a patient both *looking* at the same system, in this case the electronic health record (EHR) system. The physician views the EHR system as a tool to efficiently manage patient information, support (real-time) decision-making, and streamline administrative tasks. The patient, on the other hand, views the EHR system as a tool that focuses on accessing their medical information, communicating with healthcare providers, and managing their health in a convenient and user-friendly format.

In practice, different stakeholders may view the same system, situation, or scenario differently. This may be due to constraints, specific requirements, or specific expectations or concerns of the stakeholders. Therefore, it is very important to create multiple viewpoints for the same view.

CHAPTER 7   ARCHITECTURE REPOSITORY

Further Reading

For more information about views and viewpoints, consult *The TOGAF Standard, 10th Edition*, document number C220.

7.7. Using the Architecture Repository

The Architecture Repository is - as are most concepts described in the TOGAF Standard - positioned as an intangible object (or concept). This can make it difficult to translate the concept of the repository into an implementation suitable for everyday practice. Fortunately, the repository can be made more tangible by introducing and acquiring an architecture tool. Most architecture tools offer excellent ways to document and store essential architectural artifacts such as catalogs, diagrams, maps, and matrices. Architecture and Solution Building Blocks (the latter being responsible for providing content to the catalogs, diagrams, maps, and matrices) can also often be stored inside an architecture tool.

There are many tools on the market today, and each one offers advantages over the others. It is therefore important to think carefully about what is required of such an architecture tool (see Table 7-5).

In a small organization, it may initially seem possible to use a word processor or presentation software (or even a spreadsheet application) as a tool for maintaining the Architecture Repository. However, it will soon become apparent that these (in themselves excellent) products do not provide the capabilities that an architect would want or need. Besides the fact that these types of tools do not support a unified and globally accepted modeling language, it also means that each architecture artifact is created independently of other artifacts. This implies that each artifact must be manually tracked. As an organization and its architecture grow and take on increasingly voluminous forms, this becomes impossible to keep up with.

179

Tools that provide some sort of database construction already come close to what is needed. One of the most important things an architecture tool should provide is the ability to relate the entities used within the context of the architecture. By being able to properly establish and maintain these relationships, changes can be made quickly and accurately, without loss of context and without having to do it all over again for each individual artifact.

7.7.1. Deciding on a Tool

It was mentioned at the beginning of this chapter that the Architecture Repository is an intangible concept and is not necessarily equivalent to an architecture tool. However, many organizations use a tool under the assumption that it is.

Vendors of architecture tools offer a variety of features in their products that come close to meeting the requirements of an Architecture Repository. The use of an architecture tool is therefore essential.

Deciding which tool is best for the organization depends on several factors. Budget, for example, is one such important factor. If sufficient budget is available for the purchase of a solid architecture tool, the organization can choose a product that emerges as a market leader. By purchasing such a product, the organization will have access to all possible functionality, and the creation and management of architectural artifacts will become much easier. If there is less budget available, the organization will be forced to purchase a tool that automatically offers less functionality.

Every year, Gartner produces an overview and report of the market leaders in the field of architecture tools [17]. This report can be used to select a suitable product. The availability of an architecture tool is an absolute must in order to turn the Architecture Repository described by the TOGAF Standard into the aforementioned tangible solution.

Some considerations that can be used in the search for an appropriate tool are shown in Table 7-5. These considerations are in addition to the fact that the tool must have at least solid repository functionality.

Table 7-5. *Architecture tool functionality considerations*

Area of Functionality	Considerations
Support for a modeling language	Does the tool support the use of a modeling language (such as ArchiMate) in its entirety? Can user-defined properties be added to elements and relationships? Is the use of BPMN supported? Is it possible to create colored views?
Metamodel customization	Does the tool allow adjustments to its metamodel? Can the tool be adapted to personal or organizational needs?
Tool flexibility and stability	Does the tool allow changing one concept to another? Will previously defined relationships be maintained when changing from one concept to another? Is the tool enough when editing or creating large and bulky models?
Import/export functionality	Does the tool have import/export capabilities? Can the tool make information available to third parties (who do not work with the tool or use other tools)? Can (large) PDF files be published for use as posters and information exchange?
Support	Is the tool supported by a good help desk and user community?

CHAPTER 7   ARCHITECTURE REPOSITORY

Capturing architecture artifacts (catalogs, diagrams, maps, and matrices) in an architecture tool that provides the ability to publish them (to a website) helps increase architecture maturity. This is an important consideration when choosing a tool. It is an easy step to higher maturity.

The factors that ultimately determine the choice of an architecture tool largely depend on the situation. The same goes for the requirements for using a tool. These requirements are not easy to articulate – there is no standard set of requirements – because they are different for everyone.

The architecture tool selection described in this section should result in a product that can be used to create and maintain an Architecture Repository. At a minimum, the selected product should be able to provide a home for artifacts related to the organization's Architecture and Solutions Landscapes, Reference and Standards Libraries, and Architecture Requirements Repository. This is illustrated in Figure 7-22.

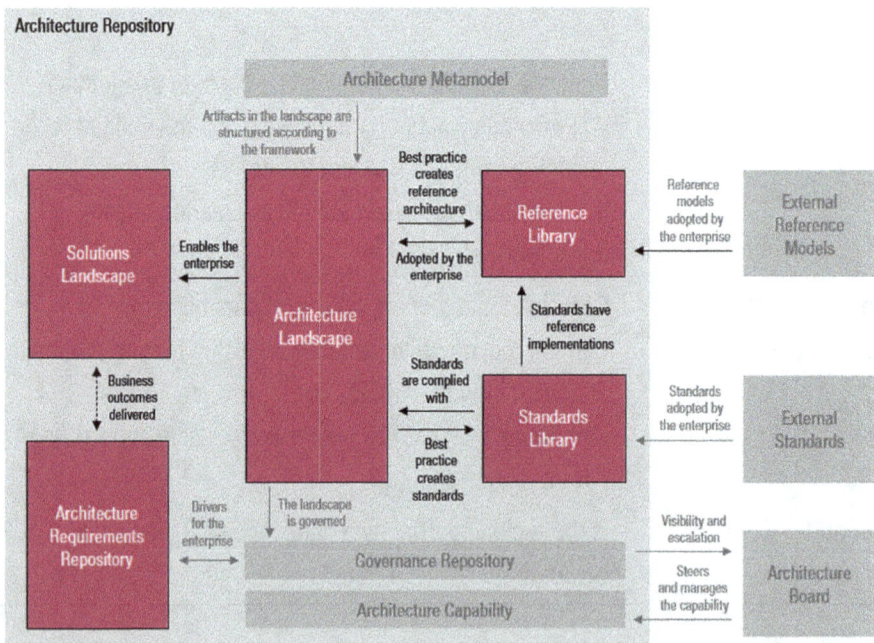

Figure 7-22. Minimum Architecture Repository components that must be represented in an architecture tool

In summary, a repository tool should offer functionality in order to support what is minimally needed to manage an Enterprise Architecture. Information that is not directly required for the Enterprise Architecture should not be included in the repository.

A well-managed repository is characterized by ruthlessly minimizing information collected and maintained [18].

In practice, however, a repository often does not work as described in the TOGAF Standard. There are few, if any, architecture tools that provide a one-size-fits-all solution that conforms to the framework's guidelines. For example, nine out of ten architecture tools do not distinguish between the Architecture Landscape and the Solutions Landscape. Of course, a good tool will allow its user to make this distinction based on a folder structure or something similar.

It's imperative to design the structure of the repository based on the architectural artifacts and the needs of the organization. This can be as simple or as complex as needed. For example, a hierarchical structure with folders and subfolders would be an excellent first step. It is important that the structure is intuitive and easy to navigate.

7.8. Summary

The Architecture Repository serves as a collection of architectures and solutions created during the architecture effort. When looking at the repository through a lens, a continuum can be used to visualize the state of the architecture work.

The TOGAF Standard recognizes three types of continua:

- Enterprise Continuum
- Architecture Continuum
- Solutions Continuum

CHAPTER 7   ARCHITECTURE REPOSITORY

The first of the three, the *Enterprise Architecture Continuum*, serves as an umbrella for the other two. The *Architecture Continuum* contains architectures that can be used as references to work toward an organization-specific architecture. There are four types of architectures in the Architecture Continuum: Foundation Architectures, Common Systems Architectures, Industry Architectures, and Organization-Specific Architectures. Each of these types of architectures contains an *increasing level of detail* as one moves through the Architecture Continuum from left to right.

The *Solutions Continuum*, the last of the three continua, contains four types of solutions. These are Foundation Solutions, Common Systems Solutions, Industry Solutions, and Organization-Specific Solutions. Again, moving from left to right provides an increasing level of detail about the solutions.

Catalogs, diagrams, maps, and matrices are used to capture solutions in the Architecture Repository. It is important to note that catalogs are *lists of building blocks* of a specific entity, type, or related types that are used for governance or reference purposes. A catalog addresses a specific entity and supplements the information about the concept with additional data. In a catalog, entities are not linked to other entities.

Diagrams and maps are *representations of architectural content* in either *graphical format* (diagrams) or *textual format* (maps). Diagrams and maps can also be used as a technique for graphically populating architectural content or for verifying the completeness of collected information. A diagram or map visualizes information captured in a catalog or matrix.

Lastly, matrices are a grid that represents the relationships between two or more model entities. The relationship between the entities – typically spread across two or more catalogs – is brought together in a matrix. The method of bringing entities together is called *cross-mapping*. Catalogs, diagrams, maps, and matrices are used to visualize building blocks. Architecture artifacts consist of one or more of these building blocks, and one or more artifacts make up an architecture deliverable.

CHAPTER 7 ARCHITECTURE REPOSITORY

When moving through the phases of the Architecture Development Method, a number of architecture artifacts may be produced. The number and composition of these artifacts largely depends on the level of customization of the TOGAF Standard. Once an organization has adapted the Enterprise Metamodel to its specific situation or requirements, an overview of the necessary architecture artifacts is formed based on this model. It is important to determine which artifacts add value to the architecture effort at hand and then create those artifacts. All other artifacts can then be ignored.

Building blocks – the components that make up an architecture artifact – are the backbone of the architecture and serve as essential, reusable components in architectural solutions. These blocks form a modular foundation, designed to be generic and adaptable, ensuring they can seamlessly integrate with other building blocks. Flexible as they are, building blocks can be defined at varying levels of detail, depending on the stage of architecture development. In the early stages, a building block might just have a name or a brief description. As development progresses, it can be broken down into supporting components, complete with detailed specifications. Building blocks serve as key references for both architecture and solution design. They are packages of functionality defined to meet the business needs across an organization. They correspond to the entities used and defined in the Enterprise Metamodel.

The process of building block definition takes place gradually as the Architecture Development Method is followed, mainly in phases A, B, C, and D.

Building blocks in the TOGAF Standard come in two flavors: Architecture Building Blocks and Solution Building Blocks. The former represents the *what* and *why*, providing high-level, technology-agnostic capabilities. The latter dives into the *how*, focusing on practical implementation using specific technologies and products.

When architecture work begins and the Architecture Development Methodology is applied during an initial iteration, Architecture Building Blocks are typically created during this round. These building blocks capture architecture requirements such as business, data, application, and technology requirements. The information stored in the building blocks from this activity can be used to direct and guide the development of Solution Building Blocks in a later phase of architecture development.

In contrast to the high-level character of Architecture Building Blocks, Solution Building Blocks are actually much more detailed in nature. In fact, these building blocks describe the entities that make up the parent Architecture Building Block. Solution Building Blocks define what products and components will implement the required functionality; they determine the actual implementation of (parts of) the architecture. Solution Building Blocks fulfil business requirements and are product- or vendor-aware, something that Architecture Building Blocks are not. Both Architecture Building Blocks and Solution Building Blocks are typically stored in the Architecture Repository.

In order to communicate the architecture – consisting of the aforementioned building blocks – the TOGAF Standard uses *views* and *viewpoints*. These two concepts address the concerns of the stakeholders involved. The concepts are similar, yet fundamentally different. A view is a *representation of a system or architecture* from the *perspective of a specific stakeholder*. It is essentially a visualization or presentation of the architecture that addresses particular concerns or interests of one or more stakeholders. A viewpoint *defines the perspective* and *rules for creating a view*. It is essentially a template or a set of guidelines for constructing views to address specific concerns. Viewpoints outline how to structure views and which stakeholders they are aimed at.

Viewpoints are like blueprints for constructing views. Multiple views can be created using a single viewpoint, but each view is tailored to specific stakeholder concerns.

In summary, a view is what you *actually present* or show to stakeholders, while a viewpoint is the *set of rules or guidelines* that dictate how you create that view. Views and viewpoints are part of the Reference Library of the Architecture Repository.

CHAPTER 7 ARCHITECTURE REPOSITORY

Mentioned at the beginning of this chapter, the Architecture Repository is an intangible concept and *is not* necessarily equivalent to an architecture tool. However, many organizations use a tool under the assumption that it is. Vendors of architecture tools offer a variety of features in their products that come close to meeting the requirements of an Architecture Repository. The use of an architecture tool is therefore essential.

Deciding which tool is best for the organization depends on several factors. The factors that ultimately determine the choice of an architecture tool largely depend on the situation. The same goes for the requirements for using a tool. These requirements are not easy to articulate – there is no standard set of requirements – because they are different for everyone.

CHAPTER 8

Architecture Governance

 This chapter is part of the GREEN reading tour. See Figure 1-1 in Chapter 1, Section 1.3, for alternative reading tours.

Architecture governance is more than just a business management tool – it's the backbone that ensures everything runs smoothly. It is best thought of as the leadership compass, guiding the organization with a deep understanding of its structure, clear direction, and the ability to fine-tune processes, all with the aim of driving the organization's strategic goals forward.

At its core, architecture governance boils down to one key mission: to streamline an organization's architectural needs into a coherent set of policies, processes, procedures, and standards. The Architecture Governance Framework makes sure the organization's vision and standards line up with real-world business demands.

CHAPTER 8   ARCHITECTURE GOVERNANCE

8.1. Managing the Architecture

Architecture governance is not a theoretical exercise. It is rooted in the day-to-day realities of business. Enterprise Architects play a key role in deploying and sustaining strategies that keep the business on track.

Architecture governance is a shared responsibility among business leaders, not just the responsibility of a CIO or CTO. Enterprise Architects, domain experts, and other key team members provide essential input. A well-structured Architecture Governance Framework helps organizations reduce costs and risks while speeding up decision-making and project delivery.

The TOGAF Standard suggests that architecture governance is intended to control and manage an organization's architecture and the Architecture Capability (the executive body behind the architecture). To this end, the TOGAF Standard has developed a framework that shows the *possible structure* of architecture governance.

I deliberately say the "possible structure" because the Governance Framework shown in Figure 8-1 serves as a reference point. The TOGAF Standard provides the framework as a handle – a starting point – indicating that it is wise to adapt the framework to the situation within one's own organization. Therefore, the framework shown here is by no means the only correct representation of an Architecture Governance Framework.

According to the TOGAF Standard, architecture governance can be captured in the following Governance Framework (Figure 8-1).

CHAPTER 8 ARCHITECTURE GOVERNANCE

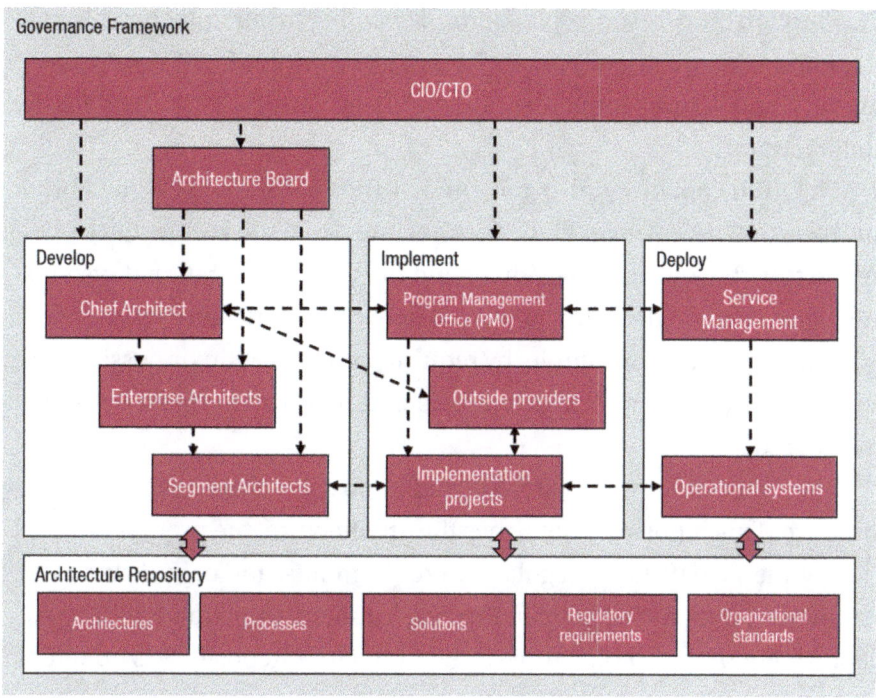

Figure 8-1. *Architecture Governance Framework*

In this Governance Framework, the overall direction is provided by the CIO/CTO. This role is responsible for the direct management of the Architecture Board and the various capabilities responsible for the development, implementation, and deployment of the architecture. These capabilities are performed by various teams within the organization.

The *development* capability (or team) consists of different architecture roles. For this capability, the framework distinguishes the roles of Chief Architect, Enterprise Architect, and Segment Architect (the former Domain Architect). The different roles are explained in more detail in Chapter 10, Section 10.2.

CHAPTER 8   ARCHITECTURE GOVERNANCE

The capability responsible for *implementing* the architecture recognizes the roles of a Program Management Office (PMO), external vendors, and various implementation projects to actually implement the architecture.

The third capability, also governed by the CIO/CTO, is responsible for the actual *deployment* of the architecture. Here the architecture interacts with service management and the various operational systems into which the architecture must fit. Where the architecture will impact existing systems - for example, by requiring modification or possibly even replacement - the connection to the operational systems must be carefully considered.

All capabilities use the Architecture Repository during the architecture initiative. This repository describes the architecture, the processes and solutions to be used, regulatory requirements, and organizational standards. These are all used in the development, implementation, and deployment of the architecture. The Architecture Repository is updated continuously during the initiative to ensure an accurate representation of the developing architecture.

The acronyms CIO and CTO stand for Chief Information Officer and Chief Technology Officer, respectively.

Looking at the first role, that of the CIO (Chief Information Officer), it's clear that this position is almost always part of the IT department. This is odd because information ownership really lies with the business side. Therefore, the CIO, who is responsible for that information, should also be on the business side.

The CTO, on the other hand – given the word "*technology*" in the name – would be perfectly at home in an IT department.

Yet nine out of ten organizations place the CIO in the IT department. The TOGAF Standard also assumes this scenario.

In addition to the existing roles of CIO and CTO, I think the framework would do well to include the role of CEO (Chief Executive Officer) in the Governance Framework.

8.2. Implementing Architecture Governance

Implementing an Architecture Board involves setting up a structured *process* and *governance body* that oversees and guides architectural decisions within an organization. This governance ensures alignment with business goals, standards, and best practices. Phase G (Implementation Governance) of the Architecture Development Method takes the approach described in this section and puts it into practice.

Implementing architecture governance boils down to the following steps. First, the Governance Framework must be established. The necessary roles and responsibilities are defined, and policies and standards are developed. The Governance Framework uses an architecture compliance process to verify that the architecture lives up to the agreements. The architecture review process is implemented, and key performance indicators are established. The entire governance process is then integrated into the Architecture Development Methodology. Next, a Communications Plan is created, and appropriate communication materials are selected or created. The governance processes are automated where possible. Finally, the governance process is subject to continuous improvement.

The steps just described are illustrated in Figure 8-2 and explained in more detail below.

CHAPTER 8   ARCHITECTURE GOVERNANCE

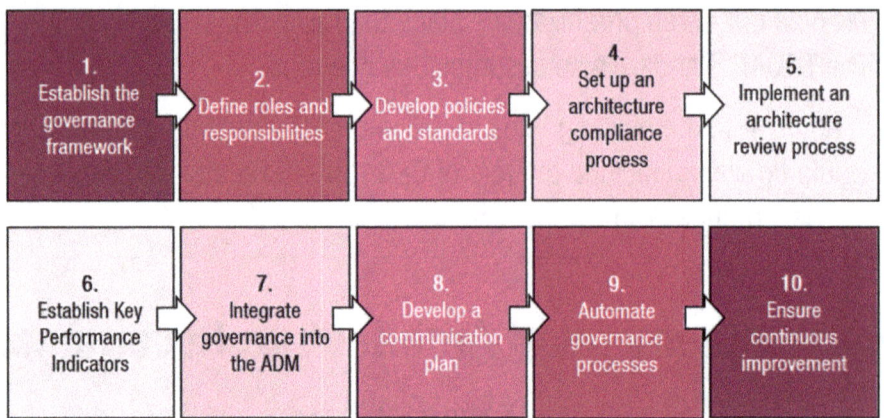

Figure 8-2. Ten steps for implementing architecture governance

Implementing architecture governance in a practical and pragmatic way can be achieved by following these steps:

1. **Establish the Architecture Governance Framework:** Ensure buy-in from stakeholders by clearly explaining the boundaries and benefits of architecture governance.

    - **Kick Off with Stakeholder Engagement:** Organize a series of workshops or meetings with key stakeholders (e.g., C-level management, IT leadership, business managers) to explain the value of architecture governance. Ensure buy-in by presenting its role in aligning IT with business goals.

    - **Create a Governance Charter:** Draft a document outlining the architecture governance scope, objectives, and guiding principles. This document should clearly define

- Which areas of the organization will be governed (business processes, data, IT systems, etc.)
- What success looks like (alignment with strategic objectives, compliance with standards)
- **Select Architecture Governance Framework Components:** Use the TOGAF Standard as a reference to decide which governance processes and structures to implement (approvals, compliance reviews, architecture change management).
- **Architecture Governance Board Setup:** Identify and invite senior stakeholders and domain architects to form the Architecture Governance Board. Typically, this includes business leaders, IT leaders, Enterprise Architects, and project managers. Define their roles and establish regular governance meetings (e.g., monthly or at key project milestones).
- **Formalize the Framework:** Document the Governance Framework. This includes governance processes (decision-making, review, and escalation) and the criteria for assessing architecture decisions.

2. **Define Governance Roles and Responsibilities:** Identify, describe, and document necessary roles and responsibilities.
    - **Organize Role Definition Workshop:** Conduct a workshop with key stakeholders to define roles (such as Architect, Business Leader, Project

CHAPTER 8   ARCHITECTURE GOVERNANCE

Manager). Identify what each role is accountable for and ensure they understand their governance responsibilities.

- **Create Role Descriptions:** Write detailed role descriptions that clearly state each person's involvement in architecture governance, including responsibilities for architecture reviews, compliance checks, and approvals.

- **Map Responsibilities:** Use a RACI matrix (see Table 8-1) to map each governance activity to roles. This ensures clarity on who makes decisions and who needs to be informed or consulted.

Table 8-1. Example RACI Matrix

Activity ▼	Architect	Business Leader	Project Manager
Establish the Architecture Governance Framework.	A	R	C
Develop governance policies and standards.	R	A	C
Set up an architecture compliance process.	AR	C	I
Implement an architecture review process.	AR	C	I
Establish key performance indicators (KPIs).	R	A	C
Integrate governance into the Architecture Development Method.	AR	C	I
Develop a Communications Plan.	AR	C	C
Automate governance processes.	A	R	R
Ensure continuous improvement.	R	A	A

CHAPTER 8 ARCHITECTURE GOVERNANCE

The RACI matrix shown in Table 8-1 is an example of what such a matrix might look like. It lists the *activities* (or processes) in the left column and shows the *roles* involved in those activities in the top row. The cells of the table contain the manner in which each of the roles is involved in the activities.

For example, when it comes to *establishing the Architecture Governance Framework*, the Architect role is *accountable* for this activity. The Business Leader is *responsible* for the activity, and the Project Manager needs to be *consulted* during the process. The four RACI responsibilities are visualized and explained in Table 8-2.

Table 8-2. The four RACI responsibilities explained

Responsibility	Description
Responsible (R)	The person who actually carries out the activity or process. Responsible to get the job done.
Accountable (A)	The person who is ultimately accountable for the activity or process being completed appropriately. Responsible person(s) is/are accountable to this person.
Consulted (C)	People who are not directly involved with carrying out the activity or process, but who are consulted. May be a stakeholder or a subject matter expert.
Informed (I)	Those who receive output from the activity or process or who have a need to stay informed.

A RACI matrix can contain any number of activities and roles. In this particular example, only one Architect role is shown. Which roles are included in the matrix depends on a number of factors, such as the topic for which the matrix is being created.

3. **Develop Governance Policies and Standards:** Form an interest group to develop policies and standards. Make sure to have them reviewed and published.

    - **Form a Policy-Drafting Group:** Select a small team of architects, business leaders, and business representatives to create architecture policies. This group should review existing organizational policies and identify gaps where new governance rules are needed (how architectural changes are approved).

    - **Create Draft Policies and Standards:** Start with high-level governance policies that apply to all areas (e.g., *All architecture changes must be reviewed by the governance board*). Next, break these into specific standards, covering technology, processes, and data handling.

    - **Review with Stakeholders:** Share draft policies and standards with business leaders and other relevant parties for feedback. Incorporate this feedback into the final documents.

    - **Publish and Communicate:** Once finalized, publish the policies and standards on a shared platform (such as the company's intranet). Host training sessions or workshops to explain these to teams involved in architecture projects.

4. **Set Up an Architecture Compliance Process:** Embed the compliance process in the organization by integrating it in the project lifecycle. Remember to include an exception handling process.

- **Develop Compliance Checklists:** For each architecture domain (Business, Data, Application, and Technology), create checklists that detail compliance requirements. These checklists should align with the governance policies and standards. See the *Further Reading* information box at the end of this section for more information.

- **Integrate Compliance into Project Lifecycle:** Add compliance checkpoints to key phases of the project lifecycle (during design reviews, before deployment). These checkpoints should mandate sign-off from the Architecture Governance Board.

- **Compliance Review Meetings:** Schedule regular compliance reviews, either as part of project status meetings or separately. During these reviews verify that projects meet the defined architecture standards. Also, identify and document any exceptions or deviations.

- **Develop Exception Handling Processes:** When compliance issues arise, have a formalized process for raising exceptions to the governance board. Ensure that a rationale for the deviation is provided and that an action plan for mitigation is agreed upon.

5. **Implement an Architecture Review Process:** Implement a formal review process for architecture documents by developing templates and documenting the outcome of the review meetings.

    - **Set Up Review Templates:** Develop standardized templates for architecture reviews. These templates should include sections for project context, compliance with Architecture Principles, risks, and proposed solutions.

- **Schedule Regular Review Meetings:** Ensure that architecture review sessions are built into the project plan at key phases (design phase, pre-deployment). Send out invitations well in advance and include all relevant stakeholders.

- **Prepare for Reviews:** Architects or project leads should prepare their review documents in advance. Use the templates to guide their preparation, focusing on key areas such as risk, compliance, and alignment with business objectives.

- **Review and Document Decisions:** During the review meeting, evaluate each architecture component against established governance standards. Record decisions, actions, and any exceptions granted.

6. **Establish Key Performance Indicators (KPIs):** Identify metrics to measure the performance of the governance activities. Use clearly defined baselines and targets and provide transparent monitoring of the performance.

    - **Define KPIs with Stakeholders:** Hold a meeting with key governance stakeholders to identify metrics that will measure the success of governance activities. KPIs could include

        - Percentage of projects compliant with the architecture framework

CHAPTER 8   ARCHITECTURE GOVERNANCE

- Number of governance reviews conducted
- Number of deviations or exceptions from standards
- Time to resolve architecture-related issues

- **Set Baselines and Targets:** Once KPIs are defined, set baseline measurements and targets for improvement (e.g., *90% of projects will be architecture-compliant by the end of the year*).

- **Create Monitoring Dashboards:** Use an architecture management tool or project management software to track KPIs in real time. Ensure that governance board members can access the dashboard to monitor progress. An example of how to monitor progress is shown in Figure 12-54 in Chapter 12, Section 12.10.2.

7. **Integrate Governance into the Architecture Development Method:** Define governance checkpoints and integrate them into the Architecture Development Method phases.

    - **Add Governance to Each ADM Phase:** For each phase of the Architecture Development Method – from the Preliminary Phase to Implementation Governance and beyond – define governance checkpoints. For example:

        - **Preliminary Phase:** Ensure Governance Framework setup is complete.
        - **Phase B (Business Architecture):** Perform reviews to validate alignment with business goals.

201

- **Governance Sign-Off at Key Milestones:** Embed formal governance reviews into project milestones. Projects should not proceed to the next phase without governance board sign-off.

8. **Develop a Communications Plan:** Identify the audience that is impacted by the architecture governance. Select, create, and use appropriate communication materials.

    - **Identify Key Audiences:** Create a list of individuals and groups impacted by architecture governance (such as project managers, developers, business stakeholders).

    - **Develop Communication Materials:** Prepare materials such as slide decks, FAQs, and email templates explaining the governance process, its benefits, and their roles. Ensure materials are tailored to different audiences (e.g., technical teams get detailed compliance information; business leaders get strategic benefits).

    - **Host Workshops and Training:** Schedule live workshops or recorded webinars to walk stakeholders through the Architecture Governance Framework, compliance expectations, and how to participate in reviews.

    The development of a Communications Plan is further detailed in Chapter 14, Section 14.3.3.

9. **Automate Governance Processes:** If possible, try to automate (parts of) the governance process such as generating reports and review notifications.

- **Select a Governance Tool:** Choose a tool that supports architecture governance, such as Enterprise Architecture management software. These tools can help automate compliance tracking, generate reports, and manage architecture documentation. See Chapter 7, Section 7.7.1, for more information on tool selection.

- **Set Up Automated Alerts and Reports:** Use the tool to set up automated notifications for governance board members when projects are due for review. Similarly, configure reports that track compliance KPIs and project alignment with Architecture Principles.

10. **Ensure Continuous Improvement:** Conduct regular meetings and ask for feedback regarding the governance process. Adjust the Architecture Governance Framework based on the feedback received.

    - **Hold Quarterly Governance Reviews:** Schedule periodic governance framework reviews with the governance board to assess the effectiveness of governance processes, based on KPIs and feedback.

    - **Gather Feedback:** Regularly solicit feedback from stakeholders, architects, and project teams on how the governance process is working. Use surveys or informal interviews to collect input.

    - **Adjust the Framework:** Based on the feedback and performance data, revise policies, roles, and governance procedures as needed. Document changes and communicate them to the organization.

By following these steps, architecture governance can be implemented effectively. It ensures alignment between the Enterprise Architecture and business objectives, while maintaining control over architectural changes and compliance.

Further Reading

For more information on architecture governance, see "The TOGAF® Leader's Guide to Establishing and Evolving an EA Capability", Chapter 6, Architecture Governance, document number G184.

The "Enterprise Architecture Capability and Governance" document of the TOGAF Standard, document number C220, provides additional information about architecture compliance.

8.3. Architecture Board

The Architecture Board is a pivotal component in an organization's Enterprise Architecture governance, ensuring alignment with strategic goals and fostering architectural best practices.

The Architecture Board – as is shown in Figure 8-1 – is responsible for managing the various architecture roles. This is *functional* rather than *hierarchical* management.

The board, composed of the previously mentioned senior management, key stakeholders, and domain experts, oversees architecture compliance, facilitates decision-making, and manages architecture-related risks. It plays a crucial role in reviewing and approving architecture deliverables, thus ensuring consistency and coherence across the organization. By doing so, the Architecture Board enhances the organization's ability to adapt and respond to changing business needs effectively.

CHAPTER 8 ARCHITECTURE GOVERNANCE

Contrary to what most people think, an Architecture Board is not a group of architects who meet regularly to discuss substantive issues.

The word *board* refers to the boardroom or board of directors. It is a direct reference to the level at which issues are discussed. Architecture issues discussed at the board level are strategic, not operational.

According to the TOGAF Standard, an Architecture Board is formed by the senior management of an organization. The ideal composition of an Architecture Board consists of C-level management, key stakeholders, and domain experts.

An Architecture Board is a cross-organization entity, which oversees the implementation of the strategy. This body should be representative of all the key stakeholders in the architecture and will typically comprise a group of executives responsible for the review and maintenance of the overall architecture [11].

The reason for this composition is that *strategic decisions* about architecture should be made in an Architecture Board. In organizations, strategic decisions are usually made by C-level management. Therefore, it is necessary to have (some form of) representation from that management at the Architecture Board level.

In practice, however, we often see that an Architecture Board consists of IT Architects. This group of individuals typically deals with substantive issues and topics related to architecture. For example, they deal with architecture designs or discuss (IT) principles and guidelines for a specific solution or solution direction.

Because organizations still view architecture as an IT discipline, in nine out of ten cases an Architecture Board is formed by IT Architects. Due to this misalignment, strategic decisions are not made within the Architecture Board. As a result, these decisions are made outside the eyes and ears of the Architecture Capability. This keeps the Architecture

Capability in a perpetual state of reactive action. It has no say in how the strategic direction is set and implemented. Consequently, architecture remains structurally behind the times.

Examining the Architecture Governance Framework outlined in the TOGAF Standard reveals that an Architecture Board is typically headed by the CIO or CTO – in other words, a C-suite executive (see Figure 8-1).

Personally, I believe that to the two C-level roles that are mentioned in the Architecture Governance Framework should be supplemented with a third, namely, the role of the CEO. When architecture is handled at the CEO level, truly strategic decisions can be made. Architecture is then a direct part of strategy execution.

8.3.1. Diverse Leadership

To illustrate how the governance of the Architecture Board differs when managed by a CIO/CTO or a CEO, two example scenarios are outlined below. The first scenario shows the impact of the Architecture Board being led by the CEO. The second scenario shows how the CIO or CTO would lead the Architecture Board. These two examples illustrate the difference in the strategic impact management can have.

8.3.1.1. Scenario 1: CEO Leadership

This first scenario assumes that the CEO is responsible for the strategic direction of the organization. If, for example, there is a need to ensure that the correct organizational structure, including associated processes, information flows, and applications, is in place due to upcoming legislation, this will have a fairly broad impact on the architecture. Not only will the architecture need to provide an appropriate organizational model, including the necessary roles and functions, but it will also need to outline a scenario for the processes used and their place within the business function model. In addition, information flows should be mapped and improved according to the strategic direction. Finally, the application landscape should be aligned with the aforementioned organizational design, processes, and information flows.

The scope of the architecture in this scenario is quite broad as it spans multiple architecture domains. The impact of the strategy needs to be mapped across the entire organization.

8.3.1.2. Scenario 2: CIO/CTO Leadership

In the scenario where the Architecture Capability is managed by the CIO or CTO, the scope of architecture work is immediately reduced compared with the scenario where it is managed by the CEO. The more strategic and therefore organization-wide nature of the architecture work is reduced because the CIO or CTO is responsible for a specific part of the strategy. For example, the CIO is only responsible for the information flow, while the CTO is responsible for the technical design and implementation of the architecture.

The situation outlined in the first scenario, where legislation comes into effect, means that in the case of CIO leadership, the architecture work is limited to the design of the information delivery. In the case of the CTO's direction, the architecture work will focus primarily on the technical side of the profession. Thus, both CIO and CTO leadership ensure that the scope of the architecture work is focused on a small part of the overall Enterprise Architecture.

An Architecture Board that is directly led by the CEO - and thus empowered to respond to strategic issues as it should - would look like the one visualized in Figure 8-3. The figure highlights the roles of the CEO, the CIO/CTO, the unit manager, and the Architecture Board itself.

CHAPTER 8   ARCHITECTURE GOVERNANCE

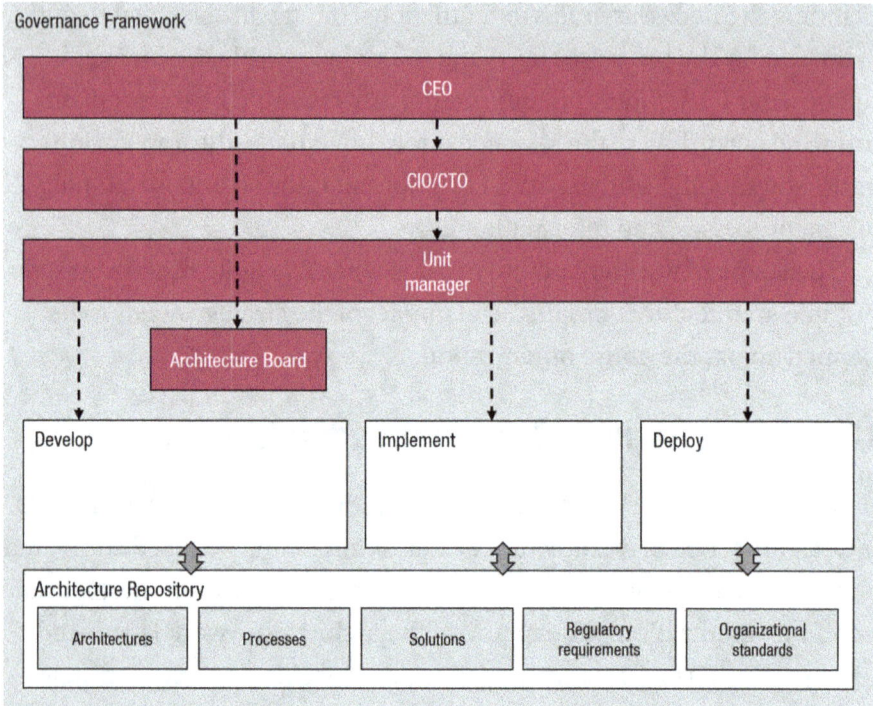

Figure 8-3. Architecture Board led by the CEO

In Figure 8-3, the interpretation of the development, implementation, and deployment capabilities has been omitted for the sake of readability. The figure shows that the CEO provides direction to both the CIO/CTO and the Architecture Board. The CIO/CTO is responsible for managing the unit manager(s). The latter role is responsible for the direct management of the various teams involved in the development, implementation, and deployment of the architecture.

As mentioned earlier, strategic decisions are made at the CEO level. By providing direct governance from the CEO to the Architecture Board, it is possible for the Architecture Board to be closely involved in strategic issues. This ensures that the strategic direction of an organization is anchored in the architecture used.

Having an Architecture Board is not just for larger organizations. Smaller organizations in particular can benefit greatly from having an Architecture Board. It is often the smaller organizations that are more pragmatic about rules, so the architecture often has to work harder to maintain certain standards.

Organizations that have a less formal organizational structure can use an Architecture Board. It is by no means necessary for the organization to have roles such as CIO, CTO, or CEO. Companies with a flatter organizational structure can also use an Architecture Board. Architecture sponsorship is critical. If someone high up in the organizational hierarchy can be made responsible for this sponsorship, that person does not necessarily have to be the CEO. What is important is that there is attention and interest from the highest levels of the organization.

8.3.2. When to Set Up an Architecture Board

An Architecture Board is essential for most (somewhat larger) organizations. In many cases, such an Architecture Board already exists, except perhaps for its form and composition. If this is not the case, certain events can trigger the creation of an Architecture Board.

Typically, the creation of an Architecture Board is triggered by one (or more) of the following events:

- The hiring of a new CIO/CTO
- Mergers or acquisitions
- Organizational restructuring, whether or not due to significant business change or rapid growth
- Recognition that the organization's strategy is not being implemented or is being implemented poorly

In many organizations, the CIO (or another C-level executive) is the executive sponsor of architecture efforts. But having a sponsor is not enough. To gain organization-wide support for the architecture effort, an Architecture Board is needed. This executive-level group is responsible for reviewing and maintaining the strategic architecture.

The Architecture Board is the primary sponsor of the architecture within the organization. But the Architecture Board itself also needs a sponsor, preferably at the highest level within the organization. However, an often-overlooked resource for Architecture Board members is the company's board of directors. Individuals in this role often have diverse knowledge of the company and its competition. Because of their significant influence on the company's vision and goals, they can be successful in validating the alignment of organizational strategy with business objectives. Having the CEO as the Architecture Board's primary sponsor would be an excellent idea.

The involvement of the Architecture Board should extend throughout the planning process and into the maintenance phase of the architecture project.

Unfortunately, many organizations still fail in their attempt to work with architecture. In nine out of ten cases, this is due to a significant lack of involvement and encouragement from senior management.

Performing architectural work depends on solid sponsorship from the highest levels of the organization. Without sponsorship, there is little or no point in doing the work. Therefore, it is important to establish sponsorship and support for architectural work before the work begins.

8.3.3. Key Responsibilities

An Architecture Board has certain responsibilities [11]. Among them are, for example, the following:

- **To Provide a Solid Basis for Decision-Making:** Decisions regarding the execution of an organization's strategy should never be made lightly. The same goes for decisions regarding architecture, for architecture plays a large and important role in realizing an organization's strategy. Having a solid basis for making these decisions will allow an organization to rely upon the Architecture Board for solid advice.

- **To Ensure Continued Flexibility of the Architecture:** Every decision made by the Architecture Board must keep in mind that the architecture must do two things in particular. On the one hand, the architecture must always strive to meet the ever-changing needs of the business. On the other hand, an organization should also want to be able to take advantage of new technologies, new ways of working, or new methods. The Architecture Board must ensure that the architecture is flexible enough to achieve these goals, while enforcing architecture compliance.

- **To Improve Architecture Maturity:** The decisions made by the Architecture Board must never allow the architecture in an organization to slip to a lower level of maturity. The Architecture Board must prevent the maturity level from dropping. Decisions on architecture issues should be made with the current and desired architecture maturity levels in mind. Decisions should be aimed at improving the level.

CHAPTER 8 ARCHITECTURE GOVERNANCE

From a more operational point of view, the Architecture Board also holds a number of responsibilities, for example:

- **To Make Sure the Architecture Is Effectively and Consistently Managed and Implemented:** Having a permanent set of members on the Architecture Board ensures a consistent approach to architecture issues. A fixed composition ensures that architecture is not approached in a different way all the time. Of course, the composition of the Architecture Board may change from time to time. One technique is to have fixed terms for members and to have these terms end at different times.

- **To Provide Advice and Guidance:** The Architecture Board has the important operational role of providing advice and direction on all types of architectural issues. Advice must be tested against the existing architecture. The direction given can be translated by experts into concrete actions to implement the architecture work.

Last but not least, the Architecture Board will need to take its responsibilities when it comes to governance. From this perspective, the Architecture Board is responsible for the following:

- **To Provide a Process for Accepting and Approving Architecture Work:** It is critical to establish a process for accepting and approving architectural work. This process should be transparent and accessible to all stakeholders. It is the role of the Architecture Board to provide and make available this process.

- **To Implement a Control Mechanism for Architecture Efforts:** All architecture work and requests for architecture work must at some point pass through a control mechanism to validate that the final execution of the work is consistent with what was proposed during the development phase. This control usually occurs during Phase G of the Architecture Development Methodology.

- **To Recognize Divergence from the Architecture and Initiatives:** Deviations from the developed architecture may occur over time. Advancing knowledge or other factors play an important role in deviations from the architecture. Identifying and responding to deviations in a timely manner is also a responsibility of the Architecture Board. This is discussed in more detail in Phase H of the Architecture Development Method.

The roles and responsibilities of the Architecture Board should be formally defined and approved by the board's lead sponsor. Ideally, this will be the CEO.

8.3.4. Composition and Size

It is recommended that the Architecture Board consist of four or five people (be careful not to exceed ten). The four or five Architecture Board members are standing members. To keep the Architecture Board at a reasonable size while still ensuring that the entire organization is represented over time, the Architecture Board membership can rotate, giving different senior managers decision-making authority and responsibility. This may be necessary if some Architecture Board members find that a lack of time prevents long-term active participation.

An organization can set up the Architecture Board so that each board member is assigned a number of stakeholders. These stakeholders are then represented at meetings. In day-to-day practice, this means that Architecture Board members need to engage with their stakeholders to get their input on various topics on the agenda.

CHAPTER 8   ARCHITECTURE GOVERNANCE

Rotating individuals within the Architecture Board is quite feasible, provided that care is taken to ensure continuity in the Architecture Board to prevent the Business Architecture from moving from one set of ideas to another. One technique to ensure rotation with continuity is to have fixed terms for members and to have the terms end at different times.

It is also recommended that the composition of the Architecture Board not be exclusively C-level management. The addition of lower-level management and possible experts from the organization, if desired and depending on the issue at hand, can ensure that the tactical and operational aspects of the architecture issue are covered. In this way, the architecture is illuminated and validated from all sides – strategic, tactical, and operational.

On several occasions, I have seen that an Architecture Board made up entirely of C-level executives was not the best solution. This was mainly because the architectural issues at hand were viewed from a too-strategic perspective.

Therefore, it is advisable – depending on the topic and the level of detail of the issue at hand – to use a healthy mix of C-level management combined with lower-level management or experts from the organization as the composition of the Architecture Board. Members drawn from lower management (think unit managers) and experts may join depending on the topic. They are not permanent members of the Architecture Board.

The addition of these additional roles can ensure that issues can be addressed in a more substantive way.

Discussing strategic issues can, in some situations, lead to too much high-level thinking. As a result, the transfer to implementation and execution capabilities may lack the information they need. This makes implementation and execution more difficult.

By involving the above roles (at the request of the Architecture Board), the architecture issue at hand can be considered and addressed from all possible angles. The issue will then receive the strategic, tactical, and even operational attention it deserves.

The size and composition of the Architecture Board may vary depending on the architectural issue at hand. Common sense in the composition of the board is always recommended. In addition, the positioning of the Architecture Capability in the organization plays an important role, as does the way the organization wants to deal with architecture. All of these factors will influence the final composition and size of the Architecture Board. Above all, ensure that the board is able to fulfill its key responsibilities.

8.3.5. How to Set Up an Architecture Board

Implementing an Architecture Board involves setting up a structured *process* and *governance body* that oversees and guides architectural decisions within an organization. This governance ensures alignment with business goals, standards, and best practices.

CHAPTER 8 ARCHITECTURE GOVERNANCE

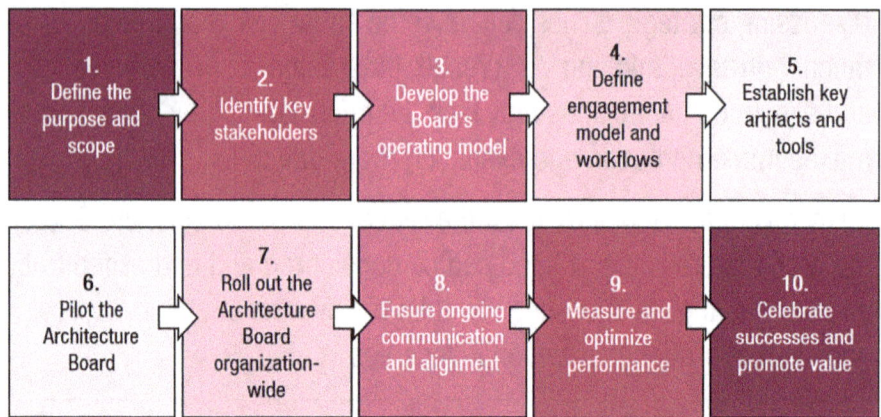

Figure 8-4. Ten steps for setting up an Architecture Board

Setting up an Architecture Board in a practical and pragmatic way can be achieved by following the steps shown in Figure 8-4:

1. **Define the Purpose and Scope:** Establish the purpose and boundaries of the Architecture Board.

    - **Identify Objectives:** Determine the key objectives of the Architecture Board, for example, ensuring architectural consistency, approving major architectural decisions, and aligning the architecture with business goals and objectives.

    - **Define Scope:** Clarify the areas of responsibility, such as specific domains (Business, Data, Application, Technology), types of decisions (project approvals, technology selections), and the levels at which the Architecture Board operates (organization-/enterprise-wide, solution specific, or at project level). This level of operations largely depends on the way the Architecture Capability is positioned in the organization.

- **Establish Governance Principles:** Set the guiding principles for decision-making, such as ensuring simplicity, promoting reusability, and adhering to industry standards.

2. **Identify Key Stakeholders:** Ensure representation from relevant stakeholders across the organization.

    - **Select Board Members:** Include representatives from key areas such as

        - **Business Stakeholders:** To ensure alignment with business objectives, it is imperative to have business representation on the Architecture Board.

        - **Enterprise Architects:** To bring in the organization-wide perspective.

        - **Segment Architects:** For specific expertise in areas like business, information, data, application, or technology.

        - **Security and Compliance Representatives:** To ensure adherence to security and regulatory requirements.

        - **IT Leaders:** To represent the overall IT strategy.

        - **Project Managers:** For project-level considerations.

    - **Establish Roles and Responsibilities:** Clearly define roles within the Architecture Board. Think, for example, of a chairperson, secretary, or decision-makers. Make sure to also define the responsibilities of each member.

3. **Develop the Board's Operating Model:** Define how the Architecture Board will function on a day-to-day basis.

    - **Set Meeting Cadence:** Decide on the frequency of meetings (e.g., weekly, biweekly) and the length of each meeting. Consider ad hoc meetings for urgent issues.

    - **Agenda Structure:** Develop a standard agenda template, including sections like

        - Review of previous decisions
        - Discussion of new proposals
        - Review of ongoing initiatives
        - Risk and issue management

    - **Decision-Making Process:** Establish how decisions will be made:
        - **Consensus-Based:** Preferable for collaborative environments.
        - **Majority Voting:** If a clear majority is needed for decisions.
        - **Escalation Process:** Define how disagreements or unresolved issues will be escalated.

4. **Define Engagement Model and Workflows:** Create a process for how the Architecture Board interacts with projects and initiatives.

    The Architecture Capability can be used to support portfolio management as well as project management. Therefore, it is advisable to clearly define its scope.

- **Submission Process:** Define how teams submit items for board review, including templates for architectural proposals, impact assessments, and decision logs.

- **Review Criteria:** Set the criteria the Architecture Board will use to assess proposals (alignment with strategic objectives, technical feasibility, Risk Management, cost/benefit analysis).

- **Feedback and Iteration:** Develop a process for providing feedback to teams, including a loop for revising and resubmitting proposals based on feedback from the Architecture Board.

- **Decision Documentation:** Ensure decisions are documented, including the rationale behind them, and stored in a central repository for future reference.

In the recent past, I worked for a government organization that regulated the healthcare industry. The organization was an amalgamation of several companies and had existed in its current form for several years before I started working there.

As professional as they thought they were in the work they were doing, the Architecture Capability left a lot to be desired.

This was largely due to management's view of architecture. They saw it – as, unfortunately, many organizations still do today – as something related to IT.

This was the main reason why the Architecture Board, as it was set up, consisted entirely of technicians. In terms of the topics on the agenda, they did not get much further than discussing technical issues, and they got pretty deep into the weeds. In fact, the Architecture Board at this organization was akin to a meeting of systems engineers, where fiery discussions about the design of a new backup solution were the highlight of the week.

Slowly but surely, I was able to get the various management teams within the organization to play a significant role in providing input on strategic issues. Little by little, I was able to put these issues on the agenda of a newly created architecture advisory group.

As the various management teams saw in practice that something was actually being done with their input, the will and commitment to these new consultations grew. Senior management participation and input grew steadily, creating a platform where strategic and architectural issues were addressed.

The new consultation was eventually renamed "Architecture Board", while the existing one took on the name "Architects' Consultation".

This way, no one lost a much-needed outlet, so there was virtually no opposition.

---

5. **Establish Key Artifacts and Tools:** Ensure that the Architecture Board has the necessary tools and documentation to function effectively.

    - **Architecture Principles and Standards:** Document and maintain a set of Architecture Principles, guidelines, and standards that the board will enforce.

- **Decision Templates:** Create standardized templates for decision requests, impact assessments, and architecture reviews.

- **Collaboration Tools:** Use a collaboration tool or platform to manage documentation, agendas, and minutes.

- **Tracking and Reporting Tools:** Implement tools to track decisions, action items, and the status of architectural reviews.

6. **Pilot the Architecture Board:** Test the board's processes and workflows in a controlled environment.

    - **Select a Pilot Project:** Choose a low-risk project or initiative to pilot the Architecture Board process.

    - **Run a Full Cycle:** Conduct a full cycle of the Architecture Board process, from submission through decision-making and feedback.

    - **Gather Feedback:** Collect feedback from both board members and the project team to identify any pain points or inefficiencies in the process.

    - **Refine the Process:** Make necessary adjustments to the workflows, tools, and engagement model based on the pilot's outcome.

7. **Roll Out the Architecture Board Organization-Wide:** Scale the Architecture Board to cover all relevant projects and initiatives.

- **Communicate the Process:** Clearly communicate the purpose, processes, and expectations of the Architecture Board to the broader organization. This may involve presentations, documentation, or training sessions.

- **Onboard Projects:** Begin onboarding new projects and initiatives, ensuring they understand how to engage with the Architecture Board.

- **Monitor and Adapt:** Continuously monitor the effectiveness of the Architecture Board. Collect metrics such as decision turnaround time, compliance with architectural standards, and feedback from project teams.

- **Iterate on the Process:** Regularly revisit the Architecture Board's processes and structure, making improvements as necessary to adapt to changing business and technological landscapes.

8. **Ensure Ongoing Communication and Alignment:** Maintain alignment between the Architecture Board, other governance bodies, and the wider organization.

    - **Regular Updates:** Provide regular updates to senior management and other stakeholders on the outcomes and value delivered by the Architecture Board.

    - **Cross-Board Collaboration:** If multiple governance boards exist (e.g., security board, innovation board), ensure cross-board communication and collaboration.

- **Continuous Education:** Keep the Architecture Board members informed about new trends, technologies, and methodologies that might impact their decisions.

9. **Measure and Optimize Performance:** Continuously evaluate and improve the effectiveness of the Architecture Board.

    - **Key Performance Indicators (KPIs):** Establish KPIs to measure the Architecture Board's effectiveness, such as decision turnaround time, adherence to architectural standards, and business value delivered. This last component, delivering business value, is often forgotten. Too often, people get caught up in the content issues, and the business value proposition takes a back seat.

    - **Feedback Loops:** Implement regular feedback loops from both board members and stakeholders to identify areas for improvement.

    - **Continuous Improvement:** Use the feedback and performance metrics to optimize processes, refine governance models, and enhance the Architecture Board's effectiveness.

10. **Celebrate Successes and Promote Value:** Highlight the achievements of the Architecture Board and ensure continued support from the organization. Keep communicating the added business value of the Architecture Board by recognizing milestones. For instance, publicize the board's successes, such as major architectural achievements, process improvements, or contributions to key business

CHAPTER 8 ARCHITECTURE GOVERNANCE

outcomes. It is also recommendable to promote the Architecture Board's value. This can be done by regularly communicating the value that the Architecture Board brings to the organization, ensuring ongoing buy-in from senior leadership and stakeholders.

By following these ten steps, a well-functioning Architecture Board can be created that provides strategic architectural oversight, aligns with business goals, and drives consistency across the organization.

8.4. Summary

This chapter has covered the topics of architecture governance and the Architecture Board. Architecture governance is intended to control and manage an organization's architecture and the Architecture Capability. The TOGAF Standard has developed a framework that shows the possible structure of architecture governance. This framework was explained in more detail and the capabilities included were highlighted.

Architecture Governance can be set up by following these ten actionable steps:

- **Establish the Architecture Governance Framework:** Ensure buy-in from stakeholders by clearly explaining the boundaries and benefits of Architecture Governance.

- **Define governance roles and responsibilities:** Identify, describe and document necessary roles and responsibilities.

- **Develop governance policies and standards:** Form an interest group to develop policies and standards.

- **Set up an architecture compliance process:** Embed the compliance process in the organization by integrating it in the project lifecycle. Remember to include an exception handling process.

- **Implement an architecture review process:** Implement a formal review process for architecture documents by developing templates and documenting the outcome of the review meetings.

- **Establish Key Performance Indicators (KPIs):** Identify metrics to measure the performance of the governance activities.

- **Integrate Governance into the Architecture Development Method:** Define governance checkpoints and integrate them into the Architecture Development Method phases.

- **Develop a Communication Plan:** Identify the audience that is impacted by the architecture governance. Select, create and use appropriate communication materials.

- **Automate governance processes:** If possible, try to automate (parts of) the governance process such as generating reports and review notifications.

- **Ensure continuous improvement:** Conduct regular meetings and ask for feedback regarding the governance process. Adjust the Architecture Governance Framework based on the feedback received.

CHAPTER 8   ARCHITECTURE GOVERNANCE

In addition to architecture governance, the topic of the Architecture Board was also discussed. An Architecture Board enhances the organization's ability to adapt and respond to changing business needs effectively. The ideal composition of an Architecture Board consists of C-level management, key stakeholders, and domain experts. The reason for this composition is that *strategic decisions* about architecture should be made in an Architecture Board.

An Architecture Board is essential for most organizations. If an Architecture Board does not exist, certain events can trigger its creation. The Architecture Board is the primary sponsor of the architecture within the organization. The board itself also needs a sponsor, preferably at the highest level within the organization. An often-overlooked resource for members is the company's board of directors. It is recommended that the composition of the Architecture Board not be exclusively C-level management.

Setting up an Architecture Board is performed by following ten actionable steps:

- **Define the Purpose and Scope:** Establish the purpose and boundaries of the Architecture Board.

- **Identify Key Stakeholders:** Ensure representation from relevant stakeholders across the organization.

- **Develop the Board's Operating Model:** Define how the Architecture Board will function on a day-to-day basis.

- **Define Engagement Model and Workflows:** Create a process for how the Architecture Board interacts with projects and initiatives.

- **Establish Key Artifacts and Tools:** Ensure that the Architecture Board has the necessary tools and documentation to function effectively.

- **Pilot the Architecture Board:** Test the board's processes and workflows in a controlled environment.
- **Roll Out the Architecture Board Organization-Wide:** Scale the Architecture Board to cover all relevant projects and initiatives.
- **Ensure Ongoing Communication and Alignment:** Maintain alignment between the Architecture Board, other governance bodies, and the wider organization.
- **Measure and Optimize Performance:** Continuously evaluate and improve the effectiveness of the Architecture Board.
- **Celebrate Successes and Promote Value:** Highlight the achievements of the Architecture Board and ensure continued support from the organization.

It is important to keep communicating the added business value of the Architecture Board by recognizing milestones.

CHAPTER 9

Architecture Capability

 This chapter is part of the GREEN and GOLD reading tours. See Figure 1-1 in Chapter 1, Section 1.3, for alternative reading tours.

This chapter deals with the concept of the Architecture Capability. It refers to an organization's ability to develop, maintain, and govern its Enterprise Architecture effectively. The Architecture Capability is crucial for achieving successful business transformation and maintaining architectural governance.

The TOGAF Standard sees the Architecture Capability as a structured approach to establishing and operating an Enterprise Architecture function within an organization. It includes key components such as architecture governance, organizational structures, processes, roles, responsibilities, and skills necessary to support the architecture practice. The capability helps organizations ensure that architectural efforts are standardized, repeatable, and aligned with corporate strategy.

A well-developed Architecture Capability enables an organization to respond effectively to change, manage complexity, and improve decision-making by providing a structured approach to Enterprise Architecture management. It ensures that architecture-related initiatives are properly governed, resourced, and aligned with business priorities.

CHAPTER 9   ARCHITECTURE CAPABILITY

9.1. The Ability to Practice Architecture

In order to establish and maintain an architecture in an organization, the creation of an appropriate capability is a must. The TOGAF Standard calls this capability the Architecture Capability. It is about an organization's *ability to practice architecture*.

A loose definition for a capability is:

The ability of an organization to do something.

The word *something* in the definition above should obviously be replaced with the specific subject for which the capability is intended or created – in this case, *to practice architecture*.

9.2. Creating the Architecture Capability

According to the TOGAF Standard, creating the Architecture Capability is no different from creating any other capability. For this reason, it is similar to creating a sustainable architecture. Therefore, the framework indicates that the creation of the Architecture Capability can be supported by going through the phases of the Architecture Development Method.

As with any business capability, the establishment of an Enterprise Architecture Capability can be supported by the TOGAF Architecture Development Method [11].

The TOGAF Standard views the Architecture Development Method as the ideal means to design and oversee the implementation of the Architecture Capability. Applying the Architecture Development Method, particularly Phase A (Architecture Vision), ensures the capability's establishment within an organization.

This capability should not be viewed as a separate phase of an architecture project nor as a stand-alone project. Rather, it should be viewed as an ongoing discipline that provides the context, environment, and resources to govern and enable the delivery of the architecture to the organization.

Implementing any capability requires designing the four architecture domains: Business, Data, Application, and Technology. Therefore, implementing the Architecture Capability follows a similar pattern to developing an average architecture, using the Architecture Development Method.

Table 9-1 shows the phases of the Architecture Development Method and the associated focus areas that must be traversed and addressed to create the Architecture Capability.

Table 9-1. *The Phases and focus areas of the ADM*

ADM Phase	Focus Areas
Phase A: Architecture Vision	Define the project, identify stakeholder concerns and business requirements, identify business goals and drivers, define scope, define constraints, perform Architecture Maturity Assessment.
Phase B: Business Architecture	Agree on a common language, define the process, identify views and viewpoints, decide on the framework, define performance metrics, agree on the governance framework.
Phases C and D: Data, Application, and Technology Architectures	Specify and govern the structure of the Architecture Repository, define requirements for architecture deliverables, define technology.

(*continued*)

Table 9-1. (*continued*)

ADM Phase	Focus Areas
Phase E: Opportunities and Solutions	Determine organizational change.
Phases F and G: Migration Planning and Implementation Governance	Govern organizational change.
Phase H: Architecture Change Management	Manage change.
Requirements Management	Manage requirements.

The creation of the Architecture Capability begins with *Phase A (Architecture Vision)*:

- **Define the Project:** This step focuses on defining the stakeholders. Stakeholders should include the roles and organizational units involved in the Architecture Capability, as well as the people who will benefit from the deliverables of the Architecture Capability. This step largely determines the positioning of the Architecture Capability in the organization. Therefore, it is a very important step.

- **Identify Stakeholder Concerns and Business Requirements:** This step generates the initial, very high-level definitions of the Architecture Capability from a business information systems and technology perspective.

- **Identify Business Goals and Drivers:** Understanding the business goals and drivers is essential to aligning the Architecture Capability with the organization.

CHAPTER 9 ARCHITECTURE CAPABILITY

Much like the first step, this step partially determines the positioning of the Architecture Capability in the organization. The exact positioning of the capability depends largely on the business goals being pursued.

- **Define Scope:** Defining the scope of the Architecture Capability cannot be omitted. It cannot be implemented without a high-level project plan of what architecture needs to be addressed in the coming period.

- **Define Constraints:** This step focuses on the organization-wide constraints that affect all architecture projects, including the Architecture Capability (which isn't a one-time project, as mentioned earlier).

- **Perform Architecture Maturity Assessment:** It is essential to conduct an initial Architecture Maturity Assessment to gain insight into the maturity of the current view of architecture within the organization. The way in which the Architecture Capability can be implemented will largely depend on the maturity level of the organization.

Moving through *Phase B (Business Architecture)*, the focus shifts to the following core components:

- **Agree on a Common Language:** Define the architectural terms and definitions that will be used in and by the organization. This will enable the use of a common language that everyone involved in the Architecture Capability will understand.

- **Define the Process:** The Architecture Development Method forms the basis for the architectural process to be used. When using the TOGAF Standard, it is essential to adapt the Architecture Development Method to the context for which it is being used. Use only those parts of the method that contribute to the creation of Architecture Capability.

- **Identify Views and Viewpoints:** Work with stakeholders to determine which views and viewpoints are important to flesh out. One viewpoint that should not be overlooked is architecture governance.

- **Decide on the Framework:** Agree on the creation of architecture deliverables and the requirements for creating these deliverables.

- **Define Performance Metrics:** Establish realistic metrics against which to measure the performance and progress of creating and governing the Architecture Capability.

- **Agree on the Governance Framework:** Formulate what the Governance Framework will look like. Draw a diagram of the framework and document it along with the other agreements.

Phases C and D (Information Systems Architectures, and Technology Architectures) perform the following activities related to the creation of the Architecture Capability:

- **Specify and Govern the Structure of the Architecture Repository:** Design or develop the metamodel to be used with the Architecture Capability. This metamodel describes the entities used by the Architecture Capability, their characteristics, and the key relationships between these entities.

CHAPTER 9   ARCHITECTURE CAPABILITY

- **Define the Requirements for the Architecture Deliverables:** Describe the requirements for being able to create, maintain, publish, distribute, and govern the architecture deliverables. A focus here is the selection of an appropriate toolset for creating and maintaining models and diagrams. The ability to publish - making the architecture artifacts findable and available - is also an important focus.
- **Define Technology:** Define the technology and infrastructure that will be used to create and operate the Architecture Capability.

Phase E (Opportunities and Solutions) focuses on determining the organizational change required to accommodate the Architecture Capability:

- **Determine Organizational Change:** Determine the organizational change that is necessary to establish and implement the Architecture Capability and how to achieve this.

The focus of *phases F and G (Migration Planning and Implementation Governance)* is to govern the implementation of the defined steps:

- **Govern Organizational Change:** The Migration Planning phase focuses on implementing the steps defined in the Business Architecture phase. This also applies to Phase G (Implementation Governance).

The Architecture Development Method cycle ends with *Phase H (Architecture Change Management)*:

- **Manage Change:** In this final phase, the focus should be on managing the changes required to establish and implement the Architecture Capability.

During the *Requirements Management Phase,* all requirements should be controlled:

- **Manage Requirements:** All requirements defined during the previous phases should be clearly articulated and aligned. This can be done by continuously following the Requirements Management Phase.

The approach described here makes it all sound and look very formal.

In practice, of course, all the steps mentioned in the Architecture Development Method are performed, but in a less deliberate way. Often, the steps and focus areas described above are reviewed and agreed upon during one or two meetings.

It is recommended that the agreements made are recorded and, if possible, formally signed off by the main sponsor of the Architecture Capability.

The applicable business goals and desired outcomes are determined during the creation of the Architecture Capability. This largely determines the Architecture Capability's position in the organization.

9.3. Positioning the Architecture Capability

Organizations that understand that Enterprise Architecture is a strategic business management tool, not a technical instrument, have a great advantage in setting the direction of the organization. When viewed as a strategic tool, architecture is directly involved in and used to translate organizational strategy into concrete implementation. This scenario is consistent with the use of the Architecture Capability as described in the Leader's Guide [3]. This guide describes *four ways* to use the Architecture Capability, which are visualized in Figure 9-1.

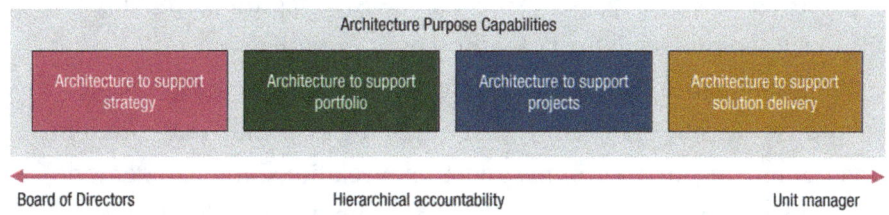

Figure 9-1. Architecture Purpose Capabilities

The four Architecture Capabilities shown in Figure 9-1 can be explained as follows:

Architecture to Support Strategy: The Architecture Capability is used to develop target architectures and create roadmaps that reflect organizational changes within the organization over a three- to ten-year period.

In this context, the Architecture Capability is used to support organization-wide change requests and multiple projects and programs and to achieve strategy execution through these projects and programs. There is ample room for the Architecture Capability to shape strategic aspects such as drivers, goals, objectives, and initiatives. Architecture used to support strategy is architecture in its most comprehensive form. The Architecture Capability would benefit greatly from being led by a CEO.

Architecture to Support Portfolio: The Architecture Capability supports change initiatives that span multiple functions, phases, and projects.

An Architecture Capability used for this purpose will primarily focus on a single portfolio. An example of a single portfolio might be migrating to a cloud environment. Within the portfolio, which has a clearly defined scope, multiple projects will require the support of the Architecture Capability. In this context, architecture is used to interpret and guide projects. The architect ensures the alignment of the projects within the portfolio and the coherence of the projects and steers their execution. The Architecture Capability is thus less of an organizational steering function than when it is primarily concerned with strategy execution.

Architecture to Support Projects: The Architecture Capability is tailored to the project delivery method of the organization.

An Architecture Capability used for this purpose typically covers one project. To stay in the context of migrating to a cloud environment, consider moving an on-premises customer relationship management system to its counterpart in a cloud environment. This migration requires the architect's involvement not only in the Application Architecture domain, but certainly in the Business and Information Architecture domains as well. In the context of supporting projects, the Architecture Capability is used to clarify the purpose and value of the project, as well as to identify requirements and monitor adherence to architecture governance. In addition, architecture supports integration and alignment across projects.

An Architecture Capability that supports portfolios and/or projects should be led by a CIO or CTO. There is still a strategic component to the architecture work that needs to be done, albeit in a slightly milder form than when the Architecture Capability is led by a CEO.

Architecture to Support Solution Delivery: The Architecture Capability supports the delivery of solutions.

An Architecture Capability used for this purpose typically covers one project or a significant portion of a project. The architecture is used to define how the change will be designed and delivered and to identify constraints, controls, and architectural requirements for the design. The architecture provides a Governance Framework for the change. Enterprise Architecture used to support solution delivery is actually similar to Solution Architecture. If the organization chooses this use of architecture, the question arises as to whether the organization would benefit more from employing a Solution Architect than an Enterprise Architect.

This last use of the Architecture Capability can be led by a unit manager. The responsibility for implementing solutions rests with this particular role.

During the creation of the Architecture Capability, it is important to define and agree on how the organization views and interprets the capability and to align it with the defined business goals and drivers. To determine where the primary focus of the capability should be, an Architecture Maturity Assessment can be performed. Such a maturity assessment helps to determine the current level of maturity of the Architecture Capability in the organization. The assessment will provide a clear indication of how the Architecture Capability should be positioned in the organization to achieve business goals. This is critical because the position of the capability is an indication of how the organization views the capability.

The position of the capability in the organization will also determine the line of reporting for the Architecture Capability. The higher the direct reporting line is in the organizational hierarchy, the more the Architecture Capability moves toward the strategic side of the profession. Figure 9-2 provides some insight into the possible placement of the Architecture Capability within the organization.

CHAPTER 9   ARCHITECTURE CAPABILITY

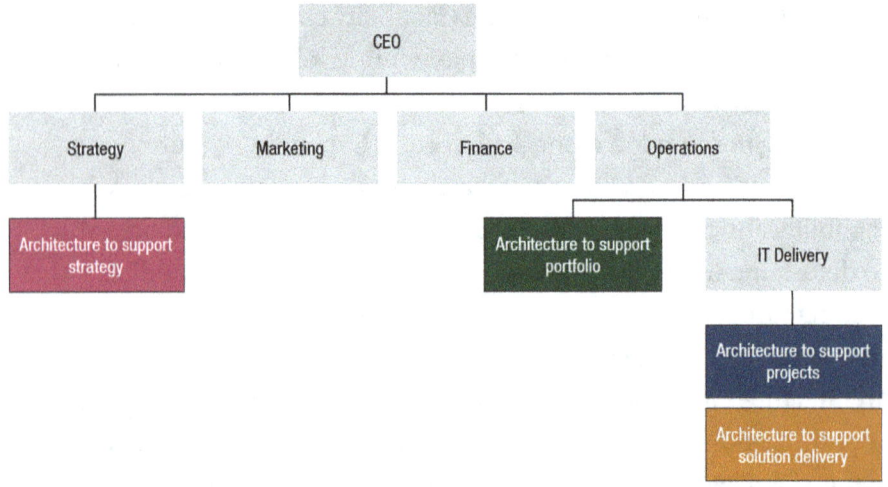

Figure 9-2. *Positioning of the Architecture Capability*

For example, Figure 9-2 shows the four Architecture Capabilities positioned on an organizational chart. The capability that uses architecture to support strategy is often positioned with a direct reporting line to the CEO or to a strategy executive. This ensures that the capability operates primarily in the strategic domain. If the architecture is used to support the portfolio, the direct reporting line will most likely be to the CIO. The architecture used in this scenario will be involved in portfolio and project management. A third way to use the Architecture Capability is to support projects. Using the architecture to structure project execution, as with using it to support the portfolio, will further reduce its scope within the organization. When architecture is used to support projects, the direct reporting line is often to the unit manager of an IT department. Finally, placing the Architecture Capability in the IT domain or department will create the smallest scope for architecture work, as it will focus primarily on IT solution delivery.

Ultimately, there is no right or wrong way to establish an Architecture Capability in an organization. Its position depends entirely on an organization's needs and desires for working with architecture.

It should be noted, however, that the further to the right the Architecture Capability is positioned - that is, toward the operational side of the organization chart in Figure 9-2 - the more the focus will be on solution-oriented issues. As a result, architecture work will be more reactive and much less proactive. Also, the goals and objectives of an organization will be represented by and in architecture work to a much lesser extent, making architecture less likely to be connected to the strategic direction of the organization. The risk here is that architecture becomes a stand-alone entity that lacks coherence with other business activities.

The Architecture Capability is directly related to the architecture roles distinguished in the TOGAF Standard. These roles are in turn related to the architecture domains.

Further Reading

For more information on the Architecture Capability, see Chapters 5, 9, 11, and 12 of "The TOGAF® Leader's Guide to Establishing and Evolving an EA Capability", document number G184.

9.4. Summary

This chapter demonstrated that in order to establish and maintain an architecture in an organization, the creation of an appropriate capability is a must. The TOGAF Standard calls this capability the Architecture Capability.

According to the framework, creating the Architecture Capability is no different from creating any other capability. For this reason, it is similar to creating a sustainable architecture. Therefore, the TOGAF Standard indicates that the creation of the Architecture Capability can be supported by going through the phases of the Architecture Development Method.

CHAPTER 9 ARCHITECTURE CAPABILITY

Establishing the capability should not be viewed as a separate phase of an architecture project or as a stand-alone project, but rather as an ongoing discipline that provides the context, environment, and resources to govern and enable the delivery of the architecture to the organization.

The creation of the Architecture Capability consists of the following steps:

- **Define the Project:** Focuses on defining the stakeholders.

- **Identify Stakeholder Concerns and Business Requirements:** Generates the initial definitions of the Architecture Capability from a business information systems and technology perspective.

- **Identify Business Goals and Drivers:** This step partially determines the positioning of the Architecture Capability in the organization.

- **Define Scope:** Defines the scope of the Architecture Capability.

- **Define Constraints:** This step focuses on the organization-wide constraints that affect the Architecture Capability.

- **Perform Architecture Maturity Assessment:** The way in which the Architecture Capability can be implemented will largely depend on the maturity level of the organization.

- **Agree on a Common Language:** Define the architectural terms and definitions that will be used in and by the organization.

- **Define the Process:** Use only those parts of the Architecture Development Method that contribute to the creation of the Architecture Capability.

- **Identify Views and Viewpoints:** Work with stakeholders to determine which views and viewpoints are important to flesh out.

- **Decide on the Framework:** Agree on the creation of architecture deliverables and the requirements for creating these deliverables.

- **Define Performance Metrics:** Establish realistic metrics against which to measure the performance and progress of creating and governing the Architecture Capability.

- **Agree on the Governance Framework:** Formulate what the Governance Framework will look like.

- **Specify and Govern the Structure of the Architecture Repository:** Design or develop the metamodel to be used with the Architecture Capability.

- **Define the Requirements for the Architecture Deliverables:** Describe the requirements for being able to create, maintain, publish, distribute, and govern the architecture deliverables.

- **Define Technology:** Define the technology and infrastructure that will be used to create and operate the Architecture Capability.

- **Determine Organizational Change:** Determine the organizational change that is necessary to establish and implement the Architecture Capability and how to achieve this.

- **Govern Organizational Change:** The Migration Planning phase focuses on implementing the Architecture Capability.

- **Manage Change:** In this final phase, the focus should be on managing the changes required to establish and implement the Architecture Capability.

- **Manage Requirements:** All requirements defined during the previous steps should be clearly articulated and aligned.

After discussing the creation and establishment of the Architecture Capability, the chapter showed what its positioning can do for an organization. This positioning is addressed in the *Identify business goals and drivers* step when establishing the Architecture Capability.

There are four possible ways of positioning the capability, each depending on the objectives of the organization:

- **Architecture to Support Strategy:** The Architecture Capability is used to develop target architectures and create roadmaps that reflect organizational changes within the organization over a three- to ten-year period.

- **Architecture to Support Portfolio:** The Architecture Capability supports change initiatives that span multiple functions, phases, and projects.

- **Architecture to Support Projects:** The Architecture Capability is tailored to the project delivery method of the organization.

- **Architecture to Support Solution Delivery:** The Architecture Capability supports the delivery of solutions.

During the creation of the Architecture Capability, it is important to define and agree on how the organization views and interprets the capability and to align it with the defined business goals and drivers. To determine where the primary focus of the capability should be, an Architecture Maturity Assessment can be performed. Such a maturity assessment helps to determine the current level of maturity of the Architecture Capability in the organization.

The position of the capability in the organization will also determine the line of reporting for the Architecture Capability. The higher the direct reporting line is in the organizational hierarchy, the more the Architecture Capability moves toward the strategic side of the profession.

CHAPTER 10

Architecture Roles, Skills, and Competencies

 This chapter is part of the GREEN and GOLD reading tours. See Figure 1-1 in Chapter 1, Section 1.3, for alternative reading tours.

The previous chapter discussed establishing the Architecture Capability. This activity is one of the first steps to be performed in the Preliminary Phase, along with the formation of an architecture team. Both activities are highly dependent on the ability to define the correct roles, skills, and competencies required to successfully establish the Architecture Capability and associated architecture team. A skills framework can be used to identify these roles and skills.

This chapter focuses on the architecture roles, skills, and competencies provided by the TOGAF Standard. The framework includes an overview of the roles used in an architecture practice and the skills associated with them. With the release of the "Architecture Roles and Skills" Series Guide, the framework now provides a more comprehensive overview of the skills associated with each role, highlighting the depth of knowledge required to successfully perform the role.

CHAPTER 10   ARCHITECTURE ROLES, SKILLS, AND COMPETENCIES

In practice, however, there still seems to be some ambiguity about job descriptions, roles, and assignments. In some situations, this leads to misunderstandings about the roles and responsibilities of architects.

10.1. Difference Between Jobs, Roles, and Assignments

To properly understand the difference between *jobs*, *roles*, and *assignments*, it is essential to clarify these terms, as they can be used in multiple, sometimes overlapping contexts. The Series Guide "Architecture Roles and Skills" defines them as follows:

- **Job:** The *formal position* someone holds, usually defined by a contract of employment. Usually the position is related to the hierarchical positioning in the organization.

- **Role:** The expected *set of activities*, typically tied to a *specific skill set*.

- **Assignment:** A *specific task* or project a person is *responsible for*, often within the scope of their job and role, though not exclusively.

Although the TOGAF Standard makes an excellent attempt to clarify the distinction between jobs, roles, and assignments, there is still little consensus about architecture roles in day-to-day practice. For example, in many organizations today, it is still unclear whether the term *Enterprise Architect* refers to a job or a role. Partly because of this continuing ambiguity, organizations are not well equipped to properly position the profession of architecture within the contours of their own organization. As a result, architecture as a discipline is still invariably viewed as part of the IT department. Being part of the IT department requires different skills and competencies than being part of a strategic team that translates business strategy into execution.

Much of the misunderstanding has to do with the naming of some architecture roles and their underlying meaning in terms of duties and responsibilities. Most ambiguity occurs when talking about architecture roles in relation to architecture domains. Take the term *Technical Architect*, for instance. Is it a job or a role? And what about an Application or Business Architect? What are their responsibilities? In an effort to eliminate confusion about jobs and roles in architecture domains, the TOGAF Standard created the Segment Architect profile. This profile describes a job, not a role or an assignment. It comes as a complete package, including a role description and assigned responsibilities.

10.2. Core Architecture Roles

The Series Guide "Architecture Roles and Skills" acknowledges that in many organizations, individuals may fulfill multiple roles, while certain roles might also be assigned to multiple individuals. The Series Guide therefore provides a practical *starting point*, showcasing the core roles essential for an Enterprise Architecture practice, which can then be tailored to meet the unique demands of the organization.

The TOGAF Standard outlines an example set of roles critical for supporting the development and governance of an organization's Architecture Landscape. These roles facilitate robust decision-making across various levels of influence.

The "Architecture Roles and Skills" Series Guide introduces three primary viewpoints: *enterprise*, *segment*, and *solution*. Figure 10-1 presents these viewpoints alongside additional perspectives, explained further below, that enhance this structure.

CHAPTER 10   ARCHITECTURE ROLES, SKILLS, AND COMPETENCIES

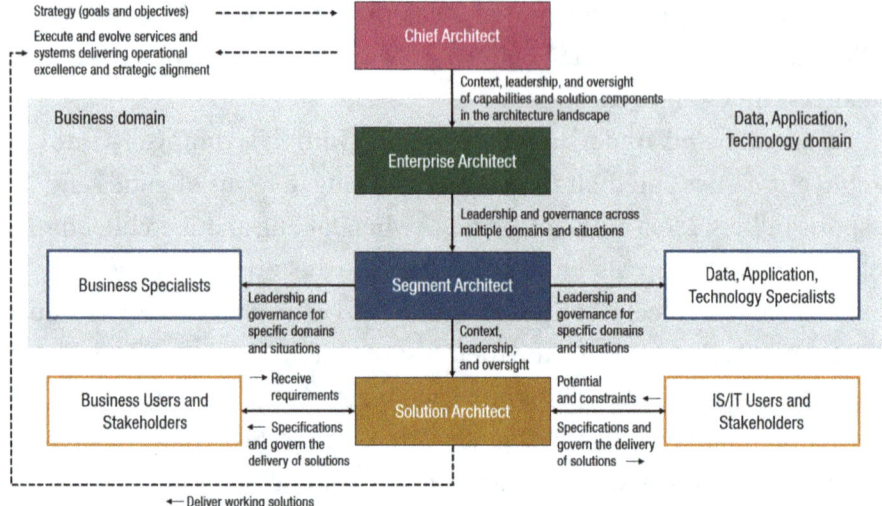

Figure 10-1. *Core architecture roles and their relationships*

The three primary viewpoints – enterprise, segment, and solution – are equal to the three core architecture roles. The TOGAF Standard does not give specific details on the *specialist*, *user*, and *stakeholder* roles. These can be found in the many different professional frameworks and associations that exist for those roles.

Figure 10-1 shows the *three core architecture roles* – supplemented by the specialized position of Chief Architect – arranged vertically to illustrate the breadth of the roles and to emphasize their interrelationship.

Starting from the Chief Architect at the top and moving down through Enterprise Architect to Segment Architect, each role brings increasing depth in expertise, leadership, and governance, tailored to distinct elements within the Architecture Landscape. This tiered approach supports balanced, well-informed decision-making. Most organizations operate with three to seven levels of management and control, and while many aim to streamline these, the TOGAF Standard's model starts with a foundational three-tier framework. This minimal setup allows flexibility to expand as needed for organizations that require additional layers of guidance and governance.

Figure 10-2 omits the additional components and relationships shown in Figure 10-1 in order to properly focus on the three primary architecture roles (with the additional specialization of the Enterprise Architect role, the Chief Architect). The dashed line around the Chief Architect and Enterprise Architect roles in Figure 10-2 is to emphasize their relationship in the sense that the Chief Architect role is a *specialization* of the Enterprise Architect role. They are *not* two separate roles like Segment Architect and Solution Architect.

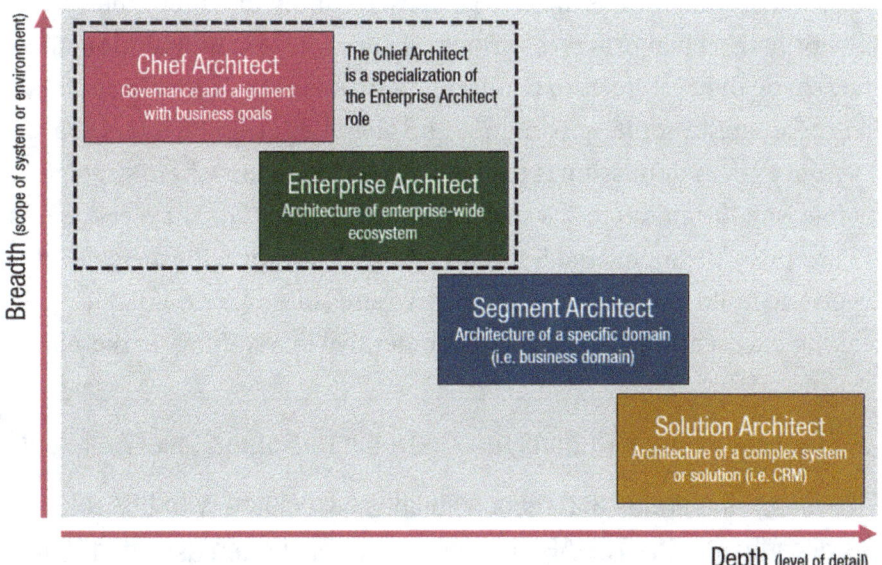

Figure 10-2. Core architecture roles

Chief Architect: The Chief Architect is a special case of an Enterprise Architect and is responsible for integrating the Enterprise Architecture view into the wider strategic and operational context of the organization. This role is performed by a seasoned, influential professional who plays a pivotal role in driving major technical and business transformations across an organization and even the broader market. Whether an internal leader or a trusted external advisor, they're responsible for building alignment and commitment at all levels, securing buy-in for impactful changes that elevate performance.

Working at the highest levels, the Chief Architect unifies executives and board members around the vision for a robust, forward-looking Architecture Landscape. They make complex technical and business concepts accessible and relevant to leadership, ensuring ideas connect seamlessly with the company's strategic and operational goals.

This role thrives in situations that are dynamic, evolving, or complex. The Chief Architect leads the charge in establishing roadmaps that guide transformative change across various business and technology domains. They shape the Architecture Landscape's structure and direction and take on accountability for its governance and alignment with business goals.

Above all, their mission is to create an environment where strategic change isn't just possible, it's sustainable and impactful. They lead teams of Enterprise, Segment, and Solution Architects, along with specialized experts, to build and manage the services and solutions essential to keeping the organization agile and competitive in a shifting marketplace.

In Europe, few organizations recognize the role of the Chief Architect.

Therefore, the duties and responsibilities associated with this role — as described in the TOGAF Standard — are most often assigned to the Enterprise Architect.

The roles of Enterprise and Chief Architect are typically combined into one.

Enterprise Architect: Seasoned experts, steering transformative change that bridges technology and business across the entire organization, ultimately bringing innovation into the market.

They don't just participate, they *lead* major initiatives, holding a seat at the table with senior decision-makers. With accountability for comprehensive business service delivery through finely tuned information systems, they demonstrate advanced expertise in both technical and business arenas. Their track record showcases substantial achievements and strong, trusted relationships with key stakeholders. Not only do they deliver, but they actively coach and mentor, setting the standard for excellence.

Enterprise Architects thrive in dynamic, complex environments where solutions are uncharted and the scope is fluid. Whether tackling evolving requirements or varied markets or integrating multiple technologies, they build roadmaps that span entire business segments. They define the guiding principles and policies that shape the architecture, ensuring cohesive governance and alignment across all initiatives.

Their core mission is to establish and enhance the framework for integration, interoperability, and governance across solution components. Leading Segment Architects, Solution Architects, and a spectrum of specialists, they shape, deliver, and sustain the solutions that keep their organization competitive in an ever-evolving market.

Segment Architect: Skilled and seasoned professionals who manage and champion the specific business or technical capabilities within a segment of the organization or the broader marketplace.

Their role is all about defining, shaping, and governing the systems, subsystems, and components that bring products and services to life. Segment Architects don't just build roadmaps, they create blueprints for the future of their segments, ensuring every technical and business option is thoroughly evaluated and effectively utilized. They've got policies and standards covered, guaranteeing every element in their segment is effective, efficient, and compliant. Collaboration is second nature for them as they work closely with other architects to ensure seamless connectivity across the entire environment.

CHAPTER 10 ARCHITECTURE ROLES, SKILLS, AND COMPETENCIES

Segment Architects excel in dynamic, evolving, or unprecedented scenarios, stepping up to tackle complex, wide-reaching solutions that keep pace with change. They oversee every development effort within their segments, ensuring that all systems and components align with purpose, comply with regulations, and follow established policies and standards.

Their mission is to create the foundation for cohesive integration, smooth interoperability, and continuous evolution across their segments. By setting up clear roles and fostering effective collaboration among technical specialists and Solution Architects, they ensure everyone works in sync to deliver components and building blocks that meet end-to-end business requirements for the organization and its clients.

Solution Architect: Also seasoned professionals who drive the delivery of new or evolved capabilities across the organization and its marketplace.

Their role is to craft and guide the solutions that align with both immediate requirements and broader organizational goals, achieving defined service levels within ambitious timelines. They break down these solutions into systems, subsystems, and components that interact seamlessly and then integrate them into a cohesive roadmap, transforming complex ideas into actionable steps through targeted Transition Architectures.

With an eye for innovation, Solution Architects make sure that each decision leverages the latest business and technical advancements. They are particularly effective in dynamic and complex situations, where solutions must be adaptable to evolving challenges. From specification through to implementation, they ensure each new solution adheres to key principles, policies, and standards, resulting in systems that are not only effective but also efficient and compliant.

Their main mission is to build a collaborative environment that supports seamless solution development, integration, and interoperability. By engaging closely with technical and business teams alike, Solution Architects clarify roles, align expectations, and streamline the delivery of each component, ultimately ensuring the end-to-end service levels are not just met but exceeded. This collaborative approach allows them to fully understand and address the diverse requirements and expectations of every stakeholder involved.

10.3. The Role of the Segment Architect

Let's take a closer look at the Segment Architect. The person who holds this *job* may very well fill the *role* of Technical Architect (based on the tasks described in the job's *assignment* and the required skills and competencies defined in the role's description). The hierarchical position within the organization is not determined by the job description, but rather by the job assignment. For example, if the job assignment states that the Segment Architect deals with technical infrastructure issues, the role description points to the Technical Architecture domain and requires certain skills and competencies that are useful when working in that particular domain. It's safe to say that this Segment Architect will most likely be part of the IT department.

Conversely, if the Segment Architect's job description states that they are primarily concerned with business issues, then all indications are that the Segment Architect works in the area of Business Architecture. The role description of the second Segment Architect is different from that of a Segment Architect working in the Technical Architecture domain. The skills and competencies required for the two roles are also different. However, both examples involve the same job description, that of a Segment Architect.

The use of *segments* also requires further explanation. This is mainly because there are two ways to look at and use segments and thus to explain the role of the Segment Architect.

The first way to look at segments is from the viewpoint or *perspective of the profession*. This is consistent with how the TOGAF Standard has articulated it. From this perspective (the *horizontal* perspective), segments, and the role of the Segment Architect, are based on the four architecture domains (Business, Data, Application, and Technology). Each of the domains represents a specialization for the Segment Architect. The specializations are Business Architect, Data Architect, Application Architect, and Technology Architect, corresponding to the four architecture domains.

A *segment* groups capabilities in a structured way, often as part of a larger portfolio or specific change program [19].

The TOGAF Standard also introduces domains, which act as key groupings of capabilities that an organization needs to manage.

A *domain* is used to reflect any useful grouping of capabilities to be managed by an organization, with four particular segment-based viewpoints. These four segments are called the business, application, data, and technology domains.

The second way to look at segments and the role of the Segment Architect is from an *organizational perspective*. This perspective (the *vertical* perspective) is not covered in the TOGAF Fundamental Content and is somewhat hidden in the "Architecture Roles and Skills" Series Guide.

In addition to the usual reference to the four architecture domains, a Segment Architect's job may also be linked to an organizational classification. This classification occurs when business operations are organized into business domains or segments (vertical pillars) rather than using a direct architecture domain classification. When architecture in an organization is considered in terms of business domains or segments instead of architecture domains, it is referred to as an *organizational* or *vertical perspective*. Figure 10-3 illustrates business domains or segments (shown vertically) compared with architecture domains (shown horizontally).

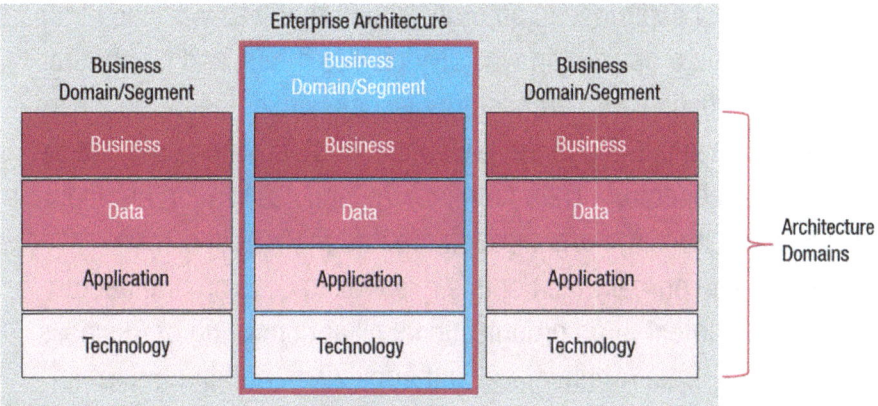

Figure 10-3. Business domains/segments versus architecture domains

---

This vertical perspective is adopted by many organizations, but it makes it difficult to assign architecture roles.

This is because each pillar in the organization actually contains multiple architecture domains. Often, Business Architecture, Data Architecture, and Application Architecture come together in one pillar. The Technology Architecture is typically separated from the other three architecture domains because, in nine out of ten cases, it is the responsibility of IT.

The application of the vertical perspective does not always support the assignment of architecture roles in the way suggested by the TOGAF Standard. Because there are so many different architecture domains in a single pillar – and usually too few architects with the necessary knowledge and expertise in these domains – pragmatic solutions are sought. Typically, this results in an organization creating new architecture roles to meet demand.

For example, I experienced an organization that was working in an agile way. The organization was divided into tribes, and one of the tribes was dealing with architecture issues within one of the pillars of the organization. Since there was no architect who could handle Business Architecture, Data Architecture, and Application Architecture at the same time, the organization created the role of a Tribe Architect. The primary task of this role was to develop architecture within the context of an organizational pillar consisting of Business Architecture, Data Architecture, and Application Architecture at an operational level.

This is just one of many examples of how the choice of organizational perspective often leads to an uncontrollable proliferation of new — and often nonsensical — architecture roles.

The horizontal perspective (based on the architecture domains) and the vertical perspective (which follows business domains or segments) are explained in the following paragraphs.

10.3.1. The Horizontal Perspective

As mentioned in Section 10.3, when viewed from the angle of the architecture profession, the TOGAF Standard recognizes three core architecture roles (with the Chief Architect being a specialization of the Enterprise Architect role). Each of the three core architecture roles is relatively easy to derive from the four Architecture Capabilities mentioned in Chapter 9. Figure 10-4 shows the four Architecture Capabilities, with the corresponding core architecture roles beneath them.

CHAPTER 10   ARCHITECTURE ROLES, SKILLS, AND COMPETENCIES

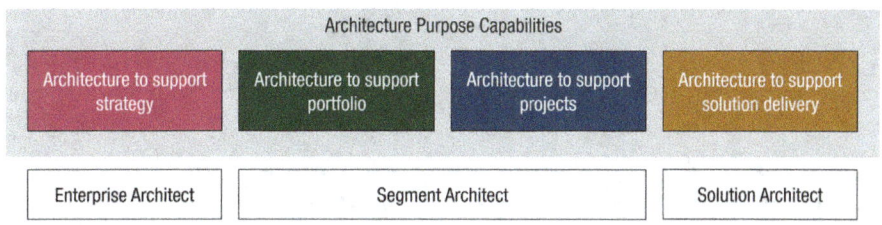

Figure 10-4. Architecture Capabilities and architecture roles

The Architecture Capability *in support of strategy* is directly related to the roles of the Chief Architect and the Enterprise Architect. It should be noted that when an organization has a Chief Architect, that role often takes on the strategic side of the discipline. The Enterprise Architect then takes on the development of the architecture for the enterprise ecosystem. If an organization is not large enough to support both roles, then often only the Enterprise Architect exists. In this scenario, the Enterprise Architect will also take on the strategic part of the Architecture Capability.

The two capabilities in the center of Figure 10-4, *in support of portfolio* and *in support of projects*, can be related to the role of the Segment Architect. This role falls into a number of specializations corresponding to the number of architecture domains: Business Architecture, Data Architecture, Application Architecture, and Technology Architecture.

Where the capability supports *solution delivery*, the role of the Solution Architect can be distinguished. Again, like the Segment Architect role, this role has multiple manifestations in the form of specializations (think Cloud, Network, and Software Architect).

Some of the specializations mentioned above that can be associated with the architecture roles by capability are shown in Figure 10-5. Each capability corresponds to an architecture role, and each architecture role corresponds to two or more specializations.

CHAPTER 10   ARCHITECTURE ROLES, SKILLS, AND COMPETENCIES

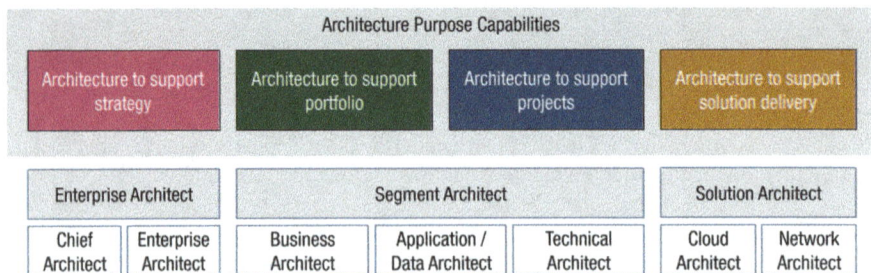

Figure 10-5. Architecture Capabilities, roles, and specializations

Figure 10-5 assumes a horizontal perspective in which architecture roles are deployed according to architecture domains. This layout fits well with the four Architecture Capabilities mentioned earlier.

Different architecture roles are required to properly implement each of the four capabilities. When architecture roles are combined into a single role, it creates ambiguity about the deployment of the capabilities and the responsibilities for their implementation and execution.

10.3.2. The Vertical Perspective

In addition to the horizontal perspective (based on the four architecture domains) of looking at architecture roles, the TOGAF Standard also describes a second way of viewing the various architecture roles: from an *organizational perspective*. This perspective is also called the vertical design of an organization. It focuses not on the four architecture domains, but on the *core activities or services* of an organization. The classification that follows is based on organizational pillars per core activity or service. For example, in the healthcare industry, this might result in the pillars of Research, Consultation, and Treatment. Each pillar then consists of a piece of Business Architecture, Data and Application Architecture, and Technology Architecture. In the context of the vertical perspective, a pillar is also called a segment and thus consists of a Segment Architecture (which is composed of the four domains mentioned earlier). Figure 10-6 illustrates the example just given.

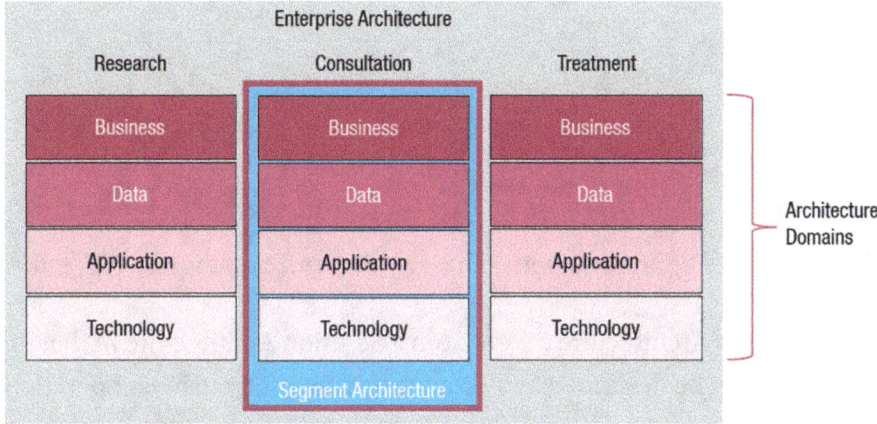

Figure 10-6. *Vertical perspective and associated architectures*

A Segment Architect is responsible for the architecture within an organizational domain (pillar), a segment of the overall Enterprise Architecture. Here, a segment is a vertical domain from an architectural perspective because it encompasses all architectural domains, but specifically from the context of an organization. This is illustrated schematically by the blue block in Figure 10-6.

When an organization adopts organizational segmentation, the work of a Segment Architect often becomes a bit more substantive and operational in nature because of the responsibility for the Application and Technology Architecture domains. This is often the domain of the Solution Architect, which explains the more operational nature of the role.

The difference between a Segment Architecture and a Solution Architecture is illustrated in Figure 10-7. For example, Consultation might be centralized, but its services are used and tasks performed in other processes – used by other departments – as well. This is highlighted by the purple lines in Figure 10-7.

CHAPTER 10   ARCHITECTURE ROLES, SKILLS, AND COMPETENCIES

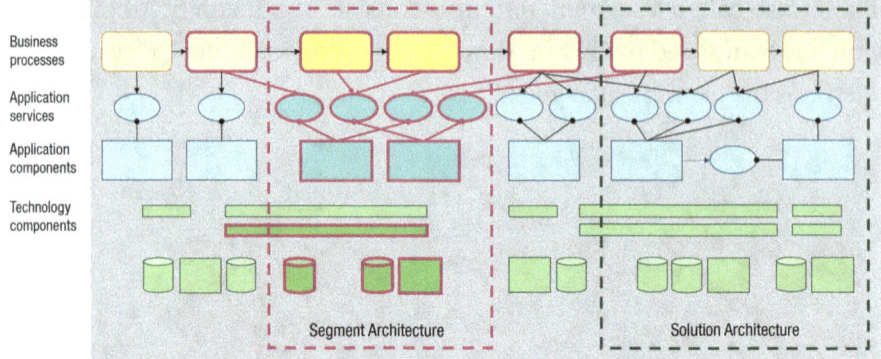

Figure 10-7. *Differences between Segment and Solution Architectures*

A Segment Architecture differs from a Solution Architecture in that the latter often can be described as a discrete and focused business operation or activity and how IT supports that operation. A Solution Architecture, in Figure 10-7 highlighted by the dark-green border, typically applies to a single project or project release.

The Governance Framework (see Chapter 8) includes the roles of Chief Architect, Enterprise Architect, and Domain Architect, but not Segment and Solution Architects.

In the "Architecture Roles and Skills" Series Guide, the term *Segment Architect* is used instead of Domain Architect, and the Solution Architect is mentioned in the context of the Architecture Capability (see Figure 10-4). The Governance Framework does not mention these roles.

Unfortunately, this type of inconsistency occurs periodically in the TOGAF Standard.

During my discussions with The Open Group, we talked about these inconsistencies. They indicated that they are gradually addressing and changing them. It is possible that the next or an updated version of the framework will be more consistent.

The primary purpose of the Segment Architect role is to focus on a particular segment of the architecture. From a horizontal perspective (based on the four architecture domains), this could be the Business Architecture, Data and Application Architecture, or Technology Architecture domain. If an organization is divided into business domains, chances are that a Segment Architect is responsible for one of these business domains. Often, all four architecture domains fall within such a domain. This is the vertical perspective. In both situations, they are called segments of the architecture.

Further Reading

Additional information about architecture roles and skills is available in the original TOGAF Series Guide "Architecture Skills Framework", document number G198, and in the updated Series Guide "Architecture Roles and Skills", document number G249.

10.4. Skills and Competencies

The previous paragraphs have explained the three core architecture roles that the TOGAF Standard prescribes for the proper execution of architecture work:

- Enterprise Architect (with the Chief Architect specialization role)
- Segment Architect (the former Domain Architect role)
- Solution Architect

CHAPTER 10   ARCHITECTURE ROLES, SKILLS, AND COMPETENCIES

Each of these roles performs its own part of the architecture work. The Enterprise Architect focuses primarily on the more *strategic aspect* of the profession, while the Segment Architect focuses on detailing and refining *the specific architecture or business domains*. The Solution Architect plays the crucial role of *developing the details* of the proposed solutions. The latter role can therefore be seen as a specialist in a particular subset of the architecture.

Obviously, the proper performance of the various architecture roles requires certain skills and competencies. These are different for each role because of their focus area. The TOGAF Standard has described the skills required for each of the architecture roles listed above in a Series Guide called the "Architecture Skills Framework". This document provides a set of roles, skills, and proficiency levels for people performing architecture work.

Later, in September 2024 to be exact, The Open Group released an update to the Series Guide. This updated Series Guide is called "Architecture Roles and Skills" (see the *Further Reading* information box).

Skills frameworks provide a view of the competency levels required for specific roles. They define the roles within a work area, the skills required by each role, and the depth of knowledge required to fulfill the role successfully.

Skills frameworks are relatively common for defining the skills required for a consultancy and/or project management assignment, to deliver a specific project or work package. They are also widely used by recruitment and search agencies to match candidates and roles.

To properly perform the work associated with the core architecture roles, it is necessary for architects to have the right set of skills and associated competencies. The TOGAF Standard uses proficiency levels for each skill. In the "Architecture Roles and Skills" Series Guide, the skill categories contain a more detailed overview of the individual skills per category. An overview of the different skill categories [19] is shown in Table 10-1.

Table 10-1. *Skill categories in alphabetical order*

Skill category	Description
Application	Application management, Application Programming Interface (API) patterns, software design principles, etc.
Architecture	Modeling, building block design, high-level design, role definition, Architecture Principle design, high-level migration planning, building block management, systems development and integration methods, Security Architecture, architecture patterns, requirements management, governance, etc.
Business	Enterprise organization knowledge, business cases, business process, strategic planning, etc.
Change	Managing business change, project management methods and tools, etc.
Data	Data analysis, data exchange, and data management.
Generic	Leadership, teamworking, inter-personal skills, etc.
Legal	Data protection laws, contract law, procurement law, fraud, etc.
Technology	Compute, storage, network infrastructure, etc.

The levels of proficiency that can be associated with each of the skills are shown in Table 10-2.

CHAPTER 10 ARCHITECTURE ROLES, SKILLS, AND COMPETENCIES

Table 10-2. *Proficiency Levels*

Level number	Name	Description
1	Background	Basic understanding, but reliant on specialists.
2	Aware	Understands and can address issues, supported by specialists.
3	Knowledgeable	Good detail enabling effective decision-making, requiring specialist input only in difficult situations.
4	Expert	Extensive experience, practical, and applied knowledge.

The "Architecture Roles and Skills" Series Guide lists all the skills in the TOGAF Standard and describes these skills in detail. The link to the relevant Series Guide can be found in the *Further Reading* information box at the end of Section 10.3.

Now, in addition to the skills mentioned above, *competencies* also play an important role. Every architect who performs one of the three core architectural roles must have not only the skills, but also the right competencies to perform the skills properly. However, a list of competencies an architect must have is exactly what is missing from the TOGAF Standard.

Competencies differ from skills in that a competency consists of abilities and knowledge in addition to skills. The three together make up a competency. Figure 10-8 illustrates the structure of a competency and the difference between competencies and skills.

CHAPTER 10   ARCHITECTURE ROLES, SKILLS, AND COMPETENCIES

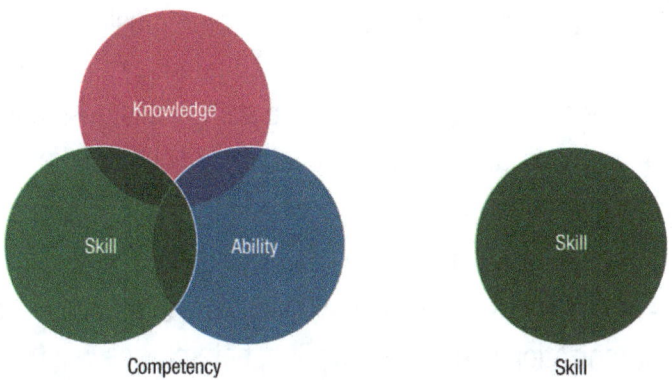

Figure 10-8. Competencies versus skills

Having clear and well-articulated competencies in addition to the required skills and associated proficiency levels is essential. Not only from the perspective of the organization, but certainly from the perspective of the architect doing the work.

In fact, I have been in a situation where I worked for an organization that predominantly (if not exclusively) viewed the profession of architecture from an IT perspective. This meant that the skills and competencies that were important to the organization and therefore required of the architect – me in this case – were radically different from the set of competencies I brought to the organization.

My skills and competencies are primarily at the strategic level, being able to translate an organization's strategy into appropriate execution. That requires very different skills compared with implementing IT Architecture.

The organization held on to its view of what it thought an Enterprise Architect should do. As a result, the scope of my work was reduced to that of the IT department. This situation eventually led me to terminate my contract with the organization.

CHAPTER 10 ARCHITECTURE ROLES, SKILLS, AND COMPETENCIES

Architecture is an essential component of sound business operations. It provides structure and insight into the coherence of everything an organization does. It is essential that the architect is positioned in the right place in the organization. This position is accompanied by the necessary competencies.

If there is a mismatch between what the architect believes is needed – based on their knowledge, skills, and experience – and how the organization sees the discipline, both parties will be left dissatisfied.

It is therefore highly recommended that a clear and mutually agreed picture of how the work will be done (from which perspective) is established prior to the start of the architecture work, so that the required skills and competencies can be aligned accordingly.

This will prevent unpleasant situations along the way.

The competencies required to properly perform architectural work vary depending on the role of the architect. In general, the following list of competencies applies to the three core architecture roles.

Strategic Thinking and Vision: This competency entails the ability to conceptualize the long-term view of enterprise systems and their alignment with organizational goals. Architects must be able to translate an organization's long-term vision into a concrete strategy. This strategy should then be turned into executable activities or initiatives that should have a place on the Architecture Roadmap.

The competency also requires architects to possess the capability to anticipate future challenges and opportunities to position the organization for success. Architects must have the foresight to help adapt the organization so that the course set can continue. It requires holistic thinking to balance business needs, technology solutions, and operational efficiency.

Strategic Decision-Making: Making informed, strategic decisions under uncertainty or complex scenarios is part of this competency. Architects must be able to make decisions based on informed arguments, even in situations that do not always present a clear picture of the future. These choices will need to be presented to decision-makers. In order for these people to make the right choice, it is important to outline several scenarios so that any risks per choice can be weighed. This means evaluating trade-offs and prioritizing initiatives based on organizational impact and ensuring alignment with organizational objectives while mitigating risks. The Architecture Alternatives and Trade-Offs method (see Chapter 14, Section 14.9) can be used to assist in this process.

Leadership and Influence: Architects need to have the capability to guide and inspire teams toward shared goals. The organization's goals need to be lived by everyone involved. Architects have an important role to play. It is important to build consensus among diverse stakeholders with varying priorities and perspectives. To achieve this, an architect must act as a trusted advisor to executives and decision-makers. The ability to demonstrate leadership contributes to the proper performance of the job.

Change Leadership: A competency closely related to *leadership and influence* is *change leadership*. This competency focuses specifically on driving organization-wide change initiatives with resilience and adaptability (also a separate competency). Part of change leadership is advocating for innovation while addressing resistance and fostering buy-in. Architects should be able to act as a mediator in case of resistance from within the organization. Building support for architectural work is essential to its proper execution. Architects must be sensitive to the culture of an organization. The likelihood of adoption is increased by supporting cultural and structural changes to accommodate new technologies and processes.

Business and Technology Alignment: Architects must be able to bridge the gap between business strategy and technology implementation. Successfully putting this competency into practice requires a solid understanding of how technology can support or enhance business processes and objectives. It also requires the ability to articulate the value of technology investments in business terms. It is important to continually align the architecture work with organizational goals.

Communication and Collaboration: Architects must demonstrate the ability to communicate complex ideas clearly to both technical and non-technical audiences. They must be able to choose the right tone and context of the conversation at all levels of the organization to get the message across clearly. This helps foster collaboration across multidisciplinary teams and departments. To do this, architects must be able to truly listen and respond constructively to the diverse viewpoints of many stakeholders.

Governance and Compliance: The *governance and compliance* competency focuses on enforcing standards, policies, and guidelines that ensure architectural consistency. Ensuring adherence to regulatory and compliance requirements is a key part of this competency. Another part is monitoring and measuring the effectiveness of architecture initiatives in accordance with the above requirements. This competency addresses the need to establish solid architecture governance. The Architecture Governance Framework described in Chapter 8 can be of assistance.

Adaptability and Resilience: Strategies typically don't change much over the course of a year. However, architects must be able to adapt them to rapidly changing technology landscapes and business environments as needed. Keeping the architecture aligned with business goals and objectives remains critical. Therefore, the ability to maintain composure and thus effectiveness in high-pressure situations is of paramount importance. One of the things that positively contributes to this competency is the ability to continuously learn to stay relevant in evolving domains. This allows architects to stay resilient.

CHAPTER 10   ARCHITECTURE ROLES, SKILLS, AND COMPETENCIES

Analytical Thinking and Problem Solving: Another competency architects must possess is the ability to break down complex challenges into manageable components. This means that architects must be able to break down a large problem or challenge into smaller, more manageable pieces. The advantage of this is that each (smaller) part can be considered and addressed as an independent part. Of course, the big picture must always be kept in mind. It is important to synthesize data and insights to identify root causes that can lead to actionable solutions. Architects need to approach problems with creativity and a structured methodology. Architecture frameworks can provide the necessary tools.

Systems Thinking: Since architecture is all about coherence between the various parts and segments of an organization, understanding the interconnectedness of components within an enterprise ecosystem is an absolute must. Architects should be able to design scalable and flexible systems and solutions that work cohesively across departments, with an eye toward the broader fit within the organization. They should focus on the sustainability and adaptability of solutions over time, which is related to the *adaptability and resilience* competency.

Table 10-3 provides an overview of the required competencies detailed above and maps them to the three core architecture roles.

CHAPTER 10 ARCHITECTURE ROLES, SKILLS, AND COMPETENCIES

Table 10-3. *Required architecture competencies per core architecture role*

Competency	Description	Core Architecture Role
Strategic thinking and vision	Ability to conceptualize the long-term view of enterprise systems and their alignment with organizational goals.	Enterprise Architect
	Capability to anticipate future challenges and opportunities to position the organization for success.	Enterprise Architect, Segment Architect
	Holistic thinking to balance business needs, technology solutions, and operational efficiency.	Enterprise Architect, Segment Architect
Strategic decision-making	Making informed, strategic decisions under uncertainty or complex scenarios.	Enterprise Architect
	Evaluating trade-offs and prioritizing initiatives based on organizational impact.	Enterprise Architect, Segment Architect
	Ensuring alignment with organizational objectives while mitigating risks.	Enterprise Architect, Segment Architect
Leadership and influence	Capability to guide and inspire teams toward shared goals.	Enterprise Architect
	Building consensus among diverse stakeholders with varying priorities and perspectives.	Enterprise Architect
	Acting as a trusted advisor to executives and decision-makers.	Enterprise Architect

(continued)

CHAPTER 10 ARCHITECTURE ROLES, SKILLS, AND COMPETENCIES

Table 10-3. (*continued*)

Competency	Description	Core Architecture Role
Change leadership	Driving organization-wide change initiatives with resilience and adaptability.	Enterprise Architect
	Advocating for innovation while addressing resistance and fostering buy-in.	Enterprise Architect, Segment Architect
	Supporting cultural and structural changes to adopt new technologies and processes.	Enterprise Architect, Segment Architect
Business and technology alignment	Competence in bridging the gap between business strategies and technology implementations.	Enterprise Architect, Segment Architect
	Strong understanding of how business processes and objectives can be supported or enhanced by technology.	Enterprise Architect, Segment Architect
	Ability to articulate the value of technology investments in business terms.	Segment Architect
Communication and collaboration	Proficiency in conveying complex ideas clearly to both technical and non-technical audiences.	Segment Architect, Solution Architect
	Fostering collaboration across multidisciplinary teams and departments.	Segment Architect, Solution Architect

(*continued*)

CHAPTER 10 ARCHITECTURE ROLES, SKILLS, AND COMPETENCIES

Table 10-3. (*continued*)

Competency	Description	Core Architecture Role
	Listening and responding constructively to diverse viewpoints.	Enterprise Architect, Segment Architect, Solution Architect
Governance and compliance	Enforcing standards, policies, and guidelines that ensure consistency in architecture.	Enterprise Architect, Segment Architect, Solution Architect
	Ensuring adherence to regulatory and compliance requirements.	Enterprise Architect, Segment Architect
	Monitoring and measuring the effectiveness of architecture initiatives.	Segment Architect, Solution Architect
Adaptability and resilience	Adjusting to rapidly changing technology landscapes and business environments.	Segment Architect, Solution Architect
	Maintaining composure and effectiveness during high-pressure situations.	Enterprise Architect, Segment Architect, Solution Architect
	Continuously learning to stay relevant in evolving domains.	Enterprise Architect, Segment Architect, Solution Architect
Analytical thinking and problem solving	Breaking down complex challenges into manageable components.	Enterprise Architect, Segment Architect, Solution Architect

(*continued*)

CHAPTER 10 ARCHITECTURE ROLES, SKILLS, AND COMPETENCIES

Table 10-3. (*continued*)

Competency	Description	Core Architecture Role
	Synthesizing data and insights to identify root causes and actionable solutions.	Solution Architect
	Approaching problems with creativity and a structured methodology.	Segment Architect, Solution Architect
Systems thinking	Understanding the interconnectedness of components within an enterprise ecosystem.	Segment Architect, Solution Architect
	Ability to design scalable and flexible systems that work cohesively across departments.	Segment Architect, Solution Architect
	Focus on sustainability and adaptability of solutions over time.	Segment Architect, Solution Architect

The competencies listed in Table 10-3 collectively ensure that architects not only design effective systems but also drive organizational transformation and deliver business value.

Each of the competencies listed in Table 10-3 corresponds to a skill listed in the "Architecture Roles and Skills" Series Guide. Figure 10-9 shows the relationships between competencies and skills.

CHAPTER 10   ARCHITECTURE ROLES, SKILLS, AND COMPETENCIES

Skill	Description	Architecture Competency									
		Strategic thinking and vision	Strategic decision-making	Leadership and influence	Change leadership	Business and technology alignment	Communication and collaboration	Governance and compliance	Adaptability and resilience	Analytical thinking and problem solving	Systems thinking
Application	Application management, Application Programming Interface (API) patterns, software design principles, etc.					X	X	X	X	X	X
Architecture	Modeling, building block design and management, high-level design, role definition, Architecture Principle design, high-level migration planning, systems development and integration methods, security architecture, architecture patterns, requirements management, governance, etc.		X	X	X	X	X	X		X	
Business	Enterprise organization knowledge, business cases, business process, strategic planning, etc.	X	X	X	X	X	X	X			
Change	Managing business change, project management methods, and tools, etc.		X	X	X	X	X	X	X		
Data	Data analysis, data exchange, and data management.								X	X	X
Generic	Leadership, teamworking, inter-personal skills, etc.		X	X		X	X				
Legal	Data protection laws, contract law, procurement law, fraud, etc.		X					X	X	X	
Technology	Compute, storage, and network infrastructure, etc.					X		X	X	X	X

Figure 10-9. Architecture competencies mapped to skills

At this time, the TOGAF Standard does not include an overview or listing of architecture competencies. Hopefully, this will be added in one of the future updates of the framework.

Further Reading

The SFIA 9 skills directory provides an overview of skills and competencies for Segment and Solution Architects:
https://sfia-online.org/en/sfia-9/all-skills-a-z

10.5. Summary

This chapter has highlighted the various architecture roles that exist according to the TOGAF Standard and are needed to properly implement the Architecture Capability. The TOGAF Standard introduces three primary views of the architecture profession: enterprise, segment, and solution. These three primary viewpoints correspond to the three core architecture roles:

- Enterprise Architect (with the Chief Architect specialization role)
- Segment Architect (the former Domain Architect role)
- Solution Architect

Each of the three roles performs its own part of the overall architecture work.

The *Chief Architect* is a special case of an Enterprise Architect and is responsible for integrating the Enterprise Architecture view into the wider strategic and operational context of the organization.

Enterprise Architects lead major initiatives and hold a seat at the table with senior decision-makers. They are accountable for comprehensive business service delivery through information systems. They demonstrate advanced expertise in both technical and business arenas.

The role of the *Segment Architect* is all about defining, shaping, and governing the systems, subsystems, and components that bring products and services to life. Segment Architects build roadmaps, create blueprints for the future of their segments, and ensure every technical and business option is thoroughly evaluated and effectively utilized.

Solution Architects are seasoned professionals who drive the delivery of new or evolved capabilities across the organization and its marketplace. Their role is to craft and guide the solutions that align with both immediate requirements and broader organizational goals, achieving defined service levels within ambitious timelines. They break down these solutions into systems, subsystems, and components that interact seamlessly and then integrate them into a cohesive roadmap, transforming complex ideas into actionable steps through targeted Transition Architectures.

There are several ways to look at an organization from an architectural perspective. The first way is from a horizontal perspective. This involves dividing the organization into the four well-known architecture domains (Business, Data, Application, and Technology). The second perspective is called the vertical perspective. This looks at the services of an organization and shows them in vertical segments. A vertical segment usually consists of one or more (or even all four) architecture domains.

Skills and competencies are required to properly perform the aforementioned architecture roles. The TOGAF Standard only describes the required skills and the required levels of proficiency. Because skills alone are not sufficient to perform architecture work or to fulfill the role of an architect, this book includes a list of ten competencies. These competencies are mapped to the skill set created by the TOGAF Standard.

CHAPTER 11

Architecture Domains

 This chapter is part of the PINK and GREEN reading tours. See Figure 1-1 in Chapter 1, Section 1.3, for alternative reading tours.

This chapter builds on the architecture roles mentioned in the previous chapter. Among other things, it identifies the Segment Architect. This role is responsible for (one of) the business, data, application, and technology domains, depending on the perspective from which the organization is viewed.

Upon reviewing the Architecture Capability described in Chapter 9, it is evident that certain architecture roles are required within an organization depending on the capability's positioning. The TOGAF Standard distinguishes three core architecture roles that relate to the Architecture Capability.

There is a role for the Chief/Enterprise Architect when the Architecture Capability is used to support strategy. The Segment Architect role is focused on supporting portfolios – and in some cases (large) projects – and the Solution Architect role is primarily focused on providing support when the Architecture Capability is used to support solutions. Regardless of the architecture role practiced, an architect will have to deal with one or more architecture domains. This chapter explains them.

CHAPTER 11   ARCHITECTURE DOMAINS

11.1. Four Architecture Domains

The previous two chapters explained the use of the Architecture Capability and described the four core architecture roles as recognized by the TOGAF Standard. The relationship between the Architecture Capability, the required architecture roles, and the architecture domains has become progressively clearer.

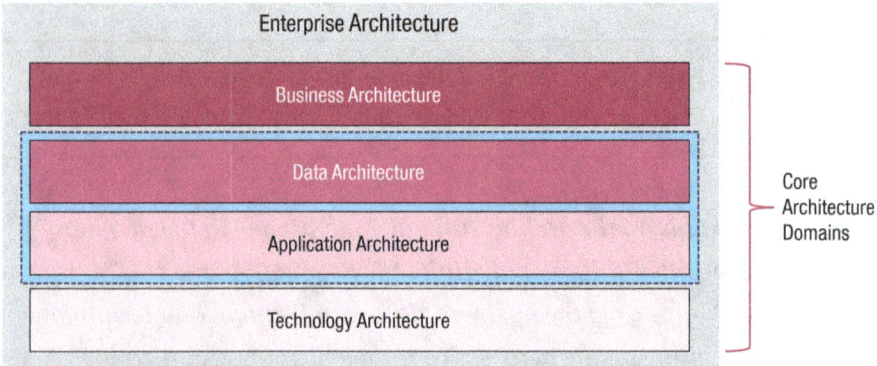

Figure 11-1. *The four core architecture domains*

Figure 11-1 illustrates the four core architecture domains identified in the TOGAF Standard:

- **Business Architecture:** Defines the business strategy, governance, organization, and key business capabilities. By articulating these elements, organizations can ensure that their IT systems support and enhance business objectives.

- **Data Architecture:** Describes the structure of an organization's logical and physical data assets and data management resources. A well-defined Data Architecture facilitates effective data governance and utilization, enabling informed decision-making.

280

- **Application Architecture:** Provides a blueprint for individual systems to be deployed, their interactions, and their relationships to core business processes. This ensures that applications are aligned with business needs and can adapt to changing requirements.

Data Architecture and Application Architecture are represented in the TOGAF Standard under the collective name of *Information Systems Architectures*. This is visualized in Figure 11-1 by the blue outlined area.

- **Technology Architecture:** Details the hardware, software, and network infrastructure required to support mission-critical applications. A robust Technology Architecture underpins the reliable and efficient operation of IT services.

The classification into four architecture domains has existed since the eighth edition of the TOGAF Standard. Due to the lack or unavailability of information on the very first versions of the TOGAF Standard, it is not possible to verify whether a different composition of architecture domains existed in the past.

What is known is that TOGAF 7 recognized only the Technology Architecture. The Business, Data, and Application Architecture domains were not yet part of this version of the framework.

CHAPTER 11   ARCHITECTURE DOMAINS

It is also interesting to note that with the release of TOGAF 8 in 2002, the classification of architecture domains was as follows:

- Business Architecture
- Information Systems Architectures
  - Data or Information Architecture
  - Application Architecture
- Technology Architecture

Note in particular the addition of "or Information" in the Information Systems Architectures subheading shown in the bulleted list above. The addition of these words disappeared in later versions of the framework.

With the gradual integration between the TOGAF Standard and the BIZBOK Guide (the latter explicitly emphasizes the importance of information), a return to the usefulness and necessity of recognizing Information Architecture is gradually taking place.

The four architecture domains mentioned above are common in architectural practice. Of course, there are methods and models that assume a slightly different division of content. For example, there is a model that divides the Business Architecture domain into two parts, *organization* and *processes*. It also assumes the existence of an Information Architecture domain instead of the Data Architecture domain. This model is called the Interoperability Model[1] and is often used in the healthcare industry (see Figure 11-2).

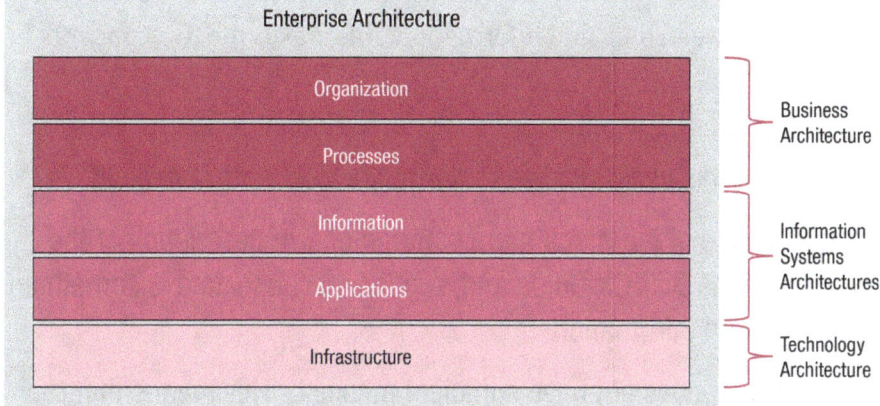

Figure 11-2. *Interoperability Model (five-layer model)*

With the exception of an additional subdivision of the Business Architecture domain, which emphasizes the distinction between the organization and the processes of the organization, the model essentially assumes the four core architecture domains as described in the TOGAF Standard.

According to the TOGAF Standard, the Information Systems Architectures consists of a Data Architecture domain and an Application Architecture domain. The Data Architecture domain recognizes the *data entity* (see Chapter 5, Section 5.1.3), but does not recognize an *information concept* – an entity that is common in the BIZBOK Guide. Contrary to the TOGAF Standard, this *Business Architecture Body of Knowledge* recognizes the importance of information. From this standard, information concepts

---

[1] The Interoperability Model distinguishes five layers of interoperability. Each layer has its own actors, concepts, and standards. Additionally, two preconditional columns apply to all layers: legislation and regulations, and security. Source: https://nictiz.nl/wat-we-doen/zorginformatiestelsel/interoperabiliteit/lagenmodel-3/

are introduced. Information concepts are described in much more detail in the BIZBOK Guide than data entities are in the TOGAF Standard. Data entities have less granularity than information concepts. Aside from the difference between data and information, this makes the two concepts incomparable.

The reason this book mentions information concepts is twofold.

First, because of the much more comprehensive description of the concept in the BIZBOK Guide and, second, because data is something different from information.

Data by itself does not have sufficient meaning. This means that raw data without context has little added value.

To give an example, the address *1 Primary Lane* is meaningless by itself. Without context, little can be done with it. Adding the context of *the President's home address* to 1 Primary Lane gives meaning to the raw data.

A second example is the row of numbers *22, 24, 26, 21,* and *23*. This is also raw data, and without context it does not say much. When context is added to this data, such as indicating that these are the expected temperatures for next week, the data becomes meaningful. It becomes information.

This yields the following equation: Data + Context = Information.

The two examples show that information (data in context) leads to a valuable concept.

CHAPTER 11 ARCHITECTURE DOMAINS

With the gradual movement of the TOGAF Standard toward more integration with other major frameworks and standards such as the BIZBOK Guide, it is quite possible that over time the Information Architecture domain will reappear in the TOGAF Standard, as it did in the 2008 eighth edition.

According to the TOGAF Standard, a complete Enterprise Architecture should address all four core architecture domains (Business, Data, Application, Technology). However, the realities of resource and time constraints often mean there is not enough time, funding, or resources to build a top-down, all-inclusive architecture description encompassing all four architecture domains.

Architecture descriptions are typically crafted with a clear purpose, driven by specific business objectives that steer architecture development. Defining the particular challenges the architecture description aims to address, and the questions it seeks to answer, is a crucial aspect of the initial phase of the Architecture Development Method.

For instance, if an architecture initiative's goal is to identify and assess technology options for achieving a specific capability, and the core business processes are not subject to change, a comprehensive Business Architecture might not be necessary. However, since Data, Application, and Technology Architectures are built upon the Business Architecture, it's essential to consider and comprehend the Business Architecture.

A complete Enterprise Architecture should address all four architecture domains (Business, Data, Application, Technology) [20].

While certain situations may necessitate developing an architecture description that doesn't encompass all four architecture domains, it is important to recognize that such an architecture cannot, by definition, be a complete Enterprise Architecture. One risk associated with this approach is a lack of consistency, leading to integration challenges. Integration may need to occur later – introducing additional costs and risks – or the architect must clearly articulate the risks and trade-offs of not developing a complete and integrated architecture, ensuring that the organization's management understands them.

11.2. Difference in Domains and Layers

Over the years, the field of architecture has matured. The evolution and application of architecture frameworks has greatly contributed to this maturation. Today, more and more organizations recognize the value of architecture. Many of these organizations employ architects to structure, unify, and standardize the deployment of information systems, the execution of business processes, and the implementation of strategies.

Architects play a critical role in establishing, applying, and evolving an organization's architecture. However, organizations are not exempt from making significant contributions. In fact, they are responsible for many areas. For instance, they must define their strategy and collaborate with architects to create feasible implementation plans. Enterprise Architecture can help organizations shape their drivers, goals, objectives, and initiatives.

An architect can do more than translate organizational strategy into execution. There are several topics that cover a wide range of areas. These topics may include organizational design, business process design and optimization, and visualization of information flows within the organization. To perform their job satisfactorily, an architect must gather information on these aspects.

Enterprise Architecture, which was considered a technology-focused field around 1990, has evolved into a mature profession over the past decades. It now covers all aspects of an organization. These aspects are referred to as *architectures* or *architecture domains*. For years, people have confused architecture domains and layers, but the two are distinct.

CHAPTER 11 ARCHITECTURE DOMAINS

Architecture layers refer to the *different levels of abstraction* within a system or Solution Architecture, typically describing how different components interact. Architecture layers thus refer to the logical divisions of a software system or application based on the *functionality* they provide. Each layer is responsible for a specific aspect of the system and communicates with adjacent layers through predefined interfaces.

Common architecture layers (see Figure 11-3) in a typical software architecture include the *presentation layer* (which manages user interaction), the *business logic layer* (which processes business rules), the *data access layer* (which interacts with databases), and the *infrastructure layer* (which supports hosting and networking). Separating the system into layers promotes modularity, maintainability, and reusability.

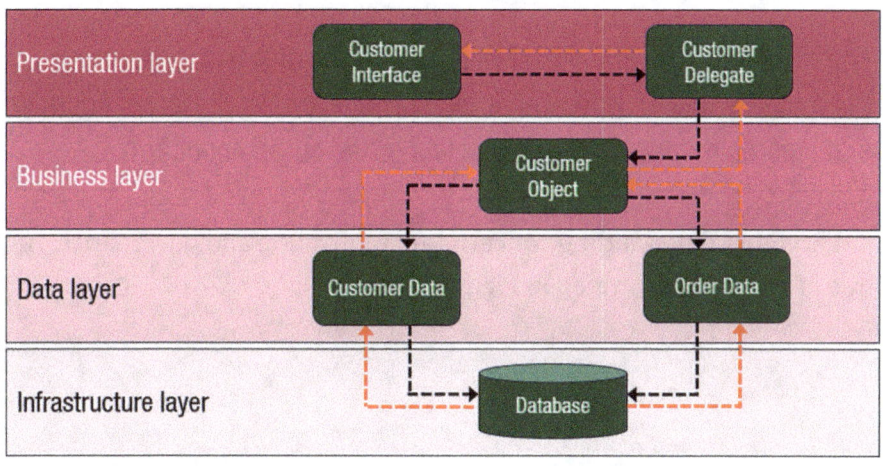

Figure 11-3. *Architecture layers*

Architecture domains, on the other hand, are broader divisions that categorize different aspects of the overall system architecture. They represent the *areas of concern or expertise* that architects need to address while developing architecture.

287

CHAPTER 11   ARCHITECTURE DOMAINS

The most common architecture domains are the ones that were mentioned earlier in this chapter: Business Architecture, Data Architecture, Application Architecture, and Technology Architecture. However, Security Architecture, for example, is also considered to be an architecture domain, although it is mostly called a *cross-cutting concern*. The reason for this is that Security Architecture cuts across all other architecture domains in an organization.

Chapter 15 discusses architecture maturity models. The TOGAF Standard mentions two well-known frameworks that provide maturity models: Architecture Capability Maturity Model (ACMM) and Capability Maturity Model Integration (CMMI).

The ACMM framework lists IT security as one of the nine elements of the maturity model. However, it focuses only on the IT part of security and ignores the fact that security is a cross-cutting concern.

Two other cross-cutting concerns are motivation and governance. All three of them are illustrated in Figure 11-4.

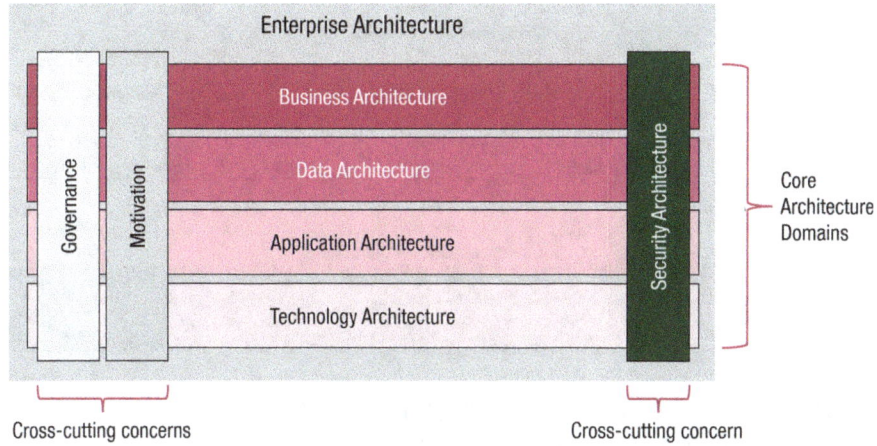

Figure 11-4. Architecture domains and cross-cutting concerns

Each architecture domain focuses on specific concerns and constraints related to its area and contributes to the overall design of the architecture. Architecture domains categorize different concerns and perspectives that must be considered during the architectural design process. Architecture layers, on the other hand, focus on organizing the components and functionality of one or more systems within the context of an architecture domain. Architecture domains and layers are complementary concepts used to create well-structured and comprehensive architectures.

11.3. Mapping Roles to Domains

The four core architecture domains are directly related to the architecture roles and specializations, which in turn can be related to the specific use of the Architecture Capability. Figure 11-5 shows the relationship between the architecture roles and specializations and the corresponding architecture domains.

CHAPTER 11　ARCHITECTURE DOMAINS

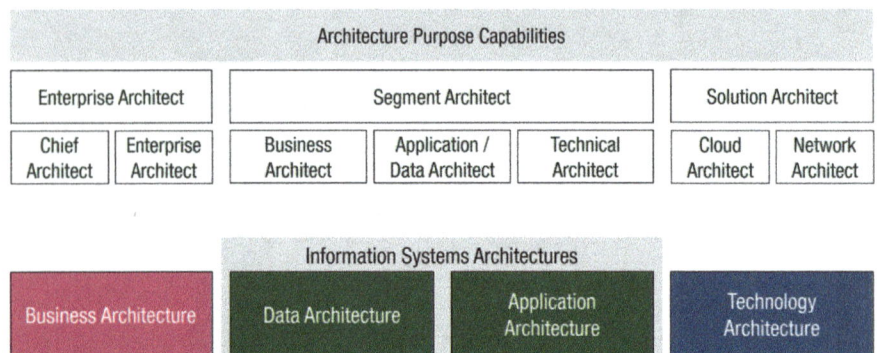

Figure 11-5. Architecture roles and specializations and the corresponding architecture domains

A standard Enterprise Architecture consists of four architecture domains. The Enterprise Architect role is responsible for the entire ecosystem and therefore has responsibility for all four architecture domains. This is shown in Figure 11-6.

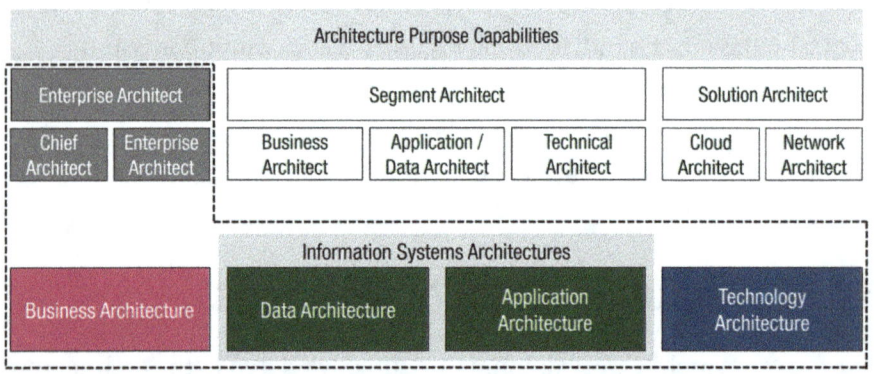

Figure 11-6. Scope of the Chief/Enterprise Architect

When an organization has a Chief Architect, they are primarily concerned with strategic issues. If an organization does not have both a Chief Architect and an Enterprise Architect, then the work of both roles is usually performed by one and the same role, that of an Enterprise Architect.

If the focus is shifted to the role of the Segment Architect, it can be seen that this role splits into four specializations: Business, Data, Application, and Technology Architect. Each of these specializations is responsible, within the context of the architecture, for its own segment or architecture domain, Business Architecture, Data Architecture, Application Architecture, and Technology Architecture, respectively. Figure 11-7 illustrates this relationship.

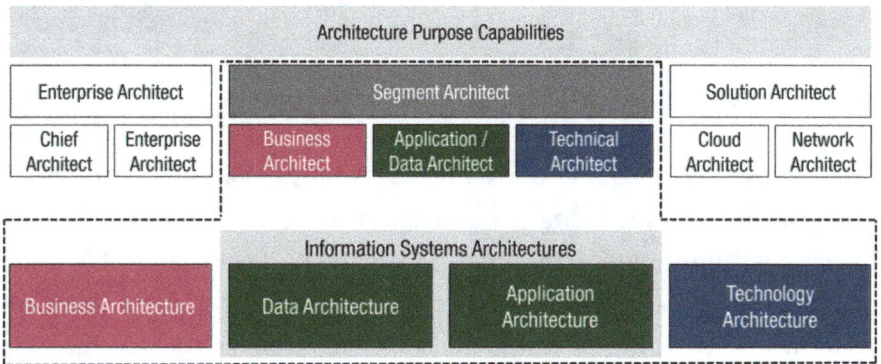

Figure 11-7. *Scope of the Segment Architect*

The Solution Architect's scope of work is typically a defined project or a significant portion of a project. This is usually a technical project. Therefore, in nine out of ten scenarios, the work of a Solution Architect will take place in the context of the Technical Architecture domain. Specializations of the Solution Architect role (such as a Cloud Architect, Network Architect, or Software Architect) are all in the Technical Architecture domain (see Figure 11-8).

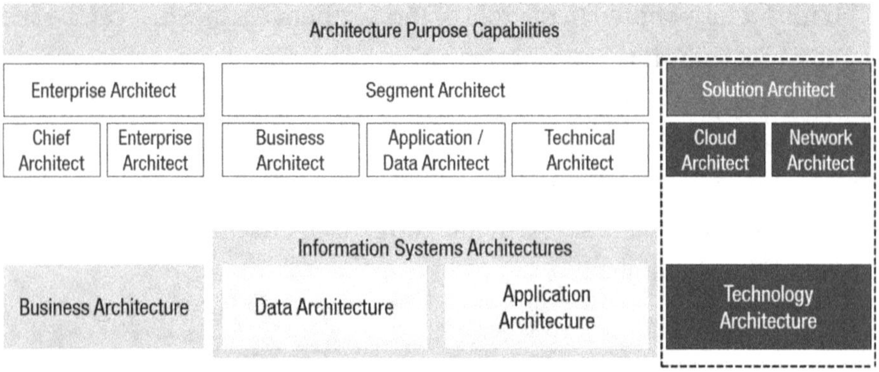

Figure 11-8. Scope of the Solution Architect

Of course, the situations and associated examples above may differ somewhat from how they are presented here in day-to-day practice. However, it is important to recognize the distinction between architecture roles and specializations. It is also important to acknowledge that each role has its own area of interest within the context of architecture. It is therefore unwise to have false expectations of the roles mentioned.

11.4. Summary

This chapter has identified and described the four core architecture roles as they appear in the TOGAF Standard:

- **Business Architecture:** Defines the business strategy, governance, organization, and key business processes
- **Data Architecture:** Describes the structure of an organization's logical and physical data assets and data management resources
- **Application Architecture:** Provides a blueprint for individual systems to be deployed, their interactions, and their relationships to core business processes

- **Technology Architecture:** Details the hardware, software, and network infrastructure required to support mission-critical applications

The four core architecture domains mentioned in this chapter are common in architectural practice. There are methodologies that assume a slightly different division of content, by dividing the Business Architecture domain into two parts, organization and processes. This method is called the Interoperability Model and is often used in the healthcare industry.

A complete Enterprise Architecture should address all four core architecture domains (Business, Data, Application, Technology). While certain situations may necessitate developing an architecture description that doesn't encompass all four architecture domains, it's important to recognize that such an architecture cannot, by definition, be a complete Enterprise Architecture.

Architecture *domains*, when compared with architecture layers, are broader divisions that categorize different aspects of the overall system architecture. They represent the areas of concern or expertise that architects need to address while developing architecture. Architecture *layers* refer to the different levels of abstraction within a system or Solution Architecture. Each layer is responsible for a specific aspect of the system and communicates with adjacent layers through predefined interfaces.

The most common architecture domains are the ones that were mentioned earlier in this chapter. However, Security Architecture, for example, is also considered to be an architecture domain, although it is mostly called a cross-cutting concern. The reason for this is that Security Architecture cuts across all other architecture domains in an organization. The same is true for motivation and governance, two other cross-cutting concerns. Each architecture domain focuses on specific concerns and constraints related to its area and contributes to the overall design of the architecture. Architecture domains categorize different concerns and perspectives that must be considered during the architectural design process.

CHAPTER 11   ARCHITECTURE DOMAINS

The four core architecture domains are directly related to the architecture roles and specializations, which in turn can be related to the specific use of the Architecture Capability.

The *Enterprise Architect* role is responsible for the entire ecosystem and therefore has responsibility for all four architecture domains.

From a horizontal perspective, the role of the Segment Architect can be divided into four specializations: Business Architect, Data Architect, Application Architect, and Technology Architect. Each of these specializations is responsible for its own segment or architecture domain.

The Solution Architect's scope of work is typically a defined project or a significant portion of a project. This is usually a technical project. Therefore, in nine out of ten scenarios, the work of a *Solution Architect* will take place in the context of the Technical Architecture domain.

It is important to recognize the distinction between architecture roles and specializations. It is also important to acknowledge that each role has its own area of interest within the context of architecture.

CHAPTER 12

Architecture Development Method

 This chapter is part of the PINK, GREEN, and BLUE reading tours. See Figure 1-1 in Chapter 1, Section 1.3, for alternative reading tours.

The Architecture Development Method is a key component of the TOGAF Standard. It is the heart of the framework as it provides a comprehensive and iterative approach for developing and managing Enterprise Architectures. It is considered a method that enables Enterprise Architects to develop and implement Enterprise Architecture.

The Architecture Development Method delivers an iterative approach by providing guidance for creating and maintaining architecture artifacts and ensuring alignment with business goals. But don't be fooled by what seems to be a waterfall approach at first sight.

The Architecture Development Method describes all the actions one can take, not what must be taken. It also does not prescribe the order of these steps. The TOGAF Standard explicitly states that the Architecture Development Method is to be tailored so that it (referring to the steps and its content) fits an organization's needs.

CHAPTER 12   ARCHITECTURE DEVELOPMENT METHOD

12.1. The Heart of the TOGAF Standard

All the theory described in the TOGAF Standard can be put into practice by executing the Architecture Development Method. This method – still too often mistaken for a process – provides tools for creating and developing a sound Enterprise Architecture.

The Architecture Development Method (see Figure 12-1) consists of ten phases, each of which has its own characteristics and provides a step-by-step approach to implementing or approaching a particular part of the architecture work. It should be noted that the activities described for each phase are *in no way prescriptive*. All steps described should be considered as *possible steps* to be taken. Depending on the situation or activities, steps may be skipped or adapted as the situation requires. The Architecture Development Method is therefore highly configurable and adaptable and can be used in almost any situation.

CHAPTER 12 ARCHITECTURE DEVELOPMENT METHOD

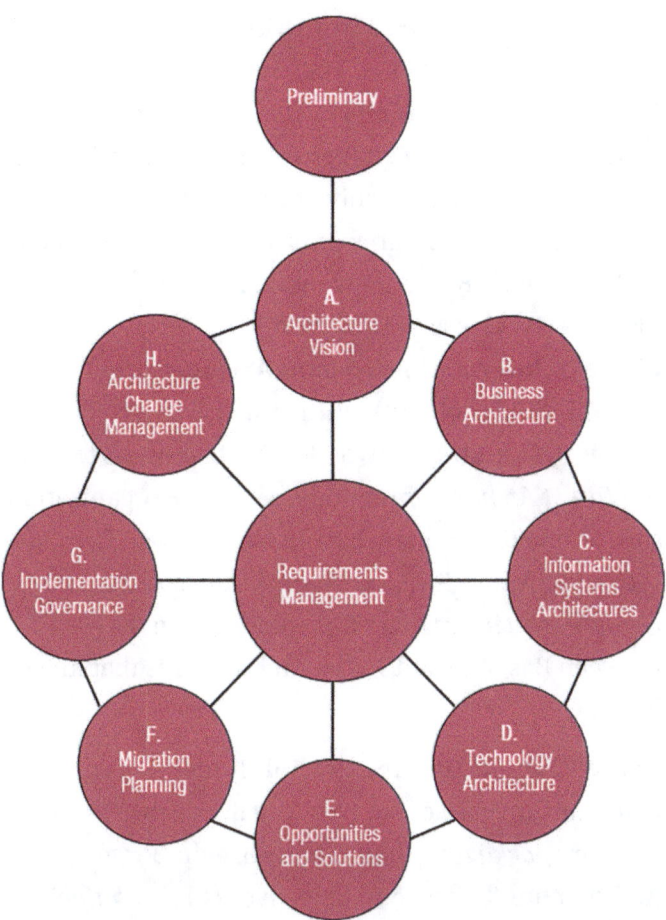

Figure 12-1. *The Architecture Development Method*

The ten phases of the Architecture Development Method each have their own distinct objectives:

- **Preliminary Phase:** The first phase of the Architecture Development Method has two unique objectives. One is to *determine the Architecture Capability* that the organization believes it needs, and the other is to *establish it* in that form.

297

- **Phase A (Architecture Vision):** The main goal of the Architecture Vision Phase is to *establish a high-level vision of the architecture* and to *ensure stakeholder alignment*. It focuses on creating a shared understanding of the architecture's goals and ensuring that the business requirements are clearly articulated and agreed upon by all stakeholders.
- **Phase B (Business Architecture), Phase C (Information Systems Architectures), Phase D (Technology Architecture):** The objectives of phases B, C, and D are twofold. The first one is to *craft the Target Architecture* to outline how the organization should function to meet business objectives. The second objective is to *identify potential components for the Architecture Roadmap*, pinpointing the gaps between Baseline and Target Architectures that must be bridged.
- **Phase E (Opportunities and Solutions):** This phase contains three key objectives. The first is to *identify the work packages* required to realize the Target Architecture. The second objective is to *determine whether an incremental approach is required* to achieve the first objective. The third and final key objective is to *define the building blocks* needed to complete the Target Architecture.
- **Phase F (Migration Planning):** The primary objective of Phase F (Migration Planning) is to *create a detailed plan* for the implementation of the architecture.

- **Phase G (Implementation Governance):** The primary objectives of Phase G include *ensuring adherence to the established architecture, managing any changes* during implementation, and *continually assessing progress* to make necessary adjustments to meet the organization's objectives.

- **Phase H (Architecture Change Management):** The key objectives of the Architecture Change Management Phase are to *continuously assess changes* in the organizational strategy, business environment, and technology landscape that may impact the architecture.

- **Requirements Management Phase:** Although not considered a separate phase, Requirements Management is a continuous activity that runs throughout the Architecture Development Method. It involves *managing and maintaining requirements* and ensuring that they are properly addressed during each phase.

During the development of architecture, it is important to be able to visualize certain components. As the saying goes, *a picture is worth a thousand words*, and visualizations are often an effective way to convey a message. Typically, an image immediately clarifies the intended change or implementation.

Artifacts are used to visualize the architecture. Several of these artifacts are created during the phases of the Architecture Development Method. It is important to remember that these artifacts must add *value* to the organization, the project, or architecture in general.

CHAPTER 12   ARCHITECTURE DEVELOPMENT METHOD

The artifacts shown in this chapter are examples based on the fictional healthcare organization that was described in Chapter 5.

The artifacts need to be adapted to your own organization, situation, and environment.

Each phase of the Architecture Development Method consists of a set of fixed components:

- Key objectives per phase
- Steps to be performed during the phase
- Outputs generated by the phase
- Approach to performing the phase

The subject matter of the phases largely determines the amount and form of information that will serve as input to the next phase.

12.2. Architecture Document Creation

During the implementation of the Architecture Development Method, a number of architecture documents are created. These documents identify the various phases of the work and describe the agreements and outline the architecture work. They do not contain specific details of the architecture or the work done in specific phases of the Architecture Development Method:

- **Request for Architecture Work:** The initiation of an architecture development cycle requires a formal document from the sponsoring organization to the architecture team. It's important that this document maintain a high-level perspective, providing high-level guidance without delving into granular details.

- **Statement of Architecture Work:** Defining the scope and approach of an architecture development cycle is critical. The Statement of Architecture Work does just that. It serves as a benchmark for measuring the success of the architecture project and can even form the basis of a contractual agreement between the provider and consumer of architecture services. It typically includes the architecture project description and scope, an overview of the architecture vision, and acceptance criteria and procedures. In addition, the Statement of Architecture Work may include the architecture project plan and schedule and formal approvals.

- **Architecture Definition Document:** The Architecture Definition Document serves as a comprehensive repository for all key architectural artifacts generated throughout a project, encompassing essential related information. It spans all architecture domains (Business, Data, Application, and Technology), while also addressing all relevant architectural states (Baseline, Transition, and Target).

- **Architecture Contract:** Architecture Contracts serve as formal agreements between development partners and sponsors that describe the deliverables, quality standards, and overall suitability of an architecture. Their successful execution depends on robust architecture governance mechanisms.

Figure 12-2 visualizes the creation of these documents.

CHAPTER 12   ARCHITECTURE DEVELOPMENT METHOD

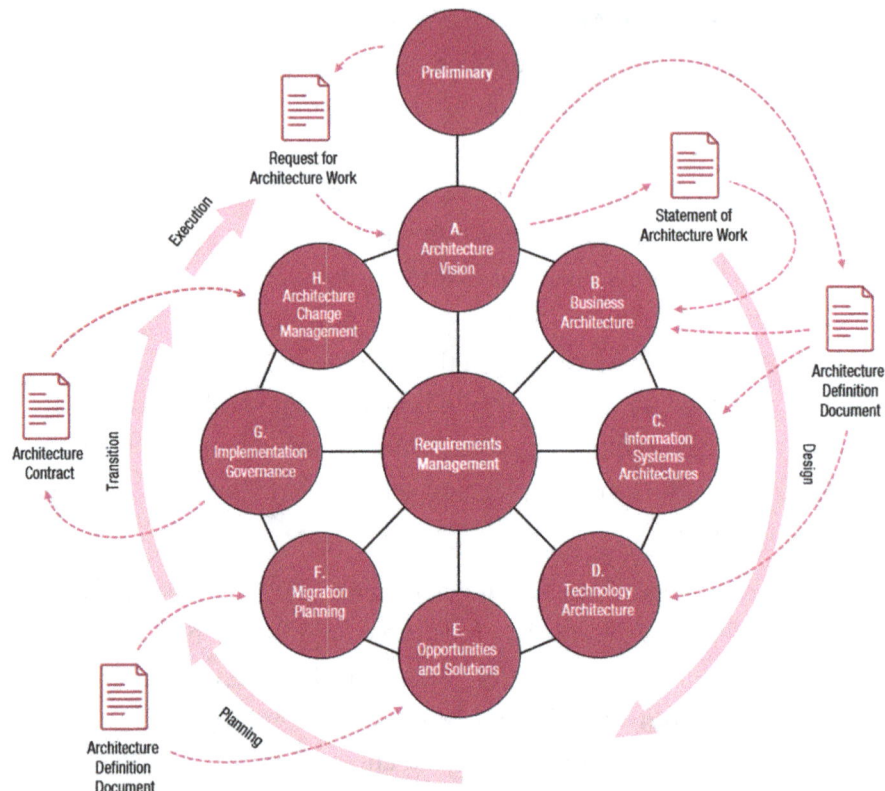

Figure 12-2. Document creation during the execution of the ADM

The Preliminary Phase initiates the architecture work and captures it in a *Request for Architecture Work* document, to the extent that it can be identified at this stage.

The Request for Architecture Work document is sent from the sponsoring organization to the architecture organization to trigger the start of an architecture development cycle [11].

During Phase A (Architecture Vision), a high-level model of the candidate building blocks is created. These are captured in a *Statement of*

CHAPTER 12   ARCHITECTURE DEVELOPMENT METHOD

Architecture Work document, which serves as input to the *design phases*, leading to the creation of the Baseline and Target Architectures. Phases B, C, and D, the Business Architecture, Information Systems Architectures, and Technology Architecture, respectively, are then completed.

The Statement of Architecture Work is created as a deliverable of Phase A and is effectively a contract between the architecting organization and the sponsor of the architecture project [11].

Reference models, viewpoints, and tools are selected, and a Baseline Architecture Description is developed for each architecture domain. This is a high-level model of the existing building blocks, reusing definitions from the Architecture Repository where available. During phases B, C and D, the *Architecture Definition Document* is created and supplemented by each Baseline Architecture Definition from the corresponding phase.

The Architecture Definition Document is the deliverable container for the core architectural artifacts created during a project and for important related information [11].

After the Baseline Architecture Description is developed, the Target Architecture Description is created. This is a more extensive process because it involves defining and developing an architecture that does not yet exist.

A view of the required building blocks is developed by creating artifacts such as catalogs, matrices, diagrams, and maps. Each building block is documented as comprehensively as possible, accompanied by a clear rationale. The decisions that led to that building block are explained. All required building blocks are identified and validated against those in the Architecture Repository. As much as possible is reused.

The reference models from the different continua are used to select standards.

Finally, the definitive building blocks selected or to be created are recorded in the Architecture Landscape. Building blocks that will serve as new standards in the future – after the Target Architecture has been implemented – are recorded in the Architecture Repository.

After the desired Target Architecture is determined, a gap analysis is performed. This determines which building blocks can be reused, which need to be built from scratch, and which will be removed from the Architecture Repository because they are no longer needed. Following the gap analysis, the Architecture Roadmap is created based on the components identified. Any deviations in the Architecture Landscape are addressed and resolved at this point.

All steps and building blocks to be created are coordinated with stakeholders through formal stakeholder reviews, after which the architecture is finalized. The *Architecture Definition Document* is now completed. In Phase E, the document is formally approved for the defined Target Architectures and includes draft Transition Architectures. The latter are approved in Phase F, where the document is finalized.

At the start of Phase E (Opportunities and Solutions), an overview of all the missing building blocks is produced. These are linked to work packages that will ultimately ensure that the missing building blocks are created. This is also when the *planning phase* begins, during which the proposed solutions are further developed. The final solution selected is documented in an *Architecture Contract*. The creation of this document initiates the *transition phase* and the governance process.

Architecture Contracts are the joint agreements between development partners and sponsors of the deliverables, quality, and fit for purpose of an architecture [11].

The *Architecture Contract* is also used during the *execution phase* where the change process is initiated. If the change process results in the need for (major) changes to the already developed architecture, a new *Request for Architecture Work* document must be created. This starts a new iteration of the Architecture Development Method.

The above description of the implementation of the Architecture Development Method is brief and concise. The following sections describe each phase in more detail.

12.3. Preliminary Phase

The Preliminary Phase is the first phase of the Architecture Development Method. It is all about getting the organization ready for architecture development. This is done by *aligning stakeholders*, *defining principles*, and *tailoring the TOGAF Standard*. The Preliminary Phase sets up the foundation for the architecture development process, by ensuring that the necessary frameworks, tools, and governance structures are in place before progressing into the more detailed architectural phases.

Skipping or inadequately executing this phase could result in misalignment between the architecture and business goals or poor management of the architecture initiative.

The Preliminary Phase is a crucial phase for ensuring that the architecture initiative is positioned for success right from the start.

12.3.1 Determining and Establishing the Architecture Capability

The Preliminary Phase of the Architecture Development Method has two distinct objectives. One is to *determine the Architecture Capability* that the organization believes it needs, and the other is to *establish it* in that form.

Essential to these objectives is to determine the organization's current architecture maturity and the desired target maturity of the capability.

CHAPTER 12 ARCHITECTURE DEVELOPMENT METHOD

It also includes assessing the organization's readiness for architectural changes and improvements.

The *first objective*, which entails the definition process, includes determining the scope of the architecture. This could be organization-wide, focused on a particular business unit, or addressing a specific concern. Part of defining the Architecture Capability is to identify the stakeholders and determine their requirements. This will inform how the architecture will be developed to meet those requirements. Tables 12-1 and 12-2 list the steps to perform the Preliminary Phase.

Table 12-1. Steps to perform the Preliminary Phase

Step to Perform	Description of Step	Output Generated by Step
Define the enterprise	By defining the organization, the scope of the impacted organization is determined, allowing architects to create an understanding of the organization and its structure.	Scope of impacted organization(s)
Identify key drivers and organizational context	Pinpointing the critical factors that influence the architecture, such as key drivers and the organizational context.	Baseline of objectives, goals, and drivers
Manage relationships between frameworks	Understanding how various management frameworks interact with each other and creating a framework to manage the architecture activities.	Enterprise Architecture Capability's structure Architecture Governance Framework
Assess Enterprise Architecture maturity	Evaluating where the organization stands in its architectural development.	Maturity assessments, identified gaps, and strategies to address them

Define the Enterprise: By defining the organization, the scope of the impacted organization is determined, allowing architects to create an understanding of the organization and its structure. Scoping the organization entails identifying stakeholders. By stakeholders, the TOGAF Standard refers to those who are most affected by and achieve most value from the architecture work, those who will see change to their work (but are otherwise not directly affected by the architecture effort), and external stakeholders outside the organization.

Make an initial assessment of the stakeholders who are or are likely to be involved in the work. Bring them together by scheduling a meeting and discussing the scope of the work during the meeting. This often leads to the recruitment of additional stakeholders. After the consultation, contact the new stakeholders and review the work and scope of the impacted organization with them as well. Over time, an overview of the actual stakeholders involved will emerge. For this particular step, it is important to understand the concept of Stakeholder Management. Since this is an essential topic that will find its way into other steps of the Architecture Development Method, it is described in more detail in Chapter 14, Section 14.3.

> **Scope the Organization:** Start by figuring out which part of the organization will be affected. This helps architects understand how the organization works and its structure. Identify who the key players are.
>
> **Identify Stakeholders:** Begin by assessing the stakeholders involved in the project. Schedule a meeting with them to discuss the work and its scope. Afterward, reach out to them and review the project scope.
>
> **Manage Stakeholder Involvement:** Understand the importance of managing the stakeholders effectively. Keep communication flowing as the project evolves, and stay flexible to ensure all relevant voices are heard and accounted for.

CHAPTER 12   ARCHITECTURE DEVELOPMENT METHOD

Identify Key Drivers and Organizational Context: Since this is the first phase of the Architecture Development Method, there are no other phases that can provide input. However, the Preliminary Phase does take input from the TOGAF Library (see Chapter 3, Section 3.4) and other architecture frameworks as appropriate.

Organizational strategies and business plans can also provide important input. These strategies may include a business and/or IT strategy, business principles and objectives, goals, and drivers. Scan these types of documents for the presence of any concrete drivers, goals, or objectives. Distill them from the documents and verify them with stakeholders.

Understand the Organizational Needs: In this first phase, there are no previous phases to provide guidance. However, useful information can be pulled from resources like the TOGAF Library and other frameworks. Also, look at the organization's business and IT strategies, goals, and plans to find important factors like drivers or objectives.

Gather and Analyze Documents: Scan business documents for key points about the organization's needs and goals. Look for concrete goals or objectives, and carefully go through them with stakeholders. Make sure to understand the true focus of the business.

Verify with Stakeholders: Once these key drivers are identified, verify and clarify them by discussing them with the relevant people in the organization. This ensures everyone is on the same page and supports the findings.

CHAPTER 12 ARCHITECTURE DEVELOPMENT METHOD

Some time ago, I worked for a hospital. This healthcare organization had done little with Enterprise Architecture prior to my arrival. There was a strategy, but it consisted mostly of stacks of documents trying to interpret the hospital's strategic direction. The pages of documents were full of information – much of it irrelevant – and were not easy for the hospital's staff to read.

I manually went through all of these strategic documents, looking for clues that said something about drivers, goals, and objectives. I spent many hours interpreting the motivational elements and summarizing them so that they could be used in the context of the architecture work.

In the end, I was able to distill a set of drivers, goals, and objectives from the strategically colored documents and visualize them in a diagram (see Figure 12-3).

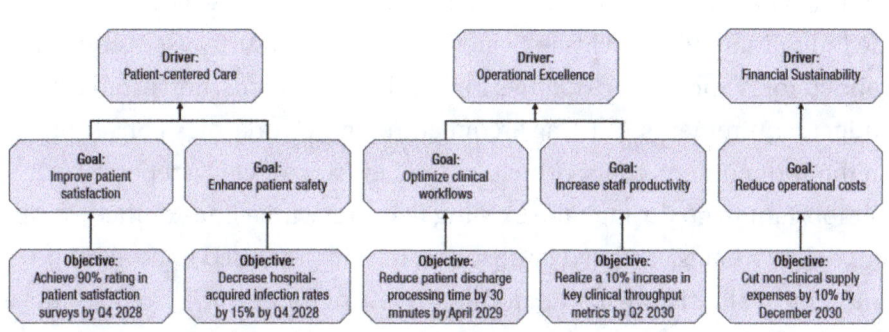

Figure 12-3. Driver/Goal/Objective Diagram

Later, I coordinated the drivers, goals, and objectives with the stakeholders (consisting mainly of senior management) to add the requirements and concerns of these stakeholders to the previously created drivers, goals, and objectives diagram. All the elements in the diagram were also given descriptions so that there would be no ambiguity as to what they meant. Over time, a Solution Concept Diagram emerged.

Today, some organizations are experimenting with using AI to do this manual and especially time-consuming work. AI tools can scan these strategic documents for keywords and then categorize them incredibly quickly. This makes it easy to find and translate an organization's strategy.

Manage Relationships Between Frameworks: In addition to the reference materials from the TOGAF Library, several types of pre-existing models for operating Enterprise Architecture can be used as a baseline for the Preliminary Phase. This may include existing maturity assessments, defined roles and responsibilities for architecture teams, and established budget requirements. Existing architecture frameworks may consist of methods or content, tools, principles, and even (parts of) a repository. It is important to go through the existing material with a fine-toothed comb and delete what is outdated or obsolete. It is not intended to be a one-to-one copy of all existing material. Other relevant frameworks, such as those used for project and portfolio management, also provide valuable insight for the Preliminary Phase.

CHAPTER 12 ARCHITECTURE DEVELOPMENT METHOD

In order to manage the architecture activities, an Architecture Governance Framework needs to be established. This framework ensures that architecture activities are aligned with organizational goals. In the Preliminary Phase, a Governance Framework focuses on how architectural materials (such as standards, models, and reports) are managed. This includes defining repository requirements and identifying governance processes (compliance, retirement, etc.). To implement this framework, existing governance models may need to be modified. Stakeholders need to evaluate current models and understand potential impacts, ensuring that all touch points and changes are identified and aligned across the organization. Bring the models to stakeholder meetings and agree on who will review the models for completeness and accuracy and when. After receiving feedback, adjust the models and have them reviewed again. Have stakeholders re-verify the accuracy of the models.

It is often forgotten that the stakeholders play a significant role in evaluating the architecture.

The decisions to be made in architecture are not the responsibility of the architect, but of the stakeholders. It's the stakeholders who make the decisions based on the architect's input.

Involving stakeholders in a timely and appropriate manner creates support for the architectural work. At this point, too, Stakeholder Management is crucial.

Review Existing Frameworks: Start by reviewing any current frameworks or models that the organization already uses, like maturity assessments, architecture roles, or budget requirements. Don't just copy these materials – scrutinize them closely and remove outdated or irrelevant parts.

Use Relevant Models: Consider other frameworks, such as those for project management, as valuable input during the Preliminary Phase.

Set Up Architecture Governance: Establish a Governance Framework that aligns architecture activities with the organization's goals. This includes managing architecture materials (standards, models, reports) and setting repository and governance requirements.

Evaluate and Adjust Governance Models: Assess the existing governance models. Involve stakeholders to identify any necessary changes and impacts, making sure everything is aligned. Discuss these changes in meetings, assign review responsibilities, and set deadlines for revising and validating the models.

Assess Enterprise Architecture Maturity: Evaluate where the organization stands in its architectural development. It is important to assess the current organization and business capabilities that are needed to understand how to work with architecture. A maturity assessment can help accomplish this.

In the "Architecture Maturity Models" Series Guide (see also the *Further Reading* information box at the end of this section), the TOGAF Standard refers to two applicable methods for measuring architecture maturity. These are the Architecture Capability Maturity Model (ACMM) framework and Capability Maturity Model Integration (CMMI) framework.

The ACMM framework has a structure in which the core components of an architecture process are represented and measured, often in relation to the organization's IT. The CMMI framework, on the other hand, was originally a standardized model for assessing and improving the processes

CHAPTER 12   ARCHITECTURE DEVELOPMENT METHOD

used in software development and other engineering disciplines. Today, the framework is also regularly used to indicate the maturity of an Enterprise Architecture because of its focus on achieving business goals.

Of course, there are alternatives to both ACMM and CMMI frameworks, and it is advisable to choose an approach that best suits the organization. Chapter 15 further details the use of architecture maturity models in combination with the TOGAF Standard.

Assess Maturity: Look at where the organization is in terms of its architecture. This means understanding what the organization is currently capable of and how ready it is to work with architecture. Knowing this helps to figure out what needs to be improved.

Use the Right Tools: The TOGAF Standard suggests using either the Architecture Capability Maturity Model framework or the Capability Maturity Model Integration (CMMI) framework. CMMI is a more appropriate framework for assessing Enterprise Architecture because of its emphasis on achieving business goals.

Choose the Best Fit: While CMMI is one option, there are other ways to assess the organization's maturity. It's important to pick the approach that works best for the situation.

Another method that helps evaluate where the organization stands in its architectural development is to understand the required readiness for change. A critical success factor for coordinating change requests lies with informing stakeholders of current architecture efforts, assessing the impact on their plans, and identifying and resolving common interests and conflicts. Chapter 14, Section 14.7, details how to apply a Business Transformation Readiness Assessment. This technique is used for evaluating and quantifying an organization's readiness to undergo change.

Returning to the objectives for the Preliminary Phase, the *second objective*, establishing the Architecture Capability, involves defining the contours of the Enterprise Architecture. This will be accomplished through the following focus areas.

CHAPTER 12   ARCHITECTURE DEVELOPMENT METHOD

Table 12-2. *Steps to perform the Preliminary Phase*

Step to Perform	Description of Step	Output Generated by Step
Establish the Enterprise Architecture team	Establishing an architecture team to perform the architecture work.	Architecture team Defined roles and responsibilities for architecture teams Constraints for architecture efforts
Set Architecture Principles	Defining the principles that will guide all architecture-related efforts.	Architecture Principles
Choose and tailor the architecture framework	Identifying and tailoring the TOGAF Standard, Architecture Development Method, and the Content Framework.	Tailored framework Tailored Architecture Development Method Tailored Content Framework

Establish the Enterprise Architecture Team: Establishing an Enterprise Architecture team is one of the first things that needs to be done to meet the requirements of working with architecture. The best results are achieved by creating a team of dedicated members. An architecture team does not have to consist only of architects. Project managers, senior management, or even other people from the organization can be part of the architecture team. However, it is important that all members are able to grasp the essence of architecture before joining the team.

The composition of the architecture team is largely related to the positioning of the Architecture Capability in the organization. If the Architecture Capability – as is still the case in many organizations – is positioned within the IT department, then the architecture team will mainly consist of one or more architecture functions, supported by technical experts. On the other hand, if the Architecture Capability is

CHAPTER 12 ARCHITECTURE DEVELOPMENT METHOD

positioned more strategically, for example, under the board of directors or a (strategic) executive, then the team may consist more of an Enterprise Architect, supplemented by members of senior management, project and portfolio managers, and other management functions.

> "Architecture teams" are different from "teams consisting of people with the job title architect". An architecture team can consist of anyone involved in architecture-related work.

Regardless of the composition of the architecture team, the roles and responsibilities of all members must be defined so that there is no ambiguity. The most effective way to do this is to use a RACI matrix (see Table 12-3).

Table 12-3. Example RACI Matrix for an Enterprise Architecture Team

Activity ▼ / Role	Enterprise Architect	Segment Architect	Solution Architect	Project Manager
Create architecture vision	AR	C	C	C
Develop Segment Architectures (Baseline and Target)	A	R	C	C
Define opportunities and solutions	A	R	R	C
Create migration plan	A	R	R	
Implement governance	AR	C	C	C
Manage architecture changes	A	R	R	C
Manage requirements	A	R	R	I

CHAPTER 12   ARCHITECTURE DEVELOPMENT METHOD

The RACI matrix shown in Table 12-3 is an example of how a matrix *might* look like when it is being set up for an architecture team. It lists the *activities* (or processes) in the left column and shows the *roles* involved in those activities in the top row. The cells of the table contain the ways in which each of the roles is involved in the activities. Chapter 8, Section 8.2, describes the structure of a RACI matrix in more detail.

Build the Architecture Team: Start by setting up the Enterprise Architecture team. This is a crucial first step when planning for architecture. The team doesn't have to consist only of architects. It could also include project managers, senior management, or anyone else in the company who gets the basics of architecture.

Define the Team Composition: The way the team is formed depends on where the Architecture Capability sits in the organization. If it's within the IT department, the team may focus on technical roles, like architects and experts. But if it's a broader, strategic initiative, the team could include senior managers and project leaders.

Clarify Roles and Responsibilities: Ensure that every team member knows their role. This removes confusion and makes things run smoothly. A great tool for this is the RACI matrix, which clarifies who's responsible, accountable, consulted, and informed.

Set Architecture Principles: Define the principles that will guide all architecture-related efforts. Architecture Principles need to be established to ensure consistency and alignment with strategic goals throughout the Architecture Development Method process. After thoroughly understanding the organizational context, it's crucial to define a tailored set of Architecture Principles that aligns with the organization's unique needs. This ensures a strong, adaptable framework that supports both governance and strategy. Defining Architecture Principles – and how they relate to and differ from Enterprise Principles – is described in more detail in Chapter 14, Section 14.2.4.

CHAPTER 12   ARCHITECTURE DEVELOPMENT METHOD

 Understand the Big Picture: Before diving into setting principles, take a moment to understand the organization's goals and context. Knowing where the company is headed helps to align the principles with the overall mission.

Define Clear Architecture Principles: These principles will guide every architectural decision. Think of them as the rules of the game, ensuring that everything that's done supports the company's strategy and vision.

Customize for the Organization: Every company is unique, so make sure the principles reflect the organization's specific needs.

Choose and Tailor the Architecture Framework: Identify and tailor the framework that will be used. To be able to create an architecture in the first place, an architecture framework and appropriate tools must be decided upon. Because each organization has unique needs, the architecture framework must be tailored to those needs, and tailoring the TOGAF Standard is therefore an absolute must.

Tailoring is done by assessing how the framework can be adapted to the organization. This includes refining the terminology so that it's clearly understood throughout the organization. To do this, an agreed-upon glossary must be created to ensure consistency in architecture discussions. Publishing the glossary so that it is always available to all stakeholders will greatly assist in achieving and maintaining consensus.

The Enterprise Metamodel must then be adapted to accurately represent the organization. This is done by defining new entities, deliverables, and artifacts that are specific to the organization.

Next, the Architecture Development Method needs to be adapted. Start by removing redundant tasks, adding organization-specific checkpoints, or integrating it with other operational processes. Finally, the Content Framework must be customized to support the organization's unique needs. This is accomplished by customizing the structure of the framework.

CHAPTER 12   ARCHITECTURE DEVELOPMENT METHOD

Having an effective tools strategy is another necessity. Creating a plan that addresses the use of both general-purpose and specialized architecture tools allows for a focus on flexibility and scalability. Techniques such as business modeling need to be incorporated to meet the diverse requirements. Be sure to formalize change control to maintain consistency across all architectural activities.

Pick the Right Framework: First, decide which architecture framework to use. A framework and tools are necessary to get started. Choose one that fits the organization's needs and customize it.

Tailor the Framework: Adjust the framework to the organization. This involves changing the terminology so everyone understands it. Create and share a glossary to keep things consistent.

Adapt the Enterprise Metamodel: Modify the Enterprise Metamodel to fit the organization by defining new entities and removing unused ones.

Modify the Architecture Development Method: Simplify or add steps to the Architecture Development Method. Remove unnecessary tasks and integrate it with existing processes.

Customize the Content Framework: Change the structure of the Content Framework to better support the organization.

Choose the Right Tools: Plan how to use both general and specialized tools. Focus on flexibility and scalability, and include techniques like business modeling.

Control Changes: Make sure there's a formal process for managing changes to keep everything consistent.

Enterprise Architecture provides a strategic view of an organization from the top down and helps different stakeholders work together effectively. It's important to recognize that Enterprise Architecture is not an

CHAPTER 12 ARCHITECTURE DEVELOPMENT METHOD

isolated effort. Enterprise Architects must consider how their frameworks interact with broader business frameworks to ensure interoperability and alignment across the organization.

The Preliminary Phase must consider business goals and ensure that the architecture adapts to different objectives. In addition, Enterprise Architecture must operate at different levels within the organization to reflect this flexibility.

Depending on the size of the organization and its budget for Enterprise Architecture, different approaches may be used to partition the architecture teams, processes, and deliverables. At times, it may be necessary to revisit the Preliminary Phase from the Architecture Vision Phase to ensure that the organization's Architecture Capability is equipped to address a specific architectural issue.

In essence, the Preliminary Phase is about building the architecture needed to support the Architecture Capability.

Further Reading

For more information on establishing and using maturity models, see the TOGAF Series Guide "Architecture Maturity Models", document number G203, and "The TOGAF® Leader's Guide to Establishing and Evolving an EA Capability" (Appendix B), document number G168.

For detailed guidance on establishing the Architecture Capability, refer to "The TOGAF® Leader's Guide to Establishing and Evolving an EA Capability", document number G168.

12.3.2 Preliminary Phase Artifacts

Table 12-4 lists the artifacts that can be created during the execution of the Preliminary Phase.

Table 12-4. Artifacts created during the Preliminary Phase

ADM Phase	Artifacts	Used Entities
Preliminary Phase	Principles Catalog	Principle

Principles Catalog: Captures Enterprise and Architecture Principles of the organization. Enterprise Principles are high-level, general guidelines that shape the overall strategy, culture, and decision-making framework of an organization. Architecture Principles describe what a *good* solution or architecture should look like.

Principles are used to evaluate and agree on an outcome for architecture decision points. Principles are also used as a tool to assist in architectural governance of change initiatives. An example of a Principles Catalog is shown in Table 12-5.

Table 12-5. *Principles Catalog*

Name	Statement	Rationale	Implications
Comply with laws and regulations	Information management processes comply with all relevant laws, policies, and regulations.	To comply with laws and regulations, relevant laws and regulations are considered in the development, procurement, and design of systems, products, and services.	• The organization must be mindful to comply with laws, regulations, and external policies regarding the collection, retention, and management of data. • Efficiency, need, and common sense are not the only drivers. Changes in the law and changes in regulations may drive changes in our processes or applications.
Unambiguous ways of working	Consistent ways of performing tasks and activities are used.	The best way of working is chosen so that work is done more efficiently. This ensures that information management is under control.	• Achieving maximum organization-wide benefit will require changes in the way we plan and manage information. • Some departments may have to concede their own preferences for the greater benefit of the entire organization. • Information management initiatives should be conducted in accordance with the organizational plan.

(continued)

Table 12-5. (*continued*)

Name	Statement	Rationale	Implications
Store information once and use it more often	Information is available from a single source.	Used information is recorded once and in one place so it can be used more often.	• To enable data sharing we must develop and abide by a common set of policies, procedures, and standards governing data management and access. • Common methods and tools for creating, maintaining, and accessing the data shared across the organization must be adopted. • The principle of data sharing will continually "bump up against" the principle of data security – under no circumstances will the data sharing principle cause confidential data to be compromised.

Table 12-5. (*continued*)

Name	Statement	Rationale	Implications
Standard products and services	Products and services are selected based on the functionality required for the task.	Duplicative capability is expensive and proliferates conflicting data.	• A capability which does not serve the entire organization must change over to the replacement organization-wide capability. • It will not be allowed to develop capabilities that are similar/duplicative of organization-wide capabilities. • Data and information used to support organizational decision-making will be standardized.

12.4. Phase A: Architecture Vision

The Architecture Vision Phase is considered the first actionable phase in the Architecture Development Method cycle. This phase plays a crucial role in setting the groundwork for subsequent architecture work, as it ensures a clear understanding of the objectives, scope, and expected outcomes of the architecture project. It builds on the initial input gathered during the previous phase, the Preliminary Phase.

12.4.1 Creating the Architecture Vision

The main goal of the Architecture Vision Phase is to *establish a high-level vision of the architecture* and to *ensure stakeholder alignment*. It focuses on creating a shared understanding of the architecture's goals and ensuring that the business requirements are clearly articulated and agreed upon by all stakeholders. It also involves obtaining initial (stakeholder) approval to proceed with creating and implementing the architecture. The end result of this work is recorded in an architecture vision document.

Important inputs to the Architecture Vision Phase are the reference materials gathered during the Preliminary Phase. These reference materials can be supplemented with existing business principles, goals, and drivers. The existing architectural documentation (framework description, architectural descriptions, baseline descriptions, etc.) is then used to populate an initial version of the Architecture Repository. For example, the capabilities needed to start the architecture work are documented as essential Architecture Building Blocks in the Architecture Landscape part of the repository. The various roles and responsibilities, preferably represented in a kind of RACI matrix, are also stored in the Architecture Repository, more specifically in the Architecture Capability section. Table 12-6 lists the steps to perform Phase A (Architecture Vision).

CHAPTER 12   ARCHITECTURE DEVELOPMENT METHOD

Table 12-6. *Steps to perform Phase A (Architecture Vision)*

Step to Perform	Description of Step	Output Generated by Step
Establish the architecture project	Obtain formal project recognition, secure endorsement from corporate leadership, and gain support and commitment from relevant line management.	Statement of Architecture Work
Identify stakeholders and determine requirements	Identify stakeholders and requirements to allow for the architecture and related work to be communicated and aligned.	Stakeholder Catalog Communications Plan
Assess readiness for business transformation	Evaluate the organization's change readiness, guiding the architecture's scope and highlighting areas requiring attention.	Business Transformation Readiness Assessment
Elaborate on business goals and Architecture Principles	Validate and refine the business goals and Architecture Principles identified in the previous phase.	Validated business objectives, goals, and drivers Validated Architecture Principles
Create a high-level architecture vision	Develop a high-level description of the architecture solution that will meet business requirements and outline how it aligns with the overall business strategy.	Solution Concept Diagram
Define the Target Architecture value propositions and KPIs	Develop a business case and define performance metrics and measures.	Value propositions and KPIs

(*continued*)

CHAPTER 12   ARCHITECTURE DEVELOPMENT METHOD

Table 12-6. (*continued*)

Step to Perform	Description of Step	Output Generated by Step
Identify the business transformation risks and mitigation activities	Identify risks and assess their level and potential frequency.	Business transformation risks Mitigating activities
Develop Statement of Architecture Work	Develop Statement of Architecture Work and secure formal approval.	Statement of Architecture Work

Establish the Architecture Project: Enterprise Architecture serves as a critical business capability, with each iteration of the Architecture Development Method ideally managed as a project within the organization's existing project management framework. In certain scenarios, architecture projects may function independently, while in others, they may be integrated into broader initiatives. Regardless of the context, it is essential to plan and oversee architectural activities using the organization's established best practices.

To ensure project success, it is vital to obtain formal project recognition, secure endorsement from corporate leadership, and gain support and commitment from relevant line management. To agree on the scope of the work – and thus define the architecture project – it is necessary to prepare a Statement of Architecture Work. This document includes the architecture *project description and scope*, the *architecture vision statement*, and the architecture *project plan and schedule*.

Initiate the Project: Recognize the architecture endeavor as a formal project within your organization's project management framework. This approach ensures structured oversight and alignment with existing processes.

Determine Project Independence: Assess whether the architecture project will operate as a standalone initiative or integrate into broader organizational programs. This decision influences resource allocation and strategic alignment.

Plan Using Best Practices: Utilize your organization's established best practices to meticulously plan and manage the architectural activities. This strategy promotes consistency and quality in execution.

Secure Formal Recognition: Obtain official acknowledgment of the project to legitimize its objectives and facilitate organizational support.

Gain Leadership Endorsement: Seek explicit approval and backing from corporate leadership to ensure the project aligns with strategic goals and receives necessary resources.

Engage Line Management: Communicate with relevant line managers to garner their support and commitment, fostering collaboration across departments.

Define Project Scope: Prepare a comprehensive Statement of Architecture Work. This document should detail the project's description, scope, and vision statement and include a project plan with a clear schedule.

Identify Stakeholders and Determine Requirements: The last step of the Preliminary Phase was to identify the stakeholders and determine their requirements. This step is continued in the Architecture Vision

Phase. Knowing who the stakeholders are ensures that the architecture and related work can be communicated and aligned. A critical part of communicating with stakeholders is the Communications Plan. Refer to Chapter 14, Sections 14.3.1 through 14.3.3, for instructions on how to perform stakeholder analysis, create a stakeholder mapping, and set up a Communications Plan.

Knowing who the stakeholders are and what requirements and concerns they have is an important part of creating the Solution Concept Diagram. This diagram is based on the organization's goals and objectives and stakeholder requirements and constraints. An example of this diagram is provided in Section 12.4.2.

Identify Stakeholders: Determine who has an interest in the architecture project. This includes individuals or groups affected by or influencing the project's outcome.

Analyze Stakeholders: Understand each stakeholder's role, concerns, and influence. This helps in prioritizing their needs and expectations.

Document Requirements: Gather and record the specific needs and constraints of each stakeholder. This ensures that the architecture aligns with organizational goals and objectives.

Develop a Communications Plan: Create a strategy for keeping stakeholders informed and engaged throughout the project. Effective communication fosters alignment and support.

Create the Solution Concept Diagram: Using the collected information, develop a visual representation of the proposed solution. This diagram should reflect organizational goals, stakeholder requirements, and any constraints identified.

Assess Readiness for Business Transformation: It is essential to evaluate and quantify an organization's preparedness for change. A key objective is to perform a Business Transformation Readiness Assessment. This assessment hinges on determining and analyzing a series of readiness factors. These factors can include the following:

- **Vision:** Does the organization have clearly defined goals and objectives for the transformation?
- **Leadership:** Can the architecture work count on strong executive sponsorship and commitment?
- **Stakeholder Engagement:** Active involvement and support from key stakeholders is paramount.
- **Governance:** Effective decision-making and accountability structures are an absolute necessity.
- **Culture and Communications:** Does the organizational culture support change and open communication?

The readiness factors mentioned above are just a few examples. Each of the factors should be assessed to determine potential risks and required interventions before executing a transformation initiative.

The insights from the Business Transformation Readiness Assessment help shape the architecture's scope, pinpoint necessary activities within the project, and identify risk areas that need addressing. Because this is a specific technique, how to conduct a Business Transformation Readiness Assessment is described in more detail in Chapter 14, Section 14.7.

Understand the Purpose: Recognize that the goal is to assess how ready the organization is for transformation.

Identify Key Factors: Focus on important areas such as vision, leadership, stakeholder engagement, governance, culture, and communication.

Evaluate Each Factor: Look at each area to find potential risks and figure out what actions might be needed before starting the transformation.

Analyze the Findings: Use the information from the assessment to define the scope of the architecture, outline necessary project activities, and highlight risk areas that need attention.

Elaborate on Business Goals and Architecture Principles:
The business objectives, goals, drivers, as well as the Architecture Principles identified in the previous phase are validated and refined in the Architecture Vision Phase. If these have already been established elsewhere in the organization, make sure they are up to date and address any areas of ambiguity. If not, reconnect with the originators of the Statement of Architecture Work to collaboratively define these fundamental elements. Once clarified, ensure that senior management provides its official approval.

Next, outline the constraints that will affect the architecture. These include both organization-wide and architecture-specific constraints, such as timelines, schedules, and available resources. Organization-wide constraints may be influenced by the foundational Enterprise and Architecture Principles identified in the Preliminary Phase or further detailed in Phase A.

Review Existing Business Objectives and Goals: Examine the current business objectives, goals, and drivers identified in earlier phases. Ensure they are up to date and accurately reflect the organization's strategic direction.

Validate and Refine Architecture Principles: Assess the Architecture Principles established previously. Confirm their relevance and applicability to the current business context. Address any ambiguities or inconsistencies.

Engage Stakeholders for Input: Collaborate with key stakeholders, including those who initiated the Statement of Architecture Work. Gather insights and feedback to ensure a comprehensive understanding of business needs.

Define Fundamental Elements Collaboratively: Work together with stakeholders to clearly articulate business objectives and Architecture Principles. Ensure these elements are well-defined and aligned with organizational goals.

Obtain Senior Management Approval: Present the refined business objectives and Architecture Principles to senior management. Seek formal approval to ensure organizational alignment and support.

Identify Organizational Constraints: Outline constraints that may impact the architecture, such as timelines, schedules, and resource availability. Consider organization-wide constraints influenced by Enterprise and Architecture Principles from the Preliminary Phase.

Document and Communicate Constraints: Clearly document identified constraints. Communicate these constraints to all relevant stakeholders to ensure transparency and informed decision-making.

Create a High-Level Architecture Vision: Develop a high-level description of the architecture solution that will meet business requirements and outline how it aligns with the overall business strategy. It usually involves preparing a business case to justify the need for architecture work, demonstrating how it will *deliver value* or *solve business challenges*.

CHAPTER 12 ARCHITECTURE DEVELOPMENT METHOD

The architecture vision document outlines the business goals, capabilities, and high-level architectural approach. It often includes a conceptual solution description and an explanation of how the proposed architecture supports the business strategy. The high-level vision can be visualized using the aforementioned Solution Concept Diagram. Although the name of this artifact may suggest a specific and detailed solution, this diagram is far from that. It simply shows the relationships between the organization's goals and objectives on the one hand and the stakeholder requirements and constraints on the other.

Understand Business Goals and Requirements: Begin by identifying your organization's key business objectives and the specific requirements that the architecture needs to address. This foundational understanding ensures that the architecture vision aligns with the overall business strategy.

Develop a Business Case: Prepare a compelling business case to justify the need for architectural work. Highlight how the proposed architecture will deliver value, address business challenges, and support strategic goals.

Outline the Architecture Vision: Create a document that articulates the high-level description of the proposed architecture solution. This should include business goals, capabilities, and a high-level architectural approach. Describe the overarching strategy for the architecture, ensuring it aligns with business goals.

Include a Conceptual Solution Description: Provide a conceptual overview of the proposed solution, detailing how it addresses the identified business requirements and supports the organization's strategy.

Visualize with a Solution Concept Diagram: Utilize a Solution Concept Diagram to illustrate the relationships between organizational goals, stakeholder requirements, and constraints. This visual representation helps stakeholders easily grasp the alignment between business objectives and the proposed architecture.

Define the Target Architecture Value Propositions and KPIs: In this step, a business case is developed that describes the architecture(s) and the organizational changes it will require. To get stakeholders on board, it is important to create value propositions. These are presented to each stakeholder group for review and approval. Another action to take during this step is to define performance metrics and accompanying measures that allow the architecture work and its results to be monitored. They should ensure that the business requirements are being met. Finally, business risk is assessed. This means that a high-level assessment of potential risks is performed, and the risks are identified and recorded in the Statement of Architecture Work. This applies to all outputs from this particular step.

Develop a Business Case: Clearly explain the proposed architecture and the organizational changes it entails.

Create Value Propositions: Identify and articulate the benefits for each stakeholder group to gain their support.

Present for Review and Approval: Share these value propositions with stakeholders and seek their feedback and endorsement.

Define Performance Metrics: Establish clear metrics to monitor the architecture's effectiveness and ensure business requirements are met.

Assess Business Risks: Conduct a high-level evaluation of potential risks, documenting them in the Statement of Architecture Work.

Identify the Business Transformation Risks and Mitigation Activities: Introducing change to an organization almost always involves some risk. This type of risk is called business transformation risk. It is an absolute must to identify the business transformation risks and assess their initial level and their potential frequency. A high-level mitigation strategy can be assigned to each risk, although this can also be done at a later stage when the Consolidated Gaps, Solutions, and Dependencies Matrix (see Section 12.8.2) is created in Phase E (Opportunities and Solutions).

CHAPTER 12 ARCHITECTURE DEVELOPMENT METHOD

Using the Risk Management technique described in Chapter 14, Section 14.8, the documented risks from the Consolidated Gaps, Solutions, and Dependencies Matrix are classified, mitigated (if possible), and otherwise managed.

Spot the Risks: Begin by identifying potential risks associated with the business transformation. These could range from employee resistance to technological challenges. Understanding what might go wrong is the first step toward prevention.

Assess the Impact: Once the risks are identified, evaluate their potential impact and how often they might occur. This assessment helps in prioritizing which risks need immediate attention and which ones can be monitored over time.

Develop Mitigation Strategies: For each identified risk, craft a high-level strategy to mitigate its impact. This might involve training programs, adopting new technologies, or revising existing processes. Detailed planning can be integrated during the creation of the Consolidated Gaps, Solutions, and Dependencies Matrix in Phase E (Opportunities and Solutions).

Document Risks: Maintain a comprehensive record of all identified risks and their corresponding mitigation strategies. This documentation will serve as a valuable reference throughout the transformation process.

Implement and Monitor: Put the mitigation strategies into action and continuously monitor their effectiveness. Be prepared to adjust the approach as needed to address any unforeseen challenges.

Develop Statement of Architecture Work: At the end of the Architecture Vision Phase, the architecture team should aim to secure formal approval of the Statement of Architecture Work and commitment to proceed to the more detailed phases of the Architecture Development Method. This typically includes securing necessary funding, resources, and time allocations.

CHAPTER 12   ARCHITECTURE DEVELOPMENT METHOD

Gaining and maintaining stakeholder engagement is crucial. The vision must clearly communicate the benefits and value to stakeholders, especially senior management and business owners. Ensuring that the architecture vision aligns with the organization's business strategy is essential to get buy-in from stakeholders. At this stage, it's also important to identify risks early on that could impact the project later, such as technological challenges or resistance from stakeholders.

Communicate Benefits Clearly: Articulate the benefits and value of the architecture work in a manner that resonates with stakeholders. Clear communication fosters buy-in and commitment.

Identify and Assess Risks: Proactively identify potential risks, such as technological challenges or stakeholder resistance. Assess their impact and develop mitigation strategies to address them effectively.

Document Resource Requirements: Detail the necessary funding, resources, and time allocations required for the project. A comprehensive resource plan is essential for informed decision-making.

Draft the Statement of Architecture Work: Compile the information gathered into a formal document. This statement should encompass the project's scope, objectives, resource requirements, risk assessment, and alignment with business strategy.

Seek Formal Approval: Present the Statement of Architecture Work to the relevant approval authority within the organization. Address any concerns raised and obtain formal approval to proceed.

Maintain Stakeholder Engagement: Continuously engage stakeholders throughout the project to ensure ongoing support and promptly address emerging issues.

12.4.2 Architecture Vision Phase Artifacts

Table 12-7 lists the artifacts that can be created during the execution of the Architecture Vision Phase.

Table 12-7. Artifacts created during Phase A (Architecture Vision)

ADM Phase	Artifacts	Used Entities
Phase A: Architecture Vision	Stakeholder Catalog	– (or stakeholder, once added to the Enterprise Metamodel)
	Value Chain Diagram	Value stream
	Solution Concept Diagram	Objective, requirement, and constraint
	Business Model Diagram	Actor, role, stakeholder (see above), contract, function, process, business capability, business service, product
	Business Capability Map	Business capability
	Value Stream Map	Value stream

Stakeholder Catalog: Identifies the stakeholders for the architecture initiative, their influence on the initiative, and their key questions, issues, or concerns that need to be addressed by the architecture framework.

Understanding the stakeholders and their requirements allows the architect to focus efforts in areas that meet the stakeholders' needs.

Due to the potentially sensitive nature of stakeholder mapping information and the fact that the Architecture Vision Phase is intended to be conducted using informal modeling techniques, no specific metamodel entities are used to generate a Stakeholder Catalog.

As I mentioned in Chapter 5, the Enterprise Metamodel allows the architect to add and remove entities from the model if they do not add value to the organization.

This is particularly useful in the context of the Stakeholder Catalog. Let me explain.

The Enterprise Metamodel provides a variety of entities that help architecture initiatives visualize the answer to questions such as why the initiative is taking place (drivers, goals, and objectives) and what needs to be done to shape the change (e.g., adjusting processes, engaging actors, and creating capabilities).

It is noteworthy that the entities needed to indicate for whom the architecture initiative is being implemented and what concerns are being addressed do not appear as such in the metamodel.

I am referring, of course, to the stakeholder entity.

This entity is a critical part of any architecture approach. The TOGAF Standard itself identifies the need for Stakeholder Management by stating that the creation of a Stakeholder Catalog is highly desirable. The Communications Plan also pays close attention to formulating a way to keep stakeholders involved in the architecture work.

The TOGAF Standard cites the potentially sensitive nature of stakeholder mapping as a reason for not including the stakeholder entity in the Enterprise Metamodel. However, in day-to-day practice it is highly desirable to be able to relate goals and objectives to stakeholders. When stakeholders are defined with appropriate roles, there is generally no risk of misuse of potentially sensitive information.

CHAPTER 12   ARCHITECTURE DEVELOPMENT METHOD

It is therefore advisable to pay special attention to creating or including an entity that represents stakeholders during the customization of the Enterprise Metamodel.

Table 12-8 shows an example of a Stakeholder Catalog. The artifact lists the stakeholders themselves and their key concerns and classifications and shows which other architectural artifacts might pique their interest or address their concerns.

Table 12-8. Stakeholder Catalog

Stakeholder	Key Concerns	Classification	Catalogs, Matrices, Diagrams, and Maps
Maria Summers (CEO)	The high-level drivers, goals, and objectives of the organization and how these are translated into an effective process to advance the business	Keep satisfied	Strategy/Goal Matrix Goal/Objective Matrix
Chris West (CIO)	Alignment of IT strategy with business goals, optimizing IT costs, and supporting business agility through technology	Key player	Goal/Objective Matrix Objective/Initiative Matrix

(*continued*)

Table 12-8. (*continued*)

Stakeholder	Key Concerns	Classification	Catalogs, Matrices, Diagrams, and Maps
James Noble (CFO)	Budget efficiency and cost control in organizational projects, long-term financial sustainability of architecture decisions	Keep informed	Application Portfolio Catalog Technology Portfolio Catalog
Anna Carter (Manager HR)	Supporting workforce needs in alignment with organizational transformation, ensuring employee adoption of new processes and technologies	Minimal effort	Business Process Map Organization/Business Process Matrix

Value Chain Diagram: The Value Chain Diagram represents the high-level processes or activities that an organization performs to deliver value to customers. Unlike the more formal Organization Map developed in Phase B (Business Architecture), this diagram emphasizes presentational impact. This diagram helps to understand how different business functions contribute to the overall value proposition. Its primary purpose is to quickly engage and align stakeholders around a specific change initiative, ensuring that all participants understand the overarching functional and organizational context of the architecture engagement.

A Value Chain Diagram typically includes *primary activities* (analogous to the core business functions of an organization) and *supporting activities*. The latter are functions that enable the primary activities.

The Value Chain Diagram has a horizontal or vertical flowchart structure. In the diagram, boxes represent the primary and supporting activities. The corresponding colors of the vertical and horizontal activities indicate the relationship between the two. The diagram should be read from left to right, showing the sequence and interdependencies between activities.

In the Value Chain Diagram as shown in Figure 12-4, activities are divided into *primary* and *support activities*. Primary activities contribute directly to the delivery of – in this case – healthcare services. In the example, the four main primary activities are:

- **Patient Admission and Triage:** Assessing and registering patients, determining urgency and care requirements.

- **Diagnosis and Treatment:** Conducting medical tests, diagnosing conditions, and providing necessary treatments or procedures.

- **Patient Care and Monitoring:** Continuous care, nursing, medication administration, and post-treatment monitoring.

- **Discharge and Follow-Up:** Ensuring proper recovery plans, follow-up consultations, and post-hospitalization care.

CHAPTER 12   ARCHITECTURE DEVELOPMENT METHOD

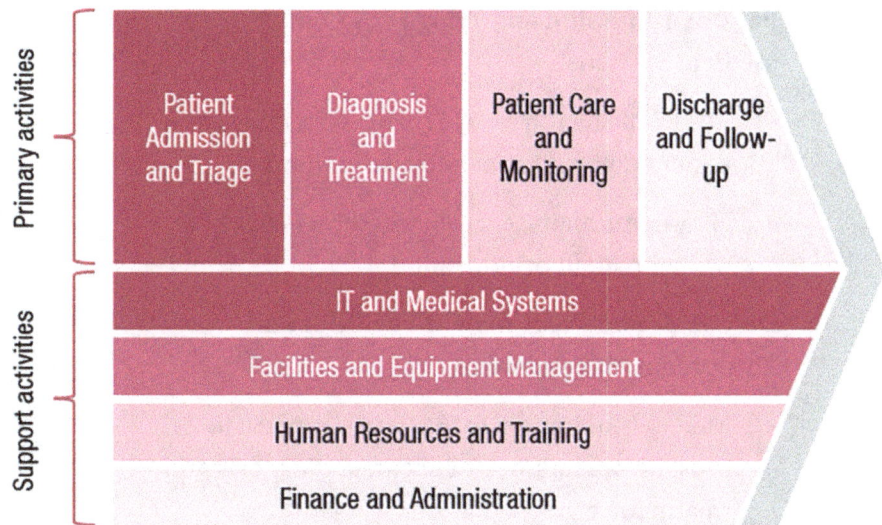

Figure 12-4. *Value Chain Diagram for a hospital*

Support activities enable primary activities to be performed efficiently. Four key support activities, shown in Figure 12-4, are:

- **IT and Medical Systems:** Managing electronic health records (EHR), hospital management software, and digital infrastructure
- **Human Resources and Training:** Recruiting, training, and managing healthcare professionals and support staff
- **Facilities and Equipment Management:** Maintaining hospital buildings, medical equipment, and supplies
- **Finance and Administration:** Budgeting, billing, insurance processing, and overall hospital governance

These activities ensure that the hospital delivers quality patient care efficiently while maintaining operational sustainability. As shown in the figure, the relationship between the primary and support activities is indicated by their matching colors:

- **Patient Admission and Triage ⇄ IT and Medical Systems**
  - Electronic health record (EHR) systems store patient data and streamline admissions.
  - IT systems manage triage workflows, ensuring proper patient prioritization.
- **Diagnosis and Treatment ⇄ Facilities and Equipment Management**
  - Medical equipment (MRI, X-ray, surgical tools) must be operational for accurate diagnosis and treatment.
  - Facility management ensures sterile, well-maintained environments for procedures.
- **Patient Care and Monitoring ⇄ Human Resources and Training**
  - Well-trained nurses, doctors, and support staff ensure effective patient care.
  - Ongoing medical training keeps staff updated on best practices and new treatments.
- **Discharge and Follow-Up ⇄ Finance and Administration**
  - Billing and insurance processing ensure proper financial clearance upon discharge.
  - Administrative teams handle scheduling for follow-ups and ensure patient records are updated.

Each support activity enables the smooth, efficient execution of its associated primary activity, ensuring the quality of the hospital's operations.

By analyzing this Value Chain Diagram, the hospital can identify areas where it can improve efficiency and reduce costs. For example, they may discover that they can improve their patient admission process by having their IT systems more efficiently manage and automate triage workflows. They may also discover that they can improve their operating rooms by investing in sterile, well-maintained environments. Analyzing a hospital's value chain provides a useful framework for identifying areas for improvement and optimizing performance. By optimizing its processes, the hospital can improve quality and create more value for its patients.

Solution Concept Diagram: This diagram offers a high-level overview of the proposed solution aimed at achieving the objectives of the architecture engagement. Unlike the more formal and detailed architecture diagrams developed in subsequent phases, this diagram serves as a preliminary *pencil sketch* of the anticipated solution at the start of the engagement.

This diagram may encapsulate key objectives, (stakeholder) requirements, and constraints for the engagement. It also may identify areas that require further investigation through formal architecture modeling. Its primary purpose is to quickly onboard and align stakeholders for a specific change initiative, ensuring that all participants comprehend the goals of the architecture engagement and how the proposed solution approach is expected to meet the organization's needs.

Business Model Diagram: A model describing the rationale for how an organization creates, delivers, and captures value. Another common name for this diagram is Business Model Canvas.

CHAPTER 12   ARCHITECTURE DEVELOPMENT METHOD

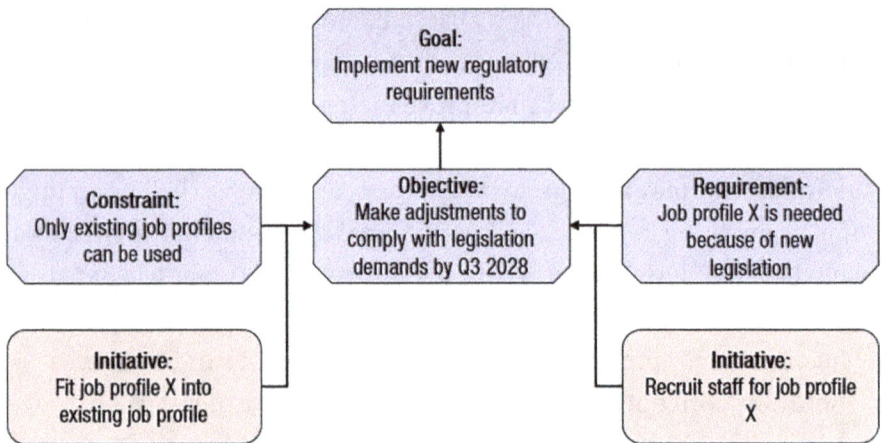

Figure 12-5. *Solution Concept Diagram*

The Business Model Diagram[1] shown in Figure 12-6 is an example of a diagram used in a hospital context [21].

---

[1] Figure 12-6 can be downloaded in a large format from https://eawheel.com/books/media/

CHAPTER 12   ARCHITECTURE DEVELOPMENT METHOD

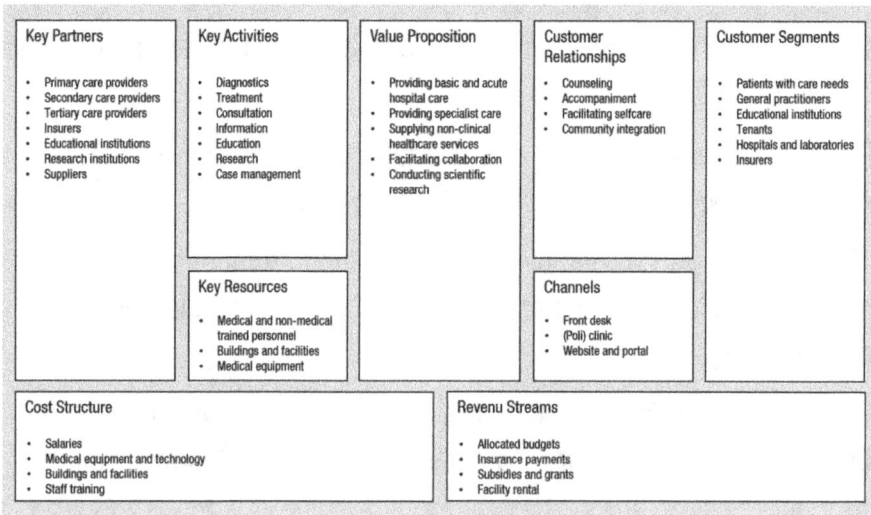

Figure 12-6. Business Model Diagram of a hospital

Note that the template of a Business Model Diagram or Canvas can contain the entities business collaboration, business interaction, business interface, resource, and value. These entities are not mentioned in the Foundation Enterprise Metamodel. It is advisable to consider them when tailoring the metamodel (as is described in Chapter 5, Section 5.2).

Business Capability Map: A diagram that shows the business capabilities that an organization needs to meet its purposes. During the Architecture Vision Phase, the diagram primarily shows capabilities related to achieving the organization's goals. It is not necessary to visualize and describe every capability the organization has at this stage.

345

CHAPTER 12 ARCHITECTURE DEVELOPMENT METHOD

> The name of this artifact does not fully align with its description in the TOGAF Standard.
>
> Please note that a *map* is a representation of architectural content in *textual format* (refer to Chapter 7, Section 7.4.2).
>
> The explanation provided by the TOGAF Standard mentions that it should actually be a *diagram* ("A diagram that shows the business capabilities").
>
> Based on the explanation, I went with showing an example of a Business Capability Diagram instead of a Business Capability Map.

The Business Capability Diagram shown in Figure 12-7 is a limited view of a Hospital Capability Diagram. This figure shows only a few capabilities for example purposes.

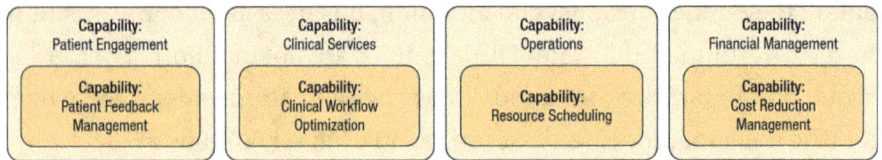

Figure 12-7. *Business Capability Diagram*

The Business Capability Diagram created during the Architecture Vision Phase is typically a high-level variant. Phase B (Business Architecture) of the Architecture Development Method updates the Business Capability Map artifact (see Section 12.5.3). During Phase B, more capability levels are added to the artifact.

Value Stream Map: A diagram representing an end-to-end collection of value-adding activities that create an overall result for a customer, stakeholder, or end user (see Figure 12-8).

CHAPTER 12   ARCHITECTURE DEVELOPMENT METHOD

Again, the name of the artifact should be Value Stream Diagram instead of Value Stream Map.

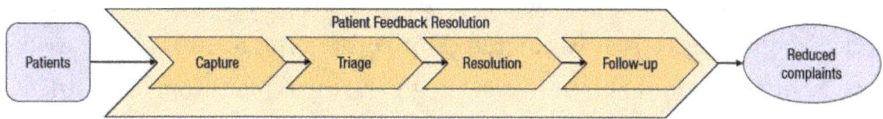

Figure 12-8. *Value Stream Diagram*

The Architecture Vision Phase sets the scope and specific objectives for the upcoming phases. It is key to ensuring that everyone involved in the project is aligned on what the architecture effort is aiming to achieve before moving forward into more detailed design and planning.

After the Architecture Vision Phase, the process moves to the Business Architecture Phase, where a more detailed view of the Business Architecture is developed, based on the high-level vision created in this phase.

 Further Reading
More information about business models, including the Business Model Diagram or Canvas, can be found in the TOGAF Series Guide "Business Models", document number G18A.

12.5. Phase B: Business Architecture

Business Architecture serves as a comprehensive framework, offering multi-dimensional views of capabilities, end-to-end value delivery, information, and organizational structure. It intricately maps the relationships among these diverse business elements, including strategies, products, policies, initiatives, and stakeholders.

At its core, Business Architecture aligns these elements with overarching business goals, providing a crucial link to other domains within the organization. Understanding Business Architecture is essential – it lays the groundwork for all architectural endeavors across various domains. It should ideally be the first architecture-related activity undertaken unless already integrated into other organizational processes, such as organizational planning, strategic business planning, or business process re-engineering.

Practically speaking, a well-defined Business Architecture is invaluable. It not only showcases the business value of subsequent architectural initiatives to key stakeholders but also illustrates the return on investment for their support and engagement in future work.

The scope of work during Phase B is primarily guided by the architecture vision articulated in Phase A. Here, the business strategy outlines the goals, drivers, and success metrics, yet it doesn't specify the path to achieving them. That's where Business Architecture comes into play, providing a roadmap in Phase B to navigate toward those defined objectives.

12.5.1 Developing Business Architecture

The objectives of Phase B (Business Architecture) are twofold. The first one is to *craft the Target Business Architecture* to outline how the organization should function to meet business objectives. Strategic drivers defined in the Architecture Vision Phase are addressed, ensuring that they align with the Statement of Architecture Work and satisfy stakeholder expectations.

The second objective is to *identify potential components for the Architecture Roadmap*, pinpointing the gaps between Baseline and Target Business Architectures that must be bridged. The key to both objectives is to reuse existing material as much as possible.

CHAPTER 12   ARCHITECTURE DEVELOPMENT METHOD

As with all phases of the Architecture Development Method, all the reference materials, external to the organization and collected from previous phases, serve as input for the next phase. Phase B (Business Architecture) is no exception to this rule.

Looking internally, the non-architectural inputs such as business principles, objectives, goals, and drivers are further refined during this phase. The capabilities defined during the Preliminary Phase (which are needed to understand how to work with architecture) undergo an assessment during Phase B. Finally, a Communications Plan is drawn up to detail how to communicate with the various stakeholders and how to address their requirements and concerns.

As for the architectural inputs, the organizational model plays an important role in describing the scope of the impacted organization. Part of the model sheds a light on the current architecture maturity level of the organization in relation to working with architecture. Then there are the roles and responsibilities for the architecture team, which are contained and visualized using a RACI matrix (see Section 12.3.1). The scope of the impacted organization is further detailed with constraints on architecture work, budget requirements, and a governance and support strategy. This last part entails the establishment of an Architecture Governance Framework in order to manage the architecture activities. Table 12-9 lists the steps to perform Phase B (Business Architecture).

CHAPTER 12 ARCHITECTURE DEVELOPMENT METHOD

Table 12-9. *Steps to perform Phase B (Business Architecture)*

Step to Perform	Description of Step	Output Generated by Step
Select reference models, viewpoints, and tools	Choose relevant models and viewpoints from the Architecture Repository to match the business goals and stakeholder interests.	Views and viewpoints Required building blocks (catalogs, matrices, and diagrams) Architecture requirements
Develop Baseline and Target Business Architecture description	Describe the current and future Business Architecture to a degree that supports the architecture vision, building on the Architecture Repository's resources as needed.	Baseline Business Architecture description Target Business Architecture description
Perform gap analysis	Confirm internal consistency, identify any gaps between baseline and target, and validate that models align with business objectives.	Gap analysis results
Define candidate roadmap components	Using the baseline, target, and gap analysis, build an initial roadmap to guide future actions and prioritization across the remaining architecture phases.	Draft Architecture Roadmap Business Architecture Roadmap components

(continued)

Table 12-9. (*continued*)

Step to Perform	Description of Step	Output Generated by Step
Resolve impacts across the Architecture Landscape	Assess the Business Architecture's wider impacts across the organization and its alignment with existing or planned projects.	Validated building blocks and other artifacts
Conduct formal stakeholder review	Revisit the architecture's initial motivation and make any necessary refinements to ensure it's fit to support other architecture domains.	Statement of Architecture Work
Finalize the Business Architecture	Standardize building blocks, fully document them, and check alignment with business goals. Finalize documentation within the Architecture Repository.	Documented building blocks and requirements
Create the Architecture Definition Document	Complete all relevant sections in the Architecture Definition Document, adding models, reports, and graphics as needed to capture a clear architectural view.	Architecture Definition Document

Select Reference Models, Viewpoints, and Tools: Start by diving into the Architecture Repository to handpick Business Architecture resources, like reference models and patterns, that align with core business drivers and address stakeholder needs (their requirements). Next, zero in on the Business Architecture viewpoints that best communicate solutions (think operations, management, finance). Lastly, choose tools that fit the job for capturing and analyzing data with the selected viewpoints. Whether they are simple documents or spreadsheets or more sophisticated modeling tools such as activity models and business process models, go with what's necessary to bring clarity and precision to the architecture.

Dive into the Architecture Repository: Begin by exploring the Architecture Repository to identify Business Architecture resources, such as reference models and patterns, that align with the organization's main business goals and meet stakeholder requirements.

Identify Suitable Viewpoints: Determine which Business Architecture viewpoints – such as those related to operations, management, or finance – will most effectively communicate the architectural solutions.

Select Appropriate Tools: Choose tools that are well suited for capturing and analyzing data within the selected viewpoints. These tools can range from simple documents and spreadsheets to more advanced modeling tools like activity models and business process models. Select the tools that will provide the necessary clarity and precision for the architectural work.

Develop Baseline and Target Business Architecture Description: Craft a baseline description of the current Business Architecture, covering as much as necessary to effectively build toward the Target Business Architecture. If an organization has existing architecture descriptions, they should be used as the basis for the baseline description. When developing the target description for the Business Architecture, focus on creating only what's necessary to support the architecture vision. The scope and level of detail to include will vary, depending on the importance of each business element in reaching the Target Architecture Vision. How much detail to include here depends on two things: first, how much of the existing structure will transition to the future architecture and, second, whether relevant architecture descriptions are already available. Utilize the Reference Library of the Architecture Repository to pinpoint the essential building blocks within the Business Architecture. If stakeholder needs call for new models, use the initial reference models found in the Reference Library of the Architecture Repository as a baseline to shape these new elements. Also, consider exploring different Target Architecture options. These are best reviewed with stakeholders using the Architecture Alternatives and Trade-Offs approach (see Chapter 14, Section 14.9) to ensure the architecture aligns with business needs.

Start with the Baseline Description: Begin by documenting the current Business Architecture. Use existing descriptions as a starting point.

Create the Target Business Architecture: Focus on what's needed to support the architecture vision. Only include what's necessary to reach the target, based on the key business areas.

Consider the Scope and Detail: The amount of detail will depend on how much of the current architecture will carry over to the future and if any existing descriptions are available.

Use Reference Models: Use the Reference Library from the Architecture Repository to identify essential building blocks for the Business Architecture.

Explore Target Architecture Options: Look at different future architecture options and review them with stakeholders. Use the Architecture Alternatives and Trade-Offs method to ensure alignment with business needs.

Perform Gap Analysis: To ensure internal consistency and accuracy within the architecture models, dive into each view to identify and resolve potential conflicts. This phase is where models are rigorously validated against core principles, objectives, and constraints, making sure they are fit for purpose. Evaluate the models for completeness, confirming that they align seamlessly with all (stakeholder) requirements. Finally, apply gap analysis to pinpoint any gaps between the Baseline and Target Architecture, as detailed in Chapter 14, Section 14.5.

CHAPTER 12   ARCHITECTURE DEVELOPMENT METHOD

Any requirement, or change in requirement, that is outside of the scope defined in the Statement of Architecture Work must be submitted to the Requirements Repository for management through the governed requirements management process (see Section 12.12).

Check for Consistency: Go through all views in the architecture model and look for any conflicts or inconsistencies that need fixing.

Validate Against Principles: Make sure the models are in line with key principles, objectives, and constraints. This helps ensure they are effective and fit for purpose.

Verify Completeness: Ensure all stakeholder needs are met. Double-check that no requirement is missed.

Identify Gaps: Compare the Baseline Architecture to the Target Architecture. Spot any differences or gaps that need addressing.

Define Candidate Roadmap Components: After establishing the Baseline and Target Architectures and analyzing the gaps, creating an Architecture Roadmap becomes essential. The results of the gap analysis lead to the creation of work packages (or initiatives) related to the architecture to be developed. These initiatives are the components to be used in the Architecture Roadmap.

Think of the initial Architecture Roadmap as the raw draft. It's here to support a more refined roadmap that will evolve during Phase E (Opportunities and Solutions) (see Section 12.8). Add the Business Architecture components to the roadmap.

Create a Draft Architecture Roadmap: Based on the gaps, start putting together the Architecture Roadmap. Think of this as the first draft – it doesn't need to be perfect yet.

Break It Down Into Work Packages: From the gap analysis, develop work packages (or initiatives) that relate to the architecture that is being developed. These work packages are what will fill out the roadmap.

Keep Improving: The roadmap will evolve, especially in the Opportunities and Solutions Phase (Phase E). So think of the initial roadmap as something that will get better over time.

Add Business Architecture: Include the Business Architecture components in the roadmap. This is crucial for aligning with business goals.

Resolve Impacts Across the Architecture Landscape: With the Business Architecture in place, the next step is to assess any potential broader impacts or ripple effects. Now's the time to look across the Architecture Landscape and examine other architecture assets to answer key questions:

- Does this Business Architecture influence any established architectures already in place?

- Have any recent adjustments or updates been made that could affect this Business Architecture?

- Are there chances to reuse or extend elements from this Business Architecture to benefit other areas of the organization?

- Could this Business Architecture affect ongoing or upcoming projects?

- Could other projects – whether planned or currently in progress – have an impact on this Business Architecture?

Discuss the results of the questions posed with the relevant stakeholders and try to determine whether or not they should lead to adjustments to the architecture. If so, update the architecture accordingly to resolve any potential impacts.

Identify Influenced Architectures: Examine existing architectures to determine if the Business Architecture impacts any of them.

Review Recent Changes: Investigate any recent modifications or updates in the organization that might affect the Business Architecture.

Seek to Reuse Opportunities: Identify elements within the Business Architecture that could be reused or extended to benefit other areas of the organization.

Evaluate Project Interactions: Assess whether the Business Architecture could influence ongoing or upcoming projects, and vice versa.

Engage Stakeholders: Discuss the findings with relevant stakeholders to determine if any adjustments to the architecture are necessary.

Update the Architecture: Implement any agreed-upon changes to address potential impacts and ensure alignment across the Architecture Landscape.

Conduct Formal Stakeholder Review: To ensure the Business Architecture Phase effectively captures and models the essence of the business, it's essential to engage core business stakeholders: executives, process owners, and subject matter experts. Their perspectives ensure that the architecture truly reflects the organization's needs and goals. Gather these stakeholders to revisit the original objectives behind the architecture work and the Statement of Architecture Work. Together, evaluate if the proposed Business Architecture aligns with and supports the upcoming work in other architecture domains. Make refinements only where necessary to keep the architecture grounded and aligned with strategic goals.

Identify Key Stakeholders: Gather a diverse group of core business stakeholders, including executives, process owners, and subject matter experts. Their insights are invaluable in shaping an architecture that truly reflects the organization's needs and goals.

Revisit Original Objectives: Together with the stakeholders, review the initial objectives outlined in the Statement of Architecture Work. This ensures everyone is aligned and focused on the intended outcomes.

Evaluate Alignment: Assess whether the proposed Business Architecture aligns with and supports upcoming work in other architecture domains. This holistic evaluation helps maintain coherence across all areas.

Refine As Necessary: Make adjustments only where needed to keep the architecture grounded and aligned with strategic goals. This approach maintains focus and prevents unnecessary complexity.

Finalize the Business Architecture: Set clear standards for each building block, using reference models from the Architecture Repository's Reference Library wherever possible. Take the time to document each block in detail (see Chapter 7, Section 7.5.2.2, for instructions on how to do this).

Once that's squared away, do a final cross-check to ensure the architecture is aligned with overarching business goals. Document the *why* behind each decision; it's vital for linking back to organizational objectives. And when it comes to documenting requirements, ensure they connect directly to specific architecture elements. Try to complete all work products as much as possible.

Set Clear Standards for Each Building Block: Begin by defining clear standards for every component of the Business Architecture. Utilize reference models from the Architecture Repository's Reference Library wherever possible to ensure consistency and reliability.

Document Each Building Block in Detail: Take the time to thoroughly document each building block.

Ensure Alignment with Business Goals: Once the building blocks are well-documented, perform a final cross-check to ensure that the architecture aligns with the organization's overarching business goals. This alignment is vital for the architecture's effectiveness and relevance.

Document the Rationale Behind Each Decision: For every decision made during this process, document the reasoning behind it. This practice is essential for maintaining a clear link between architectural choices and organizational objectives, facilitating future reviews and adjustments.

Connect Requirements to Specific Architecture Elements: When documenting requirements, ensure that each one is directly linked to specific elements within the architecture. This connection enhances traceability and ensures that all requirements are adequately addressed within the architectural framework.

Complete All Work Products Thoroughly: Strive to complete all work products as comprehensively as possible. Thorough documentation and completion of these products will serve as a solid foundation for implementation and future architectural endeavors.

Create the Architecture Definition Document: To craft a *business-oriented* Architecture Definition Document, focus on the core goals and scope relevant to the architecture. Illustrate key aspects through visuals or outputs from modeling tools to highlight the architecture's strengths. At a minimum, the document should clearly present the Baseline and Target Business Architectures. When it's ready, share it with critical stakeholders to gather feedback, ensuring refinements are made to align with organizational objectives. This iterative process is essential for achieving a comprehensive and goal-oriented architecture.

Identify Core Goals and Scope: Begin by clearly defining the main objectives and the scope of the architecture. This ensures that the Architecture Definition Document remains focused and relevant to the organization's needs.

Document Baseline and Target Architectures: Detail the Baseline and Target Business Architectures. This comparison highlights the necessary changes and developments required to achieve the goals.

Utilize Visual Aids: Incorporate diagrams, models, and other visual tools to effectively communicate key aspects of the architecture. Visual representations can simplify complex information and make the document more accessible to stakeholders.

Engage Stakeholders for Feedback: Share the Architecture Definition Document with key stakeholders to gather their insights and feedback. Their input is invaluable in ensuring that the architecture aligns with various perspectives and requirements within the organization.

Iterate and Refine: Use the feedback received to make necessary adjustments to the Architecture Definition Document. This iterative process helps in refining the architecture to better meet organizational objectives and address any concerns raised by stakeholders.

All the architecture work that needs to be done should be documented in what is called a *Statement of Architecture Work*. This document should be formally approved by the higher levels of management, in order to gain support and sponsorship for the architecture work to be done.

Since it is during this phase that actual architecture artifacts are going to be created, it is imperative to start using the Architecture Repository for documenting these artifacts. This leads to the initiation of the Enterprise

Continuum (see Chapter 7, Section 7.3.1), which is a conceptual framework and provides a lens through which to view the entire collection of architecture assets stored in the Architecture Repository. Among the assets contained in the repository are reusable building blocks, publicly available reference models, organization-specific reference models, and organization standards.

The architecture vision lays out a clear problem statement and defines the purpose of the Statement of Architecture Work – essentially, the *why* behind the architecture effort. This vision document can be enhanced with targeted views on specific architecture areas, business scenarios (if available), and an organized set of refined stakeholder requirements and concerns (which may be taken from the Solution Concept Diagram described earlier; see Figure 12-5).

12.5.2 Business Architecture Methods and Modeling Techniques

Several methods and techniques can be used to implement the steps described in the previous section. For instance, the following core Business Architecture methods are used to model the Business Architecture driven by the strategy scope from Phase A:

Applying Business Capabilities: The Business Capability Map developed in the Architecture Vision Phase acts as a high-level snapshot of the organization's core abilities, independent of its existing structure, business processes, information systems, and applications. Once outlined, these capabilities should be linked back to the specific organizational units, value streams, information systems, and strategic objectives in the architecture's scope. This mapping creates a more holistic view, enhancing the alignment and effectiveness of each area in reaching the broader goals.

CHAPTER 12 ARCHITECTURE DEVELOPMENT METHOD

Business Capability Heat Mapping: One widely used analysis technique is heat mapping, a method that provides a range of insights on core business capabilities. These insights might include maturity, effectiveness, performance, and even the value or cost each capability brings to the business. In a Business Capability Map, different colors represent specific attributes to give an at-a-glance view of each capability's status.

Take a maturity heat map, for instance. Here, a capability meeting the desired maturity level is highlighted in green, one level below target appears in yellow, and capabilities lagging by two or more levels show up in red. Colors can also reflect other conditions, for example, purple might indicate a capability that's not yet in place but highly desired or a capability with surplus funding and resources beyond what's necessary.

Remember the fictional healthcare institution described in Chapter 5 (Section 5.1.1)?

Figure 12-9 shows a portion of this institution's capabilities, using the Business Capability Heat Mapping technique. The capabilities that are slightly underperforming are shown in yellow. The capability in red is not functioning at all and requires immediate attention. The purple-colored capabilities do not yet exist and must be created (either developed or purchased).

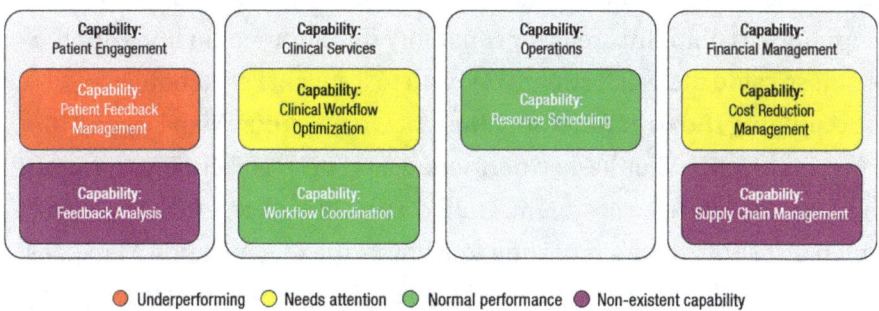

Figure 12-9. *Business Capability Heat Map*

CHAPTER 12   ARCHITECTURE DEVELOPMENT METHOD

> There is no scientific approach or calculation behind determining the maturity of a capability. Much comes down to stakeholder alignment and gut feeling. Business Capability Heat Mapping – and this is true of any type of heat mapping – requires a healthy dose of common sense.

For more information about the Business Capability Heat Mapping technique, see the *Further Reading* information box at the end of this section.

Applying Value Streams: Value streams provide valuable stakeholder context, helping them understand why certain business capabilities are critical for the organization. They set the stage by defining the *why* behind the need, while business capabilities focus on the *what* – the elements required for each phase of the value stream to deliver results.

Begin with an initial set of value stream models for the business, ideally documented in the Architecture Vision Phase. Whether working with a new or existing value stream, these models can be analyzed in the architecture's scope by heat mapping value stream stages or developing use cases that cover the entire value stream journey (for further details, see the "References" section at the end of this book).

The real power of this approach lies in mapping the relationship between each stage of a value stream and the corresponding business capabilities. This alignment allows for effective gap analysis – such as heat mapping – to pinpoint areas for capability development in line with the business value the value stream delivers to a specific stakeholder.

Applying the Organization Map: An Organization Map lays out the core organizational units, key partners, and crucial stakeholder groups that shape the enterprise ecosystem. Unlike a traditional organizational chart – which just shows who's reporting to whom – the Organization Map goes further, revealing the dynamic working relationships among these groups.

CHAPTER 12 ARCHITECTURE DEVELOPMENT METHOD

Think of it more as a network or web of connections rather than a rigid hierarchy. And while it is called a *map*, it's rather more of a *diagram*, since it is a graphical representation and not text-based.

This diagram plays a central role in Business Architecture because it sets the organizational stage for everything else in the architectural process. While capability mapping reveals *what* an organization does and value stream mapping shows *how* it delivers value to specific stakeholders, the organizational map identifies the business units or third parties that own or use these capabilities and participate in the value streams. The Organization Map shows *who* holds these capabilities and *who* is involved in these value streams, whether they're internal teams or external partners.

Together with other methods described earlier, the Organization Map clarifies which units need to be looped into the architecture effort, who to approach about specific requirements, and even how to gauge the impact of strategic decisions.

Applying Information Maps: Diving into the Business Architecture Phase starts with pinpointing the core elements that drive the business forward. Think products, customers, strategies, and so forth. With a clear view of these essentials, critical information can be mapped out in terms that resonate with the business. This mapping isn't just about listing things; it's about seeing how these elements interconnect, creating a foundational Information Map that sets the stage for the future.

The true power of this approach comes to life when matrices are built between key information concepts and business capabilities. These connections reveal the direct links between essential information on the one hand and the capabilities that turn it into real value on the other. This is fundamental to the essence of Business Architecture. These Information Maps and their ties to business capabilities aren't just one-off exercises. They'll serve as a valuable reference across future architecture phases, impacting everything from data insights to applications and infrastructure.

Applying Modeling Techniques: Beyond core techniques like capability maps, value streams, Organization Maps, and Information Maps, various modeling methods can be beneficial based on specific needs, for instance:

- **Activity Models:** Describe the organization's business activities. They detail internal data exchanges between activities and also cover interactions with external activities outside the model's scope. This versatility empowers organizations to tailor their approach, modeling only what is relevant and insightful, ensuring alignment with business goals.

The Object Management Group has developed the Business Process Modeling Notation (BPMN), a standard for business process modeling that includes a language with which to specify business processes, their tasks/steps, and the documents produced [20].

- **Use-Case Models:** Think of use-case models as a straightforward map of how a business process flows in an organization. They break down complex business processes into use cases (tasks or functions) and actors (the people or roles involved) in a way that's easy to follow. Each part of this model is visualized through use-case diagrams and detailed specs that make it simple to see who's doing what and when.
- **Logical Data Model:** The logical data model, or what some might call a class model, organizes information by defining key entities, their properties, and the acceptable values for each. Plus, it captures the connections between different entities, like using

is-a relationships for sharper analysis. For example, imagine *sedan* being a type of *car*, which is a type of *vehicle*. If there's a rule that applies to all vehicles, you can set it once, and it'll automatically apply across the board, whether you're talking about cars, trucks, or any other vehicle. Class models take this a step further by adding behaviors (methods) to each entity, giving them specific actions to perform. Typically, Application and Information Architects lean on the class diagram, while Business Architects focus on the logical data model.

Business Scenarios: This technique is complementary to the Business Architecture Phase of the Architecture Development Method. This technique evaluates, elaborates on, and/or modifies the assumptions behind an architecture project. It involves understanding and documenting the most important elements of a business scenario in successive iterations. Failing to examine all elements of a business scenario may result in an incomplete solution; however, care must be taken to avoid unnecessary repetition.

The Business Scenarios technique can be used in various phases of Enterprise Architecture development, primarily the Preliminary, Architecture Vision, and Business Architecture phases. However, it can also be used in other phases if necessary. See the *Further Reading* information box at the end of Section 12.5.3 for more information on applying the method.

> **Further Reading**
>
> For more information on the heat mapping technique, refer to the TOGAF Series Guide "Business Capabilities, Version 2", document number G211.
>
> A baseline example of Value Stream Heat Mapping can be found in the TOGAF Series Guide "Value Streams", document number G178.
>
> The BIZBOK Guide offers extensive information about heat mapping capabilities. See the chapter on capability mapping:
>
> https://www.businessarchitectureguild.org/page/BIZBOK
>
> For further guidance on using Organization Maps, see the TOGAF Series Guide "Organization Mapping", document number G206.
>
> Additional information on Information Maps is available in the TOGAF Series Guide "Information Mapping", document number G190.
>
> More information about Business Scenarios can be found in the TOGAF Series Guide "Business Scenarios", document number G176.

12.5.3 Phase B (Business Architecture) Artifacts

Table 12-10 lists the artifacts that can be created during the execution of Phase B (Business Architecture).

CHAPTER 12 ARCHITECTURE DEVELOPMENT METHOD

Table 12-10. *Artifacts created during Phase B (Business Architecture)*

ADM Phase	Artifacts	Used Entities
Phase B: Business Architecture	Organization/Actor Catalog	Organization unit, actor, role, location (when Location Catalog is not maintained)
	Driver/Goal/Objective Catalog	Organization unit, driver, goal, objective, measure (may optionally be included)
	Role Catalog	Role
	Business Service/Function Catalog	Organization unit, function, business service, application service (may optionally be included here)
	Location Catalog	Location, logical/physical application component, logical/physical technology component
	Process/Event/Control/Product Catalog	Process, event, control, product
	Contract/Measure Catalog	Business service, application service (optionally), contract, measure
	Business Capabilities Catalog	Business capability
	Business Capability Map	Business capability, entities that flesh out the business capability
	Value Stream (Stages) Catalog	Business capability, value stream

(*continued*)

CHAPTER 12 ARCHITECTURE DEVELOPMENT METHOD

Table 12-10. (*continued*)

ADM Phase	Artifacts	Used Entities
	Value Stream Map	Business capability, value stream
	Value Stream/Capability Matrix	Business capability, value stream
	Strategy/Capability Matrix	Business capability, value stream, course of action
	Capability/Organization Matrix	Business capability, value stream, organization unit
	Organization Map	Organization unit, actor, role, stakeholder (if available)
	Business Glossary Catalog	–
	Business Interaction Matrix	Organization unit, function, business service
	Actor/Role Matrix	Actor, role
	Business Footprint Diagram	Organization unit, function, business service
	Business Service/ Information Diagram	Business service, business information
	Functional Decomposition Diagram	Capability
	Product Lifecycle Diagram	Product
	Goal/Objective/Business Service Diagram	Goal, objective, business service

(*continued*)

Table 12-10. (*continued*)

ADM Phase	Artifacts	Used Entities
	Business Use-Case Diagram	Actor, role, process, function (later on, it can include data, application, and technology details)
	Organization Decomposition Diagram	Actor, role, location
	Process Flow Diagram	Process, all mappings related to the process entity
	Business Event Diagram	Event, process
	Information Map	Business information

Organization/Actor Catalog: The purpose of this catalog is to capture a definitive listing of everyone that interacts with IT, including users and owners of IT systems. The Organization/Actor Catalog can be referenced when developing requirements in order to test for completeness.

For example, requirements for an application that services customers can be tested for completeness by verifying exactly which customer types or business roles need to be supported and whether there are any particular requirements or restrictions for user types. The Organization/Actor Catalog (see Table 12-11) can also contain business roles.

Table 12-11. Organization/Actor Catalog

Organization Unit	Actor	Business Role
Board of Directors	Maria Summers	Chief Executive Officer
Finance	James Noble	Chief Financial Officer
IT Department	Chris West	Chief Information Officer
Quality and Safety	Elena Petrova	Medical Director
	Kevin O'Brien	Patient Experience Manager
Clinical Operations	Susan Delgado	Chief Nursing Officer
	Benjamin Lee	Chief Medical Information Officer
	Hannah Duarte	Charge Nurse/Unit Manager

The Organization/Actor Catalog can be visualized using a diagram as well. Figure 12-10 shows the Organization/Actor Diagram in which the organization units *Quality and Safety* and *Clinical Operations* are visualized.

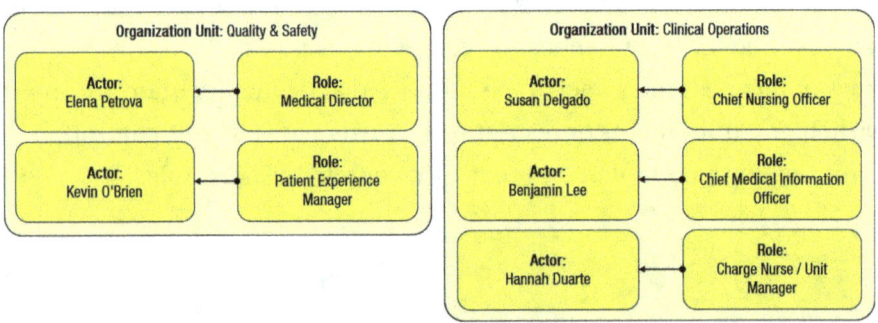

Figure 12-10. Organization/Actor Diagram

Driver/Goal/Objective Catalog: Provide a cross-organizational reference of how an organization meets its drivers in practical terms through goals, objectives, and (optionally) measures (see Table 12-12).

CHAPTER 12   ARCHITECTURE DEVELOPMENT METHOD

Publishing a definitive breakdown of drivers, goals, and objectives allows change initiatives within the organization to identify synergies across the organization (e.g., multiple organizations attempting to achieve similar objectives), which in turn allow stakeholders to be identified and related change initiatives to be aligned or consolidated.

Table 12-12. Driver/Goal/Objective Catalog

Driver	Goal	Objective
Patient-Centered Care	Improve patient satisfaction.	Achieve 90% rating in patient satisfaction surveys by Q4 2028.
	Enhance patient safety.	Decrease hospital-acquired infection rates by 15% by Q4 2029.
Operational Excellence	Optimize clinical workflows.	Reduce patient discharge processing time by 30 minutes by April 2029.
	Increase staff productivity.	Realize a 10% increase in key clinical throughput metrics by Q2 2030.
Financial Sustainability	Reduce operational costs.	Cut non-clinical supply expenses by 10% by December 2030.

Role Catalog: This artifact lists the different business roles within an organization. It can be used with other information, such as authorization levels or zones, to create a RACI matrix. The Role Catalog (see Table 12-13) is a valuable resource for identifying the impacts of organizational change management, defining job functions, and executing end-user training.

CHAPTER 12   ARCHITECTURE DEVELOPMENT METHOD

Table 12-13. Role Catalog

Role	Description
Medical Director	Oversees clinical risk, incident response, and safety programs.
Patient Experience Manager	Manages patient satisfaction and service recovery processes.
Chief Nursing Officer	Leads nursing operations, staffing, patient care quality.
Chief Medical Information Officer	Leads EHR adoption and clinical IT optimization.
Charge Nurse/Unit Manager	Coordinates shift-level operations, reports issues, escalates concerns.

Business Service/Function Catalog: The Business Service/Function Catalog provides a functional decomposition that can be filtered, reported on, and queried. It supplements graphical Functional Decomposition Diagrams.

The catalog (see Table 12-14) can be used to identify an organization's capabilities and understand the level at which governance is applied to its functions. This decomposition can identify new capabilities required to support business changes and determine the scope of change initiatives, applications, or technology components.

Table 12-14. Business Service/Function Catalog

Business Service	Description	Business Function	Capability
Patient Feedback Routing Service	Receives, categorizes, and routes patient feedback (complaints, compliments, safety concerns) to the appropriate business units for action.	Patient Experience Management	Patient Experience Management
Infection Risk Incident Escalation Service	Automatically escalates infection-related feedback or audit findings to the Infection Control Team for assessment and containment action.	Clinical Risk and Safety Oversight	Clinical Workflow Optimization Patient Feedback Management
Care Team Resource Scheduling Service	Manages automated scheduling of staff shifts, aligning personnel availability with patient load forecasts and unit needs.	Workforce Planning and Scheduling	Resource Scheduling
Cost Efficiency Analytics Service	Provides real-time and historical insights into departmental costs per treatment or service, enabling financial decision support.	Financial Performance Management	Cost Reduction Management

Since the Business Service/Function Catalog combines business services and functions, a matrix would have been a better choice. Cross-mapping between the two entities provides a clearer picture of their relationship. For this reason, a Business Service/Function Matrix is shown in Table 12-15.

CHAPTER 12   ARCHITECTURE DEVELOPMENT METHOD

Table 12-15. *Business Service/Function Matrix*

Business Service ▼	**Business Function**			
	Patient Experience Management	Clinical Risk and Safety Oversight	Workforce Planning and Scheduling	Financial Performance Management
Patient Feedback Routing Service	X			
Infection Risk Incident Escalation Service		X		
Care Team Resource Scheduling Service			X	
Cost Efficiency Analytics Service				X

Location Catalog: Documents all relevant physical or virtual locations where business operations, services, or architectural elements reside. This includes offices, clinical facilities, data centers, or cloud zones. It helps architects understand geographic dependencies, constraints (e.g., regulatory), and the distribution of capabilities, applications, or data. It supports planning for disaster recovery, latency, availability, and compliance. An example is shown in Table 12-16.

CHAPTER 12   ARCHITECTURE DEVELOPMENT METHOD

Table 12-16. *Location Catalog*

Location ID	Location Name	Type	Purpose	Applications Used
LOC-001	Central Hospital Campus	Physical Site	Inpatient care, surgery, admin HQ	EHR system (Epic), Staff Scheduling Tool
LOC-002	Data Center – Region A	Physical Site	Hosts EHR, feedback systems	Patient Feedback Repository, Infection Alert Engine
LOC-003	Telehealth Operations Hub	Virtual Location	Remote consultations and triage	Virtual Consultation Platform

The Location Catalog can also be visualized as a diagram. Figure 12-11 illustrates this.

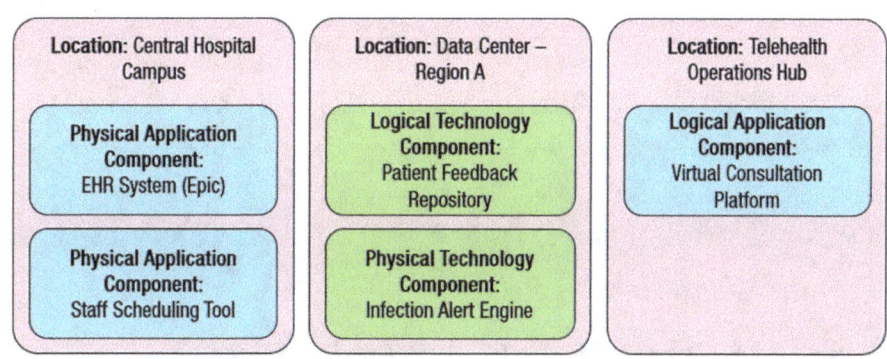

Figure 12-11. *Location Diagram*

Process/Event/Control/Product Catalog: Captures key elements of a business process, including the events that trigger it, the controls that constrain it, and the products or outcomes it delivers. This catalog (see Table 12-17) helps architects understand how business processes are governed and what inputs/outputs are involved. It supports regulatory compliance, Risk Management, and operational alignment by documenting relationships among business activities and their surrounding context.

375

CHAPTER 12 ARCHITECTURE DEVELOPMENT METHOD

Table 12-17. Process/Event/Control/Product Catalog

Process	Event	Control	Product
Patient Feedback Routing	Submission of patient survey	Feedback categorization rules	Routed feedback ticket
Infection Incident Escalation	Infection-related complaint logged	Escalation protocol (within 24 hours)	Infection control alert
Staff Schedule Generation	Weekly schedule trigger	Labor regulations, staffing ratios	Shift roster
Cost Analysis Reporting	End of quarter	Financial reporting standards	Cost performance report

Figure 12-12 shows a graphical representation of the first line from the Process/Event/Control/Product Catalog.

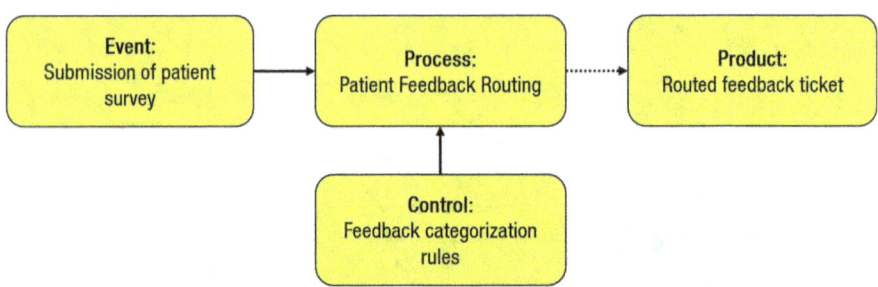

Figure 12-12. Process/Event/Control/Product Diagram

Contract/Measure Catalog: Documents the performance expectations, agreements, and service levels between business services and their consumers or providers. It helps align services with business goals by explicitly defining KPIs, SLAs, OLAs, and regulatory commitments. This catalog (see Table 12-18) supports architecture traceability by linking services to their value delivery and compliance metrics.

CHAPTER 12   ARCHITECTURE DEVELOPMENT METHOD

Table 12-18. *Contract/Measure Catalog*

Contract/ Measure Name	Applies to Service	Consumer	Provider	Measure Type	Target Value
Patient Feedback SLA	Patient Feedback Routing Service	Patient Experience Office	IT/ Experience Platform Vendor	SLA[2]	95% routed within 24 hours
Infection Escalation OLA	Infection Risk Incident Escalation	Quality and Safety Department	Clinical Risk System	OLA[3]	100% within 12 hours
Staffing Optimization KPI	Care Team Resource Scheduling Service	Nursing Administration	Scheduling System/HR	KPI	<5% manual overrides
Cost Efficiency Metric	Cost Efficiency Analytics Service	CFO/Finance	BI Platform/ Finance Analysts	KPI	Cost per case < baseline

Business Capabilities Catalog: A structured list of the core business capabilities an organization needs to fulfill its mission and strategic objectives. Each capability describes *what* the business does (not *how*)

---

[2] Service Level Agreement: a formal contract or agreement between a service provider and a client that outlines the expected level of service, performance metrics, and potential consequences if those standards aren't met.

[3] Operational Level Agreement: an internal agreement within a company that defines how different departments or teams will work together to deliver services and meet the terms of a Service Level Agreement (SLA).

CHAPTER 12   ARCHITECTURE DEVELOPMENT METHOD

and remains stable even when processes or systems change. The catalog (see Table 12-19) supports planning, impact analysis, and capability-based investment decisions. It typically includes capability name, description, maturity level, responsible organization, and related goals or services.

Table 12-19. Business Capabilities Catalog

Capability Name	Description	Maturity Level	Organization Unit	Related Goal
Patient Feedback Management	Ability to collect, route, and act on patient feedback to improve care and safety.	Ad hoc	Patient Experience Office	Improve patient satisfaction. Enhance patient safety.
Clinical Workflow Optimization	Ability to design, implement, and monitor efficient, standardized clinical paths.	Defined	Clinical Operations	Optimize clinical workflows.
Resource Scheduling	Ability to assign and manage personnel and assets based on demand forecasts.	Managed	Nursing Administration	Increase staff productivity.
Cost Reduction Management	Ability to monitor, analyze, and optimize costs without compromising quality.	Repeatable	Finance and Supply Chain	Reduce operational costs.

CHAPTER 12 ARCHITECTURE DEVELOPMENT METHOD

Business Capability Map: A map showing the business capabilities an organization needs to achieve its goals. Business capabilities can be broken down into sub-capabilities. The Business Capability Map can also be supplemented with entities that flesh out the capabilities. The Business Capabilities Catalog can be used to expand the Business Capability Map created in Phase A. Figure 12-13 provides a graphical representation of the map in the form of a Business Capability Diagram.

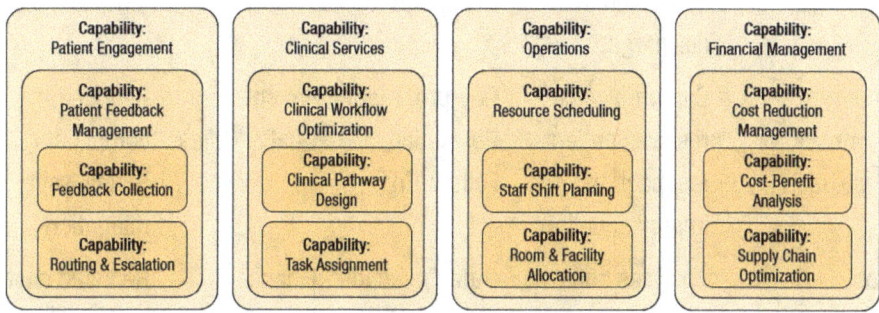

Figure 12-13. *Business Capability Diagram*

Value Stream (Stages) Catalog: A structured listing of the end-to-end collections of value-adding activities that describe *how an organization delivers value to its stakeholders*. Each entry outlines a value stream's name, description, triggering stakeholder, stages, and expected outcomes. This catalog (see Table 12-20) helps link business capabilities to customer-focused outcomes and supports strategic alignment.

379

CHAPTER 12   ARCHITECTURE DEVELOPMENT METHOD

Table 12-20. *Value Stream Catalog*

Value Stream Name	Description	Stages	Stakeholder	Outcome
Inpatient Care Delivery	Provides complete inpatient treatment from admission to discharge.	Admission, Diagnosis, Treatment, Discharge	Patients Clinicians	Safe, timely recovery Improved bed turnover
Patient Feedback Resolution	Captures and resolves patient feedback to improve quality.	Capture, Triage, Resolution, Follow-Up	Patients Quality Office	Reduced complaints Safety issue mitigation
Cost Performance Review	Analyzes and improves cost-effectiveness of clinical services.	Data Collection, Analysis, Action Plan	Finance Department Heads	Reduced cost per treatment Optimized spending

The Value Stream Catalog can be visualized by using a Value Stream Diagram for each of the listed value streams. Figure 12-14 illustrates the value stream for Patient Feedback Resolution.

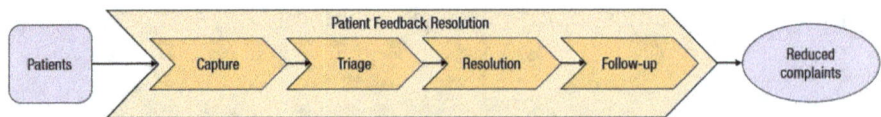

Figure 12-14. *Value Stream Diagram for Patient Feedback Resolution*

Value Stream Map: Illustrates an end-to-end collection of activities that add value and create an overall result for a customer, stakeholder, or end user.

CHAPTER 12 ARCHITECTURE DEVELOPMENT METHOD

The Value Stream (Stages) Catalog can serve as the basis for creating a Value Stream Map. The catalog can be supplemented with capabilities related to the stages of the value streams. The *Outcome* column in the Value Stream (Stages) Catalog can be replaced with a *Capabilities* column.

Value Stream/Capability Matrix: The purpose of this matrix is to link the stages of a value stream (what the organization delivers to stakeholders) to the business capabilities (what the organization needs to perform those stages). It's a key tool in capability-based planning and helps identify which capabilities support which parts of value delivery.

Table 12-21. Value Stream/Capability Matrix

Value Stream Name	Value Stream Stage	Capability
Patient Feedback Resolution	Capture	Patient Feedback Management, Digital Channel Integration
	Triage	Clinical Risk Assessment, Quality and Safety Monitoring
	Resolution	Incident Response Coordination, Clinical Workflow Optimization
	Follow-Up	Patient Communication Services, Service Quality Evaluation

The Value Stream/Capability Matrix, as shown in Table 12-21, can be visualized by using a Value Stream/Capability Diagram. Figure 12-15 illustrates the value stream for Patient Feedback Resolution. Using the *Business Capability Heat Mapping* technique (see Section 12.5.2) on this artifact reveals capability gaps, redundancies, and areas for improvement.

CHAPTER 12 ARCHITECTURE DEVELOPMENT METHOD

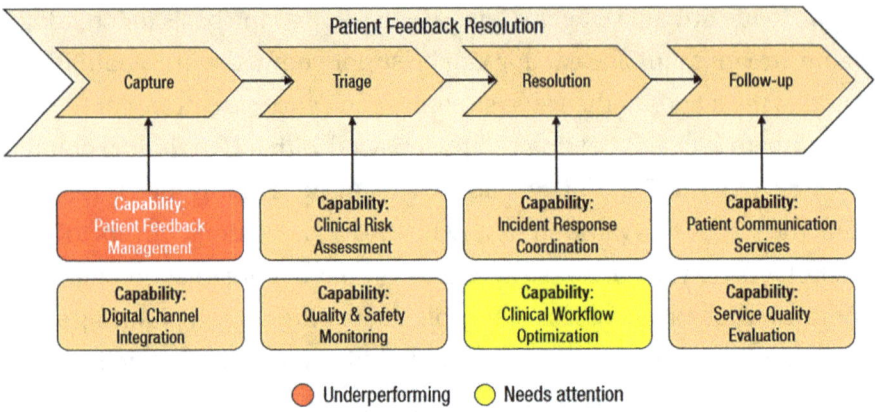

Figure 12-15. Value Stream/Capability Diagram for Patient Feedback Resolution

Strategy/Capability Matrix: Shows the capabilities required to support specific strategy statements. It aligns the organization's strategic goals with the business capabilities needed to realize them. It helps stakeholders see how strategic intents translate into organizational capabilities, ensuring traceability and focus in the architecture.

This matrix (see Table 12-22) supports capability-based planning, enabling prioritization of investments and architectural change by showing where critical capability development is required to meet strategic goals.

Table 12-22. Strategy/Capability Matrix

Strategic Goal	Capability	Priority
Improve patient satisfaction	Patient Engagement	High
Enhance patient safety	Patient Engagement	High
Optimize clinical workflows	Clinical Services	Medium
Increase staff productivity	Operations	Medium
Reduce operational costs	Financial Management	Low

CHAPTER 12 ARCHITECTURE DEVELOPMENT METHOD

Capability/Organization Matrix: Lists the organization elements that implement each capability. The matrix (see Table 12-23) maps business capabilities to the organizational units responsible for delivering or supporting them. It helps clarify ownership, accountability, and potential gaps or overlaps in capability delivery.

Table 12-23. Capability/Organization Matrix

Capability ▼	Organization Unit			
	Patient Experience Office	Quality and Safety	Clinical Operations	Finance
Patient Feedback Management	Primary	Contributor	Contributor	
Clinical Workflow Optimization			Primary	
Resource Scheduling			Contributor	
Cost Reduction Management				Primary

Organization Map: A map showing the relationships between the primary entities that make up the organization, its partners, and stakeholders. The map (see Table 12-24) lists the organization unit entities and their types, as well as their relationships to other entities on the map.

CHAPTER 12 ARCHITECTURE DEVELOPMENT METHOD

Table 12-24. Organization Map

Organization Unit	Type	Related to
Patient Experience Office	Business unit	Healthcare Institution
Quality and Safety Department	Business unit	Healthcare Institution Ministry of Health
Clinical Operations	Business unit	Healthcare institution Medical Equipment Supplier
Finance	Business unit	Healthcare Institution Insurance Partner Training Institute Medical Equipment Supplier
Nursing Administration	Business unit	Healthcare Institution Training Institute
Ministry of Health	Partner	Quality and Safety Department
Medical Equipment Supplier	Supplier	Clinical Operations Finance
Training Institute	Supplier	Nursing Administration Finance
Insurance Partner	Partner	Finance

The Organization Map can also be visualized using an Organization Diagram, as illustrated in Figure 12-16.

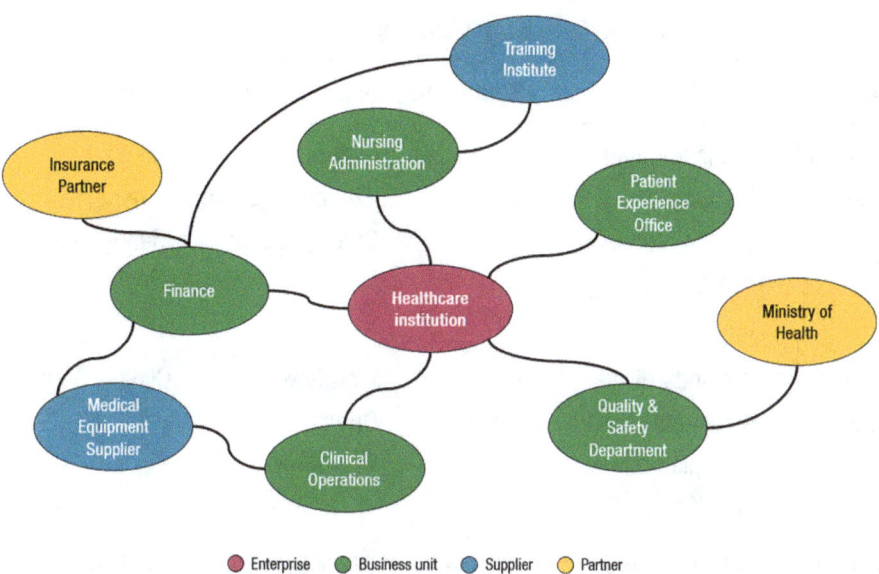

Figure 12-16. Organization Diagram

Business Glossary Catalog: A structured list of business terms and their unambiguous definitions that are used throughout the organization. This ensures semantic consistency across business processes, applications, and data models. The glossary (see Table 12-25) is essential for bridging the communication gap between business and IT stakeholders by standardizing the terminology used in requirements, reports, services, and interfaces. This artifact evolves throughout the Architecture Development Method.

CHAPTER 12   ARCHITECTURE DEVELOPMENT METHOD

Table 12-25. *Business Glossary Catalog*

Term	Definition	Synonyms	Usage Context	Capability
Feedback Case	A documented instance of patient feedback, complaint, or compliment.	Feedback Ticket	Patient Experience Portal, Quality Review	Patient Experience Management
Clinical Pathway	A standardized, evidence-based care plan for specific conditions.	Care Protocol	Workflow Optimization, EHR Task Flows	Clinical Services
Cost per Patient Day	Total operating cost divided by the number of patient days in a time period.	Per Diem Cost	Financial KPIs, Departmental Cost Reviews	Financial Management

Business Interaction Matrix: This matrix models the interactions between business roles and functions. It helps identify collaboration points, dependencies, and responsibility relationships within an organization. The matrix (see Table 12-26) is particularly useful for clarifying who does what and how organizational units engage with business capabilities.

CHAPTER 12 ARCHITECTURE DEVELOPMENT METHOD

Table 12-26. *Business Interaction Matrix*

Business Role ▼	Business Function			
	Patient Feedback Routing	Infection Risk Escalation	Resource Scheduling	Cost Efficiency Analysis
Patient Experience Manager	Performs	Not Involved	Not Involved	Not Involved
Infection Control Officer	Receives/Acts On	Performs	Not Involved	Not Involved
Nursing Administration	Not Involved	Informed	Performs	Supports
Financial Analyst	Not Involved	Not Involved	Not Involved	Performs

Actor/Role Matrix: Shows which actors perform which roles to support the definition of security and skill requirements (see Table 12-27).

Personally, I think this is one of the least common artifacts you create as an architect.

The reason is that the same information can easily be requested from the HR department.

It is questionable whether there is added value in keeping track of actors and roles in the form of an Actor/Role Matrix in the Architecture Repository. This creates an additional source of information that belongs with the HR department.

Understanding actor-to-role relationships is key to defining training needs, user security settings, and organizational change management.

CHAPTER 12   ARCHITECTURE DEVELOPMENT METHOD

Table 12-27. Actor/Role Matrix

Actor	Role
Maria Summers	Chief Executive Officer
James Noble	Chief Financial Officer
Chris West	Chief Information Officer
Elena Petrova	Medical Director
Kevin O'Brien	Patient Experience Manager
Susan Delgado	Chief Nursing Officer
Benjamin Lee	Chief Medical Information Officer
Hannah Duarte	Charge Nurse/Unit Manager

Business Footprint Diagram: This diagram illustrates the connections among business goals, organizational units, business functions, and services. It also maps these functions to the technical components that deliver the necessary capabilities.

The diagram provides clear traceability between technical components and the business goals they satisfy, demonstrating ownership of the identified services.

A Business Footprint Diagram only shows the key facts that link organizational unit functions to delivery services. They are used as communication platforms for senior-level (CxO) stakeholders. Figure 12-17 only shows the key entities and has removed the relations between them for better readability.

CHAPTER 12　ARCHITECTURE DEVELOPMENT METHOD

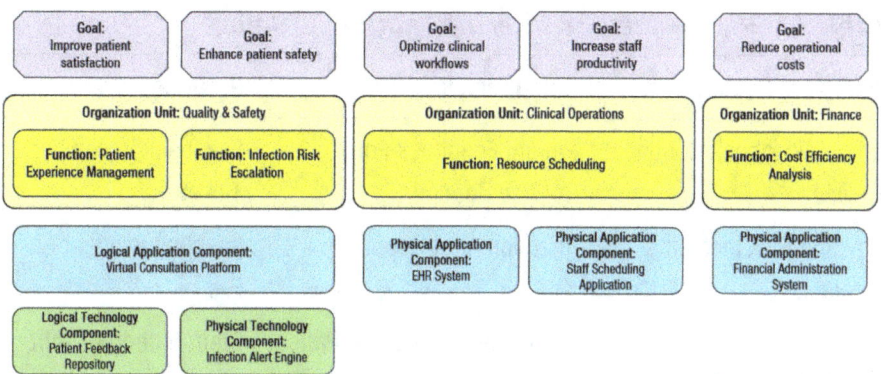

Figure 12-17. Business Footprint Diagram

Business Service/Information Diagram: Illustrates how business services consume or produce business and data objects (information). It shows the relationship between services (what the business offers) and the information they rely on or generate, helping ensure that business processes, applications, and data are aligned. This diagram (see Figure 12-18) is valuable during Phase B (Business Architecture) and Phase C (Data Architecture) to clarify which services handle specific information types and to identify potential redundancies, gaps, or integration needs.

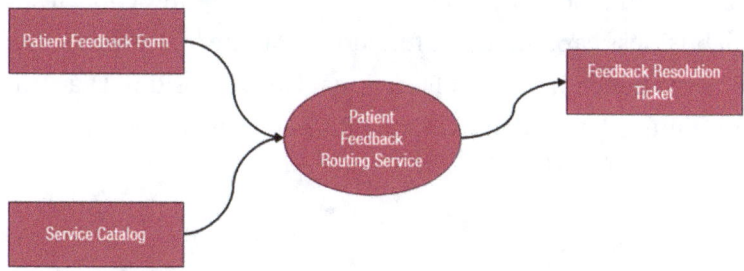

Figure 12-18. Business Service/Information Diagram

The Business Service/Information Diagram can also be visualized in tabular format using a Business Service/Information Catalog (see Table 12-28).

CHAPTER 12 ARCHITECTURE DEVELOPMENT METHOD

Table 12-28. *Business Service/Information Catalog*

Business Service	Consumes	Produces
Patient Feedback Routing Service	Patient Feedback Form, Service Catalog	Feedback Resolution Ticket
Infection Risk Incident Escalation	Infection Alert Report	Infection Incident Case File
Care Team Resource Scheduling Service	Staff Availability Data, Patient Census	Staff Schedule, Shift Alerts
Cost Efficiency Analytics Service	Cost per Service Data, Treatment Outcomes	Departmental Cost Report

Functional Decomposition Diagram: The purpose of the Functional Decomposition Diagram (See Figure 12-19) is to break down high-level business capabilities or functions into their constituent sub-functions. It helps architects and stakeholders understand *what* an organization does, supporting capability-based planning and aligning architecture with strategic objectives. This diagram provides a single-page overview of organizational functionality relevant to the architecture scope. It shows top-level business capabilities or functions and the logical grouping of sub-functions beneath them. A Functional Decomposition Diagram is very similar to a multi-level Business Capabilities or Function Catalog.

CHAPTER 12   ARCHITECTURE DEVELOPMENT METHOD

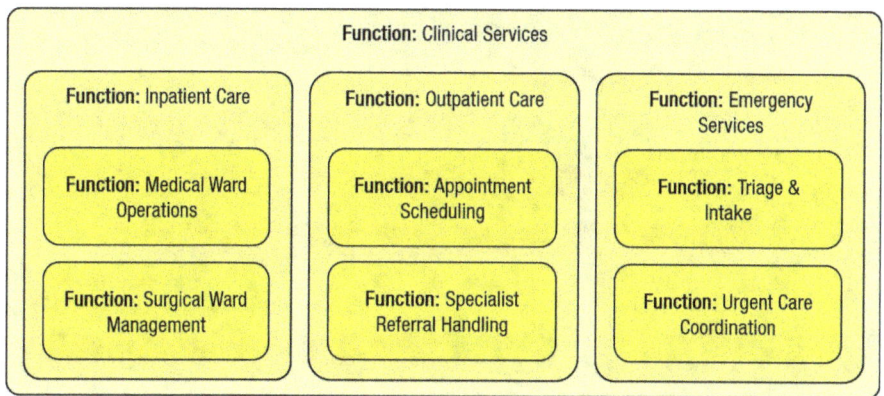

Figure 12-19. Functional Decomposition Diagram (using business functions)

Product Lifecycle Diagram: This diagram illustrates the stages a product or service goes through from inception to retirement. It provides a structured view of how products relate to business capabilities and processes. This allows architects to assess dependencies, transitions, and impacts. The diagram usually maps products to lifecycle phases, such as *concept, development, operation*, and *retirement*. This diagram is useful for managing portfolios, understanding transitions, and planning change initiatives. Each product is tracked individually, often with time-bound markers, enabling strategic alignment with organizational goals.

Figure 12-20 shows that the Patient Feedback Application will be operational until the end of Q2 2028. After that, it will be retired. Its successor, the Patient Feedback Application v2, is under development and will become operational at the start of Q3 2028. The Procurement Application will retire in Q2 2030. It will be replaced by the Supply Chain Management Application. Development of the replacement application will begin in Q3 2029.

CHAPTER 12 ARCHITECTURE DEVELOPMENT METHOD

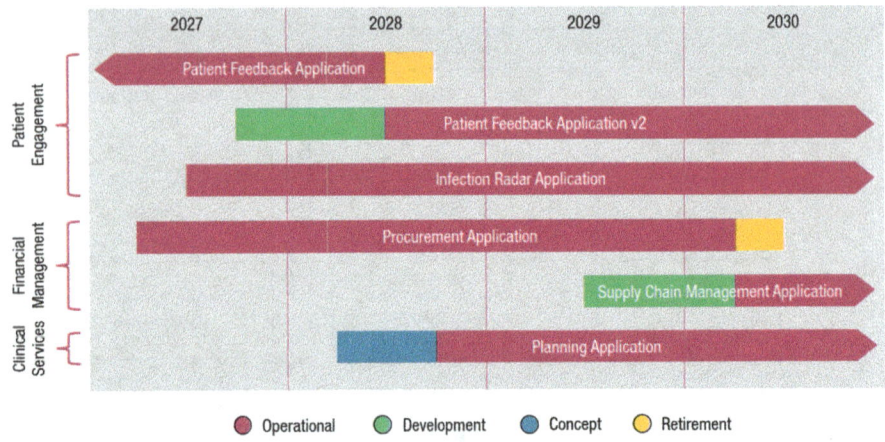

Figure 12-20. Product Lifecycle Diagram

As with most diagrams, the Product Lifecycle Diagram can also be visualized as a catalog. This is illustrated in Table 12-29.

Table 12-29. Product Lifecycle Catalog

Product Name	Lifecycle Stage	Timeframe	Related Capability
Patient Feedback Application	Operational	Q1 2025 to Q2 2028	Patient Engagement
Infection Radar Application	Operational	Q2 2027 to Q4 2035	Patient Engagement
Procurement Application	Operational	Q1 2027 to Q4 2029	Financial Management
Patient Feedback Application v2	Development	Q3 2027 to Q2 2028	Patient Engagement

(continued)

CHAPTER 12 ARCHITECTURE DEVELOPMENT METHOD

Table 12-29. (*continued*)

Product Name	Lifecycle Stage	Timeframe	Related Capability
Supply Chain Management Application	Development	Q2 2029 to Q1 2030	Financial Management
Planning Application	Concept	Q1 2028 to Q3 2028	Clinical Services
Patient Feedback Application	Retirement	Q3 2028	Patient Engagement
Procurement Application	Retirement	Q2 2030	Financial Management

Goal/Objective/Business Service Diagram: Visually shows how business goals are supported by objectives and how those objectives are realized through specific business services. This helps trace strategic intent through measurable targets to operational functionality. This Goal/Objective/Business Service Diagram (See Figure 12-21) is a derivative of the Business Footprint Diagram (Figure 12-17).

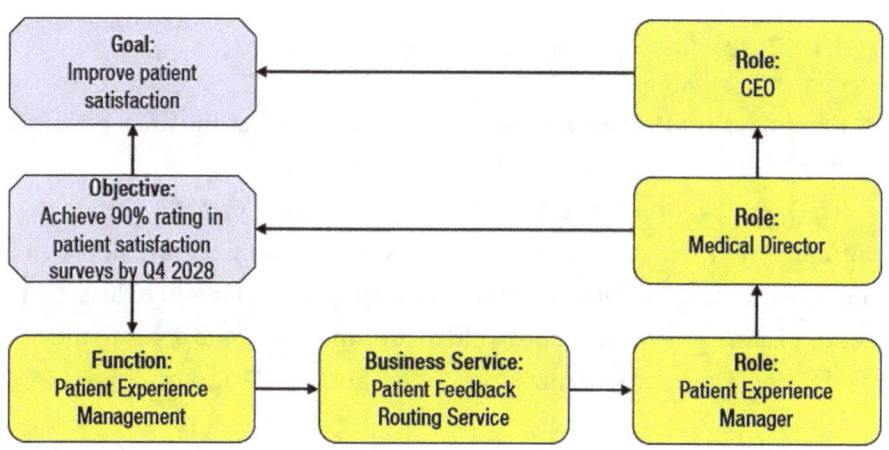

Figure 12-21. *Goal/Objective/Business Service Diagram*

CHAPTER 12   ARCHITECTURE DEVELOPMENT METHOD

Business Use-Case Diagram: Represents the interactions between business actors (roles) and business services or processes. It clarifies system boundaries and stakeholder interests by highlighting who uses what. Business Use-Case Diagrams are useful for identifying requirements and stakeholder concerns. It can also be used to map roles to business services, functions, or processes. As the architecture evolves, use cases can progress from the business level to include data, application, and technology details.

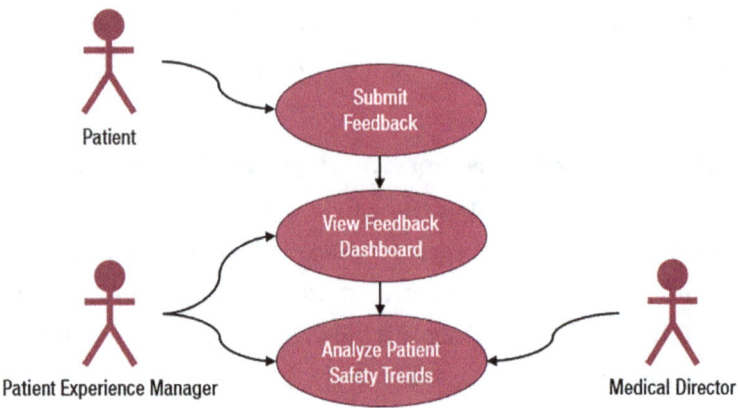

Figure 12-22. Business Use-Case Diagram

Organization Decomposition Diagram: Describes the links between actor, roles, and location within an organization tree.

An Organization Decomposition Diagram should provide a chain of command of owners and decision-makers in the organization. Although it is not the intent of the Organization Decomposition Diagram to link goal to organization, it should be possible to intuitively link the goals to the stakeholders from the Organization Decomposition Diagram.

CHAPTER 12   ARCHITECTURE DEVELOPMENT METHOD

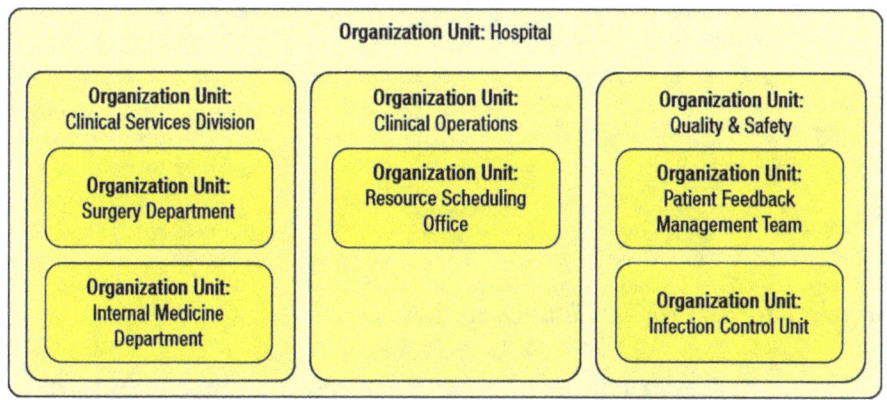

Figure 12-23. *Organization Decomposition Diagram*

Process Flow Diagram: The purpose of a Process Flow Diagram is to depict all mappings related to the *process* metamodel entity.

Process Flow Diagrams illustrate the sequential flow of control between activities and may use swimlane techniques to depict the ownership and execution of process steps. For instance, the application supporting a process step can be displayed as a swimlane.

In addition to showing a sequence of activities, these diagrams can detail the controls that apply to a process, the events that trigger or result from the completion of a process, and the products generated from process execution.

Process Flow Diagrams are useful for elaborating on the architecture with subject matter experts because they allow experts to describe *how the job is done* for a particular function.

CHAPTER 12   ARCHITECTURE DEVELOPMENT METHOD

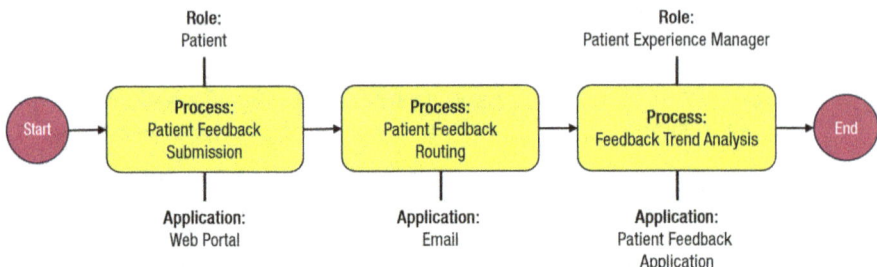

Figure 12-24. Process Flow Diagram

Business Event Diagram: Shows how business events trigger business processes, functions, or services and their resulting outcomes. A business event is an occurrence, either internal or external, that requires a response from the organization. This diagram (See Figure 12-25) helps model an organization's reactive behavior, which is essential for aligning processes with unpredictable or time-sensitive stimuli. Events may originate from users, systems, or environmental changes and often lead to process initiation, decision-making, or notifications.

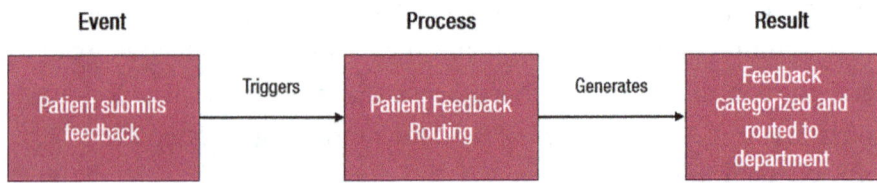

Figure 12-25. Business Event Diagram

Information Map: This map shows a collection of information concepts and their relationships to one another. It provides a structured view of an organization's key information concepts and how they relate to business capabilities, value streams, and processes. Information concepts reflect the business vocabulary. The Information Map (see Table 12-30) helps identify what information is critical to the organization, where it flows, and how it connects to decision-making and operational

CHAPTER 12   ARCHITECTURE DEVELOPMENT METHOD

performance. The Information Map organizes information into categories and shows relationships between concepts to support traceability, governance, and integration.

Table 12-30. Information Map

Information Concept	Information Concept Category	Definition	Related Information Concepts
Patient Feedback	Transactional	Input submitted by patients about their experience, including complaints, compliments, and safety concerns.	Patient ID, Feedback Category, Department
Feedback Category	Reference	Standard classification of feedback types such as hygiene, wait time, or staff behavior.	Patient Feedback
Scheduled Staff Roster	Transactional	A daily or weekly assignment list showing which staff are scheduled for shifts.	Staff Member, Shift Time, Unit
Staff Availability	Reference	Availability status of individual staff based on contracts, leave, or training.	Scheduled Staff Roster

The *transactional* category of an information concept refers to information generated by business activities or events. This type of information is dynamic, time-sensitive, and event-driven. It is frequently updated or changed and tied to a specific process or workflow.

CHAPTER 12   ARCHITECTURE DEVELOPMENT METHOD

The *reference* category assigned to information concepts refers to stable, reusable, and governed information. It supports categorization, validation, and consistent interpretation across processes. This type of information changes slowly, if at all, over time. It can be used for lookup, classification, or validation.

Phase B (Business Architecture) provides critical inputs to other phases of the Architecture Development Method. The insights gained about the business help guide decisions in subsequent phases:

- Business Architecture informs how data and applications should be structured to support business processes during Phase C (Information Systems Architectures).

- The Technology Architecture (established in Phase D (Technology Architecture)) will be aligned with the business needs defined in Phase B.

- Projects and solutions are identified in Phase E (Opportunities and Solutions), based on the business requirements from the Business Architecture Phase.

It is critical to devote sufficient time and attention to Phase B because it is the foundation for all subsequent architectural work.

12.6. Phase C: Information Systems Architectures

The purpose of the Information Systems Architectures Phase is to develop a blueprint of the information systems that support the Business Architecture. It identifies the key information systems required to implement the business capabilities and defines the Data and Application Architectures. The architects assess the existing systems and determine the gaps that need to be addressed. Key deliverables from this phase include data models and application portfolios.

Phase C is typically divided into two distinct yet interconnected architectures:

- **Data Architecture:** Focuses on the organization's data assets, data management practices, and governance of data. It describes how data is stored, managed, and utilized across the organization.

- **Application Architecture:** Focuses on the individual applications needed by the organization and how they interact with each other and with users. This includes the design of application components and their relationships, ensuring they align with the business goals.

The TOGAF Standard refers to the BIZBOK Guide for the development of Business Architecture. The BIZBOK Guide addresses the concept of information and plays an important role in understanding the relationship between business processes and functions and the use of information enabled by applications. The information concept thus assumes a role between Business Architecture and Data Architecture.

Unfortunately, this concept is not yet sufficiently reflected in the Fundamental Content of the TOGAF Standard.

The framework explains the concept of information mapping in the Series Guide "Information Mapping" [22] and bases the creation of an Information Map on the content as described in the BIZBOK Guide [5].

The two architectures that make up Phase C (Information Systems Architectures) are detailed in the following sections.

12.6.1 Developing Data Architecture

The objectives of Phase C (Data Architecture) are consistent with Phase B (Business Architecture). The first objective is to *craft the Target Data Architecture* to outline how the organization should function to meet business objectives. For example, identifying the data entities and their relationships and defining data storage, handling, and integration approaches.

The second objective is to *identify potential Data Architecture components for the Architecture Roadmap*, pinpointing the gaps between Baseline and Target Data Architectures that must be bridged. The key to both objectives is to reuse existing material as much as possible.

Table 12-31. Steps to perform Phase C (Data Architecture)

Step to Perform	Description of Step	Output Generated by Step
Select reference models, viewpoints, and tools	Choose relevant models and viewpoints from the Architecture Repository to match the business goals and stakeholder interests.	Views and viewpoints Required building blocks (catalogs, matrices, and diagrams) Architecture requirements
Develop Baseline and Target Data Architecture description	Describe the current and future Data Architecture to a degree that supports the architecture vision, building on the Architecture Repository's resources as needed.	Baseline Data Architecture description Target Data Architecture description

(continued)

CHAPTER 12   ARCHITECTURE DEVELOPMENT METHOD

Table 12-31. (*continued*)

Step to Perform	Description of Step	Output Generated by Step
Perform gap analysis	Confirm internal consistency, identify any gaps between baseline and target, and validate that models align with business objectives.	Gap analysis results
Define candidate roadmap components	Using the baseline, target, and gap analysis, build an initial roadmap to guide future actions and prioritization across the remaining architecture phases.	Data Architecture Roadmap components
Resolve impacts across the Architecture Landscape	Assess the Data Architecture's wider impacts across the organization and its alignment with existing or planned projects.	Validated building blocks and other artifacts
Conduct formal stakeholder review	Revisit the architecture's initial motivation and make any necessary refinements to ensure it's fit to support other architecture domains.	Updated Statement of Architecture Work
Finalize the Data Architecture	Standardize building blocks, fully document them, and check alignment with business goals. Finalize documentation within the Architecture Repository.	Documented building blocks and requirements
Update the Architecture Definition Document	Complete all relevant sections in the Architecture Definition Document, adding models, reports, and graphics as needed to capture a clear architectural view.	Updated Architecture Definition Document

The steps defined in Table 12-31 are the same as those in the previous phase, Phase B (Business Architecture). Because phases B, C, and D all deal with creating specific parts of the Enterprise Architecture (Business Architecture, Data Architecture, and Application Architecture, respectively), the steps don't really differ between the three phases. The minor differences between the phases are described below. For a detailed description of all the steps, refer to Table 12-9 in Section 12.5.1.

Select Reference Models, Viewpoints, and Tools: Refer to the Architecture Repository to select Data Architecture resources, like reference models and patterns, that align with core business drivers and address stakeholder needs (requirements). Zero in on the Data Architecture viewpoints.

Access the Architecture Repository: Begin by exploring the organization's Architecture Repository to identify available resources pertinent to Data Architecture.

Identify Relevant Reference Models and Patterns: Within the repository, locate Data Architecture reference models and patterns that resonate with the organization's business objectives and stakeholder expectations.

Select Appropriate Viewpoints: Choose Data Architecture viewpoints that effectively address and represent the concerns of stakeholders, ensuring comprehensive coverage of their requirements.

Evaluate Tools and Techniques: Assess the tools and techniques available in the repository to determine which ones best support the selected models and viewpoints, facilitating efficient architecture development.

Align with Business Drivers: Ensure that the selected models, viewpoints, and tools are in harmony with the organization's core business drivers, promoting coherence between Data Architecture and business strategy.

Document the Selection Process: Maintain clear documentation of the choices made and the rationale behind them, providing transparency and a reference for future architectural endeavors.

CHAPTER 12   ARCHITECTURE DEVELOPMENT METHOD

Develop Baseline and Target Data Architecture Description:
Develop a baseline description of the current Data Architecture, covering as much as necessary to effectively build toward the Target Data Architecture. Use existing architecture descriptions if available, since they should be used as the basis for the baseline description. When developing the target description for the Data Architecture, focus on creating only what's necessary to support the architecture vision. Use the Architecture Repository Reference Library to locate the essential building blocks within the Data Architecture. As suggested in the previous phase, if stakeholder needs require new models, use the initial reference models found in the Reference Library of the Architecture Repository as a baseline to shape these new elements.

Consider exploring different Target Architecture options and review them with stakeholders using the Architecture Alternatives and Trade-Offs approach (see Chapter 14, Section 14.9) to ensure the architecture aligns with business needs.

Understand the Current Data Architecture: Take stock of all existing data assets, systems, and processes. Document how data flows within the organization and identify any existing issues or gaps.

Gather Existing Documentation: Locate any current architecture descriptions or data flow diagrams. Use these documents as a foundation to build the Baseline Data Architecture.

Define the Future Data Architecture Needs: Clarify the organization's goals and how data supports these objectives. Determine what changes or improvements are necessary in the Data Architecture to meet future demands.

Use Available Resources: Consult the Architecture Repository Reference Library to find standard building blocks and best practices. Incorporate these elements to ensure consistency and efficiency in the Target Data Architecture.

Develop the Target Data Architecture: Design the Target Data Architecture that aligns with the organization's vision. Focus on creating components that are essential to support business objectives without overcomplicating the design.

Explore Different Options: Consider various architectural approaches and solutions. Evaluate the pros and cons of each option in the context of the organization's needs.

Engage Stakeholders for Feedback: Present the proposed Target Data Architecture options to key stakeholders. Discuss potential trade-offs and gather input to ensure the architecture aligns with business requirements.

Refine and Finalize the Data Architecture: Incorporate feedback from stakeholders to adjust the design. Ensure that the final architecture is practical, scalable, and aligned with the organization's strategic goals.

Perform Gap Analysis: This phase is where models are rigorously validated against core principles, objectives, and constraints, making sure they are fit for purpose. To ensure internal consistency and accuracy within the architecture models, dive into each view to identify and resolve potential conflicts. Evaluate the models for completeness, confirming that they align seamlessly with all stakeholder requirements. Finally, apply gap analysis to pinpoint any gaps between the Baseline and Target Architecture, as detailed in Chapter 14, Section 14.5.

Validate Models Against Core Principles: Begin by thoroughly reviewing each model to ensure they adhere to the organization's fundamental principles, objectives, and constraints. This step is crucial to confirm that the models are appropriate and aligned with the overarching goals.

Ensure Internal Consistency: Next, examine each view within the architecture models to identify any inconsistencies or conflicts. Addressing these issues is essential for maintaining the accuracy and reliability of the models.

Confirm Completeness: After resolving internal conflicts, assess the models to verify that they comprehensively meet all stakeholder requirements. This ensures that no critical elements are overlooked.

Conduct Gap Analysis: Finally, perform a gap analysis to detect any discrepancies between the Baseline and Target Architectures. This involves identifying areas where the current architecture falls short of meeting the target state.

Define Candidate Roadmap Components: After establishing the Baseline and Target Architectures and analyzing the gaps, it is essential to create components for the Architecture Roadmap. The results of the gap analysis lead to the creation of work packages (or initiatives) related to the architecture to be developed. These initiatives are the components to be used in the Architecture Roadmap. In this step, the Data Architecture components are added to the draft roadmap created in the previous phase. A more refined roadmap is developed during Phase E (Opportunities and Solutions) (see Section 12.8).

Review the Gap Analysis Results: Begin by thoroughly examining the outcomes of the gap analysis. This will help in understanding the specific areas where the current architecture falls short in meeting the Target Architecture.

Identify Necessary Initiatives: For each identified gap, determine the initiatives or work packages required to address and close these gaps. These initiatives should be clearly defined and directly related to achieving the Target Architecture.

Add to the Draft Architecture Roadmap: Compile all the identified initiatives related to Data Architecture into the Draft Architecture Roadmap. This roadmap should outline the sequence and priority of each work package, providing a clear path from the Baseline to the Target Architecture.

Prepare for Further Refinement in Phase E: Recognize that this draft is not final. The Architecture Roadmap will undergo further refinement during Phase E (Opportunities and Solutions). This phase focuses on identifying and evaluating potential solutions, ensuring that the initiatives are feasible and aligned with business objectives.

Resolve Impacts Across the Architecture Landscape: Look across the Architecture Landscape and examine other architecture assets to answer key questions, such as the following:

- Does the Data Architecture influence any established architectures, and vice versa?
- Are there chances to reuse or extend elements from this Data Architecture to benefit other areas of the organization?

- Could this Data Architecture impact current or upcoming projects, and vice versa?

Discuss the results of the questions posed with the relevant stakeholders and try to determine whether or not they should lead to adjustments to the architecture. If so, update the architecture accordingly to resolve any potential impacts.

Review Existing Architectures: Examine current architectural assets to identify interactions between the Data Architecture and other established architectures.

Engage Stakeholders: Discuss the findings with relevant stakeholders to determine if adjustments to the architecture are necessary.

Update the Data Architecture: If changes are needed, revise the architecture to address potential impacts and ensure alignment across the organization.

Conduct Formal Stakeholder Review: It's essential to engage key stakeholders to ensure the Data Architecture Phase effectively captures and models the essence of the organization. The stakeholder perspectives ensure that the architecture truly reflects the organization's needs and goals. Gather these stakeholders to revisit the original objectives behind the architecture work and the Statement of Architecture Work. Evaluate if the proposed Data Architecture aligns with and supports the upcoming work in other architecture domains. Make refinements where necessary, based on stakeholder feedback.

Assemble Key Stakeholders: Identify and bring together individuals who have a vested interest in the architecture's success.

Revisit Original Objectives: Collaboratively review the initial goals set for the architecture work to ensure alignment.

Examine the Statement of Architecture Work: Assess this foundational document to confirm that the Data Architecture aligns with its directives.

Evaluate Alignment with Other Domains: Determine if the proposed Data Architecture supports and integrates well with forthcoming efforts in other architectural areas.

Collect Stakeholder Feedback: Encourage stakeholders to provide insights and perspectives on the current architectural approach.

Implement Necessary Refinements: Based on the feedback received, make adjustments to the Data Architecture to better meet organizational objectives.

Finalize the Data Architecture: Set clear standards for each building block, using reference models from the Architecture Repository's Reference Library wherever possible. Meticulously document each block (see Chapter 7, Section 7.5.2.2, for instructions on how to do this).

Afterward, do a final cross-check to ensure the architecture is aligned with overarching business goals. Document the *why* behind each decision; it's vital for linking back to organizational objectives.

Set Clear Standards for Each Building Block: Begin by defining clear standards for every component of the Data Architecture. Utilize reference models available in the Architecture Repository's Reference Library to ensure consistency and industry alignment.

Document Each Building Block Thoroughly: For each component, create detailed documentation.

Cross-Check Alignment with Business Goals: After documenting, perform a comprehensive review to ensure that the Data Architecture aligns with the organization's overarching business goals. This alignment is vital for the architecture's effectiveness and relevance.

Document the Rationale Behind Each Decision: For every decision made during this process, document the reasoning behind it. This practice is essential for maintaining a clear connection between your Data Architecture and the organization's objectives, facilitating future reviews and adjustments.

Update the Architecture Definition Document: Update the Architecture Definition Document by presenting the Baseline and Target Data Architectures. Share the document with the key stakeholders to gather feedback, ensuring refinements are made to keep the architecture aligned with organizational objectives. This iterative process is essential for achieving a comprehensive and goal-oriented architecture.

Share with Key Stakeholders: Distribute the updated Architecture Definition Document to essential stakeholders. Their insights are invaluable for ensuring the architecture meets organizational needs.

Gather and Incorporate Feedback: Collect feedback from stakeholders and refine the document accordingly. This iterative process helps in addressing concerns and improving the architecture.

Align with Organizational Objectives: Ensure that the updated architecture aligns with our organization's goals and strategies. This alignment is key to delivering value.

Similar to Phase B (Business Architecture), all the architecture work that needs to be done should be documented in what is called a *Statement of Architecture Work*. This document should be formally approved by the higher levels of management, in order to gain support and sponsorship for the architecture work to be done.

The Data Architecture Phase is key in ensuring the data infrastructure of an organization is designed to meet business goals efficiently, is secure, and is scalable for future needs.

12.6.2 Phase C (Data Architecture) Artifacts

Table 12-32 lists the artifacts that can be created during the execution of Phase C (Data Architecture).

Table 12-32. Artifacts created during Phase C (Data Architecture)

ADM Phase	Artifacts	Used Entities
Phase C: Data Architecture	Data Entity/Data Component Catalog	Data entity, logical/physical data component
	Data Entity/Business Function Matrix	Data entity, function (optionally, data entity relationship to owning organization unit)
	Application/Data Matrix	Logical application component, data entity
	Conceptual Data Diagram	Data entity
	Data Dissemination Diagram	Data entity, business service, logical/physical application component
	Data Security Diagram	Role, organization unit, logical/physical application component
	Data Migration Diagram	Data entity, location
	Data Lifecycle Diagram	Data entity, process

Data Entity/Data Component Catalog: This catalog identifies and maintains a list of the data used across the organization, relating data entities to data components showing how data entities are structured. Its purpose is to identify all data used in the organization and to encourage effective data use. This artifact (see Table 12-33) is based on the Information Map created previously (see Section 12.5.3). The Information Map defined information concepts. According to the TOGAF Standard, this entity is comparable to the logical data component entity. Mapping these concepts to logical or physical data components results in the Data Entity/Data Component Catalog.

CHAPTER 12   ARCHITECTURE DEVELOPMENT METHOD

Table 12-33. Data Entity/Data Component Catalog

Data Entity	Data Component	Business Service
Patient Feedback Record	Patient Feedback	Patient Feedback Routing Service
	Feedback Category	
Staff Schedule	Scheduled Staff Roster	Care Team Resource Scheduling Service
	Staff Availability	

The Data Entity/Data Component Catalog, when visualized graphically, resembles the illustration in Figure 12-26 (which uses only the *Patient Feedback Record* data entity).

Figure 12-26. Data Entity/Data Component Diagram

Data Entity/Business Function Matrix: The purpose of the Data Entity/Business Function Matrix is to depict the relationship between data entities and business functions within the organization. Business functions are supported and realized by business processes. The mapping of the data entity to the business function entity enables the organization to gain insight into the assigned ownership of data entities. The matrix illustrates where data is being used in the organization.

CHAPTER 12   ARCHITECTURE DEVELOPMENT METHOD

In my experience, it is much more common for an organization to want insight into what data is being used by what processes. In practice, mapping data entities to business functions will be less common than mapping data entities to processes.

In most cases, people in an organization are very familiar with the processes they perform on a daily basis. The mapping of data entities to processes will therefore provide a more helpful picture than the mapping to business functions. The latter entity generally paints a much more general picture than the process entity.

In this sense, the creation of a Data Entity/Business Process Matrix offers more advantages than the matrix proposed in the TOGAF Standard.

A Data Entity/Business Function Matrix (or Data Entity/Business Process Matrix) supports a gap analysis to determine whether any data entities are missing and need to be created. For illustrative purposes, Table 12-34 lists both business functions and processes.

Table 12-34. Data Entity/Business Function and Process Matrix

Data Entity ▼	Business Function		Business Process	
	Patient Experience Management	Workforce Planning and Scheduling	Patient Feedback Routing	Staff Schedule Generation
Patient Feedback Record	X		X	
Staff Schedule		X		X

CHAPTER 12 ARCHITECTURE DEVELOPMENT METHOD

Application/Data Matrix: Shows the relationship between applications (application components) and the data entities they create, use, update, or delete. It helps identify data duplication, integration needs, and application ownership over specific data. This matrix supports decisions about data centralization, application rationalization, and system interfaces. The matrix (see Table 12-35) typically lists data entities (rows) and applications (columns), with indicators like C (Create), R (Read), U (Update), and D (Delete) to show interactions.

Table 12-35. Application/Data Matrix

Data Entity ▼	Application Component			
	Patient Feedback Application	Infection Radar Application	Planning Application	Supply Chain Management Application
Patient Feedback Record	C, R, U, D			
Infection Incident Report	R	C, R, U		
Staff Schedule			R	
Department Cost				C, R, U

Conceptual Data Diagram: The primary purpose of the Conceptual Data Diagram (see Figure 12-27) is to illustrate the relationships between critical data entities within an organization. It is developed to address the concerns of business stakeholders.

It shows logical views of the relationships between critical data entities within the organization. It is largely comparable to the *Information Map* mentioned in Section 12.5.3.

CHAPTER 12   ARCHITECTURE DEVELOPMENT METHOD

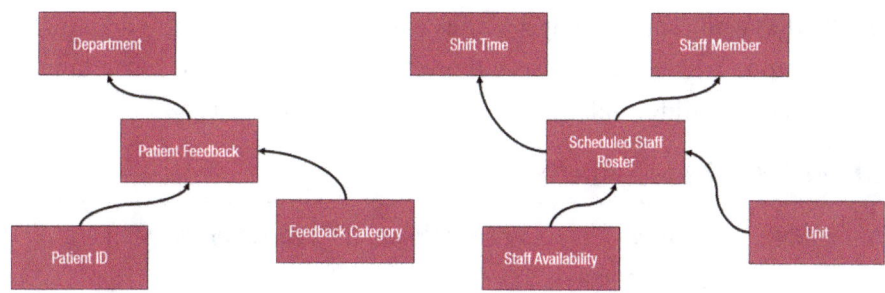

Figure 12-27. Conceptual Data Diagram

Data Dissemination Diagram: Shows the relationship between data entities, business services, and application components. The diagram (see Figure 12-28) shows how the *logical data components* are to be *physically realized by application components*. This allows effective sizing to be carried out and refined. Moreover, by assigning business value to data, an indication of the business criticality of application components can be gained.

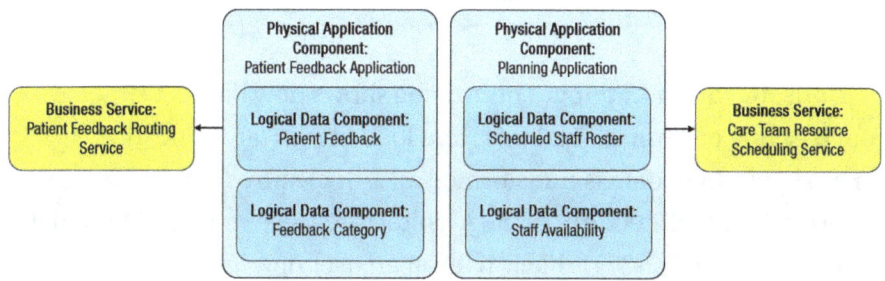

Figure 12-28. Data Dissemination Diagram

Data Security Diagram: Illustrates which data is accessed by which roles, organizational units, and applications. This information can be presented as a diagram or a catalog (see Table 12-36). The catalog displayed here maps data entities to security classifications and illustrates the access roles and protection mechanisms applied to each. Additional entities, such as business functions, business services, and locations, may be added as well. The catalog specifies which roles or systems can read, write, or update specific data sets, thereby helping to ensure compliance with privacy, confidentiality, and regulatory requirements.

Table 12-36. Data Security Catalog

Data Entity	Security Classification	Access Roles	Protection Mechanism
Patient Feedback Form	Internal Use	Patient Experience Manager	Secure web access, pseudonymization
Infection Incident Log	Confidential	Quality and Safety	RBAC, immutable audit trail, limited access via VPN
Staff Credential Database	Restricted	HR Admin, Compliance Officer	Encrypted, multi-factor access

Data Migration Catalog: This catalog shows how data is extracted from baseline or source location(s) and loaded into target location(s). It may show where data is transformed and/or cleansed on the way. In Phase C (Data Architecture) of the Architecture Development Method, the catalog (see Table 12-37) is likely to be at an overview level. Later, it may be elaborated to show which source data items map to which target data items. For illustration purposes, a Data Migration Diagram is also shown (Figure 12-29), visualizing the first row of the catalog.

CHAPTER 12　ARCHITECTURE DEVELOPMENT METHOD

Table 12-37. *Data Migration Catalog*

Source System	Target System	Data Entities Migrated	Transformation Activities
Patient Feedback Application	Patient Feedback Application v2	Patient feedback, feedback categories	Reformat text fields, reclassify codes
Manual Excel Reports	Quality Management System	Infection-related incident logs	Normalize dates, map legacy IDs
Archived Survey Data (CSV)	Analytics Data Lake	Patient satisfaction scores, free text	Clean nulls, tokenize free text
Intranet Forms Repository	Service Recovery Workflow Tool	Feedback escalation requests	Convert form fields to structured JSON

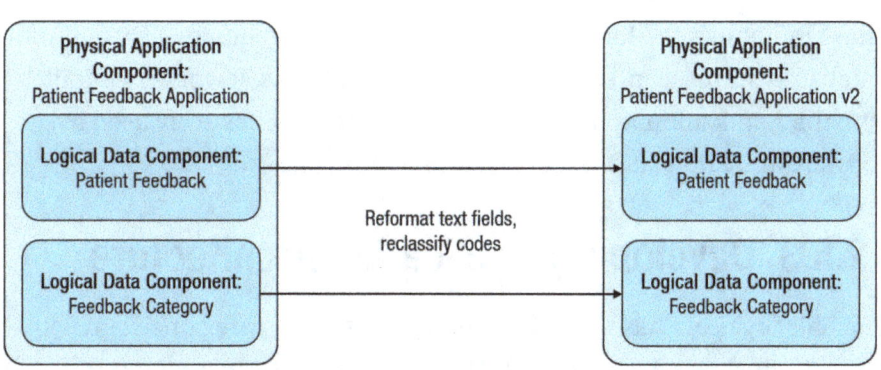

Figure 12-29. *Data Migration Diagram showing the first row of the Data Migration Catalog*

Data Lifecycle Diagram: The Data Lifecycle Diagram is an essential part of managing business data throughout its lifecycle from conception until disposal within the constraints of the business process. The diagram (see Figure 12-30) shows how a data entity progresses through various stages in its lifecycle. Typically, this includes the stages of creation, modification, use, storage, archival, and deletion.

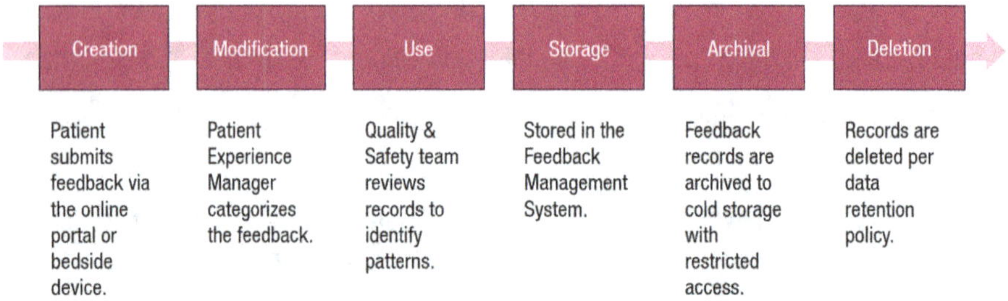

Creation	Modification	Use	Storage	Archival	Deletion
Patient submits feedback via the online portal or bedside device.	Patient Experience Manager categorizes the feedback.	Quality & Safety team reviews records to identify patterns.	Stored in the Feedback Management System.	Feedback records are archived to cold storage with restricted access.	Records are deleted per data retention policy.

Figure 12-30. *Data Lifecycle Diagram*

Phase B (Business Architecture) provided the business context and requirements for the data described in this phase. The Data Architecture Phase influences the infrastructure and technology choices, especially for databases, storage, and data management platforms. It works in parallel with the Application Architecture to ensure data is effectively used by software and applications.

12.6.3 Developing Application Architecture

The objectives of Phase C (Application Architecture) are consistent with Phase B (Business Architecture). The first objective is to *craft the Target Application Architecture* to outline how the organization should function to meet business objectives.

The second objective is to *identify potential Application Architecture components for the Architecture Roadmap*, pinpointing the gaps between Baseline and Target Application Architectures that must be bridged. The key to both of these goals is the same as Phase B: reuse existing material as much as possible.

Table 12-38. *Steps to perform Phase C (Application Architecture)*

Step to Perform	Description of Step	Output Generated by Step
Select reference models, viewpoints, and tools	Choose relevant models and viewpoints from the Architecture Repository to match the business goals and stakeholder interests.	Views and viewpoints Required building blocks (catalogs, matrices, and diagrams) Architecture requirements
Develop Baseline and Target Application Architecture description	Describe the current and future Application Architecture to a degree that supports the architecture vision, building on the Architecture Repository's resources as needed.	Baseline Application Architecture description Target Application Architecture description
Perform gap analysis	Confirm internal consistency, identify any gaps between baseline and target, and validate that models align with business objectives.	Gap analysis results
Define candidate roadmap components	Using the baseline, target, and gap analysis, build an initial roadmap to guide future actions and prioritization across the remaining architecture phases.	Application Architecture Roadmap components

(continued)

Table 12-38. (continued)

Step to Perform	Description of Step	Output Generated by Step
Resolve impacts across the Architecture Landscape	Assess the Application Architecture's wider impacts across the organization and its alignment with existing or planned projects.	Validated building blocks and other artifacts
Conduct formal stakeholder review	Revisit the architecture's initial motivation and make any necessary refinements to ensure it's fit to support other architecture domains.	Updated Statement of Architecture Work
Finalize the Data Architecture	Standardize building blocks, fully document them, and check alignment with business goals. Finalize documentation within the Architecture Repository.	Documented building blocks and requirements
Update the Architecture Definition Document	Complete all relevant sections in the Architecture Definition Document, adding models, reports, and graphics as needed to capture a clear architectural view.	Updated Architecture Definition Document

The steps defined in Table 12-38 are the same as those in Phase B (Business Architecture). Because phases B, C, and D all deal with creating specific parts of the Enterprise Architecture (Business Architecture, Data Architecture, and Application Architecture, respectively), the steps don't really differ between the three phases. The minor differences between the phases are described below. For a detailed description of all the steps, refer to Table 12-9 in Section 12.5.1.

Select Reference Models, Viewpoints, and Tools: Refer to the Architecture Repository to select Application Architecture resources, like reference models and patterns, that align with core business drivers and address stakeholder requirements and constraints. Zero in on the Application Architecture viewpoints.

Check the Architecture Repository: Start by looking into the Architecture Repository. This is where all the organization's architecture resources are stored.

Find Application Architecture Resources: Within the repository, search for materials related to Application Architecture. These might include reference models and patterns that have been used before.

Match Resources to Business Goals: Review these materials to see which ones align with the main business objectives. Choose the ones that best support what the business aims to achieve.

Consider Stakeholder Needs: Think about the requirements and limitations of stakeholders. Ensure the selected resources address their concerns and fit within any constraints they have.

Select Appropriate Viewpoints: Identify the Application Architecture viewpoints that are most relevant. These perspectives will help in understanding and communicating the architecture effectively.

Develop Baseline and Target Application Architecture Description: Develop a baseline description of the current Application Architecture, covering as much as necessary to effectively build toward the Target Application Architecture:

- Use existing architecture descriptions if available.
- Focus on creating only what's necessary to support the architecture vision.

- Use the Architecture Repository Reference Library to locate the essential building blocks within the Application Architecture.

- Explore different Target Architecture options and review them with stakeholders using the Architecture Alternatives and Trade-Offs approach.

Develop the Baseline Application Architecture: Begin by documenting the current state of the organization's Application Architecture. If existing architecture descriptions are available, utilize them to save time and ensure consistency. Focus on capturing only the details necessary to support the architecture vision, avoiding unnecessary complexity.

Identify Essential Building Blocks: Leverage the Architecture Repository Reference Library to locate the fundamental components within the Application Architecture. This resource will help to identify existing building blocks that can be reused or adapted for the Target Architecture.

Explore Target Architecture Options: Consider various options for the Target Application Architecture. Engage stakeholders in reviewing these alternatives using the Architecture Alternatives and Trade-Offs approach. This collaborative process ensures that the selected architecture aligns with business objectives and stakeholder expectations.

Perform Gap Analysis: This step validates and refines architecture models to ensure accuracy, alignment, and completeness. The models are rigorously tested against principles, objectives, and constraints to confirm they are suitable for purpose. Each architecture view is examined for consistency and conflicts, while completeness is verified against stakeholder requirements. A gap analysis is then performed to identify differences between the Baseline and Target Architecture, as outlined in Chapter 14, Section 14.5.

Validate models against core principles: Begin by thoroughly reviewing each model to ensure they adhere to the organization's fundamental principles, objectives, and constraints. This step is crucial to confirm that the models are appropriate and aligned with the overarching goals.

Ensure internal consistency: Next, examine each view within the architecture models to identify any inconsistencies or conflicts. Addressing these issues is essential for maintaining the accuracy and reliability of the models.

Confirm completeness: After resolving internal conflicts, assess the models to verify that they comprehensively meet all stakeholder requirements. This ensures that no critical elements are overlooked.

Conduct gap analysis: Finally, perform a gap analysis to detect any discrepancies between the Baseline and Target Architectures. This involves identifying areas where the current architecture falls short of meeting the target state.

Define Candidate Roadmap Components: After establishing the Baseline and Target Architectures and analyzing the gaps, it is essential to create components for the Architecture Roadmap. The results of the gap analysis lead to the creation of work packages (or initiatives) related to the architecture to be developed. These initiatives are the components to be used in the Architecture Roadmap. In this step, the Application Architecture components are added to the draft roadmap created in Phase B (Business Architecture).

Identify the Baseline and Target Architectures: Start by understanding where the architecture currently stands (Baseline) and where it needs to go (Target). This will provide a foundation for everything.

Analyze the Gaps: Figure out the differences between the Baseline and Target. What's missing? What needs improvement?

Define Initiatives: From the gap analysis, come up with initiatives that will help bridge the gaps. These are the action points for developing the architecture.

Add Application Architecture Components: Now add the Application Architecture elements to the Architecture Roadmap.

Resolve Impacts Across the Architecture Landscape: Look across the Architecture Landscape and examine other architecture assets to answer key questions, such as the following:

- Does the Application Architecture influence any established architectures, and vice versa?
- Are there chances to reuse or extend elements from this Application Architecture to benefit other areas of the organization?
- Could this Application Architecture impact current or upcoming projects, and vice versa?

Discuss the results of the questions posed with the relevant stakeholders and try to determine whether or not they should lead to adjustments to the architecture. If so, update the architecture accordingly to resolve any potential impacts.

CHAPTER 12   ARCHITECTURE DEVELOPMENT METHOD

Look at the Whole Architecture Landscape: Check if the Application Architecture impacts other parts of the business and how other architectures might influence it.

Ask Key Questions: Does this Application Architecture affect other existing architectures, or vice versa? Can we reuse parts of this Application Architecture elsewhere? How could this Application Architecture affect ongoing or future projects?

Discuss These Findings with the Relevant People: Make sure to talk through the answers with stakeholders and decide if changes are needed in the architecture.

Adjust the Architecture If Necessary: If there are any issues or improvements identified, update the architecture to address them.

Conduct Formal Stakeholder Review: It's essential to engage key stakeholders to ensure the Application Architecture Phase effectively captures and models the essence of the organization. The stakeholder perspectives ensure that the architecture truly reflects the organization's needs and goals. Gather these stakeholders to revisit the original objectives behind the architecture work and the Statement of Architecture Work. Evaluate if the proposed Application Architecture aligns with and supports the upcoming work in other architecture domains. Make refinements where necessary, based on stakeholder feedback.

Gather the Right People: Start by bringing together the key stakeholders. These are the people whose input is critical for making sure the architecture reflects the organization's real needs.

Check the Original Goals: Review the initial objectives and the Statement of Architecture Work. This will help ensure the work aligns with what was originally planned.

Align with Other Domains: Evaluate whether the proposed Application Architecture fits with and supports the upcoming work in other areas of architecture.

Refine Based on Feedback: Make changes based on the feedback from stakeholders to ensure the architecture is on track and effective.

Finalize the Application Architecture: Set clear standards for each building block, using reference models from the Architecture Repository's Reference Library wherever possible. Document each block in detail (see Chapter 7, Section 7.5.2.2).

Do a final cross-check to ensure the architecture is aligned with overarching business goals. Document the *why* behind each decision; it's vital for linking back to organizational objectives.

Set Standards for Each Building Block: Start by defining clear standards for every component. Use reference models from the Architecture Repository's Reference Library whenever possible. This provides a solid foundation to build upon.

Document Each Building Block: Record detailed information about each building block.

Align with Business Goals: Do a final check to make sure the architecture supports the overall business objectives.

Explain the Decisions: Write down the reasoning behind each architectural decision. This is crucial for connecting each decision back to the organization's goals.

CHAPTER 12   ARCHITECTURE DEVELOPMENT METHOD

Update the Architecture Definition Document: Update the Architecture Definition Document by presenting the Baseline and Target Application Architectures. Share the document with the key stakeholders to gather feedback, ensuring refinements are made to keep the architecture aligned with organizational objectives. This iterative process is essential for achieving a comprehensive and goal-oriented architecture.

Update the Architecture Definition Document: Add details about both the Baseline and Target Application Architectures.

Share with Key Stakeholders: Distribute the updated document to those who matter most in the decision-making process.

Gather Feedback: Collect input from stakeholders to ensure the document aligns with the organization's goals.

Refine the Application Architecture: Use the feedback to make necessary adjustments, ensuring continuous alignment with business objectives.

Iterate the Process: This feedback loop is key to refining the architecture and keeping it relevant as the business evolves.

Similar to the previous phase, all the architecture work that needs to be done should be documented in the *Statement of Architecture Work*.

Phase C (Application Architecture) focuses on defining a structured, cohesive, and efficient application landscape that supports business processes and objectives while considering technical, strategic, and operational requirements. The Application Architecture Phase ensures that the applications are interoperable and can exchange data effectively.

12.6.4 Phase C (Application Architecture) Artifacts

Table 12-39 lists the artifacts that can be created during the execution of Phase C (Application Architecture).

Table 12-39. *Artifacts created during Phase C (Application Architecture)*

ADM Phase	Artifacts	Used Entities
Phase C: Application Architecture	Application Portfolio Catalog	Application service, logical/physical application component
	Interface Catalog	Logical/physical application component
	Application/Organization Matrix	Logical application component, organization unit, actor, business service
	Role/Application Matrix	Logical application component, role
	Application/Function Matrix	Logical application component, function, business service
	Application Interaction Matrix	Application service, logical/physical application component
	Application Communication Diagram	Logical/physical application component, data entity
	Application and User Location Diagram	Logical/physical application component, location
	Application Use-Case Diagram	Actor, role, logical application component
	Enterprise Manageability Diagram	Logical/physical application component, logical technology component

(continued)

CHAPTER 12 ARCHITECTURE DEVELOPMENT METHOD

Table 12-39. (*continued*)

ADM Phase	Artifacts	Used Entities
	Process/Application Realization Diagram	Logical application component, process
	Software Engineering Diagram	Physical application component (composed of packages, modules, services, and operations)
	Application Migration Diagram	Logical/physical application component, staging areas (plateaus), logical technology component
	Software Distribution Diagram	Physical application component, physical technology component, location

Application Portfolio Catalog: A structured list of all the software applications used within the organization. It includes information on the purpose, lifecycle status, and how the applications support business capabilities and processes. This catalog (see Table 12-40) helps assess redundancy, alignment, interoperability, and risk in the application landscape. It supports gap analysis, rationalization, and planning in later phases.

CHAPTER 12 ARCHITECTURE DEVELOPMENT METHOD

Table 12-40. Application Portfolio Catalog

Application Name	Business Function Supported	Purpose	Lifecycle Status	Interfaces
Patient Feedback Application	Patient Experience Management	Collects and routes patient feedback	Production	EHR, CRM
Planning Application	Workforce Planning and Scheduling	Staff rostering	Development	HR System, Alert System
Workflow Management Application	Workforce Planning and Scheduling	Shift optimization	Production	HR System, Alert System
Financial Administration System	Financial Performance Management	Financial reporting	Production	ERP, BI Tool
Supply Chain Management Application	Financial Performance Management	Cost analytics and financial reporting	Development	ERP, BI Tool

Creating an Application Portfolio Catalog provides a comprehensive overview of all the applications used within an organization. It also reveals where applications with similar functionality are being used. For example, consider situations in which application components are used to run reports, take screenshots, or store and process customer data. It's more common than one might think for an organization to have several application components in use that provide virtually the same functionality.

Without a complete view of all application components, problems and challenges are likely to arise. For instance, different departments may request various applications when the desired functionality is often already available within the organization. Since the availability of the requested functionality cannot be readily verified, requests for new application components are often approved too quickly. This eventually leads to an overpopulation of the application landscape and the need for an organization to rationalize its applications, a process that can take several years.

In the absence of an Application Portfolio Catalog, a second problem that can arise is that costs are incurred that are not fully visible. For instance, an organization may discover that it is spending a lot of money on software licenses, yet the breakdown of those costs is unclear. The lack of an Application Portfolio Catalog hinders the ability to get a clear picture of license costs. Often, it is only after creating an Application Portfolio Catalog that it becomes apparent that license costs are driven by duplicate application functionality. Removing application components with duplicate (or sometimes triple or quadruple!) functionality from the application landscape can significantly reduce license costs.

The Application Portfolio Catalog provides an overview of the organization's applications. To take full advantage of the catalog, additional information should be added to it.

For example, consider adding confidentiality, integrity, and availability (CIA) information to the artifact. This provides a valuable overview of the information used by applications and the additional security measures required for compliance. When visualized as a diagram, the catalog provides a clear, comprehensible overview of the application landscape and its security-related points of attention.

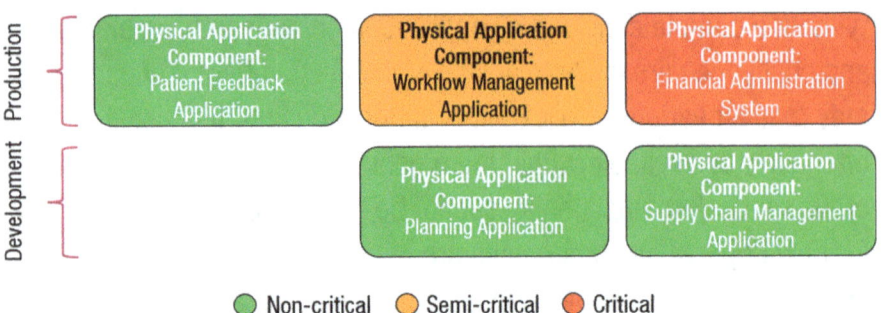

Figure 12-31. Application Portfolio Diagram with CIA score

Of course, the Application Portfolio Catalog can be enriched with various information components. For example, consider adding the location property to the catalog to indicate whether applications are on-premises or in the cloud. Another option is to add the capabilities automated by the applications to the list. This will show where work needs to be done to eliminate manual activities. Displaying that information through color views transforms the catalog into a powerful tool.

Interface Catalog: The purpose of the Interface Catalog is to scope and document the interfaces between applications to enable the overall dependencies between applications to be scoped as early as possible.

Applications will create, read, update, and delete data within other applications; this will be achieved by some kind of interface, whether via a batch file that is loaded periodically, a direct connection to another application's database, or some form of API or web service.

The mapping of one application component entity to another is an important step as it enables understanding the degree of interaction between applications. It allows an organization to understand the number and types of interfaces between applications and the degree of duplication of interfaces between applications.

CHAPTER 12   ARCHITECTURE DEVELOPMENT METHOD

The Interface Catalog displayed in Table 12-41 only shows four columns. This is done to improve readability. A complete catalog contains additional columns, including *data exchanged, protocol, frequency,* and *security.*

Table 12-41. Interface Catalog

Interface ID	Source System	Target System	Interface Type
IF-001	Patient Feedback Application	Quality Management Dashboard	REST API
IF-002	Patient Feedback Application	Nursing Admin Portal	REST API
IF-003	Nursing Admin Portal	Patient Feedback Application	REST API
IF-004	Workflow Management Application	Nursing Admin Portal	REST API
IF-005	Financial Administration System	Cost Analytics Dashboard	SQL View

Application/Organization Matrix: This matrix shows the relationship between applications and organizational units within the organization. Organizational units perform business operations. Some of these operations and services will be supported by applications. Mapping the application component to organization units provides insight into which applications are used by which organization units. This allows an organization to understand the application support requirements of the business services and processes carried out by a particular organizational unit. An example is shown in Table 12-42.

CHAPTER 12   ARCHITECTURE DEVELOPMENT METHOD

Table 12-42. *Application/Organization Matrix*

Application Component ▼	Organization Unit			
	Patient Experience Office	Quality and Safety	Clinical Operations	Finance
Patient Feedback Application	X	X	X	
Workflow Management Application			X	
Financial Administration System				X

Role/Application Matrix: The Role/Application Matrix illustrates the relationship between applications and the business roles that utilize them within an organization.

People within an organization interact with applications. During these interactions, they assume a specific role to perform a task. The matrix assigns application usage to specific roles within the organization. This matrix (see Table 12-43) is essential to any transition to role-based computing. Consequently, it is primarily used in the context of security.

Table 12-43. *Role/Application Matrix*

Role ▼	Application Component		
	Patient Feedback Application	Workflow Management Application	Financial Administration System
Patient Experience Manager	X		
Nursing Administration		X	
Financial Analyst			X

CHAPTER 12 ARCHITECTURE DEVELOPMENT METHOD

Application/Function Matrix: Demonstrates the relationship between applications and business functions within the organization. Business functions are performed by organizational units. Some of the business functions and services will be supported by applications. The Application/Function Matrix (see Table 12-44) demonstrates which applications are used by which business functions.

Table 12-44. *Application/Function Matrix*

Application Component ▼	**Business Function**		
	Patient Feedback Routing	Resource Scheduling	Cost Efficiency Analysis
Patient Feedback Application	X		
Workflow Management Application		X	
Financial Administration System			X

Application Interaction Matrix: The purpose of the Application Interaction Matrix is to illustrate the communication relationships between applications. The matrix (see Table 12-45) shows the equivalent of the Interface Catalog in the form of a map of application interactions. It is a two-dimensional table with logical or physical application components in both the rows and columns. The application service entity is displayed in the intersecting cells.

Table 12-45. Application Interaction Matrix

Application Component ▼	Application Component			
	Patient Feedback Application	Quality Management Dashboard	Workflow Management Application	Nursing Admin Portal
Patient Feedback Application		Push safety alerts		Submit complaints
Quality Management Dashboard	Read feedback data		View escalation	
Workflow Management Application		Escalate issue		Show nurse tasks
Nursing Admin Portal	Submit patient input	Report hygiene issues	View shift plan	

Application Communication Diagram: Illustrates how different applications interact by showing the flow of data between them across interfaces. It details which applications exchange information and which protocols or mechanisms are used. Optionally, it also details which business services or processes those exchanges support.

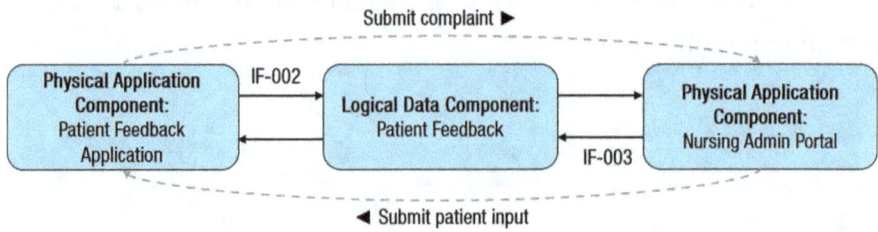

Figure 12-32. Application Communication Diagram

CHAPTER 12 ARCHITECTURE DEVELOPMENT METHOD

Application and User Location Diagram: Shows where applications are used in relation to user groups and locations. It helps visualize the distribution of application services and their geographic dependencies. This diagram clearly depicts business locations where business users typically interact with applications. It can also be used to illustrate the location where the application infrastructure is hosted.

In practice, this diagram will rarely be used. Adding the application location to the Application Portfolio Catalog eliminates the need for an Application and User Location Diagram.

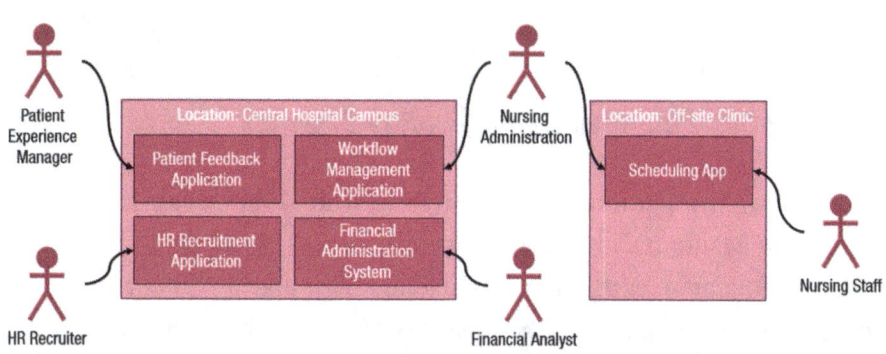

Figure 12-33. Application and User Location Diagram

Application Use-Case Diagram: Illustrates how business users (actors) interact with application components through specific application services to support business processes or capabilities. It describes who uses what, how, and for what purpose at the application layer. The Application and User Location Diagram can be expanded to include the application services used by users (actors) and applications.

Enterprise Manageability Diagram: This diagram is essentially a filter of the Application Communication Diagram. It illustrates how the organization's applications will be monitored, controlled, and supported

437

to ensure reliable operation in alignment with business needs. The diagram identifies management services, such as monitoring, alerting, and logging. It defines the relationships between system elements and their manageability controls and supports the operational governance of the architecture. Analysis can reveal duplication, gaps, and opportunities in an organization's IT service management operation.

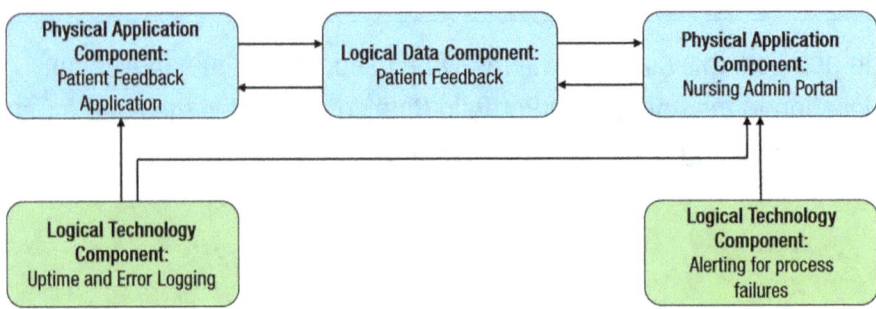

Figure 12-34. Enterprise Manageability Diagram

Process/Application Realization Diagram: The Process/Application Realization Diagram shows how a business process is supported by one or more application components. The diagram maps the execution of a process to the applications that enable or automate it, highlighting integration points and data flows.

The Process/Application Realization Diagram enhances the Application Communication Diagram by adding any sequencing constraints. It can also pinpoint process efficiency improvements that could reduce interaction traffic between applications.

CHAPTER 12   ARCHITECTURE DEVELOPMENT METHOD

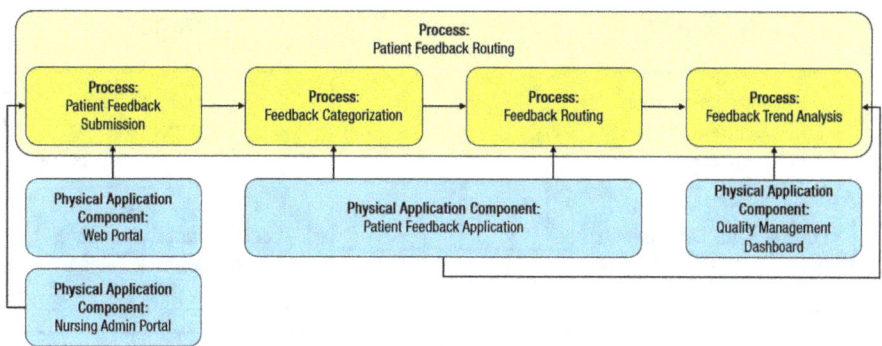

Figure 12-35. Process/Application Realization Diagram

Software Engineering Diagram: From a development perspective, it breaks applications into packages, modules, services, and operations. It enables more detailed impact analysis when planning migration stages and analyzing opportunities and solutions. It is ideal for application development and management teams when working with complex environments.

Since none of the three core architecture roles (see Chapter 10, Section 10.2) creates this artifact, an example is omitted here.

Application Migration Diagram: Identifies application migration from baseline to target application components. It enables a more accurate estimation of migration costs by showing precisely which applications and interfaces need to be mapped between migration stages.

It would identify temporary applications, staging areas, and the infrastructure required to support migrations (e.g., parallel run environments and others).

CHAPTER 12 ARCHITECTURE DEVELOPMENT METHOD

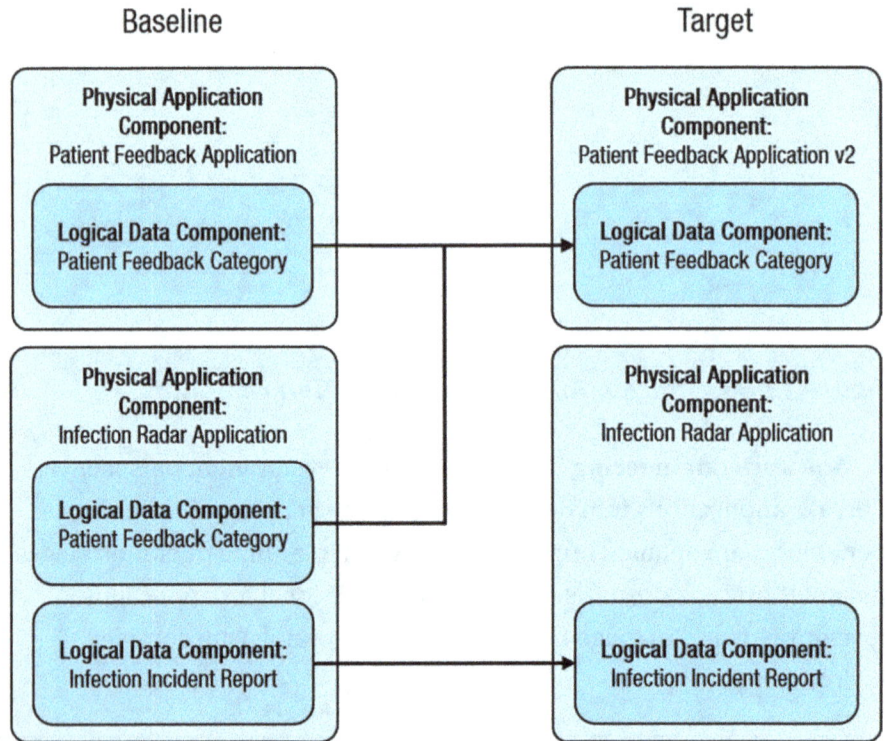

Figure 12-36. *Application Migration Diagram*

Software Distribution Diagram: This diagram illustrates the distribution of physical applications across physical technology and its location. It is useful for system upgrades or application consolidation projects.

This diagram largely follows the Application and User Location Diagram. A similar result can be achieved by adding application locations to the Application Portfolio Catalog. This eliminates the need for a Software Distribution Diagram.

CHAPTER 12 ARCHITECTURE DEVELOPMENT METHOD

Phase C (Information Systems Architectures) builds on the understanding of the business goals and objectives established in Phase B (Business Architecture).

Once the information systems architecture is defined, it serves as an input to the next phase (Phase D (Technology Architecture)), which focuses on the underlying technology infrastructure required to support the applications and data.

Any gaps identified in Phase C will drive planning and decisions (see Section 12.8) about how to implement or update systems in later phases.

In summary, the Information Systems Architectures Phase is essential for defining the structure of data and applications that enable the organization's Business Architecture. It helps bridge the gap between the business strategy and the technology infrastructure by specifying the systems that manage data and processes critical to business functions.

12.7. Phase D: Technology Architecture

Phase D (Technology Architecture) focuses on defining the technology infrastructure necessary to support the Business, Data, and Application Architectures defined in the previous phases of the Architecture Development Method. This phase, which may follow the Application Architecture (Phase C), is geared toward determining the most suitable technology components, platforms, and services needed to enable efficient operation of the organization's applications. In Phase D, architects identify hardware, software, and network infrastructure requirements and determine technology standards to ensure consistency, scalability, and security. The goal is to create a well-defined Technology Architecture that aligns with the broader organizational goals while ensuring interoperability and future readiness.

12.7.1 Developing Technology Architecture

The objectives of Phase D (Technology Architecture) – as well as those of Phase C (Data and Application Architectures) – are consistent with Phase B (Business Architecture). The first objective is to *craft the Target Technology Architecture* to outline how the organization should function to meet business objectives.

This seems like a logical first step, but in nine out of ten organizations, IT is viewed as a separate entity with its own planning and strategy. The idea behind the TOGAF Standard is to link the architecture work that takes place within each architecture domain to the business goals.

The Technology Architecture defines the IT infrastructure, systems, and services required to run the organization's applications and meet the business needs.

The second objective is to *identify potential Technology Architecture components for the Architecture Roadmap,* locating the gaps between Baseline and Target Technology Architectures that need to be addressed. It is important to reuse existing material as much as possible.

CHAPTER 12　ARCHITECTURE DEVELOPMENT METHOD

Table 12-46. *Steps to perform Phase D (Technology Architecture)*

Step to Perform	Description of Step	Output Generated by Step
Select reference models, viewpoints, and tools	Choose relevant models and viewpoints from the Architecture Repository to match the business goals and stakeholder interests.	Views and viewpoints Required building blocks (catalogs, matrices, and diagrams) Architecture requirements
Develop Baseline and Target Technology Architecture description	Describe the current and future Technology Architecture to a degree that supports the architecture vision, building on the Architecture Repository's resources as needed.	Baseline Technology Architecture description Target Technology Architecture description
Perform gap analysis	Confirm internal consistency, identify any gaps between baseline and target, and validate that models align with business objectives.	Gap analysis results
Define candidate roadmap components	Using the baseline, target, and gap analysis, build an initial roadmap to guide future actions and prioritization across the remaining architecture phases.	Technology Architecture Roadmap components

(*continued*)

CHAPTER 12 ARCHITECTURE DEVELOPMENT METHOD

Table 12-46. (*continued*)

Step to Perform	Description of Step	Output Generated by Step
Resolve impacts across the Architecture Landscape	Assess the Technology Architecture's wider impacts across the organization and its alignment with existing or planned projects.	Validated building blocks and other artifacts
Conduct formal stakeholder review	Revisit the architecture's initial motivation and make any necessary refinements to ensure it's fit to support other architecture domains.	Updated Statement of Architecture Work
Finalize the Technology Architecture	Standardize building blocks, fully document them, and check alignment with business goals. Finalize documentation within the Architecture Repository.	Documented building blocks and requirements
Update the Architecture Definition Document	Complete all relevant sections in the Architecture Definition Document, adding models, reports, and graphics as needed to capture a clear architectural view.	Updated Architecture Definition Document

The steps defined in Table 12-46 are the same as those in Phase B (Business Architecture). Phases B, C, and D all deal with creating Baseline and Target Architectures. Therefore, the steps don't really differ between the three phases. The minor differences between the phases are described below. For a detailed description of all the steps, refer to Table 12-9 in Section 12.5.1.

Select Reference Models, Viewpoints, and Tools: Refer to the Architecture Repository to select Technology Architecture resources, like reference models and patterns, that align with core business drivers and address stakeholder requirements and constraints. Focus on the Technology Architecture viewpoints.

Explore the Architecture Repository: Begin by examining the Architecture Repository. This is the go-to resource for Technology Architecture materials, including reference models and patterns.

Align with Business Drivers: Identify the core business drivers of the organization. Understanding these will help choose resources that support and enhance these key objectives.

Address Stakeholder Needs: Consider the requirements and constraints of the stakeholders. Selecting resources that meet their expectations ensures the architecture will be well-received and effective.

Focus on Technology Architecture Viewpoints: Pay special attention to viewpoints specific to Technology Architecture. These perspectives will help to understand and address the technical aspects of the architecture.

Develop Baseline and Target Technology Architecture Description: Develop a baseline description of the current Technology Architecture, covering as much as necessary to effectively build toward the Target Technology Architecture:

- Use existing architecture descriptions if available.
- Focus on creating only what's necessary to support the architecture vision.
- Use the Architecture Repository Reference Library to locate the essential building blocks within the Technology Architecture.

- Explore different Target Architecture options and review them with stakeholders using the Architecture Alternatives and Trade-Offs approach.

Understand the Baseline Technology Architecture: Gather existing documentation and resources that detail the organization's current technology infrastructure. Focus on collecting only the information necessary to support the architecture vision, avoiding unnecessary details.

Identify Essential Building Blocks: Utilize the Architecture Repository Reference Library to locate and understand the fundamental components within the current Technology Architecture.

Explore Potential Target Architectures: Investigate various options for the organization's Target Technology Architecture. Consider different technologies, structures, and strategies that could be implemented.

Evaluate and Compare Options with Stakeholders: Engage with key stakeholders to review the potential Target Architectures. Use the Architecture Alternatives and Trade-Offs approach to assess the advantages and disadvantages of each option.

Select the Most Suitable Target Architecture: Based on stakeholder feedback and thorough evaluation, choose the Target Technology Architecture that best aligns with the organization's goals and needs.

Document the Baseline and Target Architectures: Clearly document both the Baseline and Target Technology Architectures. Ensure that this documentation is accessible to all relevant parties within the organization.

Perform Gap Analysis: The step focuses on ensuring that architecture models are accurate, consistent, and aligned with objectives and stakeholder needs. Each view is reviewed to resolve conflicts,

CHAPTER 12   ARCHITECTURE DEVELOPMENT METHOD

completeness is verified, and a gap analysis is carried out to highlight differences between the Baseline and Target Architecture, as outlined in Chapter 14, Section 14.5.

Validate models against core principles: Begin by thoroughly reviewing each model to ensure they adhere to the organization's fundamental principles, objectives, and constraints. This step is crucial to confirm that the models are appropriate and aligned with the overarching goals.

Ensure internal consistency: Next, examine each view within the architecture models to identify any inconsistencies or conflicts. Addressing these issues is essential for maintaining the accuracy and reliability of the models.

Confirm completeness: After resolving internal conflicts, assess the models to verify that they comprehensively meet all stakeholder requirements. This ensures that no critical elements are overlooked.

Conduct gap analysis: Finally, perform a gap analysis to detect any discrepancies between the Baseline and Target Architectures. This involves identifying areas where the current architecture falls short of meeting the target state.

Define Candidate Roadmap Components: After establishing the Baseline and Target Architectures and analyzing the gaps, create components for the Architecture Roadmap. Use the results of the gap analysis to create work packages (initiatives) related to the architecture to be developed. These initiatives are to be used in the Architecture Roadmap. The Technology Architecture components are added to the draft roadmap created in Phase B (Business Architecture).

Review Baseline and Target Architectures: Begin by thoroughly understanding both the Baseline and Target Architectures. This comprehension sets the foundation for identifying necessary changes.

Conduct a Gap Analysis: Examine the differences between the Baseline and Target Architectures. This analysis will highlight the gaps that need to be addressed to achieve the desired Target Architecture.

Identify Required Work Packages: Based on the gap analysis, determine specific initiatives or projects – referred to as work packages – that will bridge these gaps. These work packages represent the actionable steps needed for development.

Develop the Architecture Roadmap: Organize and sequence the identified work packages into a coherent Architecture Roadmap. This roadmap serves as a strategic plan, outlining the path from the Baseline Architecture to the Target Architecture.

Incorporate Technology Components: Integrate relevant Technology Architecture components into the Draft Architecture Roadmap. This integration ensures that technological considerations are aligned with business objectives and are part of the overall transition strategy.

Resolve Impacts Across the Architecture Landscape: Look across the technology landscape and examine other architecture assets to answer key questions, such as the following:

- Does the Technology Architecture influence any established architectures, and vice versa?

- Are there chances to reuse or extend elements from this Technology Architecture to benefit other areas of the organization?

- Could this Technology Architecture impact current or upcoming projects, and vice versa?

CHAPTER 12   ARCHITECTURE DEVELOPMENT METHOD

Discuss the results with the relevant stakeholders and determine whether or not they should lead to adjustments to the architecture. If so, update the architecture accordingly.

Review the Technology Landscape: Begin by examining the current Technology Architecture. Identify its components, integrations, and dependencies within the organization.

Identify Interdependencies: Determine if the Technology Architecture affects existing architectures or if established architectures influence the Technology Architecture.

Spot Reuse Opportunities: Look for elements within the Technology Architecture that can be reused or extended to benefit other areas of the organization.

Assess Project Impacts: Evaluate whether the Technology Architecture impacts current or upcoming projects, and vice versa.

Engage Stakeholders: Discuss the findings with relevant stakeholders to understand their perspectives and gather additional insights.

Decide on Adjustments: Collaborate with stakeholders to determine if the identified impacts warrant changes to the architecture.

Update the Technology Architecture: If adjustments are necessary, revise the architecture to reflect the agreed-upon changes, ensuring alignment with organizational objectives.

Conduct Formal Stakeholder Review: It's essential to engage key stakeholders to ensure the Technology Architecture Phase effectively captures and models the essence of the organization. The stakeholder perspectives ensure that the architecture truly reflects the organization's needs and goals. Gather these stakeholders to revisit the original objectives behind the architecture work and the Statement of Architecture Work. Evaluate if the proposed Technology Architecture aligns with and supports the upcoming work in other architecture domains. Make refinements where necessary, based on stakeholder feedback.

Schedule a Review Meeting: Coordinate a time and place that accommodates all identified stakeholders. Ensure the meeting is set up well in advance to allow for maximum participation.

Revisit Original Objectives: At the beginning of the meeting, remind everyone of the initial goals behind the architecture project and review the Statement of Architecture Work. This sets a clear context for the discussion.

Present the Proposed Technology Architecture: Share the current state of the Technology Architecture with the stakeholders. Use diagrams, documents, and presentations to clearly convey the information.

Collect Feedback: Encourage stakeholders to share their thoughts on the architecture. Focus on whether it aligns with the organization's objectives and supports future work in other architectural areas.

Analyze Feedback: After the meeting, review all input received. Identify common themes, concerns, and suggestions that emerged during the discussion.

Refine the Technology Architecture: Make necessary adjustments to the Technology Architecture based on the feedback. Ensure that these changes enhance alignment with organizational goals and support other architectural domains.

Document Changes: Clearly record all modifications made to the architecture. This documentation will be valuable for future reference and for maintaining transparency.

Communicate Updates: Share the revised Technology Architecture with all stakeholders. Highlight the changes made and explain how stakeholder feedback was incorporated.

Seek Final Approval: Request confirmation from stakeholders that the revised architecture meets the organization's needs and is ready for implementation.

Finalize the Technology Architecture: Set clear standards for each building block, using reference models from the Architecture Repository's Reference Library wherever possible. Document each block in detail (see Chapter 7, Section 7.5.2.2).

Do a final cross-check to ensure the architecture is aligned with overarching business goals. Document the *why* behind each decision; it's vital for linking back to organizational objectives.

Establish Clear Standards for Each Component: Define explicit standards for every building block within the Technology Architecture. Utilize reference models from the Architecture Repository's Reference Library to ensure consistency and leverage established best practices.

Document Each Component Thoroughly: For every building block, provide detailed documentation. This should include its purpose, functionality, relationships, and any other pertinent information.

Ensure Alignment with Business Goals: Conduct a comprehensive review to verify that the Technology Architecture supports and advances the overarching business objectives. This alignment is crucial for the architecture's effectiveness and relevance.

Record the Rationale Behind Decisions: Document the reasoning for each architectural decision made. This practice is vital for maintaining traceability and demonstrating how each choice contributes to achieving organizational goals.

Update the Architecture Definition Document: Update the Architecture Definition Document by presenting the Baseline and Target Technology Architectures. Share the document with the key stakeholders to gather feedback, ensuring refinements are made to keep the architecture aligned with organizational objectives. This iterative process is essential for achieving a comprehensive and goal-oriented architecture.

Share with Key Stakeholders: Distribute the updated document to relevant stakeholders, including business leaders, IT teams, and other pertinent parties, to gather their insights and feedback.

Incorporate Feedback for Alignment: Revise the document based on the feedback received to ensure the architecture remains in sync with organizational objectives and addresses any concerns raised.

Iterate for Comprehensive Technology Architecture: Repeat the process as necessary, refining the architecture to achieve a thorough and goal-oriented framework that supports the organization's mission and vision.

Similar to the previous phase, all the architecture work that needs to be done should be documented in the *Statement of Architecture Work*.

Building on the information gathered in the previous phases, the Technology Architecture Phase focuses on defining the infrastructure and technology required to support the information systems. Architects consider factors such as hardware, software, networking, security, and integration requirements.

In summary, the Technology Architecture Phase is where the organization defines the technical building blocks and infrastructure necessary to support its business, application, and data needs. It serves as a bridge between the conceptual designs created in earlier phases and the actual implementation of technology solutions.

12.7.2 Phase D (Technology Architecture) Artifacts

Table 12-47 lists the artifacts that can be created during the execution of Phase D (Technology Architecture).

Table 12-47. Artifacts created during Phase D (Technology Architecture)

ADM Phase	Artifacts	Used Entities
Phase D: Technology Architecture	Technology Standards Catalog	Technology service, logical/physical technology component
	Technology Portfolio Catalog	Technology service, logical/physical technology component
	Application/Technology Matrix	Logical/physical application component, application service, logical/physical technology component
	Environments and Locations Diagram	Logical/physical technology component, logical/physical application component, location, actor
	Platform Decomposition Diagram	Logical/physical technology component
	Processing Diagram	Logical/physical technology component, logical/physical application component, actor, location
	Networked Computing/Hardware Diagram	Logical/physical application component, logical/physical technology component, function
	Network and Communications Diagram	Logical/physical technology component

Technology Standards Catalog: The catalog outlines the approved technologies, products, and standards that support infrastructure, platforms, and networks used for delivering solutions. It helps ensure interoperability, maintainability, and compliance across systems. The catalog (see Table 12-48) provides a snapshot of the organization's standard technologies that can be or are deployed.

Table 12-48. Technology Standards Catalog

Standard Name	Technology Type	Version	Status	Comments
Red Hat Enterprise Linux	OS/Server Platform	10	Approved	Standard for clinical server deployments
PostgreSQL	Database Management System	17.5	Approved	Preferred for analytics and cost reporting modules
HL7 FHIR	Interoperability API	R4 R6	Approved Under Review	Required for patient data exchange with EHR systems

Technology Portfolio Catalog: This catalog identifies and maintains a list of all technology in use across the organization. This includes hardware, infrastructure software, and application software. An agreed-upon technology portfolio supports the lifecycle management of technology products and versions, as well as the definition of technology standards.

The Technology Portfolio Catalog (see Table 12-49) establishes the foundation for the remaining matrices and diagrams. It is typically the starting point of the Technology Architecture Phase. Technology registries and repositories (such as a CMDB application) provide input for this catalog from baseline and target perspectives.

Table 12-49. *Technology Portfolio Catalog*

Technology Component	Technology Type	Version	Lifecycle Status
Epic EHR System	Application Platform	2025	Active
ServiceNow Feedback Portal	Web Platform	Zurich Q4 2025	Active
Power BI Reporting	Analytics Tool	2.143	Active

The Technology Portfolio Catalog can be expanded to include additional columns, such as the organizational unit using the technology and the specific owner (business role) of the technology component.

Application/Technology Matrix: Shows the relationships between applications and the technology platforms (infrastructure and middleware) that support them. It helps identify technical dependencies, opportunities for reuse, and potential risks due to outdated or unsupported technologies. It should be aligned with and complement one or more Platform Decomposition Diagrams. An example of an Application/Technology Matrix is shown in Table 12-50.

Table 12-50. *Application/Technology Matrix*

| Application Component ▼ | Technology Component | | | | |
	Epic EHR	Web Platform	Azure Cloud	PostgreSQL	Power BI
Patient Feedback Application		X	X	X	
Workflow Management Application	X				X
Planning Application		X		X	
Supply Chain Management Application			X		X

Environments and Locations Diagram: This diagram depicts the *physical* and *virtual* locations in which application and technology components operate. It helps architects understand deployment constraints, operational risks, disaster recovery requirements, and regional dependencies. An example is shown in Table 12-51.

Table 12-51. Environments and Locations Diagram

Location/Environment	Type	Purpose
Central Hospital Campus	Physical Site	Inpatient care, surgery, admin HQ
Outpatient Clinic	Physical Site	Ambulatory care and diagnostics
Data Center – Region A	Physical Site	Hosts EHR
Telehealth Operations Hub	Virtual Environment	Remote consultations and triage
Azure Cloud Environment	Virtual Environment	Hosts patient feedback platform and analytics dashboards

The Environments and Locations Diagram is essentially a filter of the Location Catalog that was created during Phase B (see Table 12-16 in Section 12.5.3).

Platform Decomposition Diagram: This artifact illustrates the structure of technology platforms and their interrelationships. The diagram (see Figure 12-37) provides an overview of the organization's technology platform and covers all aspects of the infrastructure platform. It typically includes servers, operating systems, networks, and databases *grouped into physical and logical platforms*. The diagram can be expanded to map the technology platform to the appropriate application components.

CHAPTER 12　ARCHITECTURE DEVELOPMENT METHOD

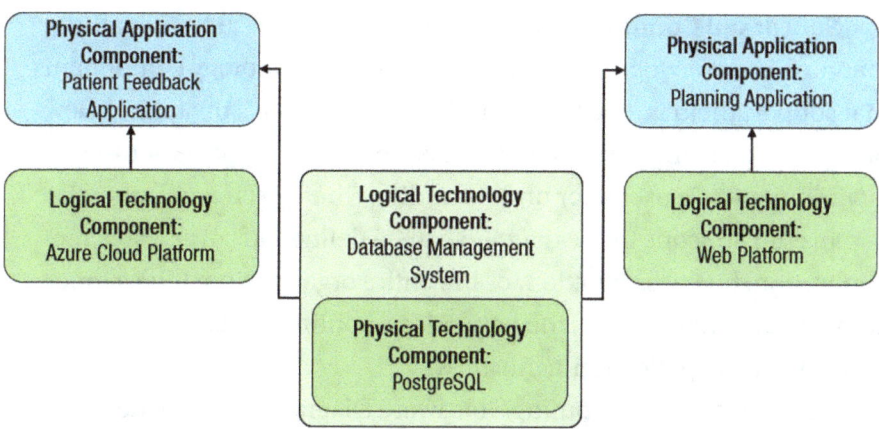

Figure 12-37. Platform Decomposition Diagram

Processing Diagram: The diagram illustrates how components are organized and deployed on a technology platform. These groups are called deployment units and typically represent business capabilities, services, applications, or technology components. The diagram helps identify which components should be packaged together and how the deployment units connect and interact with each other.

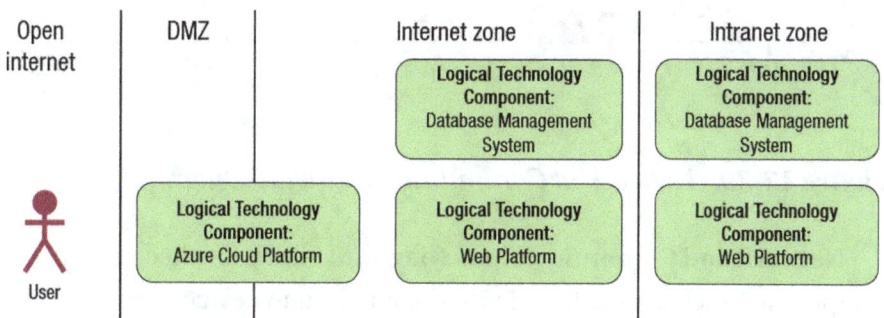

Figure 12-38. Processing Diagram

457

CHAPTER 12　ARCHITECTURE DEVELOPMENT METHOD

Networked Computing/Hardware Diagram: Documenting the mapping between logical applications and the technology components that support them is essential. This diagram illustrates the logical view of application components within a distributed network computing environment. It is useful for understanding where each application is deployed. The scope of the diagram can be defined to cover a specific application, business function, or the entire organization. If developed at the organizational level, the network computing landscape can be depicted in an application-agnostic way.

The Networked Computing/Hardware Diagram is a more detailed version of the Processing Diagram. The logical technology components shown in the Processing Diagram (see Figure 12-38) have been replaced by logical application components (as shown in Figure 12-39).

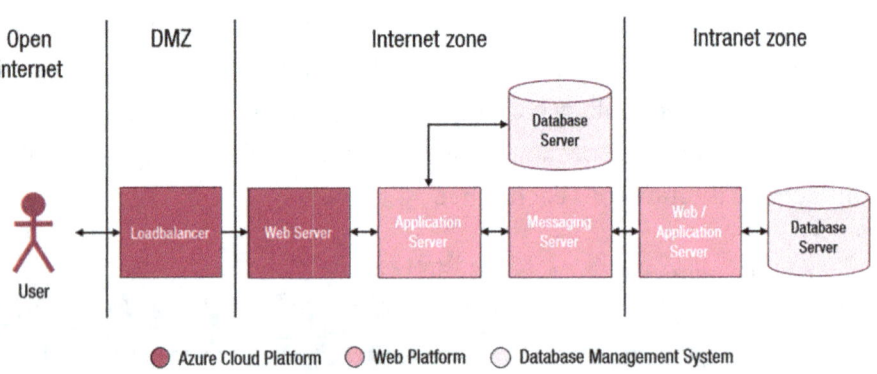

Figure 12-39. Networked Computing/Hardware Diagram

Network and Communications Diagram: Shows how technology components such as servers, clients, databases, and devices are connected across networks, emphasizing infrastructure, boundaries, and communication paths, not data formats. An example is shown in Figure 12-40.

CHAPTER 12 ARCHITECTURE DEVELOPMENT METHOD

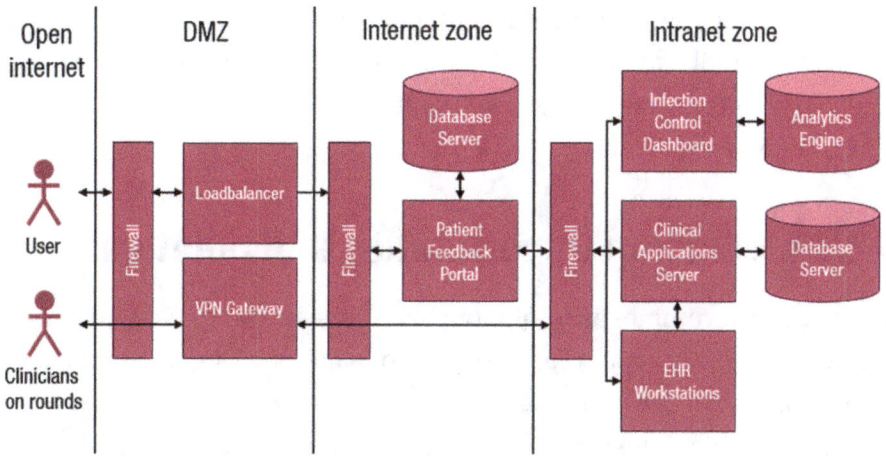

Figure 12-40. *Network and Communications Diagram*

Since the preceding phase, Phase C (Information Systems Architectures) (consisting of Data and Application Architectures), defines the applications that need technology support, the Technology Architecture must align with these architectures. Furthermore, the Technology Architecture serves as a foundation for Phase E (Opportunities and Solutions), where the focus is on identifying and prioritizing the implementation of architectural solutions.

12.8. Phase E: Opportunities and Solutions

The fifth alphabetized phase of the Architecture Development Method focuses on identifying and prioritizing the potential solutions that can deliver the capabilities required to achieve the Target Architecture. Phase E (Opportunities and Solutions) is primarily intended to identify projects or programs (in the form of work packages or initiatives) that will ultimately implement the Target Architecture. It is used to determine the most effective ways to realize the architecture vision and deliver business value.

Phase E focuses on evaluating potential implementation approaches – such as buy, build, or reuse – and assessing dependencies, opportunities, and constraints to create a feasible Architecture Roadmap (detailed in 12.8.3) for implementation.

12.8.1 Defining Opportunities and Solutions

Phase E (Opportunities and Solutions) has three key objectives. The first is to *identify the work packages* required to realize the Target Architecture. The second objective is to *determine whether an incremental approach is required* to achieve the first objective. The third and final key objective is to *define the building blocks* needed to complete the Target Architecture.

Input from the previous phases of the Architecture Development Method is required to identify the work packages (or initiatives) that will lead to the implementation of the Target Architecture. Phase E creates a first complete version of the Architecture Roadmap based on the gap analysis and candidate Architecture Roadmap components from phases B, C, and D.

During the execution of Phase E, the architect determines whether an incremental approach is required to reach the Target Architecture. If so, one or more Transition Architectures must be identified that will deliver ongoing business value.

The third objective of Phase E is to define the overall Solution Building Blocks (see Chapter 7, Section 7.5.2.5) to complete the Target Architecture based on the Architecture Building Blocks (see Chapter 7, Section 7.5.2.4). During this phase, the Architecture Building Blocks are further detailed so that clearly specified Solution Building Blocks can be defined. Table 12-52 lists the steps to perform Phase E (Opportunities and Solutions).

Table 12-52. *Steps to perform Phase E (Opportunities and Solutions)*

Step to Perform	Description of Step	Output Generated by Step
Determine/confirm key corporate change attributes	Align the architecture with an organization's business culture. This involves creating an Implementation Factor Catalog and assessing the transition capabilities of organizational units.	Implementation Factor Catalog
Determine business constraints for implementation	Identify business drivers that could compromise the Target Architecture. If there are any constraints, try to resolve them.	
Review and consolidate gap analysis results from phases B–D	Consolidate and integrate gap analysis results from the Business, Information Systems, and Technology Architectures.	Consolidated Gaps, Solutions, and Dependencies Matrix
Consolidate and reconcile interoperability requirements	Identify and address constraints that may hinder interoperability.	
Refine and validate dependencies	Analyze dependencies by grouping related activities.	Work Package Portfolio Map
Confirm readiness and risk for business transformation	Review the Business Transformation Readiness Assessment to inform the Architecture Roadmap and implementation strategy.	Updated Consolidated Gaps, Solutions, and Dependencies Matrix

(*continued*)

CHAPTER 12   ARCHITECTURE DEVELOPMENT METHOD

Table 12-52. (*continued*)

Step to Perform	Description of Step	Output Generated by Step
Formulate Implementation and Migration Strategy	Develop an Implementation and Migration Strategy by selecting a strategic approach and applying methodologies to effectively implement the Target Architecture while mitigating identified risks.	Implementation and Migration Strategy
Identify and group major work packages	Assess and address gaps in business capabilities in order to enable the organization to effectively plan and implement necessary changes aligned with strategic objectives.	Work Package Portfolio Map
Identify Transition Architecture(s)	Identify Transition Architecture(s) to provide structured, value-driven steps toward achieving the Target Architecture, tailored to the organization's implementation strategy and capacity for change.	Required building blocks (catalogs, matrices, and diagrams) Architecture Definition Increments Table Transition Architecture(s)
Create Architecture Roadmap and Implementation and Migration Plan	Create a structured approach to transition from Baseline to Target Architectures through a detailed Architecture Roadmap and Implementation and Migration Plan.	Architecture Roadmap Implementation and Migration Plan

CHAPTER 12 ARCHITECTURE DEVELOPMENT METHOD

Determine/Confirm Key Corporate Change Attributes:
Implementing Enterprise Architecture effectively requires aligning it with the organization's unique business culture. A crucial step in this process is developing an *Implementation Factor Catalog*. This catalog serves as a repository for decisions related to architecture implementation and migration. Additionally, it's essential to assess the transition capabilities of the involved organizational units, considering their culture and abilities, as well as evaluating the organization's overall culture and skill sets.

Documenting the outcomes of these assessments in the Implementation Factor Catalog is vital. For organizations with a well-established Enterprise Architecture practice, this step may be straightforward. However, establishing the catalog is necessary to maintain a record of decisions made, ensuring a structured and adaptable approach to architecture implementation.

The Implementation Factor Catalog is an invaluable tool for documenting elements that influence the architecture's Implementation and Migration Plan. This catalog should encompass a comprehensive list of factors, detailed descriptions, and the resulting actions or constraints to consider during planning.

Typically, these factors include risks, issues, and assumptions. It is also advisable to include dependencies, actions, and impacts. By meticulously documenting these elements, organizations can ensure a more structured and effective approach to implementing and migrating their architecture.

One could argue that the Implementation Factor Catalog and the Consolidated Gaps, Solutions, and Dependencies Matrix are artifacts because they are a catalog and a matrix.

Although catalogs, diagrams, maps, and matrices are generally referred to as artifacts, they must contain architectural entities to be called an artifact (see Chapter 7, Section 7.4).

Since both the Implementation Factor Catalog and the Consolidated Gaps, Solutions, and Dependencies Matrix do not contain entities, they are not actual artifacts.

Evaluate Company-Wide Culture and Skills: Don't forget to look at the organization as a whole. You'll need to assess the overall culture and the skills available in the company to support the architecture implementation.

Create an Implementation Factor Catalog: Develop an Implementation Factor Catalog. This is where the elements that influence the architecture's Implementation and Migration Plan (like culture and available skills) are documented. It acts as a roadmap for the whole process. Assess the Implementation Factor Catalog.

Document Everything: Record the results of these assessments. Keeping detailed notes is critical. It ensures that things stay on track and can be easily referred back to previous decisions.

Stay Adaptable: Understand that if the organization is new to Enterprise Architecture, creating this catalog might feel like a big step. But it's necessary to keep everything organized and adaptable for future changes.

Determine Business Constraints for Implementation: It is important to identify any business drivers that could compromise the implementation of the Target Architecture. If there are any constraints, try to resolve them by discussing them with the stakeholders. Work out the best possible solution by acting as a mediator, utilizing the *Architecture Alternatives and Trade-Offs technique*, as outlined in Chapter 14, Section 14.9.

Identify Potential Constraints: Start by understanding the business drivers that might limit how to implement the Target Architecture. Think about anything that could cause challenges.

Engage Stakeholders: Have conversations with key stakeholders to see if any constraints can be resolved. Clear communication is key here.

Collaborate on Solutions: Act as a mediator, bringing stakeholders together to find the best solutions. Use techniques like the Architecture Alternatives and Trade-Offs method to guide the decision-making.

Review and Consolidate Gap Analysis Results from Phases B-D:
A crucial aspect of this process is the consolidation and integration of gap analysis results from the Business, Information Systems, and Technology Architectures. By assessing these gaps, organizations can identify potential solutions and understand interdependencies, facilitating the creation of a *Consolidated Gaps, Solutions, and Dependencies Matrix* (see Section 12.8.2). This matrix is instrumental in pinpointing Solution Building Blocks that can address multiple gaps and their corresponding Architecture Building Blocks.

The development of Solution Building Blocks is informed by candidate roadmap components identified in phases B, C, and D.

To effectively consolidate the gap analysis results from phases B, C, and D, it's advisable to compile them into a single, comprehensive list. This list should include the *identified gaps* alongside *potential solutions* and their *dependencies*. A recommended approach for determining these dependencies involves utilizing a set of views, such as the Business Interaction Matrix, the Data Entity/Business Function Matrix, and the Application/Function Matrix. These tools help in relating elements across different architecture domains, ensuring a holistic perspective.

Once the Consolidated Gaps, Solutions, and Dependencies Matrix is documented, the next step is to rationalize it by reorganizing the gap list and grouping similar items together. During this process, it's essential to refer to the Implementation Factor Catalog and review the implementation factors. Any additional factors identified should be incorporated into the Implementation Factor Catalog, ensuring a comprehensive and up-to-date reference.

Collect Gap Results: Gather the gap analysis findings from the Business, Information Systems, and Technology Architectures (phases B, C, and D).

Combine and Organize Gaps: Create a single, easy-to-read list that includes the gaps, possible solutions, and any dependencies between them.

Create a Gap–Solution Matrix: Use a Consolidated Gaps, Solutions, and Dependencies Matrix to show the gaps, solutions, and dependencies clearly.

Identify Building Blocks: Use the matrix to figure out which Solution Building Blocks can solve multiple gaps.

Check Dependencies: Use specific views (such as the Business Interaction Matrix and the Data Entity/Business Function Matrix) to map out dependencies. This will provide a complete view of how everything connects.

Build Architecture Roadmap: Use the solutions identified to create an Architecture Roadmap for addressing gaps, making sure to align everything with the broader architecture.

Consolidate and Reconcile Interoperability Requirements: Review the architecture vision, Target Architectures, Implementation Factor Catalog, and the Consolidated Gaps, Solutions, and Dependencies Matrix, to identify and address constraints that may hinder interoperability. This proactive approach minimizes conflicts arising from reused Solution

Building Blocks, commercial off-the-shelf products, and third-party service providers. When conflicts are identified, two primary strategies can be employed: either developing a building block that transforms or translates between conflicting components or modifying the specifications of the conflicting building blocks themselves. This ensures a cohesive and efficient architectural framework that aligns with organizational objectives.

Understand the Requirements: Start by looking over key documents like the architecture vision, Target Architectures, the Implementation Factor Catalog, and the Consolidated Gaps, Solutions, and Dependencies Matrix. These will help identify any potential issues or gaps that might make things hard to integrate smoothly.

Identify Problems: Check for anything that could cause problems between different systems, such as reused solutions, third-party tools, or products.

Find Solutions: If issues are found, do one of two things:

Create a solution to convert or link conflicting systems.

Change the systems or their specs to make them work better together.
Ensure Alignment: This process ensures everything works well together and supports the company's overall goals.

Refine and Validate Dependencies: To refine initial dependencies and identify constraints on Implementation and Migration Plans, it's essential to consider key dependencies. Analyzing them helps determine the implementation sequence and the necessary coordination. Grouping related activities forms a foundation for establishing projects, and examining these projects can reveal logical increments of deliverables. This approach also aids in scheduling the delivery of these increments. Documenting the assessment of these dependencies should be part of the Architecture Roadmap and any required Transition Architectures. Addressing dependencies is fundamental to effective migration planning.

CHAPTER 12   ARCHITECTURE DEVELOPMENT METHOD

The *Work Package Portfolio Map*, in conjunction with the *Architecture Roadmap* (both artifacts are detailed in Section 12.8.3), provides a cohesive method for mapping interdependencies between work packages. In particular, the Work Package Portfolio Map provides tools for grouping work packages. The Architecture Roadmap visualizes these different groups of work packages and also shows their interrelationships and dependencies in relation to the organizational goals. Refer to Figure 12-47 for an example of an Architecture Roadmap.

Refine Dependencies: Start by figuring out which tasks depend on each other. This will help to understand the order to work on things and where to coordinate.

Group Related Tasks: Group tasks that go together. This makes it easier to manage the project and plan delivery in manageable chunks.

Schedule and Plan: With the groups of tasks, plan when each group of work will be done. This helps avoid confusion and ensures timely delivery.

Document Dependencies: Write down the findings about these dependencies. This should be included in the Architecture Roadmap to help guide the migration process.

Use the Work Package Portfolio Map: This tool helps to see how activities are connected. It shows the relationship between different activities and aligns them with the organization's goals.

Confirm Readiness and Risk for Business Transformation: The *Business Transformation Readiness Assessment* plays a pivotal role in determining whether and to what extent an organization is able to adapt in accordance with the changes required as a result of architecture work. During Phase A (Architecture Vision), the Business Transformation Readiness Assessment is initiated. By reviewing its findings during Phase E (Opportunities and Solutions), organizations can gauge their preparedness

CHAPTER 12   ARCHITECTURE DEVELOPMENT METHOD

for transformation initiatives. This assessment directly influences the Architecture Roadmap and the Implementation and Migration Strategy, ensuring they are grounded in the organization's current state and readiness levels.

Identifying, classifying, and mitigating risks associated with the transformation effort is crucial. Documenting these risks in the Consolidated Gaps, Solutions, and Dependencies Matrix provides a structured approach to address potential challenges, facilitating a smoother transition toward the Target Architecture.

Assess Transformation Readiness: Start by evaluating whether the organization is ready for change, especially concerning the architecture work. This is done through a Business Transformation Readiness Assessment, which kicks off in Phase A.

Review Findings: Revisit the findings from the readiness assessment to gauge how prepared the organization is for transformation initiatives.

Link to the Architecture Roadmap: Use the assessment results to shape the Architecture Roadmap and Implementation and Migration Strategy, ensuring that these plans reflect the current state and readiness.

Identify Risks: Spot any risks that could arise during the transformation. Document these risks in the Consolidated Gaps, Solutions, and Dependencies Matrix.

Mitigate Risks: Address the identified risks systematically, making the transition smoother and more aligned with the Target Architecture.

Formulate Implementation and Migration Strategy: Implementing a Target Architecture requires a comprehensive Implementation and Migration Strategy to guide the process and structure any necessary Transition Architectures. The initial step involves determining a strategic approach to implement solutions or capitalize on opportunities. Three fundamental approaches include:

- **Greenfield:** Initiating a completely new implementation
- **Revolutionary:** Undertaking a radical change, such as a complete system overhaul
- **Evolutionary:** Adopting a strategy of gradual convergence, like parallel operations or phased introductions of new capabilities

Subsequently, it's essential to establish a strategic direction that addresses and mitigates risks identified in the Consolidated Gaps, Solutions, and Dependencies Matrix. Common implementation methodologies encompass:

- **Quick Wins:** Achieving immediate benefits through targeted initiatives
- **Achievable Targets:** Setting realistic goals that are attainable within the current constraints
- **Value Chain Method:** Enhancing the entire value chain to improve overall performance

These methodologies, along with identified dependencies, form the foundation for creating work packages. This process culminates in the agreement on the Implementation and Migration Strategy for the organization.

Decide on the Approach: Before anything, figure out how to start the implementation of the Target Architecture. There are three basic options: Greenfield, Revolutionary, and Evolutionary.

Plan the Direction: Next, set clear goals that focus on fixing the issues found in the assessment. Think about quick wins, achievable targets, and the entire value chain.

Build the Action Plan: Now, organize the plan by creating small, manageable work packages. These tasks will guide the strategy and ultimately lead to successful implementation.

Identify and Group Major Work Packages: To effectively address identified gaps in business capabilities, key stakeholders, planners, and Enterprise Architects should assess these deficiencies as outlined in the architecture vision and Target Architecture. Utilizing tools such as the Consolidated Gaps, Solutions, and Dependencies Matrix, along with the Implementation Factor Catalog, enables logical grouping of activities into work packages. This approach ensures that each gap is addressed with appropriate solutions, whether through new development, existing products, or purchasable solutions. Additionally, classifying current systems as *mainstream, contain,* or *replace* helps in planning their future roles within the organization. By decomposing work packages into capability increments and analyzing them concerning business transformation issues and strategic implementation approaches, organizations can effectively group these packages into portfolios and projects, considering dependencies and strategic goals. A key artifact that supports this method is the Work Package Portfolio Map (see Section 12.8.3).

Group Activities Into Work Packages: Use tools like the Consolidated Gaps, Solutions, and Dependencies Matrix, along with the Implementation Factor Catalog, to organize the activities into clear groups. This helps ensure that each gap is addressed with the right solution – whether building new systems, improving existing ones, or buying ready-made solutions.

Classify Current Systems: Evaluate the current systems. Decide if they need to be kept, improved, or replaced. This helps plan how they will fit into the future of the organization.

Break Work Packages Into Smaller Pieces: Split the work into manageable tasks that align with business goals and transformation efforts. This makes it easier to organize into portfolios and projects, taking into account dependencies and strategic priorities.

Use a Visual Tool: Use a visual map, like a Work Package Portfolio Map, to clearly show how everything fits together and how each piece will be completed.

Identify Transition Architectures: Implementing a Target Architecture often necessitates an incremental approach, introducing one or more Transition Architectures. These Transition Architectures serve as clear *milestones* on the roadmap toward the Target Architecture, each delivering measurable business value. It's important to note that the intervals between successive Transition Architectures need not be uniform.

The development of Transition Architectures should be grounded in the preferred implementation strategy; the Consolidated Gaps, Solutions, and Dependencies Matrix; the project and portfolio listings; and the organization's capacity for change. Creating an *Architecture Definition Increments Table* (see Section 12.8.2) supports these activities. Identifying challenging activities is crucial; unless compelling reasons dictate otherwise, it's advisable to schedule these after implementing activities that more readily provide missing capabilities.

Identify Transition Architectures: When moving toward a Target Architecture, it's smart to take small steps. This means introducing one or more Transition Architectures along the way. These steps act as milestones, each delivering tangible benefits. Record them in the Architecture Definition Increments Table.

Map Out the Road to Success: Make sure the Transition Architectures align with the overall strategy. Use tools like the Consolidated Gaps, Solutions, and Dependencies Matrix and the Architecture Definition Increments Table. Take into account the project portfolio and the organization's ability to adapt.

Focus On the Tough Stuff: Identify the hardest tasks first. But don't try to tackle them all at once! It's better to start with easier tasks that build up the necessary capabilities and schedule the tougher activities for later.

Create Architecture Roadmap and Implementation and Migration Plan: By consolidating work packages and Transition Architectures into a Draft Architecture Roadmap, organizations can outline a clear timeline for progression. This roadmap not only informs the Implementation and Migration Plan but also frames the migration planning in Phase F of the Architecture Development Method (see Section 12.9). It's essential that each identified Transition Architecture and corresponding work packages have a well-defined set of outcomes, demonstrating how their selection and timing contribute to realizing the Target Architecture.

The Draft Architecture Roadmap should mirror the level of detail found in the Architecture Definition Document developed during phases B, C, and D. If significant additional detail is required before implementation, it may indicate a transition to a different level of architecture.

The Implementation and Migration Plan (see Table 12-53) must detail the activities necessary to achieve the Architecture Roadmap, serving as the foundation for migration planning in Phase F. The plan's detail should align with that of the Architecture Roadmap, sufficiently identifying the projects and resources required to realize the roadmap.

Table 12-53. Implementation and Migration Plan

Phase	Timeframe	Work Packages	Key Milestones	Dependencies
1	Q2 2028 to Q4 2028	Implement new capability for Patient Engagement.	Implemented new capability	–
		Implement new capability for Financial Management.	Implemented new capability	–
2	Q3 2028 to Q2 2030	Improve capability for Clinical Services.	Improved performance of capability	Staff training complete
		Improve capability for Financial Management.	Improved performance of capability	–
3	Q4 2028 to Q4 2030	Implement and improve processes for Patient Engagement.	Improved and new process	Implemented new capability for Patient Engagement
		Improve processes for Financial Management.	Improved process	Improved capability for Financial Management

When developing the Implementation and Migration Plan, consider various approaches, such as a data-driven sequence where applications that create data are implemented first, followed by those that process the data. A thorough understanding of the dependencies and lifecycle of existing Solution Building Blocks (see Chapter 7, Section 7.5.2.5) is crucial for an effective plan.

Finally, update the architecture vision and Architecture Definition Document with any additional relevant outcomes from this phase to ensure all documentation reflects the current state and planned progression.

Draft the Architecture Roadmap: Start by consolidating the work packages and Transition Architectures into the Draft Architecture Roadmap. This gives a clear timeline for moving forward.

Link to Implementation and Migration: This roadmap will guide the Implementation and Migration Plan. It's key in Phase F.

Set Clear Outcomes: Each work package and Transition Architecture should have clear outcomes. These outcomes help show how each piece fits into the bigger picture of the Target Architecture.

Mirror Architecture Documents: The Draft Architecture Roadmap should reflect the detail level found in the Architecture Definition Document (from phases B, C, and D). If more detail is required, this may indicate a need for a new level of architecture planning.

The Opportunities and Solutions Phase is a bridge between defining the architecture and planning how to implement it, focusing on identifying and evaluating solution options and setting the stage for successful migration and implementation.

12.8.2 Phase E (Opportunities and Solutions) Artifacts

The Opportunities and Solutions Phase evaluates and selects the most appropriate solutions to address the gaps identified in the previous phases. Architects identify potential architecture options, perform a risk assessment, and recommend appropriate solutions to stakeholders. They consider both

CHAPTER 12 ARCHITECTURE DEVELOPMENT METHOD

internal development and external sourcing options. The output of this phase is a set of architecture specifications and implementation plans.

Table 12-54 lists the artifacts that can be created during the execution of Phase E (Opportunities and Solutions).

Table 12-54. Artifacts created during Phase E (Opportunities and Solutions)

ADM Phase	Artifacts	Used Entities
Phase E: Opportunities and Solutions	Project Context Diagram	Work package, organization unit, function, business service, process, logical application component, data entity, logical technology component
	Benefits Diagram	Stakeholder, once added to the metamodel, business capability, business service, product
	Implementation Factor Catalog	Risk
	Consolidated Gaps, Solutions, and Dependencies Matrix	Gap, work package
	Architecture Definition Increments Table	–
	Work Package Portfolio Map	Work package, goal, objective
	Architecture Roadmap	Work package, goal, objective

CHAPTER 12 ARCHITECTURE DEVELOPMENT METHOD

Project Context Diagram: This diagram shows the scope of one or more work packages (initiatives) that will be implemented as part of a broader transformation roadmap, depending on the size of the project. The diagram (see Figure 12-41) links the work packages to the organizational units, functions, services, processes, applications, data, and technology that will be affected by the project. It is a valuable tool for project portfolio management and project mobilization.

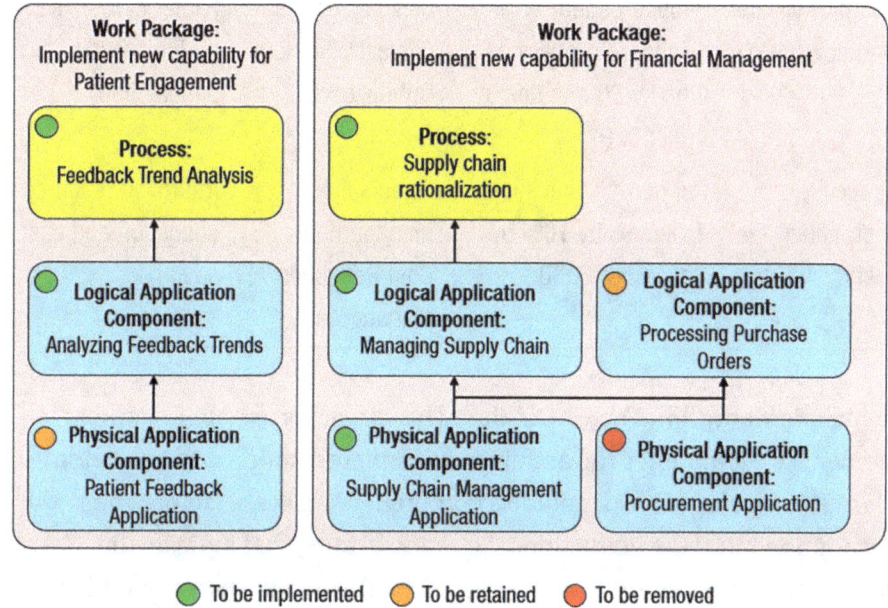

Figure 12-41. Project Context Diagram

Benefits Diagram: Shows opportunities identified in an architecture definition, classified according to their relative size, benefit, and complexity. This diagram, or catalog (see Table 12-55), can be used by stakeholders to make selection, prioritization, and sequencing decisions on identified opportunities.

CHAPTER 12   ARCHITECTURE DEVELOPMENT METHOD

Table 12-55. Benefits Catalog

Goal	Objective	Capability	Benefit
Improve patient satisfaction	Achieve 90% rating in patient satisfaction surveys by Q4 2028.	Patient Engagement Patient Feedback Management	Higher satisfaction ratings and service trust
Optimize clinical workflows	Reduce patient discharge processing time by 30 minutes by April 2029.	Clinical Services Clinical Workflow Optimization	Faster care delivery and improved satisfaction
Reduce operational costs	Cut non-clinical supply expenses by 10% by December 2030.	Financial Management Cost Reduction Management	Improved resource allocation and cost savings

Implementation Factor Catalog: This artifact is used to document elements that influence the architecture Implementation and Migration Plan. This catalog should include a comprehensive list of factors, their descriptions, and the deductions (i.e., conclusions) that indicate the corresponding actions or constraints to be considered when creating the migration plan. By systematically documenting these factors, a more robust and adaptable Implementation and Migration Plan can be developed, ensuring alignment with organizational objectives and external conditions.

The Implementation Factor Catalog ensures that each gap between the Baseline and Target Architectures is addressed with appropriate solutions. The catalog is used to effectively address identified gaps (during phases B, C, and D) by evaluating the deficiencies as outlined in the architecture vision and Target Architecture. The use of techniques such as the Implementation Factor Catalog helps with creating logical groups of activities and combine them into work packages.

CHAPTER 12 ARCHITECTURE DEVELOPMENT METHOD

In the *Deduction column* of Table 12-56, the contours of the work packages begin to take shape. For example, the inference that there is a need to *train personnel to use the new application* could result in a work package that includes the activity of *providing training options for personnel*. Another inference that the *new application will automate most application services* is likely to result in the creation of a work package containing the activity to *automate application services*.

Table 12-56. Implementation Factor Catalog

Factor Category	Factor	Description	Deduction
Effect (impact)	Change in applications	Replace HR Recruitment Application A with Application B to streamline storage of candidate information.	Need for personnel training to use new application. New application has major personnel savings and should be given priority.
Dependency	Consolidation of application services	Multiple information streams will be consolidated.	New application will automate most application services.
Action	Introduction of new HR Recruitment Application		

The Implementation Factor Catalog may also contain constraints that could hinder interoperability. To address these constraints, the same approach of creating work packages containing activities to address these constraints can be used.

479

Begin creating an Implementation Factor Catalog by listing elements that could affect the implementation of the architecture. Common factors include risks, assumptions, dependencies, and actions and effects. Create a table with the following columns:

- **Factor Category:** The type of factor (e.g., risk, assumption)
- **Factor:** A brief title or name for the factor
- **Description:** A concise explanation of the factor
- **Deduction:** Steps needed to address or manage the factor

Next, enter the type of factor and provide a short, descriptive name. Offer a succinct explanation, keeping it clear and to the point. List the necessary steps to address the factor, using bullet points for clarity.

To illustrate the steps outlined in the approach above, the first factor listed in Table 12-56 mentions a change in applications. The category assigned to the factor is Effect (impact). The description expands on the briefly stated factor by adding that HR Recruitment Application A will be replaced by HR Recruitment Application B to streamline the storage of candidate information. The deduction leads to the assumption that staff will need to be trained to use the new application and that the implementation of the application will lead to significant staff savings and should therefore be given priority. The bulleted lists are omitted in Table 12-56 to save space.

Using an Implementation Factor Catalog forces careful consideration of factors that could potentially affect the Implementation and Migration Plan.

CHAPTER 12   ARCHITECTURE DEVELOPMENT METHOD

Consolidated Gaps, Solutions, and Dependencies Matrix: This matrix is used to group the gaps that were identified during the gap analysis that took place during phases B, C, and D of the Architecture Development Method. The technique of using this matrix allows architects to assess potential solutions and their dependencies to the discovered gaps. The matrix is particularly useful for creating work packages. These sets of grouped activities will eventually be used to form the items on the Architecture Roadmap that will be created during Phase E (Opportunities and Solutions) and finalized during Phase F (Migration Planning).

During phases B, C, and D of the Architecture Development Method, the gap analysis results are consolidated and documented in the Consolidated Gaps, Solutions, and Dependencies Matrix. This matrix is instrumental in pinpointing Solution Building Blocks that can address multiple gaps and their corresponding Architecture Building Blocks (see Chapter 12, Section 12.8.1).

A Consolidated Gaps, Solutions, and Dependencies Matrix can be created by meticulously recording and consolidating the gaps. For each identified gap, propose a possible solution. Identify Solution Building Blocks that can effectively bridge these gaps. Link each Solution Building Block to the relevant Architecture Building Block(s) that it will modify or enhance. Note any interdependencies among the gaps, solutions, and architecture components. This will ensure a clear understanding of how changes may impact various parts of the architecture. Cluster similar gaps and solutions to streamline the implementation process. This grouping helps identify patterns and potential shared solutions.

Refer to the Implementation Factor Catalog to understand the factors that influence implementation, such as organizational constraints, resources, and timelines.

CHAPTER 12   ARCHITECTURE DEVELOPMENT METHOD

As new factors emerge during the rationalization process, add them to the catalog. Keeping this catalog current ensures that all relevant considerations are accounted for in future implementations.

Because the Consolidated Gaps, Solutions, and Dependencies Matrix lists gaps, their potential solutions, and dependencies, it is an excellent resource for defining Transition Architectures. These Transition Architectures serve as clear *milestones* on the roadmap to the Target Architecture, each providing measurable business value. The development of Transition Architectures should be based on the Consolidated Gaps, Solutions, and Dependencies Matrix. Figure 12-42 illustrates how the input from the Consolidated Gaps, Solutions, and Dependencies Matrix can be used to visualize the Baseline, Transition, and Target Architectures using a Plateau Planning.

***Figure 12-42.** Plateau Planning of Baseline, Transition, and Target Architectures*

In Figure 12-42, the gaps from the business and application domains (see items #1 and #2 in Table 12-57) are visualized using the Transition Architecture Plateau (the middle block). Item #3 in Table 12-57, a gap from the information domain, forms the Target Architecture Plateau.

Table 12-57. Consolidated Gaps, Solutions, and Dependencies Matrix

#	Architecture Domain	Gap	Potential Solution(s)	Dependencies
1	Business	New candidate profile processing process	Design new process for candidate profile processing (Business Interaction Matrix, Process Flow Diagram).	Replacement of HR Recruitment Application A
2	Application	New application workflow mechanism	Implement HR Recruitment Application B (Process/Application Realization Diagram, Application/Function Matrix).	
3	Information	Consolidated information storage	Develop consolidated information flows (Data Entity/Business Function Matrix, Application/Data Matrix).	

Section 12.8.3 details the creation of an Architecture Roadmap. A Plateau Planning, as illustrated in Figure 12-42, is a useful tool for this process.

Architecture Definition Increments Table: Allows architects to detail a series of Transition Architectures in tabular form. The table outlines the state of the architecture at specific times during its development.

The idea behind creating an Architecture Definition Increments Table (Table 12-58) is to list the projects and their assigned incremental deliverables across the Transition Architectures. Another way to achieve the same result is to create a slightly modified Plateau Planning Diagram (see Figure 12-43).

CHAPTER 12   ARCHITECTURE DEVELOPMENT METHOD

Table 12-58. Architecture Definition Increments Table

Project	Transition Architecture 1	Transition Architecture 2	Transition Architecture 3
Project A: Implement new HR Recruitment Application	Identified missing capabilities	Implemented HR Recruitment Application B	Eliminated HR Recruitment Application A
Project B: Automate capabilities through HR Recruitment Application		Automated capability Inflow Support	Automated capability Personnel Information Management

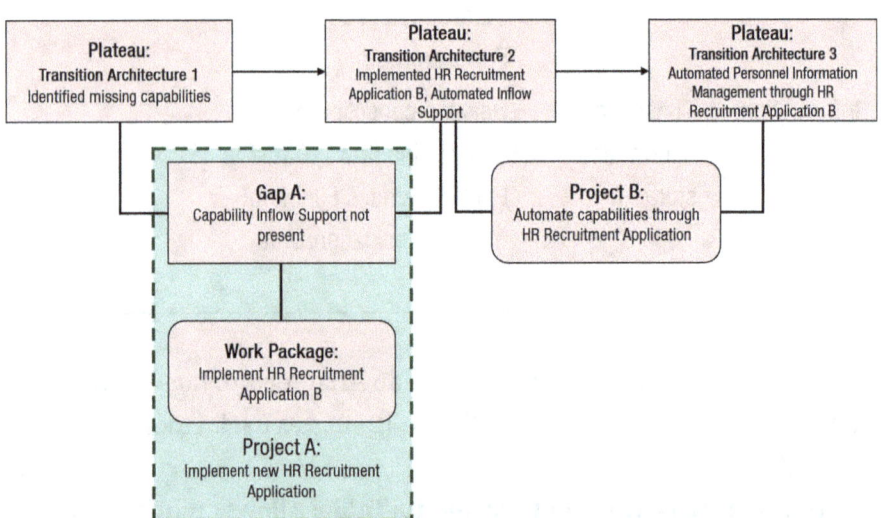

Figure 12-43. Plateau Planning Diagram showing projects instead of gaps and work packages

Instead of visualizing the gaps and work packages separately (as shown on the left in Figure 12-43), the project names are used to detail the activities required to close each gap (shown on the right). This is because projects often consist of one or more work packages. In turn, work packages contain the activities to close gaps between architecture states.

484

For instructions on how to create a Work Package Portfolio Map and an Architecture Roadmap, refer to the next section, "Creating an Architecture Roadmap".

12.8.3 Creating an Architecture Roadmap

The Architecture Roadmap is a pivotal deliverable within the TOGAF Framework, serving as a strategic guide that outlines the transition from the current Baseline Architecture to the desired Target Architecture. This roadmap is incrementally developed throughout phases E and F and informed by readily identifiable roadmap components from phases B, C, and D within the Architecture Development Method, ensuring a comprehensive and cohesive plan.

The Architecture Roadmap not only serves as a planning tool but also as a communication instrument, conveying the strategic direction and planned initiatives to stakeholders across the organization. By providing a visual representation of the transformation journey, it fosters understanding and buy-in from all involved parties. Moreover, the roadmap's iterative development allows for flexibility, enabling adjustments in response to changing business environments or emerging technologies, thereby ensuring the architecture remains aligned with the organization's evolving needs.

By detailing individual work packages (also called initiatives) and their sequencing, the Architecture Roadmap provides a clear timeline for implementing architectural changes, aligning each step with the organization's strategic objectives. It shows the progression from the Baseline Architecture to the Target Architecture.

The Architecture Roadmap highlights the business value of individual work packages at each stage. Transition Architectures necessary to effectively realize the Target Architecture are identified as intermediate steps.

CHAPTER 12   ARCHITECTURE DEVELOPMENT METHOD

In Phase E (Opportunities and Solutions), the Architecture Roadmap is refined to identify and prioritize work packages that address capability gaps and leverage opportunities for improvement. This phase ensures that each work package is evaluated against business goals, resource availability, and potential risks, facilitating informed decision-making. The outcome is a well-structured plan that balances immediate needs with long-term objectives, setting the stage for effective implementation in subsequent phases.

Before creating the Architecture Roadmap, a diagram such as Figure 12-44 is often created. This figure shows the phases that an architecture may be in (Baseline, Transition, and Target). Each phase – also known as a *plateau* – implies that there is work to be done to get to that phase. The gap between phases is represented by the *gap* entity.

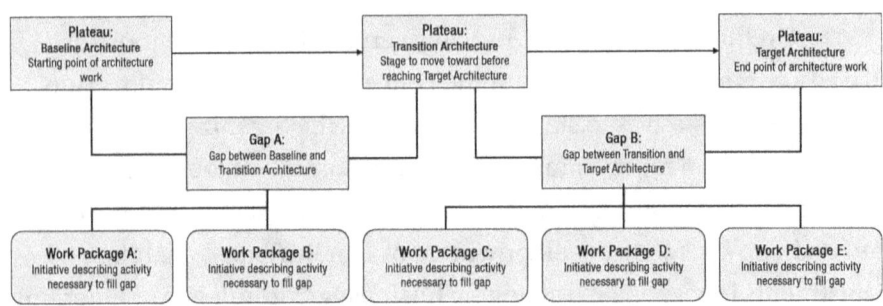

Figure 12-44. *From Baseline through Transition to Target Architecture*

To denote the work required to move from one phase to the next, the work package entity is used. These work packages represent the activities needed to bridge the corresponding gap.

High-level work packages are similar to initiatives. A cluster, set, or grouping of multiple work packages can be considered an initiative. These clusters of work packages often form a project, program, or even a portfolio.

A work package generally does not describe an ongoing activity such as a business process. The subject of the work package is performed once and produces a well-defined end result. This can be a goal or an objective

CHAPTER 12   ARCHITECTURE DEVELOPMENT METHOD

or the activities necessary to fill the gap between two architecture states. A work package can be used to model tasks within a project, entire projects, programs, or entire portfolios. In an agile context, a work package can be used to model the work done in an agile iteration (e.g., a sprint) or in a higher-level increment [9].

The TOGAF Standard defines a work package as a set of actions aimed at achieving one or more objectives:

A set of actions identified to achieve one or more objectives for the business. A work package can be a part of a project, a complete project, or a program [2].

Conceptually, a work package is similar to a business process in that it consists of a series of related tasks aimed at producing a well-defined end result. In terms of content, a work package can be said to be a *unique and one-time process*; it performs a series of activities that lead to an end result. A work package can be described in much the same way as a process. Figure 12-45 illustrates the similarities (and differences) between a work package and a process.

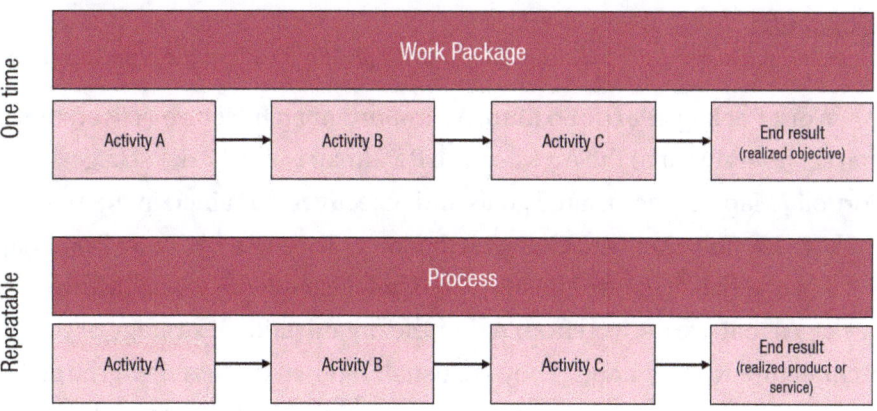

Figure 12-45. *Similarities and differences between work packages and processes*

Work packages translate the previously defined initiatives into concrete, actionable steps. Figure 12-44 illustrates the operation and use of work packages. It shows an overview of (from left to right) the Baseline Architecture, Transition Architecture, and the Target Architecture. Between each set of architecture states (Baseline and Transition, and Transition and Target) is a gap visible. Each of the gaps can be bridged by performing the activities listed in the corresponding work packages.

The TOGAF Standard references the work package entity within the Foundation Enterprise Metamodel. However, this entity is not used in any artifacts.

Therefore, I would like to propose an addition to the TOGAF Standard regarding this point.

It is important to have a thorough understanding of the initiatives (work packages) that make their way into the Architecture Roadmap. For this reason, I suggest adding two new artifacts to be produced during Phase E of the Architecture Development Method.

These are a Work Package Portfolio Map and an Architecture Roadmap.

Work Package Portfolio Map: An important architecture deliverable that serves as input to the Architecture Roadmap is the Work Package Portfolio Map. The associated goals and objectives contained in the Work Package Portfolio Map are elements that recur in the Architecture Roadmap.

A Work Package Portfolio Map consists of an overview of all work packages that are derived from either the organizational strategy or the architecture work at hand. They are usually shown clustered or grouped together. A Work Package Portfolio Map is similar in structure and visualization to an Application Portfolio Catalog. Table 12-59 shows an example of a Work Package Portfolio Map.

Table 12-59. Work Package Portfolio Map

Work Package Name	Associated with Objective(s)	Associated with Goal(s)
Work Package A: Fit job profile X into existing job profile	Objective A: Make adjustments to comply with legislation demands by Q3 2028.	Goal A: Implement new regulatory requirements.
Work Package B: Recruit staff for job profile X	Objective A: Make adjustments to comply with legislation demands by Q3 2028.	
	Objective B: Eliminate overlapping job profiles by 2027.	Goal B: Efficiently use human resources.
Work Package C: Distribute updated job profiles	Objective B: Eliminate overlapping job profiles by 2027.	
Work Package D: Update compensation application	Objective C: Re-evaluate job compensation by Q4 2028.	
Work Package E: Update job information	Objective C: Re-evaluate job compensation by Q4 2028.	

A Work Package Portfolio Map is the foundation of the Architecture Roadmap. The goals and objectives contained in this artifact are recurring elements. For governance purposes, it is important to link the initiatives from the Work Package Portfolio Map to the items in the Architecture Roadmap. This allows stakeholders and senior management to track the architecture's progress and performance against the organization's strategy. Without initiatives tied to organizational goals and objectives, this is virtually impossible to measure.

CHAPTER 12 ARCHITECTURE DEVELOPMENT METHOD

For example, Table 12-59 shows that Work Packages A and B are linked to Objective A, which, in turn, is linked to Goal A. Work Package B is also linked to Objective B. The second goal, Goal B, is composed of two objectives (B and C). Objective B is composed of Work Packages B and C, whereas Objective C has Work Packages D and E linked to it.

If the Work Package Portfolio Map was to be visualized (by creating a Work Package Portfolio Diagram), the result would be as shown in Figure 12-46.

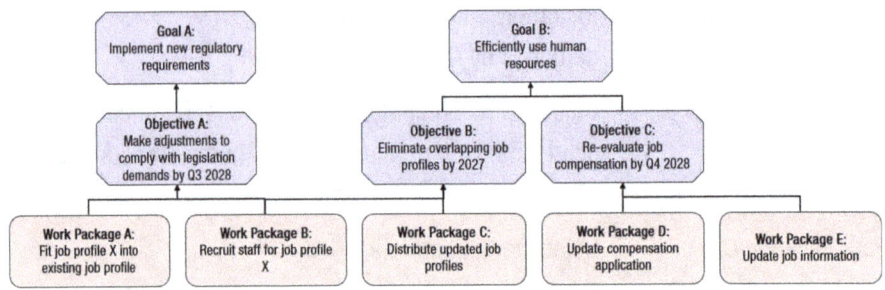

Figure 12-46. Work Package Portfolio Diagram

Linking the work packages (or initiatives) to the objectives and goals is an absolute must, because this allows the architecture activities to be tracked back to organizational goals and objectives.

During Phase F (Migration Planning), it is important to map the work packages needed to implement the architecture to the ongoing projects in the organization (refer to Section 12.9.1 for additional information). This prevents things from being done twice and can be used as a kind of consistency check.

To ensure the relationship between work packages and projects, it is advisable to add an extra column to the Work Package Portfolio Map. This column should contain the projects that are related to the work packages.

Architecture Roadmap: An Architecture Roadmap consists of initiatives (or work packages). These entities are the embodiment of the activities that need to be performed to realize an organization's strategy.

CHAPTER 12 ARCHITECTURE DEVELOPMENT METHOD

Phases B, C, and D of the Architecture Development Method develop architectures for the business, data, application, and technology domains. Each of these architectures creates components that find their way onto the roadmap. The form in which they appear on the roadmap is that of a work package or initiative. Either are plannable entities that contain a *description* of the activity, together with a *start* and *end date*.

Creating an Architecture Roadmap is an important part of developing an Enterprise Architecture. The development of a roadmap enables an organization to align the created initiatives with the defined goals and corresponding objectives. In effect, a roadmap represents the realization of the organization's strategy, visualized in concrete and defined steps.

A roadmap should never be created without linking business goals and objectives to the initiatives [9].

Goals and objectives, key components of the organization's strategy, are defined using SMART[4] criteria. This makes them specific and therefore understandable, realistic (or relevant), and, above all, measurable. The latter is particularly important for translating goals and objectives into initiatives by developing the architecture.

Once the organization's goals and objectives are defined, initiatives can be created by following the phases of the Architecture Development Method. When these initiatives are implemented, they achieve (part of) the objectives. Therefore, it is important to have a clear understanding of the initiatives so that the appropriate governance can be applied at the right time.

As mentioned earlier, initiatives are visualized using the work package entity. A work package contains a description of the initiative and has

---

[4] Specific, Measurable, Achievable, Realistic, and Timebound (SMART).

a start and end date. Work packages can be associated with additional information such as status, cost, or ownership. The entities are plotted over time in a roadmap using architecture tools.

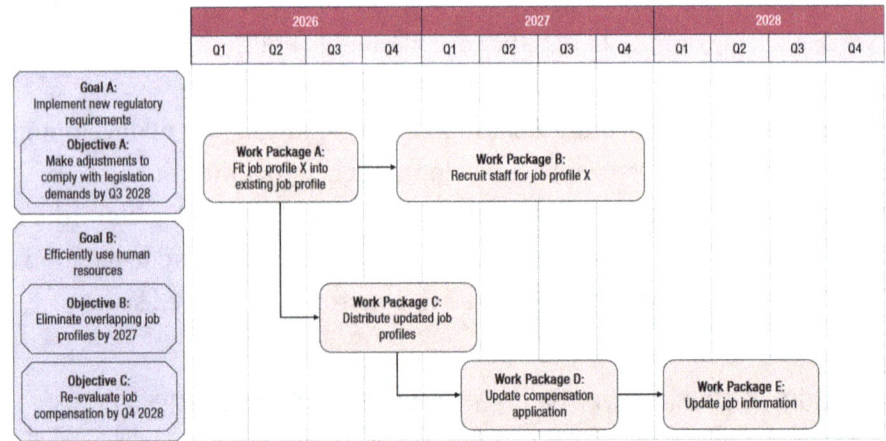

Figure 12-47. Architecture Roadmap containing work packages

Using a good architecture tool, it is relatively easy to create an Architecture Roadmap. This artifact shows which work packages are related to which goals or intended objectives. Additionally, owners and (progress) states can be assigned to the work packages. Thus, an Architecture Roadmap can be used to monitor the progress of the implementation of the organization's strategy.

A roadmap detailed with properties provides a wealth of information for an organization. For example, it can be used to visualize the cost of achieving the goals and objectives. By assigning the cost of executing a work package as an attribute to the element, the roadmap can show how this relates to the realization of the objective(s). Having an overview of the costs per work package can help prioritize projects or programs.

An Architecture Roadmap is created by processing the work packages from the Work Package Portfolio Map (see Table 12-59) and plotting them as activities over time. This is done by positioning the goals and/or objectives on the left side of the roadmap (vertically) and the work packages from the Work

CHAPTER 12 ARCHITECTURE DEVELOPMENT METHOD

Package Portfolio Map in the timeline area (horizontally). The roadmap can be enriched by adding additional properties to the work packages. Examples include owners of the activities (organization units or business roles), potential costs, and progress states (e.g., not started, pending, completed, canceled).

By adding these additional properties to the work packages, a broad spectrum of additional options becomes available. For instance, by adding the business roles as work package owners to the work packages, it becomes possible to visualize who is responsible for what. This can be helpful in monitoring the progress of the organization's goals and objectives. Color views are a powerful way to visually illustrate specific parts of a diagram – in this particular case, the progress of a project. An example of a color view that highlights the progress is shown in Figure 12-48.

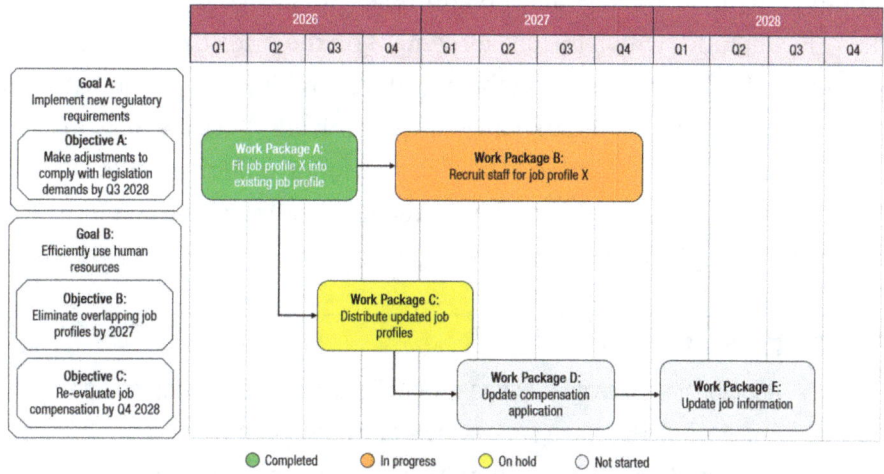

Figure 12-48. *Color view of an Architecture Roadmap*

The primary purpose of an Architecture Roadmap is to be able to plot work packages over time so that the organization has a good picture of what activities are needed to realize the desired Target Architecture. By cleverly using the translation of Transition Architectures into work packages, the intermediate steps can also be visualized. An Architecture Roadmap is a powerful tool in the context of Enterprise Architecture.

Phase E (Opportunities and Solutions) focuses on transforming the architecture vision and Architecture Roadmap into actionable project solutions. This phase identifies potential work packages, including Transition Architectures, and evaluates them against business goals, financial feasibility, and resource availability. By aligning opportunities with the Target Architecture, Phase E establishes a clear connection between strategic objectives and tactical implementation.

A critical aspect of Phase E is prioritizing and sequencing projects based on their value, dependencies, and organizational constraints. The phase ensures that all proposed solutions address identified gaps and risks while optimizing resource use.

Stakeholder engagement plays a central role, as their input ensures alignment with business goals and addresses practical considerations. Phase E concludes with a formal agreement on the prioritized set of projects and solutions, bridging the high-level architecture vision with the practical execution needed to realize the organization's strategic goals.

Phase E (Opportunities and Solutions) is followed by the Migration Planning Phase, where a detailed Architecture Roadmap and schedule for implementing the solutions are developed.

Further Reading
Additional articles on creating a Work Package Portfolio Map and an Architecture Roadmap are available at the following URLs:
https://eawheel.com/wheel/execute/roadmap/work-package-portfolio-map/
https://eawheel.com/wheel/execute/roadmap/architecture-roadmap/
For more information about the Architecture Roadmapping process, visit the following URL:
https://eawheel.com/blog/2024/04/architecture-roadmapping/

12.9. Phase F: Migration Planning

In this phase, the architects focus on creating a detailed plan for implementing the selected solutions. A step-by-step approach for transitioning from the baseline state to the target state architecture is developed. This includes defining Transition Architectures and identifying critical milestones, resources, and timelines. The Migration Planning document provides the roadmap for the transformation journey.

12.9.1 Creating a Migration Planning

The primary objective of Phase F (Migration Planning) is to create a detailed plan for the implementation of the architecture. This involves identifying the projects and initiatives (the aforementioned work packages) that will help the organization move from the Baseline Architecture to the Target Architecture. The aim is to:

- **Prioritize projects** and activities that are part of the transition.
- **Assess dependencies and risks** between projects.
- **Plan resource allocation** (budget, people, technology) for these projects.
- **Develop a migration strategy** that ensures business continuity.

Crucial inputs for this phase are the Target Architecture and the Architecture Roadmap components. These components include the gaps identified between the Baseline and Target Architectures, opportunities for improvement, and solutions. They were developed during phases B–D and were translated to work packages (initiatives) in Phase E. The Architecture Roadmap gets finalized during Phase F (Migration Planning). Other key inputs to this phase are the Implementation Governance Model, which will

CHAPTER 12　ARCHITECTURE DEVELOPMENT METHOD

guide the architecture governance during the migration process. Last but not least, the insights from previous phases on potential impacts and risks also play an important role. Table 12-60 lists the steps to perform Phase F (Migration Planning).

Table 12-60. Steps to perform Phase F (Migration Planning)

Step to Perform	Description of Step	Output Generated by Step
Confirm management framework interactions	Align the Implementation and Migration Plan with an organization's management frameworks.	Updated Implementation and Migration Plan
Assign business value to work packages	Define and measure business value for work packages, identify corresponding projects, assess associated risks, and estimate business value.	Updated Work Package Portfolio Map Business Value and Risk Matrix
Estimate resource requirements, project timings, and availability/delivery vehicle	Allocate resources and estimate costs for each project increment, distinguishing between capital and operational expenses, and identify cost-saving opportunities.	
Prioritize migration projects through cost/benefit assessment and risk validation	Determine the sequence of projects based on business priorities, technical dependencies, resource availability, and risk factors.	Risk Management Plan

(continued)

Table 12-60. (*continued*)

Step to Perform	Description of Step	Output Generated by Step
Confirm Architecture Roadmap	Update the Architecture Roadmap, incorporating any necessary Transition Architectures.	Transition Architecture State Evolution Table Finalized Architecture Roadmap
Complete Implementation and Migration Plan	Develop a detailed timeline with clear milestones for each transition stage.	Finalized Implementation and Migration Plan
Complete architecture development cycle	If the Architecture Capability's maturity permits, craft an Implementation Governance Model. Document lessons learned.	

Confirm Management Framework Interactions: Coordinating the Implementation and Migration Plan with an organization's management frameworks is crucial for achieving concrete business outcomes. Typically, four key frameworks must collaborate closely:

- **Business Planning** conceives, directs, and provides the necessary resources for activities aimed at achieving specific business objectives.
- **Enterprise Architecture** structures and contextualizes all organizational activities, delivering tangible business outcomes.

- **Project/Portfolio Management** coordinates, designs, and builds business systems that deliver the desired outcomes.
- **Operations Management** integrates, operates, and maintains deliverables that achieve the business objectives.

The Implementation and Migration Plan will impact the outputs of each of these frameworks and must be reflected within them. During this step, it's essential to understand the organization's frameworks and ensure that the plans are coordinated and summarized within each framework's plans. Identify the people in the organization who are responsible for the frameworks listed above. These will typically be department heads, as well as portfolio or project managers. Present the Implementation and Migration Plan to them, and make sure that the steps described in it are adopted in the other frameworks, where applicable. Conversely, if there are things described in the other frameworks that may affect the Implementation and Migration Plan, they need to be incorporated.

The outcome of this step may be that the Implementation and Migration Plan becomes part of a different plan produced by another framework, with Enterprise Architecture participation.

Understand Existing Frameworks: First, get to know the main management frameworks the organization uses. These include Business Planning, Enterprise Architecture, Project/Portfolio Management, and Operations Management.

Coordinate Plans: Make sure the Implementation and Migration Plan fits well with these frameworks. The steps in the plan should be reflected in theirs, and vice versa. If other frameworks have plans that could affect the Implementation and Migration Plan, be sure to include them.

Engage the Right People: Identify the heads of the departments and project managers responsible for these frameworks. These are the people who need to be involved in making sure the plans line up.

Present the Plan: Share the Implementation and Migration Plan with the department heads and make sure it's aligned with their frameworks. They should adopt the necessary parts of the Implementation and Migration Plan, where applicable.

Update and Finalize: As a result, the Implementation and Migration Plan might be integrated into other framework plans, especially those involving Enterprise Architecture. Ensure everyone is on the same page to move forward smoothly.

Assign Business Value to Work Packages: Begin by defining what constitutes business value within the organization and determining how to measure it. Then, apply this understanding to each project and its increments.

Utilize the work packages (refer to the Work Package Portfolio Map) to identify projects for inclusion in the Implementation and Migration Plan. These projects will be further developed in subsequent steps of Phase F. Adjustments to the Architecture Roadmap and Architecture Definition Document may be necessary to accommodate these projects and their increments. Update the Work Package Portfolio Map by mapping the work packages to the projects.

Next, assign risks to the projects and their increments by consolidating risks identified in the Consolidated Gaps, Solutions, and Dependencies Matrix from Phase E.

Finally, estimate the business value for each project using the Business Value and Risk Matrix (see Section 12.9.2).

Assessing business value effectively involves creating a matrix that evaluates both business value and risk indices. The business value index should encompass criteria such as adherence to principles, financial contribution, strategic alignment, and competitive positioning. Conversely, the risk index should consider factors like project size and complexity, technological aspects, organizational capacity, and potential failure impacts. Assigning appropriate weights to each criterion is essential.

Define Business Value: Start by figuring out what business value means for the organization and how to measure it. Then, apply this to each project and its phases.

Identify Projects: Use the Work Package Portfolio Map to spot which projects should be included in the Implementation and Migration Plan. These projects will be developed further in the next steps of Phase F.

Update Architecture Deliverables: Adjust the Architecture Roadmap and Definition Document to fit the new projects and their stages. Also, update the Work Package Portfolio Map to link the projects with their work packages.

Assess Risks: Look at the risks from the Consolidated Gaps, Solutions, and Dependencies Matrix in Phase E, and apply these to the projects and their phases.

Estimate Business Value: Finally, calculate and visualize the business value for each project using the Business Value and Risk Matrix.

CHAPTER 12   ARCHITECTURE DEVELOPMENT METHOD

Estimate Resource Requirements, Project Timings, and Availability/Delivery Vehicle: This process entails providing initial cost estimates, which should be distinctly categorized into capital expenditures (to establish the capability) and operational and maintenance costs (to sustain the capability). Identifying opportunities to offset expenses by decommissioning existing systems when delivering new or enhanced capabilities is also crucial. Assigning the required resources to each activity and aggregating them at both the project increment and overall project levels ensures a structured and efficient approach to Enterprise Architecture development. Most of the activities that need to be performed during this step of Phase F will be performed by a project manager or someone in a similar role. It is important to stay involved here.

Estimate Resources: Start by estimating how much the project will cost. Break the costs into two parts: one for the initial setup (capital expenses) and another for ongoing operations (maintenance).

Look for Savings: Think about ways to save money. For example, could using old systems be stopped and replaced by new ones?

Assign Resources: Ensure each task has the right people and resources assigned to it. Tally up all the resources needed for each step of the project.

Involve Project Management: This step is mainly managed by the project manager or someone in charge. Stay involved and make sure everything is going as planned.

Prioritize Migration Projects Through Cost/Benefit Assessment and Risk Validation: This step consists of four sub-steps. The first is to *assess the business value and cost*. This is done by determining the net benefit of Solution Building Blocks delivered by projects. This involves a thorough analysis to ensure that the anticipated benefits justify the associated costs. By doing so, organizations can make informed decisions that align with their strategic objectives.

The second step is about *risk mitigation*. Identifying and mitigating risks is paramount. One of the tools that can aid the architect is creating a comprehensive review of potential risks associated with project deliverables. By proactively addressing these risks, organizations can update project plans with pertinent risk-related insights, ensuring that all potential challenges are accounted for and managed effectively.

Third, *building consensus among stakeholders* is critical to successfully prioritizing projects. The TOGAF Standard provides criteria that incorporate elements from the draft Architecture Roadmap, developed in Phase E, as well as individual stakeholder agendas. This holistic approach ensures that even projects with modest immediate benefits but critical long-term deliverables receive appropriate prioritization.

A side-by-side comparison of the requirements, desires, and concerns of all stakeholders involved can be used to prioritize. In some cases, it may be necessary to re-engage with stakeholders to address some of the concerns. It may also be useful to look at the sum of the wants and needs with the stakeholders and work together to determine the best prioritization. Stakeholders may include the following groups:

- **Architecture Team:** Ensures technical feasibility and alignment with the overall architectural vision
- **Business Stakeholders:** To prioritize projects based on business needs
- **Project Managers:** To ensure that the migration aligns with broader program or project plans
- **Finance and Resource Planners:** To ensure that budgetary and resource constraints are considered

CHAPTER 12   ARCHITECTURE DEVELOPMENT METHOD

Formal risk assessment review is the essential fourth step. This step involves reviewing and revising the risk assessment, if necessary. It ensures a comprehensive understanding of residual risks associated with project prioritization and the projected funding line, enabling organizations to proceed with confidence. This is where a Risk Management Plan is created. This plan addresses potential risks identified during migration planning, with proposed mitigation strategies. A Risk Management Plan contains at least three basic components:

- **Document Information:** Things like the project name, date, and author of the document, including the version number.

- **Approvals:** A list of the dates, names, and signatures of the people who approved the approach described in the document.

- **Approach:** A detailed description of how risks will be managed. It also includes a summary of the tools that will be used and an outline of the information that will be recorded. Perhaps more importantly, it defines the metrics that will be used to classify risks. Risk tolerance is also described in the document.

CHAPTER 12   ARCHITECTURE DEVELOPMENT METHOD

Assess Business Value and Costs: Start by evaluating how much value each project brings versus its cost. This involves understanding the expected benefits of each project and comparing them to the costs involved. This helps ensure projects contribute to the broader business goals.

Mitigate Risks: Risk Management is critical. Identify any risks tied to the projects and plan ahead to manage them. It's helpful to create a detailed review to highlight potential challenges and ensure that these risks are properly handled in the project plans. Use the Risk Management technique for this.

Build Stakeholder Consensus: Next, it's essential to get buy-in from all key players. Use criteria based on long-term goals and individual stakeholder needs to guide this process. Regular discussions might be needed to address any concerns and align priorities.

Review Risk Assessment: Finally, revisit and update the risk assessment as needed. This ensures that all risks are accounted for and the necessary mitigation strategies are in place, allowing the organization to move forward with confidence.

Confirm Architecture Roadmap: To ensure a successful implementation, it's essential to update the *Architecture Roadmap*, incorporating any necessary Transition Architectures. This involves reviewing current progress to determine appropriate intervals between Transition Architectures, considering factors such as business value increments, capability enhancements, and associated risks. Once these capability increments are finalized, consolidating deliverables by project will lead to a refined Architecture Roadmap. In practice, this means updating the work packages that were mapped to the goals and objectives in the first version of the Architecture Roadmap. This meticulous process

CHAPTER 12   ARCHITECTURE DEVELOPMENT METHOD

is crucial for coordinating the development of multiple concurrent architecture instances. Utilizing a *Transition Architecture State Evolution Table* (see Section 12.9.2) can effectively illustrate the proposed states of Domain Architectures at varying levels of detail.

If the implementation approach has evolved due to the confirmation of implementation increments, it's imperative to update the Architecture Definition Document accordingly. This update may involve setting project objectives and aligning projects and their deliverables with the Transition Architectures.

Update the Architecture Roadmap: Begin by reviewing and updating the Architecture Roadmap. Add any necessary Transition Architectures.

Review Progress: Assess how things are going and decide the right gaps for each Transition Architecture, considering business needs, new capabilities, and potential risks.

Finalize Changes: Once the needed improvements are clear, group the deliverables by project. This helps refine the roadmap.

Track Progress: Update work packages that align with the original roadmap goals.

Use Artifacts: A helpful artifact like the Transition Architecture State Evolution Table can show architecture changes at different stages.

Complete Implementation and Migration Plan: A thorough capture of all external dependencies is essential, ensuring they are incorporated into the plan. Additionally, a comprehensive assessment of resource availability is conducted to guarantee that the necessary assets are in place for successful implementation. Detailed project plans are often embedded within the overarching Implementation and Migration Plan, providing a clear roadmap for execution.

Identify All External Dependencies: List everything that could affect the project and make sure it's included in the plan.

Check Resource Availability: Ensure all needed resources (people, tools, etc.) are ready to go.

Create a Detailed Project Plan: Develop a clear roadmap for how things will happen, including key milestones.

Complete Architecture Development Cycle: Transitioning from architecture development to realization is a pivotal step to be performed. At this juncture, and if the Architecture Capability's maturity permits, crafting an Implementation Governance Model becomes essential. Documenting lessons learned during architecture development is crucial; these insights should feed into the governance processes in Phase H, thereby enhancing the Architecture Capability.

The Architecture Roadmap and the Implementation and Migration Plan should mirror the detail level of the Architecture Definition Document from phases B, C, and D. If subsequent phases demand more granularity, it indicates a shift to a different architectural tier. In such cases, initiating another Architecture Development Method cycle at a more detailed level might be necessary.

Architecture Development Cycle: Moving from architecture creation to real-world application is a crucial step. At this stage, if the team's skills allow, it's time to develop a plan for overseeing implementation (governance model).

Document Insights: Keep track of key lessons learned during development. These should be used to improve future governance processes in Phase H, which will make the architecture stronger.

Details Matter: The Architecture Roadmap and Implementation and Migration Plan should be as detailed as the Architecture Definition Documents from earlier phases (B, C, D). If more detail is needed later on, that's a sign to start a new cycle at a more detailed level.

In short, during Phase F (Migration Planning), the Transition Architectures are further developed. These are intermediate architectures that serve as a stepping stone between the Baseline and Target Architectures. They help manage risk and ensure the organization is making incremental progress. Phase F focuses on creating a detailed and practical implementation plan (the Implementation and Migration Plan) to guide the organization through its architectural transformation. It ensures that the transition is phased, manageable, and aligned with business needs, minimizing risks and maximizing value.

The Implementation and Migration Plan serves as input to Phase G (Implementation Governance), where the execution of the plan will be governed, monitored, and adjusted as needed.

12.9.2 Phase F (Migration Planning) Artifacts

Phase F of the Architecture Development Method focuses primarily on creating an Implementation and Migration Plan. Output from Phase E serves as input for Phase F. Activities include assessing the dependencies, costs, and benefits of various migration projects within the organization. Several artifacts greatly aid in realizing the Implementation and Migration Plan.

Table 12-61 lists the artifacts that can be created during the execution of Phase F (Migration Planning).

Table 12-61. *Artifacts created during Phase F (Migration Planning)*

ADM Phase	Artifacts	Used Entities
Phase F: Migration Planning	Transition Architecture State Evolution Table	Work package, organization unit, function, business service, process, logical/physical application component, data entity, logical/physical technology component
	Business Value and Risk Matrix	Work package, risk

Transition Architecture State Evolution Table: As mentioned in the previous section, it is essential to include any Transition Architectures in the Architecture Roadmap to ensure a successful architecture implementation. In order to do this, the current progress needs to be reviewed to determine appropriate intervals (and corresponding actions) between Transition Architectures. In practice, this means updating the work packages that were mapped to the goals and objectives in the first version of the Architecture Roadmap. This process is crucial for coordinating the development of multiple concurrent architecture instances. Utilizing a *Transition Architecture State Evolution Table* can effectively illustrate the proposed states of domain or Segment Architectures at varying levels of detail.

Figure 12-49 shows the proposed Solution Building Blocks in the cells under the Transition and Target Architecture headings. In this case, HR Recruitment Application A, HR Recruitment Application B, and Database

Management System are the three Solution Building Blocks used in the example. The reason they are called Solution Building Blocks rather than Architecture Building Blocks is because they all describe the specific applications in detail. Architecture Building Blocks would refer to high-level descriptions of products or processes.

Domain	Service	Transition Architecture 1	Transition Architecture 2	Target Architecture
Applications	Application Services	HR Recruitment Application A (replace)	HR Recruitment Application B (transition)	HR Recruitment Application B (new)
Technology	Database Management Services	Database Management System (retain)	Database Management System (retain)	Database Management System (retain)

🔴 To be replaced 🟠 In transition 🟢 In place (new or retained)

Figure 12-49. Transition Architecture State Evolution Table

Remember the difference between Architecture Building Blocks and Solution Building Blocks? They were described in Chapter 7, Sections 7.5.2.4 and 7.5.2.5, respectively.

Architecture Building Blocks typically describe a required capability and shape the specification of Solution Building Blocks. For example, an organization may require a Human Resources Management capability that is supported by a number of Solution Building Blocks, such as an organization unit, a business service, an application, and a data entity.

The Architecture Building Block does not specify a particular product, process, or service, so it remains high-level. Solution Building Blocks, on the other hand, do.

CHAPTER 12   ARCHITECTURE DEVELOPMENT METHOD

The Transition Architecture State Evolution Table shows the Transition Architectures in the top row. To the left, the corresponding architecture domain or segment is shown, along with the service it affects. For example, looking at the first row in Figure 12-49, it is evident that HR Recruitment Application A needs to be replaced by HR Recruitment Application B. This observation is recognized as the first transition to be made and is therefore called Transition Architecture 1.

In Transition Architecture 2, HR Recruitment Application B is introduced to the landscape. It requires the application services mentioned in the Service column to be transitioned from HR Recruitment Application A. Finally, in Transition Architecture 3, the new application (HR Recruitment Application B) is fully implemented and operational, and HR Recruitment Application A no longer exists in the landscape.

Looking at the second row of the Transition Architecture State Evolution Table, it is clear that the Database Management System exists in the Baseline Architecture (because it is retained during Transition Architecture 1). It also exists in Transition Architectures 2 and 3, which means it will also exist in the Target Architecture. The database management services supported HR Recruitment Application A until it was replaced. Later, during Transition Architectures 2 and 3, they support HR Recruitment Application B.

A Transition Architecture State Evolution Table can be created by making a table with two main columns labeled *Domain* and *Service*. These columns serve as the basis for categorizing the different domains and their related services within the architecture. For each relevant Transition Architecture, add an additional column to the right of the *Service* column. Label these columns according to the respective Transition Architecture names or identifiers. This setup allows for documenting changes specific to each Transition Architecture.

In the *Domain* column, list each pertinent architecture domain (e.g., Business, Data, Application, Technology). In the *Service* column adjacent to each domain, specify the corresponding service or component associated with that domain.

For each service listed, move across the row to the columns representing the Transition Architectures. In each cell, provide a brief description of the planned changes or developments for that service within the context of its respective Transition Architecture. This could include modifications or upgrades (in transition), deprecations (to be removed), or the introduction of new functionalities (new or retained).

Repeat this process for every domain and its associated services that are part of the project scope. This comprehensive approach ensures that all aspects of the architecture are considered and documented.

A Transition Architecture State Evolution Table provides a lot of information about the architecture work to be done. It demands a lot of the architect(s) who create the table, as it must be done meticulously for each project. The pros outweigh the cons because using an artifact such as the Transition Architecture State Evolution Table clearly details the necessary steps to be performed per architecture domain (or segment) and service for each Transition Architecture.

Business Value and Risk Matrix: Effective assessment of business value can be achieved by creating what the TOGAF Standard calls a Business Value and Risk Matrix. The artifact has two dimensions: a *business value* index on the y-axis and a *risk* index on the x-axis. The business value index includes criteria such as compliance, strategic alignment, and competitive position, while the risk index includes factors such as size and complexity, technology, organizational capacity, and impact of failure. Assigning individual weights to each criterion is critical.

Establishing decision criteria *before* populating the diagram is essential. Senior management should develop and approve the indices, criteria, and their respective weights to ensure alignment with organizational goals.

CHAPTER 12 ARCHITECTURE DEVELOPMENT METHOD

The Business Value and Risk Matrix in Figure 12-50 shows that projects A, B, C, and D deliver the highest business value. Projects D and H pose the greatest risk. However, since Project H delivers less business value than Project D, it is more likely to be canceled.

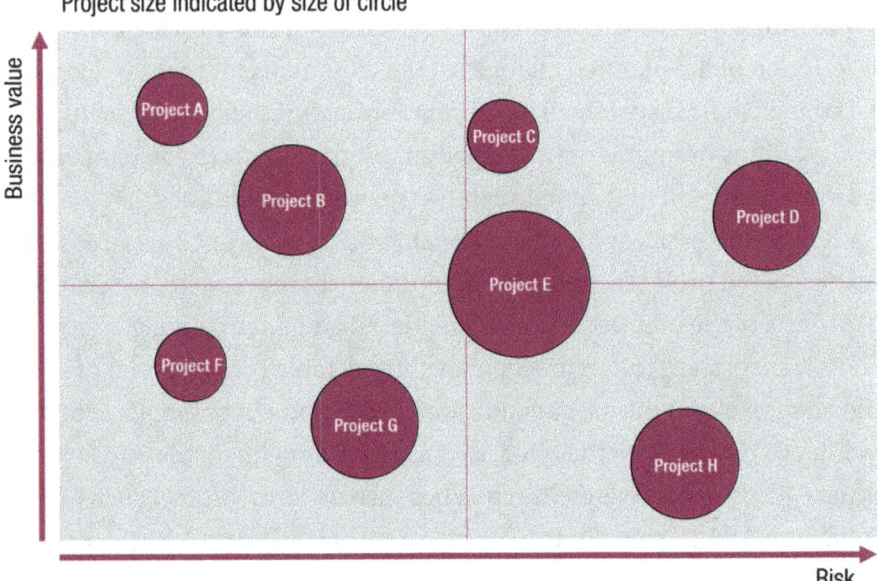

Figure 12-50. Business Value and Risk Matrix

The project size depends on the number of work packages associated with the project. Therefore, it is important to update the Work Package Portfolio Map with the projects. The Work Package Portfolio Map artifact (see Chapter 12, Section 12.8.3) can then be used to determine the project size.

The colors indicate the presence of (potential) risks.

The Business Value and Risk Matrix is used in Phase F (Migration Planning) to assign business value to work packages.

This matrix is created by identifying the criteria that determine a project's contribution to an organization. Common factors include compliance with regulations, alignment with strategic goals, and enhancement of competitive position.

It's also important to determine the factors that assess a project's potential challenges. These may include project size and complexity, technology involved, organizational capacity, and potential failure impact.

Recognize that not all criteria carry equal importance. Assign appropriate weights to each factor to reflect its significance in the assessment.

Engage senior management to review and approve the defined indices and criteria and their respective weights. This ensures the assessment aligns with the organization's goals and priorities.

Then, plot each project in the matrix based on its evaluated business value and risk. This visual representation will help compare projects and make informed decisions. Use colors to indicate whether a project is at risk, potentially at risk, or not at risk with regard to execution.

12.10. Phase G: Implementation Governance

Implementation Governance is about establishing the mechanisms to oversee the execution of the architecture project. It is where the architectural vision transitions into operational reality. Phase G involves defining the organizational structure, roles, and responsibilities to ensure that the architecture is implemented as intended. Architects work closely with project management teams to monitor progress and address any deviations from the plan. The result is a Governance Framework that helps manage risk and ensures alignment with the architecture vision. This phase ensures that the solutions developed align with the defined architecture and adhere to established standards. It's the stage where meticulous planning meets execution, and governance plays a pivotal role in overseeing this transformation.

12.10.1 Establishing Architecture Governance

The primary objectives of Phase G include *ensuring adherence to the established architecture, managing any changes* during implementation, and *continually assessing progress* to make necessary adjustments to meet the organization's objectives.

Key activities include establishing an Architecture Governance Framework that defines roles and responsibilities, monitoring the progress of implementation projects, and promptly addressing any issues that arise. This structured approach ensures that the implementation is aligned with the organization's strategic goals and the architectural blueprint.

Setting up and implementing an Architecture Governance Framework is explained in more detail in Chapter 8. Phase G (Implementation Governance) leverages this. Table 12-62 lists the steps to perform Phase G.

Table 12-62. Steps to perform Phase G (Implementation Governance)

Step to Perform	Description of Step	Output Generated by Step
Confirm scope and priorities	Identify and address gaps within the existing architecture. Review the Work Package Portfolio Map to determine which initiatives lead to optimal value extraction.	Architecture Contract
Identify deployment resources and skills	Identify the tools needed and the skills to use them.	
Guide development of solutions deployment	Map out the scope of the project. Define the strategic requirements, as well as the various change requests needed to implement the architecture.	Change requests

(continued)

Table 12-62. (*continued*)

Step to Perform	Description of Step	Output Generated by Step
Perform architecture compliance reviews	Regularly assess the implementation activities to ensure that they are consistent with the architectural standards and principles.	Compliance assessments
Implement business and IT operations	Implement business and IT service delivery, skills development, and training.	Architecture-compliant solutions
Perform post-implementation review	Examine the end result of the architecture work. If there are deviations, submit change requests to correct them.	Publication of reviews

Confirm Scope and Priorities: This step identifies and addresses gaps within the existing architecture. It involves a thorough gap analysis to pinpoint deficiencies and determine the necessary Solution Building Blocks to bridge these gaps. Solution Architects play a pivotal role in this process, meticulously defining how each Solution Building Block will be implemented. It's essential to recognize that multiple projects may target similar capabilities; therefore, Solution Architects must ensure optimal value extraction from these initiatives by leveraging existing investments. The Work Package Portfolio Map again plays an important role in providing insight into the initiatives.

To implement this first step, it's important to engage with key stakeholders and ensure alignment between the architectural plans and the development team's priorities. Set up a collaborative meeting or workshop with the development management team to confirm scope and priorities. This is essential because it fosters direct communication, where all parties can align their views and expectations. Before the meeting, review any prior architectural work, including the architecture vision, business requirements, and the broader deployment objectives. This gives context for the discussion and ensures everyone is on the same page regarding strategic goals.

Confirm the Scope and Priorities: Start by identifying gaps in the current architecture. This involves a detailed gap analysis to find where improvements are needed. Once the gaps are clear, figure out what Solution Building Blocks are required to fill them. Solution Architects lead this effort. Keep in mind that different projects might target the same goals, so it's crucial to extract the most value by leveraging past investments. The Work Package Portfolio Map can provide useful insights here.

Align with Stakeholders: It's important to make sure the architectural plans match the development team's priorities. Set up a workshop or meeting with key stakeholders to confirm scope and priorities. This direct communication helps ensure everyone is on the same page. Before the meeting, review existing architectural documents like the architecture vision and business requirements to give everyone context and align expectations with strategic goals.

Identify Deployment Resources and Skills: In the context of project resources, it's essential to ensure that development teams are well versed in the overall Enterprise Architecture deliverables and the expectations specific to their development and implementation projects. This includes identifying the system development methods required for solutions development.

The use of modeling language tools will go a long way in getting everyone involved to speak the same language.

Evaluate the readiness of the development team to implement the architecture. This includes assessing their current skill sets and competencies, tools, and processes and determining if any gaps need to be addressed before proceeding. Identify any dependencies between the architecture deployment and other ongoing initiatives or projects. Use the updated Work Package Portfolio Map for this. Make sure to understand if any existing work must be completed or aligned with the deployment.

Identify Skills and Resources: Before jumping into development, make sure the team understands the big picture of Enterprise Architecture. This includes knowing the expected deliverables and the development methods that fit the project. Using modeling tools helps everyone get on the same page and communicate clearly.

Check Readiness: Assess the team's current skills, competencies, tools, and processes. Look for gaps that could prevent them from successfully implementing the architecture. If there are any skill or tool shortages, address them before moving forward.

Identify Dependencies: Check if the architecture rollout depends on other ongoing projects. Use an updated Work Package Portfolio Map to track these dependencies. Be clear on any existing tasks that must align with the deployment to avoid delays.

Guide Development of Solutions Deployment: To perform this step, it is necessary to formulate a project recommendation. This should be done for each individual implementation. To do this, the scope of the project should be mapped out and accompanied with an impact analysis.

It's important to clarify what will and won't be included. Ensure to define clear boundaries for what will be part of the deployment in this phase. Include features, functionalities, systems, and business processes.

Engage with the stakeholders to determine which features or systems should be prioritized. This might involve evaluating which components will deliver the most immediate business value or address the highest-priority business needs.

Together with the development management team, create a priority list. This can be based on business urgency, technical dependencies, available resources, or potential risks. The *Business Value and Risk Matrix* comes in handy at this point (refer to Section 12.9.2).

CHAPTER 12   ARCHITECTURE DEVELOPMENT METHOD

The strategic requirements (from an architectural perspective) should be defined, as well as the various change requests needed to implement the architecture. This means managing any changes or deviations from the original architecture plan. Performing change management involves evaluating the impact of changes and ensuring that they are appropriately documented and approved.

Be sure to document and confirm both the functional and non-functional scope clearly. Share this document with stakeholders to ensure there are no misunderstandings about what will be deployed.

Update the Architecture Repository with the latest architecture work.

Create a Project Recommendation: Start by outlining the project, including what's in and out of scope. Be sure to clearly identify what features, systems, and processes are involved.

Engage with Stakeholders: Collaborate with key stakeholders to determine the most important features and systems to focus on based on business needs.

Prioritize Tasks: Work with the architecture development team to create a list of tasks, ranking them by business urgency, technical dependencies, and risks.

Set Strategic and Change Management Requirements: Define the key requirements for the project and anticipate any needed changes to the original plan. Ensure changes are evaluated, documented, and approved.

Document Scope and Share: Clearly document both the functional and non-functional scope of the deployment, and share with stakeholders for alignment.

Update the Architecture Repository: Don't forget to update the Architecture Repository with the latest architecture work for consistency and clarity.

Perform Architecture Compliance Reviews: As the architecture is implemented, it is important to keep track of whether or not it is going according to plan. One of the things that needs to be done during this step is to perform an ongoing review of each building block from an architecture compliance perspective. Compliance checking involves regularly assessing the implementation activities to ensure that they are consistent with the architectural standards and principles established in earlier phases. In other words, it ensures that the building blocks are being implemented in the agreed manner. If not, a new change request must be issued to address the situation.

Define what success looks like for this phase, whether it's specific performance metrics, completion of key deliverables, or meeting certain quality standards. Agree on metrics for success. This will help ensure that everyone is aligned on the desired outcomes.

Before proceeding, confirm that all stakeholders agree with the defined scope, priorities, and resources. In other words, get them to formally sign off on the work to be done. All architecture compliance audits are reviewed and coordinated with relevant stakeholders. This is crucial for ensuring there's shared ownership and accountability for the deployment. Document all the decisions made during the meeting/workshop and create a formal deployment plan. This should include the scope, priorities, milestones, timelines, resource plans, and success metrics.

Finally, tracking the progress of the implementation against established metrics and performance indicators to ensure that it meets time, cost, and quality objectives is another important part of this step. This particular part is described in more detail in Section 12.10.2.

Review the Architecture's Progress: As the architecture is being built, regularly check if it aligns with the initial plan. This includes reviewing each component (building block) for compliance with the standards set in earlier stages.

Define Success: Set clear goals for this phase, whether it's meeting performance targets, completing key tasks, or ensuring quality. Agree on these success metrics to keep everyone on the same page.

Get Stakeholder Approval: Before moving forward, confirm that all stakeholders agree with the plan, priorities, and resources. They should formally approve the work to be done.

Document and Plan: After the stakeholder meeting, document the decisions made and create a formal plan. This should include the scope, milestones, timelines, resources, and success measures.

Track Progress: Regularly monitor the project to ensure it's meeting its goals in terms of time, cost, and quality. Keep an eye on the metrics to ensure everything stays on track.

Implement Business and IT Operations: The next step in Phase G is to execute the deployment projects. These include the implementation of business and IT service delivery, skills development and training, and the publication of communication documentation. It is important to publish new baseline architectures to the Architecture Repository and update other affected repositories, such as operational configuration management stores (such as a CMDB, if one exists).

Deploy Business and IT Operations: The first thing to do is to roll out the deployment projects. This includes setting up how the business and IT services will be delivered, plus ensuring the teams have the right skills through training.

Publish Documentation: Next, it's key to share any important communication materials. This includes publishing updated architectures to the Architecture Repository. Make sure to also update other systems that manage configurations, like a CMDB, if there is one.

Keep Everything Updated: Lastly, don't forget to regularly check and refresh all related documents and repositories to stay on top of any changes.

Perform Post-implementation Review: Once the previously conceived architecture has been implemented, it is important to complete the work. To do this, post-implementation reviews should be performed. These reviews examine the end result of the architecture work. If there are deviations from the original Implementation and Migration Plan, change requests should be submitted to correct the deviations. At this point, Phase H (Architecture Change Management) is initiated.

The actual completion of the work is done by publishing the reviews performed and formally closing the projects. The involvement of project managers is desirable for this last step.

One other important factor to consider throughout Phase G is Risk Management. This technique is used to identify and mitigate risks associated with the implementation to minimize disruptions and ensure successful delivery. Risk Management is further detailed in Chapter 14, Section 14.8.

Do a Post-implementation Review: Once the architecture is set up, take time to evaluate how everything turned out.

Check for Differences: Compare the final result to the original plan. If there are discrepancies, request changes to fix them.

Start Architecture Change Management: This marks the beginning of Phase H (Architecture Change Management).

Wrap Up the Project: Finalize everything by sharing the review results and formally closing the project. It's helpful if project managers are involved in this step.

Phase G (Implementation Governance) serves as the bridge between meticulously crafted designs and their real-world deployment. It serves as a framework for overseeing the architecture implementation, ensuring compliance, managing changes, and engaging stakeholders effectively. It is essential for achieving the intended outcomes of the architectural vision laid out in the previous phases of the Architecture Development Method.

Through effective governance and control mechanisms, organizations can confidently move forward, knowing that their architectural vision is being realized in a controlled and aligned manner. Implementation governance is critical for ensuring that the architecture delivers the desired business value and that any issues are addressed promptly.

In essence, Phase G is where the rubber meets the road. It helps to minimize risks, manage stakeholder expectations, and ensure alignment with the organization's strategic goals.

12.10.2 Architecture Performance Monitoring

Monitoring the performance of the implementation of the architecture – in particular the implementation of the Target Architecture – can be done by creating dashboards and/or graphs. These tools allow not only the architects but all stakeholders to monitor and adjust the architecture development as needed as the implementation progresses.

There are two main areas that should be addressed when measuring progress. These are tracking the completion of work packages and measuring the achievement of goals and objectives.

12.10.2.1. Measuring Progress in Completing Work Packages

During the development of the architecture, several work packages need to be executed and completed in order to realize the Target Architecture. Most of the work packages consist of one or more activities that need to be performed.

It is necessary to track the progress of these activities to get a good idea of the progress of the architecture development. This can be achieved by measuring the completion of the work packages containing the activities. In order to help determine their progress, a percentage can be assigned to each activity within a work package. Executing all activities normally leads to the completion of the work package. The method of assigning percentages should be applied to each individual activity for which progress is to be measured.

Let's say that the execution of a work package involves four separate activities. Each activity can be assigned one-fourth of the total, leading up to 25% per activity.

Once these four activities have been successfully performed, the completion of the work package has been realized for 100% (see Figure 12-51). This is of course a very basic approach as it assumes that a proportional percentage is assigned to all activities of the work package to be completed.

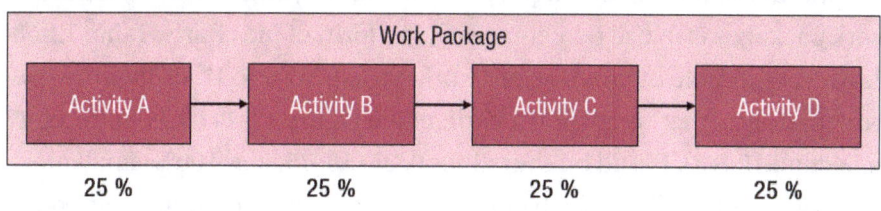

Figure 12-51. Basic method for assigning percentages to activities

CHAPTER 12   ARCHITECTURE DEVELOPMENT METHOD

A more refined method involves assigning percentages based on, for example, the duration of each activity. When the more refined method (Figure 12-52) is used, for example, assigning a percentage based on the duration of each activity, the distribution of percentages is often quite different. To illustrate, consider a work package consisting of four activities, A, B, C, and D. Activity A takes one hour to complete. Activities B and C each take two hours. Activity D has a lead time of three hours.

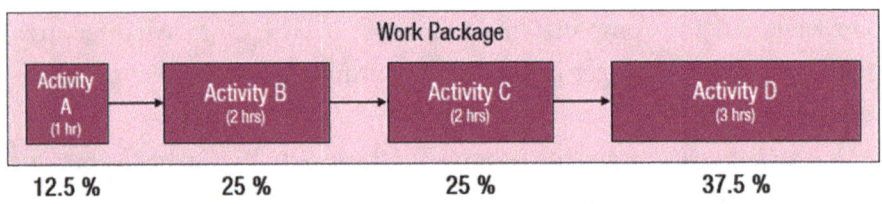

Figure 12-52. More refined method of assigning percentages to activities

Using the more refined method of assigning percentages in proportion to the duration of the activities, the percentages for activities A–D would be 12.5, 25, 25, and 37.5%, respectively. Thus, assigning percentages based on the more refined method provides a more detailed picture of progress than using the basic concept of proportional percentages.

When the two methods are applied to a work package in which three of the four activities are completed, the basic method yields a rate of 75%. The more refined method shows that completion of the first three activities accounts for 62.5%. The latter method, therefore, provides a more realistic picture of the progress being made.

Another common technique for charting the progress of individual work packages is to use percentages in the form of pie charts. Table 12-63 shows an example of progress per work package. Table 12-63 lists the five work packages that were shown earlier (see Figure 12-44 and Figure 12-47 in Section 12.8.3). Based on the values 0, 25, 50, 75, and 100%, small pie charts are used to illustrate the progress of *each individual work package*.

Table 12-63. *Progress Summary per Work Package*

Work Package	Description	Progress
Work Package A	Fit job profile X into existing job profile.	● (100%)
Work Package B	Recruit staff for job profile X.	◕ (75%)
Work Package C	Distribute updated job profiles.	◐ (50%)
Work Package D	Update compensation application.	○ (0%)
Work Package E	Update job information.	○ (0%)

A third way to track progress is to use spider or bar charts. These types of charts use percentages to plot the current state against the desired end state, 100%. Which form of visualization is chosen depends largely on personal preference. None of the forms described is more or less useful than the others.

12.10.2.2. Measuring Achievement of Goals and Objectives

Aside from measuring the progress of the completion of individual work packages and their activities, it is also very useful to track the progress being made on the realization of the organization's goals and objectives.

As mentioned in Section 12.8.3, the Architecture Roadmap contains an overview of the initiatives that need to be deployed in order to realize the Target Architecture. The initiatives are related to the organization's goals and objectives.

Using the same approach as described for charting progress for each work package, progress toward goals and objectives can also be charted.

To do this, it is important to assign a percentage to the progress of each initiative (work package). By using the approach described in Section 12.10.2, a progress percentage can be assigned to each individual work package. Add the percentages of all the work packages related to a specific goal or objective, and divide by the number of work packages. The resulting number can be used as the average progress percentage per goal or objective.

As an example, suppose an Architecture Roadmap contains two goals to be realized (see Figure 12-53). Goal A has two work packages (A and B) assigned to it. Goal B is linked to work packages C, D, and E. Work Package A is 100% complete and Work Package B is 75% complete. Goal B is related to Work Package C, which is only 50% complete and currently on hold. The activities belonging to work packages D and E have not yet started, so the percentage for these work packages is 0%.

CHAPTER 12 ARCHITECTURE DEVELOPMENT METHOD

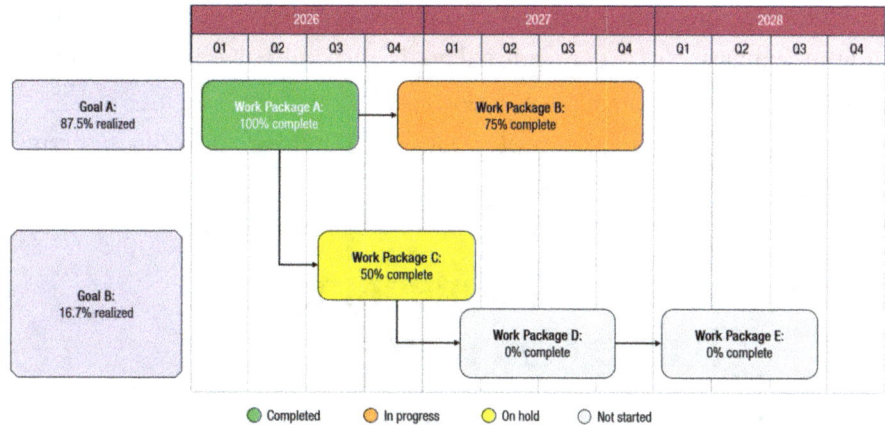

Figure 12-53. *Architecture Roadmap showing work package progress percentages*

The result of averaging work packages A and B is 87.5% (100% + 75% / 2). The same calculation for work packages C, D, and E yields 16.7% (50% + 0% + 0% / 3). Thus, goals A and B are 87.5% and 16.7% realized, respectively. Figure 12-54 illustrates the above example.

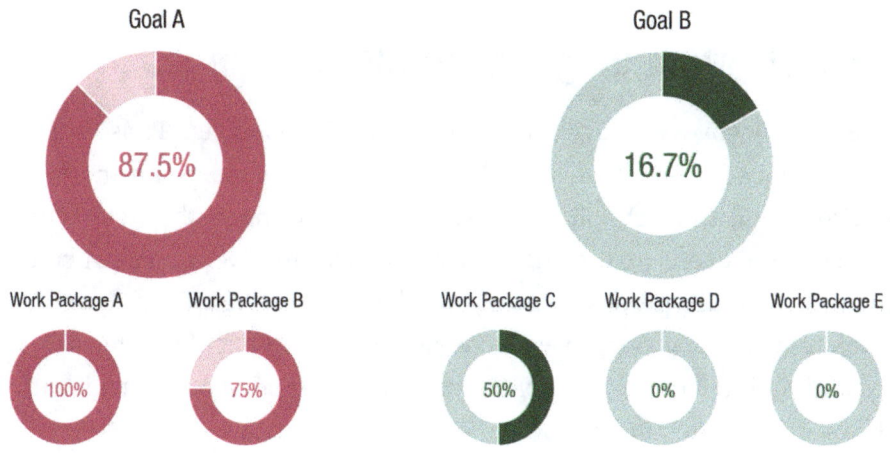

Figure 12-54. *Goal realization progress*

Understanding the progress of both the completion of work packages and the realization of organizational goals and objectives is important to both the architecture team and senior management. The method described here is relatively simple. However, it helps to quickly understand the progress of both concepts.

12.11. Phase H: Architecture Change Management

Architecture is not static, and the environment in which it operates evolves over time. The Architecture Change Management Phase is concerned with managing changes to the architecture throughout its lifecycle. It includes assessing the impact of changes, defining procedures for making changes, and ensuring that the architecture remains aligned with business goals.

Phase H (Architecture Change Management) of the Architecture Development Method is critical because it ensures that the architecture remains relevant and aligned with business needs as they evolve over time.

12.11.1 Managing Architecture Changes

The key objectives of the Architecture Change Management Phase are to *continuously assess changes* in the organizational strategy, business environment, and technology landscape that may impact the architecture. It is essential to ensure that there are processes and governance structures in place to handle proposed changes to the architecture effectively. Therefore, it is necessary to evaluate the potential impact of changes on existing architecture components, ensuring that modifications align with the overall architectural vision.

When implementing approved changes, it is important to facilitate the process while minimizing disruption to ongoing operations. Stakeholders need to be kept informed about changes to the architecture, ensuring that they understand the implications and benefits.

CHAPTER 12　ARCHITECTURE DEVELOPMENT METHOD

After implementing changes, it is vital to review the architecture to validate that it continues to meet business requirements and that the changes have been successfully integrated. Table 12-64 lists the steps to perform Phase H (Architecture Change Management).

Table 12-64. Steps to perform Phase H (Architecture Change Management)

Step to Perform	Description of Step	Output Generated by Step
Establish value realization process	Implement a change management process to govern modifications to the architecture in order to align changes with business value and maintain architectural integrity.	Architecture Compliance Report
Deploy monitoring tools	Deploy tools to monitor the business and technology changes that could impact the architecture, as well as the maturity of the Architecture Capability.	
Manage risks	Identify, classify, and mitigate risk(s) before business transformation.	Risk Management Plan
Provide analysis for architecture change management	Analyze information from monitoring tools, stakeholder feedback, and performance reviews.	

(*continued*)

Table 12-64. (*continued*)

Step to Perform	Description of Step	Output Generated by Step
Develop change requirements to meet performance targets	Identify and define necessary modifications to the architecture.	Changes to the architecture framework and principles (minor changes) New Request for Architecture Work (major changes)
Manage governance process	Develop policies and guidelines that outline how changes should be proposed, reviewed, and approved.	
Activate the process to implement change	Create a structured rollout plan with timelines, dependencies, and milestones.	Architecture Contract

Establish Value Realization Process: Establishing a robust value and change management process is essential for governing the evolution of an Enterprise Architecture post-deployment. This process delineates the specific scenarios that permit modifications to the Enterprise Architecture or its components and outlines the procedures for implementing such changes. Additionally, it defines the conditions that trigger a new cycle of the Architecture Development Method to develop a refreshed architecture.

The architecture change management process is intricately linked to the organization's Architecture Governance Framework (see Chapter 8) and the management of the Architecture Contract between the architecture team and business stakeholders. During Phase H, it is imperative for the

CHAPTER 12   ARCHITECTURE DEVELOPMENT METHOD

governance body to establish clear criteria to assess whether a change request necessitates a simple architecture update or the initiation of a new Architecture Development Method cycle.

An Architecture Compliance Report plays a crucial role in this process by evaluating whether proposed changes adhere to the current architecture. In cases of non-compliance, exemptions may be granted if justified appropriately. For changes with significant architectural impact, a comprehensive strategy should be devised to manage their effects.

An Architecture Compliance Report should state whether the change is compliant to the current architecture. If it is non-compliant, an exemption may be granted with valid rationale. If the change has high impact on the architecture, then a strategy to manage its impact should be defined [20].

Providing universal guidelines for establishing these criteria is challenging due to the varying risk appetites of organizations. The iterative nature of the Architecture Development Method allows an organization to continue to mature its governance body. Each time the Architecture Development Method is used, the organization is able to take a small step forward toward a higher maturity level (the use of maturity models is explained in Chapter 15). Over time, this leads to the development of criteria that are tailored to the organization's specific needs.

Define When to Change: Clearly outline the situations that allow for modifications to the Enterprise Architecture or its parts.

Set Up Procedures: Develop step-by-step methods for making these changes when the defined situations occur.

Know When to Restart: Identify the conditions that require starting a new cycle of the Architecture Development Method to create an updated architecture.

531

Deploy Monitoring Tools: This step entails deploying tools to monitor the business and technology changes that could impact the architecture, as well as the maturity of the Architecture Capability. In practice, tools such as a change management application are often used for this in combination with project management methods such as PRINCE2 or service management methods such as ITIL. To monitor the maturity of the Architecture Capability, maturity models provide a convenient method.

Choose the Right Tools: Select tools that can track changes in the business and technology environments. Change management applications are commonly used for this purpose.

Integrate with Management Methods: Combine these tools with established project management methods like PRINCE2, or service management frameworks such as ITIL, to ensure comprehensive oversight.

Assess Architecture Maturity: Utilize maturity models to evaluate and monitor the development of the Architecture Capability. These models provide a structured way to measure progress and identify areas for improvement.

Manage Risks: Embarking on an architecture or business transformation inherently involves risks. Identifying, classifying, and mitigating these risks before initiation is crucial for effective tracking throughout the transformation process.

Continuous risk monitoring is performed during Phase H with respect to changes to the architecture. The Architecture Change Management Phase identifies risks associated with new requirements, technologies, or business changes. It implements risk response mechanisms as part of change management. Risk Management is further detailed in Chapter 14, Section 14.8.

Monitor Continuously: Keep a constant watch on the transformation process, as new risks can emerge unexpectedly, such as those from mergers or acquisitions.

Establish Governance: Set up a governance framework to formally accept and manage risks, ensuring they are addressed appropriately throughout the transformation.

Adapt to Changes: During the Architecture Change Management Phase, identify risks associated with new requirements, technologies, or business changes, and implement appropriate responses as part of change management.

Provide Analysis for Architecture Change Management: During this step, it is important to gather information from monitoring tools, stakeholder feedback, and performance reviews. Based on the information gathered, various scenario analyses and impact assessments can be performed to understand the potential impact of the changes to the architecture.

Collaboration with key stakeholders is very important. Work with business units, IT, and external partners to validate the results of the analyses. Then present the findings in a format that decision-makers can use to drive action. Useful formats include executive summaries and heat maps.

Remember that change analysis is not a one-time activity. It needs to be an ongoing process. Therefore, it is important to create feedback loops. This will allow the organization to establish a mechanism where architectural changes are continually reviewed and improved.

Gather Information: Begin by collecting data from various sources such as monitoring tools, stakeholder feedback, and performance reviews. This comprehensive information forms the foundation for informed decision-making.

Perform Scenario Analyses and Impact Assessments: Utilize the gathered data to conduct various scenario analyses and impact assessments. This helps in understanding the potential effects of proposed changes on the architecture and operations.

Collaborate with Key Stakeholders: Engage with business units, IT teams, and external partners to validate the results of your analyses. Their insights and perspectives are invaluable in ensuring the accuracy and feasibility of proposed changes.

Present Findings to Decision-Makers: Compile the analysis into actionable insights and present them in formats that facilitate decision-making. Executive summaries and heat maps are effective tools for conveying complex information succinctly.

Establish Continuous Feedback Loops: Recognize that change analysis is not a one-time activity. Implement mechanisms for ongoing review and improvement of architectural changes. Continuous feedback loops enable the organization to adapt proactively to emerging challenges and opportunities.

Develop Change Requirements to Meet Performance Targets: This step refers to identifying and defining the necessary modifications to the Enterprise Architecture to ensure it continues to support the organization's performance goals. It involves evaluating how well the current architecture meets business and IT performance targets. Ensuring that the proposed changes align with business strategies and deliver measurable value is another thing to consider.

An important input to this step are the *gap analysis results* that were created during the Architecture Development Method phases B, C, and D. The gaps described can be used to indicate what changes are needed to comply with the proposed architecture.

Changes may be made to the architecture framework itself, including the Architecture Principles. If this is the case, it may be necessary to update the Enterprise Metamodel and/or the Content Framework so that the architecture developed from that point forward reflects these changes. These modifications are often referred to as *minor changes*. If larger or more complex changes to the architecture are required, a new Request for Architecture Work must be created. This document is used to define and describe the *major changes* that trigger a new cycle of the Architecture Development Method.

Essentially, developing change requirements ensures that the architecture remains agile, continuously evolving to meet business and technical performance expectations.

Too often, organizations are left with an Enterprise Architecture that fit yesterday's organization, but may not provide sufficient capabilities to meet the needs of today's and tomorrow's organization.

In many cases, the architecture still fits, but the underlying solutions may not, and some changes are required. The Enterprise Architect must be aware of these change requirements and view them as an essential part of the ongoing renewal of the architecture [20].

CHAPTER 12   ARCHITECTURE DEVELOPMENT METHOD

Use Gap Analysis Results: Use the results from previous analyses (specifically from phases B, C, and D of the Architecture Development Method) to pinpoint discrepancies between the current architecture and desired outcomes.

Identify Necessary Changes: Based on the gap analysis, determine the specific modifications required to align the architecture with performance targets. Ensure these changes are in harmony with the organization's business strategies and are expected to deliver measurable benefits.

Update Architecture Framework: If the identified changes affect foundational elements like Architecture Principles, update relevant documents such as the Enterprise Metamodel and Content Framework to reflect these adjustments.

Document Major Changes: For significant or complex modifications, prepare a Request for Architecture Work document. This formalizes the proposed changes and may initiate a new cycle of the Architecture Development Method.

Manage Governance Process: To manage the change governance process, it is advisable to define who approves changes to the architecture and what criteria will be used to evaluate them. A necessary step is to develop policies or guidelines that outline how changes should be proposed, reviewed, and approved. Involve the Architecture Board in this process, as it can help to monitor changes. Governance should always be aligned with organizational standards, regulatory requirements, and best practices. Use scorecards or dashboards to track governance effectiveness and make necessary adjustments.

Identify Decision-Makers: Determine who will be responsible for approving changes to the architecture. This could be an individual or a group, such as an Architecture Board.

Set Evaluation Criteria: Define clear criteria that will be used to assess proposed changes. This ensures that every change is evaluated consistently and objectively.

Develop Policies and Guidelines: Create comprehensive policies or guidelines that outline the procedures for proposing, reviewing, and approving changes. This provides a structured approach and sets clear expectations for all stakeholders.

Engage the Architecture Board: Involve the Architecture Board in the governance process. Their role is to oversee and monitor changes, ensuring that each modification aligns with the organization's strategic objectives and architectural principles.

Align with Standards and Regulations: Ensure that the governance process adheres to organizational standards, complies with regulatory requirements, and incorporates industry best practices. This alignment helps maintain consistency and legal compliance.

Monitor and Adjust: Utilize tools like scorecards or dashboards to track the effectiveness of the governance process. Regularly review these metrics and make necessary adjustments to improve and adapt to evolving organizational needs.

Activate the Process to Implement Change: Once all the necessary changes to the architecture have been formally approved, it is essential to create a structured rollout plan with timelines, dependencies, and milestones. This will help ensure that all stakeholders (executives, IT teams, end users) understand the changes and their role in the process. Communicating with key stakeholders has proven to be essential in a number of different areas, and this is no exception. Communication is key.

Implementing changes is best done in a phased approach. This will almost always minimize disruption and allow for adjustments if needed. Be sure to continually track the results of the change and refine the implementation approach as needed.

Get Approval for Changes: Before initiating any modifications, ensure that all proposed changes have received formal approval from the relevant decision-makers.

Create a Detailed Rollout Plan: Develop a comprehensive plan that outlines the timeline, identifies dependencies, and sets clear milestones. This plan will serve as a roadmap for the implementation process.

Inform All Stakeholders: Communicate the details of the upcoming changes to everyone involved, including executives, IT teams, and end users. Clearly explain their roles and responsibilities to ensure alignment and understanding.

Implement Changes in Phases: Adopt a phased approach to roll out the changes. This strategy helps minimize disruptions and provides opportunities to make adjustments as necessary.

Monitor and Adjust: Continuously track the outcomes of each phase. Use the insights gained to refine the implementation strategy and address any issues promptly.

Each of the steps described in this phase requires continuous engagement, data-driven decision-making, and an iterative mindset. The key is not just to define a process but to ensure it remains flexible, responsive, and aligned with evolving business needs.

In short, Phase H (Architecture Change Management) is about identifying the drivers for change. This means identifying internal and external factors that require change, such as shifts in business strategy, technological advances, or regulatory requirements. It is essential to formulate change proposals that detail the rationale, expected benefits,

CHAPTER 12   ARCHITECTURE DEVELOPMENT METHOD

and impact of the changes. The proposed changes must then go through a thorough analysis process. This is where risk, cost, and impact on existing architectures are determined.

The next step is to follow established governance processes to approve or reject the proposed changes, always involving relevant stakeholders. A detailed plan for implementing the approved changes is then developed. This plan must include timelines, resources, and responsibilities. During the implementation of the changes, it is imperative to ensure effective communication and coordination among the teams. After the changes have been implemented, the results must be monitored and feedback gathered. The feedback is necessary to make informed adjustments as needed.

The importance of architecture change management is that it enables the organization to adapt quickly to market or technology changes, ensuring that the architecture can support evolving business needs. Change management provides a method for systematically evaluating and managing change. As a result, organizations can reduce the risks associated with architecture changes.

And, just as importantly, architecture change management helps maintain alignment between the architecture and business goals, ensuring that the architecture delivers maximum business value.

12.12. Requirements Management

The Requirements Management Phase plays a vital role in ensuring that the architecture development process aligns with the business needs and stakeholder expectations.

As the name already suggests, the Requirements Management Phase is all about identifying requirements. This entails gathering and documenting requirements from stakeholders, which ensures a comprehensive understanding of their needs and expectations. These requirements then need to be managed so they can be made available when other phases of the Architecture Development Method call on them. It is important to establish a system for tracking and managing requirements throughout the entire Architecture Development Method cycle. This includes categorizing, prioritizing, and versioning requirements as needed.

Requirements need to be continuously refined and clarified based on feedback and evolving business needs. This may involve engaging with stakeholders to validate and adjust requirements as necessary. Make sure to verify that the requirements align with the overall architecture vision and objectives by assessing how changes in requirements might impact the architecture. Sharing requirements with relevant stakeholders will ensure transparency and buy-in. Clear communication helps to manage expectations and reduce misunderstandings.

12.12.1 Capture, Manage, and Refine Requirements

Although not considered a separate phase, Requirements Management is a continuous activity that runs throughout the Architecture Development Method. It involves *managing and maintaining requirements* and ensuring that they are properly addressed during each phase. As the architecture evolves, new requirements may emerge, and existing requirements may change. Effective requirements management ensures that the final architecture meets all the necessary criteria.

The Requirements Management Phase is designed to capture, manage, and refine the requirements that inform and drive the architecture development process throughout the Architecture Development Method cycle. This ensures that the architecture is aligned with the business goals and objectives, something that cannot be stressed enough.

The key objective of this phase is to make sure that the requirements management process is available and applied during all phases of the Architecture Development Method. This phase ensures that all requirements that are identified during an Architecture Development Method cycle – or even during a specific phase of a cycle – are thoughtfully managed.

CHAPTER 12　ARCHITECTURE DEVELOPMENT METHOD

Table 12-65 lists all the steps to be performed during the Requirements Management Phase and during other phases of the Architecture Development Method.

Table 12-65. Steps to Perform Requirements Management

No.	Requirements Management Phase Step	Architecture Development Method Phase Step
1		Identify and document requirements.
2	Establish baseline requirements and document them in the Requirements Repository.	
3	Monitor baseline requirements.	
4		Identify new and changed requirements.
5	Identify changed requirements and record priorities.	
6		Assess impact of changed requirements on current and previous phases.
7		Implement requirements arising from Phase H.
8	Update the Architecture Requirements Repository with information relating to the changes requested.	
9		Implement change in the current phase.
10		Assess and revise gap analysis for past phases.

CHAPTER 12   ARCHITECTURE DEVELOPMENT METHOD

1. **Identify and Document Requirements:** Phase A (Architecture Vision) outlines the architecture to be developed. All requirements from the previous phase are recorded in the Requirements Repository of the Architecture Repository (see Figure 12-55).

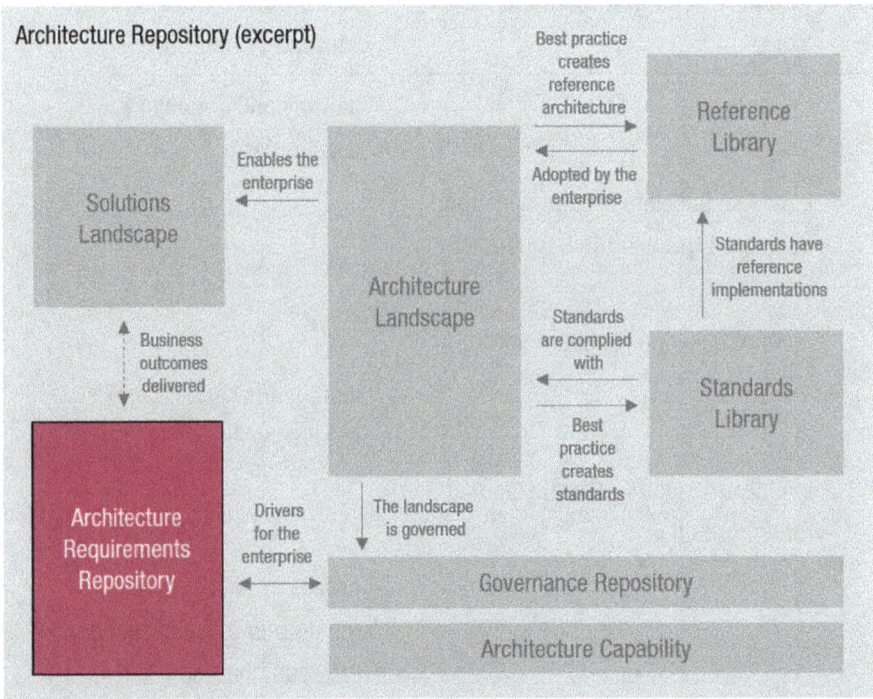

Figure 12-55. Excerpt of the Architecture Repository, showing the Requirements Repository

2. **Establish Baseline Requirements:** Performing one or more iterations of the Architecture Development Method during phases B, C, and D identifies the requirements for the Business, Data, Application, and Technology Architectures to be created. During Phase E (Opportunities and Solutions), the

interoperability requirements for the architecture segments are identified. All of the requirements from the above phases are also recorded in the Requirements Repository of the Architecture Repository (see Figure 12-55).

3. **Monitor Baseline Requirements:** This step involves the continuous tracking and management of the established requirements that serve as a foundation for the architecture. This process ensures that the requirements remain relevant, valid, and aligned with business objectives throughout the architecture development lifecycle. Requirements are actively tracked to detect any changes due to evolving business needs, regulations, or technological advancements.

4. **Identify New and Changed Requirements:** Based on the priorities that have been set, requirements should be added, or existing requirements should be changed. As the priorities change, the requirements should reflect those changes. It is a constant game of re-evaluating and updating requirements as business priorities are adjusted.

5. **Identify Changed Requirements and Record Priorities:** When moving through the Architecture Development Method, it is important to identify changed requirements. They must be properly prioritized by the architect working on a particular phase. This involves validating not only the

priorities but also the requirements with the relevant stakeholders. Be aware of any conflicts that may arise. They must be managed through the phases of the Architecture Development Method to a successful conclusion.

6. **Assess Impact of Changed Requirements:** If requirements change, either during the current phase or the previous phase, their impact must be assessed. It could mean that a change needs to be implemented. This change could be implemented during the current iteration of the Architecture Development Method or postponed to a later iteration.

7. **Implement Requirements Arising from Phase H:** During Phase H, any predefined changes to the architecture are implemented. In some cases, this may result in the creation of new or updated requirements. If these requirements are then validated and prioritized, they can be made available to new iterations of the Architecture Development Method and used in future architecture changes.

8. **Update the Architecture Requirements Repository:** After all requirements changes have been prioritized and validated, the Architecture Requirements Repository must be updated to contain the correct requirements.

9. **Implement Change in the Current Phase:**
   This is the point at which all changes to the architecture are implemented. At this point, no specific action is required from the Requirements Management Phase.

10. **Assess and Revise Gap Analysis for Past Phases:**
    During the gap analysis performed in phases B–D, requirements known as *gap requirements* (identifying the gaps between the Baseline and Target Architectures) may have been identified. This step ensures that they are addressed, documented, and recorded in the Architecture Requirements Repository and that the Target Architecture is revised accordingly.

Managing requirements is essential during architecture development because it ensures that the architecture remains aligned with business objectives, reduces risks, and supports successful project execution.

12.12.2 Requirements Management Artifacts

The Requirements Management Phase recognizes only a single essential deliverable (see Table 12-66). This is a comprehensive list of documented requirements, the Requirements Catalog. This catalog is often accompanied by a requirements management plan that outlines how requirements will be tracked and managed. Together, these two outputs ensure that updates to the architecture vision and models are based on refined requirements.

CHAPTER 12   ARCHITECTURE DEVELOPMENT METHOD

Table 12-66. Artifacts created during Requirements Management

ADM Phase	Artifacts	Used Entities
Requirements Management	Requirements Catalog	Requirement, objective, goal, assumption, constraint, gap

Requirements Catalog: The Requirements Catalog captures things that the organization needs to do to meet its objectives. It is therefore extremely important to link requirements to business goals or objectives whenever possible. This ensures that requirements can always be traced back to the realization of business goals based on organizational strategy.

Requirements generated from architecture engagements are typically implemented through change initiatives identified and scoped during Phase E (Opportunities and Solutions). Requirements can also be used as a quality assurance tool to ensure that a particular architecture is fit for purpose (i.e., the architecture can meet all identified requirements). An example of a Requirements Catalog is shown in Table 12-67.

Table 12-67. Requirements Catalog

Requirement	Goal
A centralized digital feedback platform needs to be made available that enables real-time collection and routing of patient satisfaction and safety-related feedback to appropriate departments.	Improve patient satisfaction. Enhance patient safety.
Clinical workflows must be digitized and integrated into the hospital's EHR system, allowing automated role-based task assignments.	Optimize clinical workflows. Increase staff productivity.
A cost-tracking dashboard must be implemented that links departmental spending to patient outcomes, enabling quarterly reviews of cost-effectiveness.	Reduce operational costs.

CHAPTER 12   ARCHITECTURE DEVELOPMENT METHOD

Requirements management is not a one-time activity. It is continuous throughout the Architecture Development Method cycle. As the architecture evolves, requirements may change, and ongoing management ensures that the architecture remains relevant and effective.

The Requirements Management Phase in the Architecture Development Method is essential for aligning the architecture with business goals, ensuring that stakeholder needs are met, and maintaining the integrity of the architecture throughout its lifecycle.

12.13. The Architecture Development Method in Short

A lot happens during the implementation of the Architecture Development Method. This is not surprising since it is the heart of the TOGAF Standard. Although the Architecture Development Method is called an approach, it is really a logical method that groups key activities to clarify their relationships and information flow. Each phase of the Architecture Development Method consists of a number of steps (process steps) that *can* be performed.

I say *can* here deliberately, because it is certainly not the case that all the steps of the Architecture Development Method must be implemented. This is something that is often misunderstood.

One of the most common complaints or (negative) comments about the use of the TOGAF Standard is that many architects assume that everything in the Architecture Development Method has to be implemented. This is completely wrong and misguided.

The whole idea behind the TOGAF Standard – and therefore the Architecture Development Method – is to provide tools and a description of all the actions that can potentially be performed when the organization or architecture work requires or needs them.

Tailoring the framework is paramount, and it is not for nothing that the very first chapter of the TOGAF Standard refers directly to tailoring.

I would therefore like to reiterate that the Architecture Development Method in no way prescribes that all the steps must be performed. Even the order of the steps is not fixed.

In fact, the Architecture Development Method provides an iterative approach (see Chapter 13.3) that can be run over and over again, where skipping steps (e.g., when using an agile approach) is actually supported and even encouraged.

This chapter explained in detail what can happen during each phase of the Architecture Development Method. To make things easier, here is a brief summary of each phase and the corresponding steps to be performed, including a short explanation of each step.

Preliminary Phase:

- **Define the Enterprise:** By defining the organization, the scope of the impacted organization is determined, allowing architects to create an understanding of the organization and its structure.

- **Identify Key Drivers and Organizational Context:** Pinpointing the critical factors that influence the architecture, such as key drivers and the organizational context.

- **Manage Relationships Between Frameworks:** Understanding how various management frameworks interact with each other and creating a framework to manage the architecture activities.

- **Assess Enterprise Architecture Maturity:** Evaluating where the organization stands in its architectural development.

- **Establish the Enterprise Architecture Team:** Establishing an architecture team to perform the architecture work.

- **Set Architecture Principles:** Defining the principles that will guide all architecture-related efforts.

- **Choose and Tailor the Architecture Framework:** Identifying and tailoring the TOGAF Standard, Architecture Development Method, and the Content Framework.

Phase A (Architecture Vision):

- **Identify Stakeholders and Determine Requirements:** Identify stakeholders and requirements to allow for the architecture and related work to be communicated and aligned.

- **Assess Readiness for Business Transformation:** Evaluate the organization's change readiness, guiding the architecture's scope and highlighting areas requiring attention.

- **Elaborate on Business Goals and Architecture Principles:** Validate and refine the business goals and Architecture Principles identified in the previous phase.

- **Create a High-Level Architecture Vision:** Develop a high-level description of the architecture solution that will meet business requirements and outline how it aligns with the overall business strategy.

Phase B (Business Architecture),
Phase C (Information Systems Architectures),
Phase D (Technology Architecture):

- **Select Reference Models, Viewpoints, and Tools:** Choose relevant models and viewpoints from the Architecture Repository to match the business goals and stakeholder interests.

- **Develop Baseline and Target Business Architecture Description:** Describe the current and future Business, Data, Application, and Technology Architecture to a degree that supports the architecture vision, building on the Architecture Repository's resources as needed.

- **Perform Gap Analysis:** Confirm internal consistency, identify any gaps between baseline and target, and validate that models align with business objectives.

- **Define Candidate Roadmap Components:** Using the baseline, target, and gap analysis, build an initial roadmap to guide future actions and prioritization across the remaining architecture phases.

- **Resolve Impacts Across the Architecture Landscape:** Assess the Business, Data, Application, and Technology Architecture's wider impacts across the organization and its alignment with existing or planned projects.

CHAPTER 12 ARCHITECTURE DEVELOPMENT METHOD

- **Conduct Formal Stakeholder Review:** Revisit the architecture's initial motivation and make any necessary refinements to ensure it's fit to support other architecture domains.

- **Finalize the Business, Data, Application, and Technology Architecture:** Standardize building blocks, fully document them, and check alignment with business goals. Finalize documentation within the Architecture Repository.

- **Create the Architecture Definition Document:** Complete all relevant sections in the Architecture Definition Document, adding models, reports, and graphics as needed to capture a clear architectural view.

Phase E (Opportunities and Solutions):

- **Determine/Confirm Key Corporate Change Attributes:** Align the architecture with an organization's business culture. This involves creating an Implementation Factor Catalog and assessing the transition capabilities of organizational units.

- **Determine Business Constraints for Implementation:** Identify business drivers that could compromise the Target Architecture. If there are any constraints, try to resolve them.

- **Review and Consolidate Gap Analysis Results from Phases B–D:** Consolidate and integrate gap analysis results from the Business, Information Systems, and Technology Architectures.

- **Consolidate and Reconcile Interoperability Requirements:** Identify and address constraints that may hinder interoperability.

- **Refine and Validate Dependencies:** Analyze dependencies by grouping related activities.

- **Confirm Readiness and Risk for Business Transformation:** Review the Business Transformation Readiness Assessment to inform the Architecture Roadmap and implementation strategy.

- **Formulate Implementation and Migration Strategy:** Develop an Implementation and Migration Strategy by selecting a strategic approach and applying methodologies to effectively implement the Target Architecture while mitigating identified risks.

- **Identify and Group Major Work Packages:** Assess and address gaps in business capabilities in order to enable the organization to effectively plan and implement necessary changes aligned with strategic objectives.

- **Identify Transition Architecture(s):** Identify Transition Architecture(s) to provide structured, value-driven steps toward achieving the Target Architecture, tailored to the organization's implementation strategy and capacity for change.

- **Create Architecture Roadmap and Implementation and Migration Plan:** Create a structured approach to transition from Baseline to Target Architectures through a detailed Architecture Roadmap and Implementation and Migration Plan.

Phase F (Migration Planning):

- **Confirm Management Framework Interactions:** Align the Implementation and Migration Plan with an organization's management frameworks.

- **Assign Business Value to Work Packages:** Define and measure business value for work packages, identify corresponding projects, assess associated risks, and estimate business value.

- **Estimate Resource Requirements, Project Timings, and Availability/Delivery Vehicle:** Allocate resources and estimate costs for each project increment, distinguishing between capital and operational expenses, and identify cost-saving opportunities.

- **Prioritize Migration Projects Through Cost/Benefit Assessment and Risk Validation:** Determine the sequence of projects based on business priorities, technical dependencies, resource availability, and risk factors.

- **Confirm Architecture Roadmap:** Update the Architecture Roadmap, incorporating any necessary Transition Architectures.

- **Complete Implementation and Migration Plan:** Develop a detailed timeline with clear milestones for each transition stage.

- **Complete Architecture Development Cycle:** If the Architecture Capability's maturity permits, craft an Implementation Governance Model. Document lessons learned.

CHAPTER 12   ARCHITECTURE DEVELOPMENT METHOD

Phase G (Implementation Governance):

- **Confirm Scope and Priorities:** Identify and address gaps within the existing architecture. Review the Work Package Portfolio Map to determine which initiatives lead to optimal value extraction.

- **Identify Deployment Resources and Skills:** Identify the tools needed and the skills to use them.

- **Guide Development of Solutions Deployment:** Map out the scope of the project. Define the strategic requirements, as well as the various change requests needed to implement the architecture.

- **Perform Architecture Compliance Reviews:** Regularly assess the implementation activities to ensure that they are consistent with the architectural standards and principles.

- **Implement Business and IT Operations:** Implement business and IT service delivery, skills development, and training.

- **Perform Post-implementation Review:** Examine the end result of the architecture work. If there are deviations, submit change requests to correct them.

Phase H (Architecture Change Management):

- **Establish Value Realization Process:** Implement a change management process to govern modifications to the architecture in order to align changes with business value and maintain architectural integrity.

CHAPTER 12 ARCHITECTURE DEVELOPMENT METHOD

- **Deploy Monitoring Tools:** Deploy tools to monitor the business and technology changes that could impact the architecture, as well as the maturity of the Architecture Capability.

- **Manage Risks:** Identify, classify, and mitigate risk(s) before business transformation.

- **Provide Analysis for Architecture Change Management:** Analyze information from monitoring tools, stakeholder feedback, and performance reviews.

- **Develop Change Requirements to Meet Performance Targets:** Identify and define necessary modifications to the architecture.

- **Manage Governance Process:** Develop policies and guidelines that outline how changes should be proposed, reviewed, and approved.

- **Activate the Process to Implement Change:** Create a structured rollout plan with timelines, dependencies, and milestones.

Requirements Management:

- **Identify and Document Requirements:** Record all requirements from the Architecture Vision Phase in the Requirements Repository of the Architecture Repository.

- **Establish Baseline Requirements:** Document the baseline requirements created during phases B, C, D, and E in the Requirements Repository of the Architecture Repository.

- **Monitor Baseline Requirements:** Continuously track and manage the established requirements.

- **Identify New and Changed Requirements:** Add or change new and existing requirements.

- **Identify Changed Requirements and Record Priorities:** Identify and prioritize changed requirements. This step involves validating the requirements with the relevant stakeholders. Resolve any conflicts that may arise.

- **Assess Impact of Changed Requirements:** Changes to requirements might implicate a change to the architecture. Determine whether to implement new changes during the current or later iterations of the Architecture Development Method.

- **Implement Requirements Arising from Phase H:** Make new or updated requirements available for future architecture changes.

- **Update the Architecture Requirements Repository:** Record all changed, prioritized, and validated requirements in the Architecture Requirements Repository.

- **Implement Change in the Current Phase:** No specific action is required from the Requirements Management Phase.

- **Assess and Revise Gap Analysis for Past Phases:** Identify gap requirements and revise the Target Architecture as needed.

The summary of the steps per phase of the Architecture Development Method are visualized in Figure 12-56.

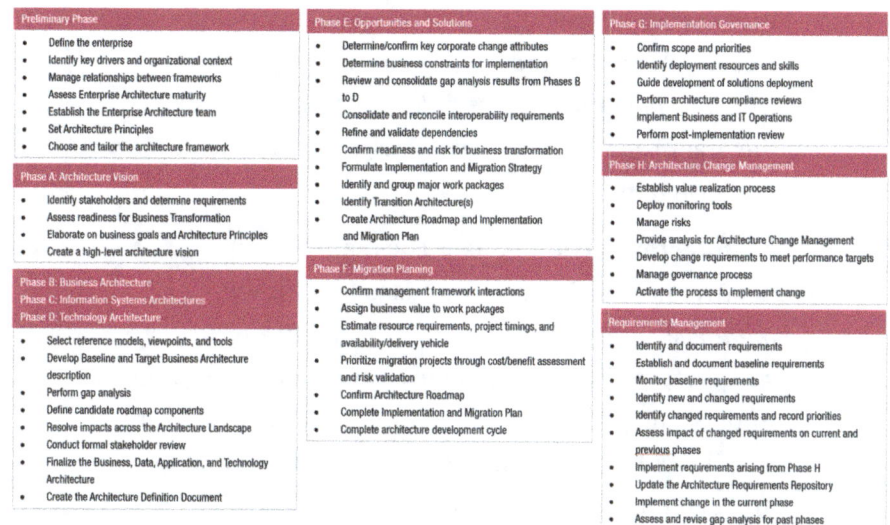

Figure 12-56. Summary of the steps of the Architecture Development Method

A large, poster-size version of Figure 12-56 can be downloaded from https://eawheel.com/books/media/.

12.14. Summary

This chapter has covered the Architecture Development Method, the heart of the TOGAF Standard. Often confused with a process, the Architecture Development Method is a logical method that groups key activities to clarify their relationships and information flow. It is a key component of the TOGAF Standard as it provides a comprehensive and iterative approach for developing and managing Enterprise Architectures.

CHAPTER 12   ARCHITECTURE DEVELOPMENT METHOD

The Architecture Development Method consists of ten phases, each with its own distinct objectives:

- **Preliminary Phase:** The first phase of the Architecture Development Method has two unique objectives. One is to *determine the Architecture Capability* that the organization believes it needs, and the other is to *establish it* in that form.

- **Architecture Vision Phase:** The main goal of the Architecture Vision Phase is to *establish a high-level vision of the architecture* and to *ensure stakeholder alignment*. It focuses on creating a shared understanding of the architecture's goals and ensuring that the business requirements are clearly articulated and agreed upon by all stakeholders.

- **Phase B (Business Architecture), Phase C (Information Systems Architectures), Phase D (Technology Architecture):** The objectives of phases B, C, and D are twofold. The first one is to *craft the Target Architecture* to outline how the organization should function to meet business objectives. The second objective is to *identify potential components for the Architecture Roadmap*, pinpointing the gaps between Baseline and Target Architectures that must be bridged.

- **Phase E (Opportunities and Solutions):** This phase contains three key objectives. The first is to *identify the work packages* required to realize the Target Architecture. The second objective is to *determine whether an incremental approach is required* to achieve

CHAPTER 12   ARCHITECTURE DEVELOPMENT METHOD

the first objective. The third and final key objective is to *define the building blocks* needed to complete the Target Architecture.

- **Phase F (Migration Planning):** The primary objective of Phase F (Migration Planning) is to *create a detailed plan* for the implementation of the architecture.

- **Phase G (Implementation Governance):** The primary objectives of Phase G include *ensuring adherence to the established architecture, managing any changes* during implementation, and *continually assessing progress* to make necessary adjustments to meet the organization's objectives.

- **Phase H (Architecture Change Management):** The key objectives of the Architecture Change Management Phase are to *continuously assess changes* in the organizational strategy, business environment, and technology landscape that may impact the architecture.

- **Requirements Management Phase:** Although not considered a separate phase, Requirements Management is a continuous activity that runs throughout the Architecture Development Method. It involves *managing and maintaining requirements* and ensuring that they are properly addressed during each phase.

The Architecture Development Method describes all the actions one *can* take, not what *must* be taken. It also does not prescribe the order of these steps. The TOGAF Standard explicitly states that the Architecture Development Method is to be tailored so that it (the steps and its content) fits an organization's needs.

CHAPTER 12   ARCHITECTURE DEVELOPMENT METHOD

The Architecture Development Method delivers its iterative approach by providing a step-by-step guide for creating and maintaining architecture artifacts and ensuring alignment with business goals.

This chapter has reviewed all the phases of the Architecture Development Method. Each phase has been explained in detail, and examples of each artifact to be created have been provided.

CHAPTER 13

Applying the ADM

 This chapter is part of the GREEN and BLUE reading tours. See Figure 1-1 in Chapter 1, Section 1.3, for alternative reading tours.

The Architecture Development Method describes a structured approach to achieving a sound Enterprise Architecture in a variety of situations and scenarios. The step-by-step approach – not to be confused with a waterfall approach – allows architects to address each sub-area of an architecture challenge in a very focused manner.

The Architecture Development Method can be used in combination with a variety of architectural styles, and the iterative nature of the method makes it possible to respond effectively to different situations. This makes the Architecture Development Method ideal for supporting architecture development. Together with the various applications of the method, the Architecture Development Method provides a solid approach for addressing any architectural issue.

CHAPTER 13 APPLYING THE ADM

13.1. Various Architecture Development Method Approaches

The Architecture Development Method can be adapted to a number of different scenarios and situations. Its tailorability makes it an easy method to adopt when dealing with different process styles or even specialized architectures such as Security Architecture. Because of its adaptability, the Architecture Development Method can be applied to almost any situation for any organization.

There are three specific scenarios that require further explanation:

- **Using the TOGAF Standard with Different Architectural Styles:** This section explains how the TOGAF Standard can be used and adapted to different architectural styles.

- **Applying Iteration to the Architecture Development Method:** This section discusses the concept of iteration and provides an overview of how it can be applied to the Architecture Development Method.

- **Architecture Partitioning:** This section shows how partitions can be used to structure and simplify architecture development.

There are, of course, many other possible applications of the Architecture Development Method. However, the situations or specific applications just mentioned generally cover most of them.

13.2. Using the TOGAF Framework with Different Architecture Styles

The TOGAF Standard is an exceptionally versatile framework, carefully crafted to accommodate a wide variety of architectural styles (e.g., Service-Oriented Architectures, Microservices Architectures, or Zero-Trust Architectures). These styles vary in focus, form, techniques, materials, subject matter, and historical context. As a comprehensive framework, the TOGAF Standard is designed to be adaptable to different environments, allowing organizations to tailor it to their unique architectural needs.

Within an organization's architectural landscape, it's common to encounter architectural work developed in multiple styles. The TOGAF Standard ensures that the requirements of each stakeholder are appropriately addressed, taking into account the perspectives of others and the Baseline Architecture.

When using the TOGAF Standard to support a particular architectural style, architects must consider the unique characteristics of that style. The first step is to identify these unique characteristics. Next, architects determine how to address these characteristics within the framework. This typically involves adapting models, views, and tools, rather than making significant changes to the TOGAF Framework.

During phases B, C, and D, architects are expected to select relevant architectural resources – including models, viewpoints, and tools – to accurately describe the architecture domains and ensure that stakeholder concerns are addressed. Depending on the unique features, different architectural styles may introduce new elements, emphasize existing ones, adjust notations, and focus attention on specific stakeholders or their concerns. Addressing these unique features often requires extending the Enterprise Metamodel and Content Framework, using specific notations or modeling techniques, and identifying appropriate viewpoints.

For example, a predominant architectural style may cause architects to revisit the Preliminary Phase to modify the Architecture Capability or to address special features within a single Architecture Development Method cycle. Style-specific reference models and maturity models are commonly used tools to assist architects in this endeavor.

Throughout its life, the TOGAF Framework has evolved to encompass different architectural styles, address key challenges faced by practitioners, and demonstrate its adaptability within defined contexts.

13.3. Applying Iteration to the Architecture Development Method

The Architecture Development Method of the TOGAF Standard emphasizes flexibility and adaptability in the architecture development process. Iteration plays a key role in ensuring that the architecture evolves in response to changing requirements, feedback, and context. The Architecture Development Method is iterative in nature, meaning that it can be revisited as needed. Architects continually evaluate and refine the architecture as they move through the phases, ensuring that it remains relevant and effective. The end result is a comprehensive and well-aligned Enterprise Architecture that supports the organization's business goals.

CHAPTER 13   APPLYING THE ADM

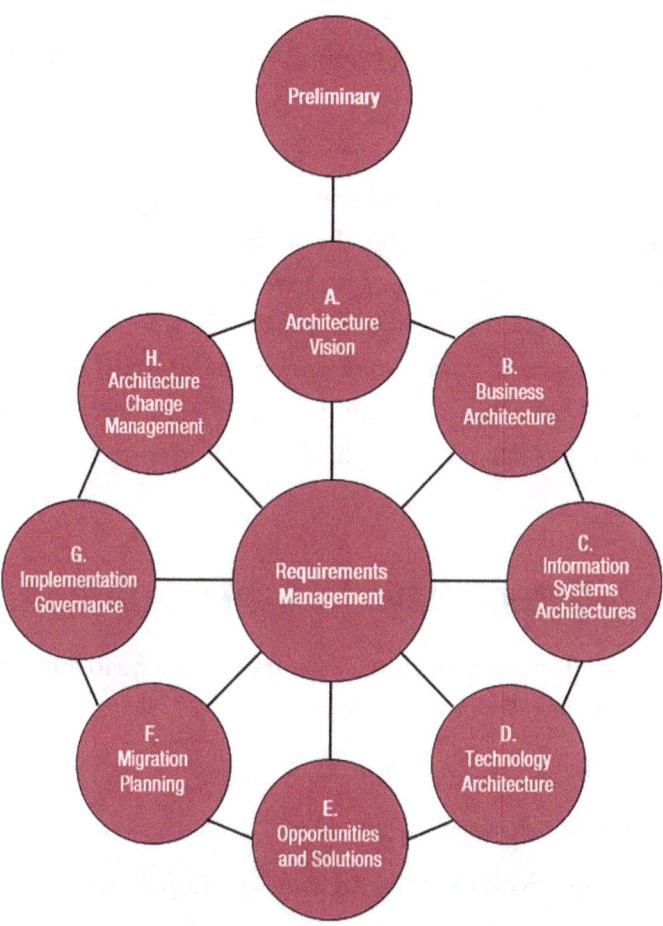

Figure 13-1. *The Architecture Development Method graphic*

It is still quite common to misunderstand the concept of the Architecture Development Method, especially regarding its iterative nature. The Architecture Development Method graphic (Figure 13-1) is frequently misinterpreted as a linear, waterfall-style process. In reality, it is a logical method that groups key activities to clarify their relationships and information flow. The well-known graphic is a stylized representation showing these information flows, and it is *not a representation of an activity sequence.*

CHAPTER 13   APPLYING THE ADM

The Architecture Development Method should not be understood as a process model [18].

It's crucial to recognize that whenever the Enterprise Architecture team engages in any activity within the Architecture Development Method's scope, they are executing a phase and contributing to the Enterprise Architecture Landscape. For instance, when an architect works on roadmap development, they are engaging in Phase E (Opportunities and Solutions). This requires consuming mandatory inputs and producing mandatory outputs, a principle applicable to all Architecture Development Method phases.

13.3.1. Different Ways to Apply Iteration

According to the TOGAF Standard, there are three ways to apply iteration to the Architecture Development Method:

- Iteration across the whole ADM cycle
- Iteration within an ADM cycle
- Iteration to manage the Architecture Capability

These approaches reflect different levels and ways in which the Architecture Development Method can be tailored based on an organization's needs or specific architectural context.

Iteration Across the Whole Architecture Development Method Cycle: This type involves executing *the entire Architecture Development Method cycle multiple times* for different levels of architecture or across different scope definitions (e.g., organization-level versus project-level architectures).

Applying iteration to the whole Architecture Development Method cycle is often used when a comprehensive Architecture Landscape is developed. This can be done in layers or when working in an agile or incremental development approach, where the entire process is applied iteratively to build upon previous cycles.

After completing an Architecture Development Method cycle for a specific architectural layer (e.g., Business Architecture), the process is applied again for other layers (Information Systems Architectures and/or Technology Architecture).

The key difference between applying iteration to the whole Architecture Development Method cycle and other ways of applying iteration is that this method of iteration is typically used to address multiple layers or versions of an architecture, such as transitioning from a high-level Enterprise Architecture to more detailed architectures.

Iteration Within an Architecture Development Method Cycle: In this approach, iteration occurs *between two or more phases* of the Architecture Development Method cycle.

This type is used for moving back and forth between different phases, revisiting previous phases as new information or insights emerge.

Applying iteration between two or more phases of the Architecture Development Method cycle is helpful when these phases have interdependencies or when new insights from one phase necessitate reworking a previous phase. Think, for example, about managing the interrelationship between Business Architecture, Information Systems Architectures, and Technology Architecture. Also, applying iteration within an Architecture Development Method cycle can be used to converge on a detailed Target Architecture when higher-level architecture does not exist to provide context and constraint.

The three most common groups of iteration within an Architecture Development Method cycle are called *Architecture Development, Transition Planning,* and *Architecture Governance*. Figure 13-2 illustrates these three iteration groups.

CHAPTER 13   APPLYING THE ADM

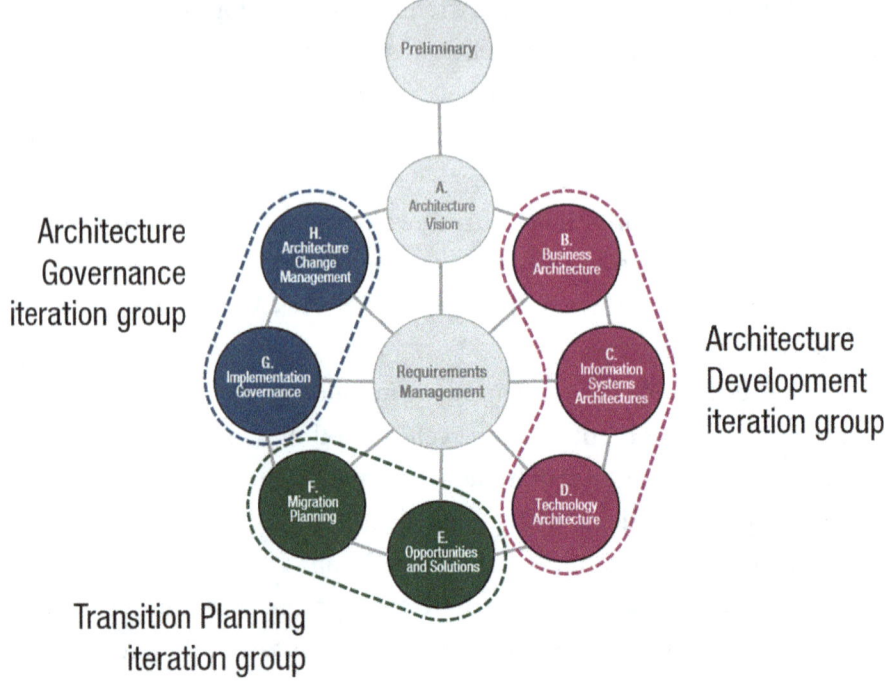

Figure 13-2. Three most common iteration groups

The first of the three iteration groups is called the *Architecture Development iteration group*. This particular iteration is commonly used for developing the Baseline and Target Architectures during phases B, C, and D.

The second scenario where iteration occurs between two or more phases is to arrive at an executable Architecture Roadmap or Implementation and Migration Plan. This type of iteration is triggered when the implementation details and scope of change demand a change or reprioritization of stakeholder requirements. Also, there may be changes in business strategy, new requirements may emerge, or issues may arise that require re-evaluation of earlier decisions. A common name for this particular iteration group is the *Transition Planning iteration group*.

CHAPTER 13 APPLYING THE ADM

The third situation that calls for applying iteration within an Architecture Development Method cycle is to support governance of change activities when moving toward a Target Architecture. This specific iteration group is called the *Architecture Governance iteration group*.

The key difference between applying iteration to two or more phases and other ways of applying iteration is that this form of iteration *involves moving backward or looping between distinct phases* to refine the architecture as it evolves across the lifecycle. This type of iteration is focused on deepening the activities within one or more phases without affecting the overall flow of the Architecture Development Method.

Consider the following scenario. An organization has two goals: implement new regulatory requirements and efficiently use human resources. In order to achieve both goals, certain architectural things need to be accomplished. In this scenario, it is assumed that there are two initiatives (Initiative A and Initiative B), each associated with a specific goal, that are needed to achieve the described goals (see Figure 13-3).

Figure 13-3. *Goals and related initiatives*

Each initiative requires its own approach and adaptation to the organization's setup.

Initiative A indicates that new employees with the required profile must be hired. As a result, HR will need to invest time and effort in finding and hiring the right people. This will require, among other things, the Inflow Support capability.

Initiative B involves updating the application responsible for determining the appropriate compensation for the job profiles. The job profiles need to be re-evaluated. This will require an investment in new software. The amount of the investment is equal to the investment that HR spends on recruiting (in terms of internal staff hours).

During phases B and C of the Architecture Development Method, the two possible scenarios described above are developed. Both scenarios are presented to the CIO during Phase F (Migration Planning). Since the available budget can only be spent once, the CIO must choose between the two proposed scenarios. The CIO uses the Architecture Alternatives and Trade-Offs method to determine the final direction (see Figure 13-4).

CHAPTER 13 APPLYING THE ADM

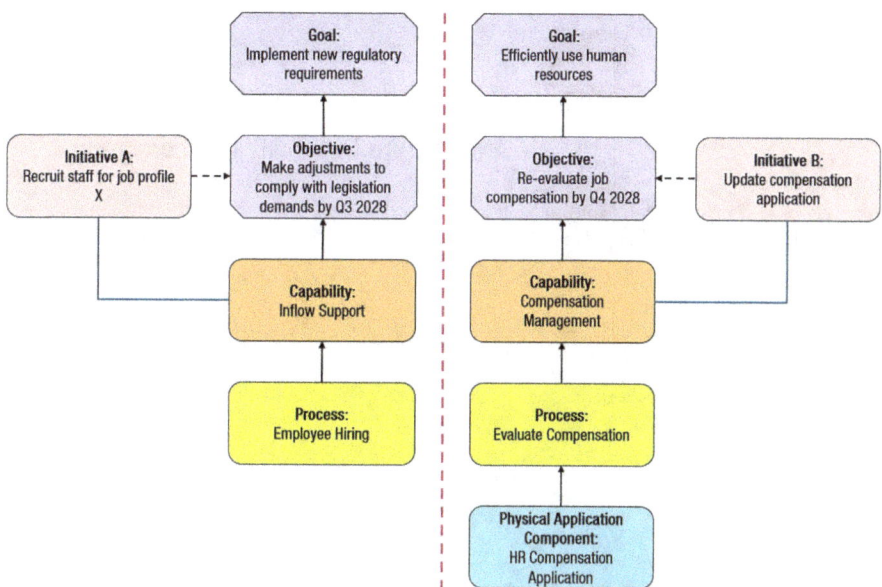

Figure 13-4. *Two possible scenarios created when applying the Architecture Alternatives and Trade-Offs method*

For the selected scenario, another *Architecture Development iteration cycle* is initiated to further refine the organizational implications.

Iteration to Manage the Architecture Capability: This type involves executing *the Preliminary Phase and Phase A* in order to adjust (parts of) the Architecture Capability.

The third type of iteration concerns itself solely with the Architecture Capability and is therefore fittingly called the *Architecture Capability iteration group* (see Figure 13-5).

571

CHAPTER 13   APPLYING THE ADM

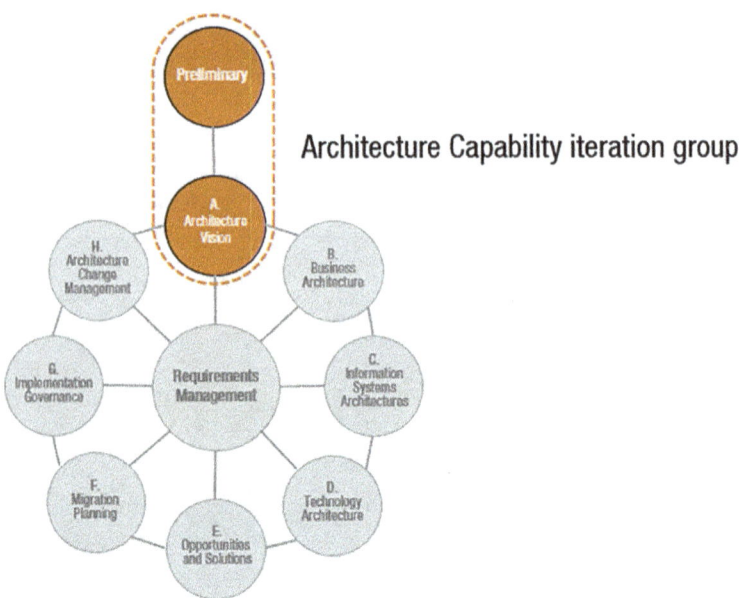

Figure 13-5. *Architecture Capability iteration group*

The iteration of the Preliminary Phase and Phase A (Architecture Vision) may be triggered by projects wanting to (re-)establish aspects of the Architecture Capability identified in Phase A (Architecture Vision). A second example is when projects require a new iteration of the Preliminary Phase to adjust the organization's Architecture Capability. This need may arise from the identification of new or changed requirements for the Architecture Capability as a result of a change request in Phase H (Architecture Change Management).

In short, applying iteration to the Preliminary Phase and Phase A (Architecture Vision) is primarily used to process changes to the Architecture Capability. Iteration between Architecture Development Method phases means revisiting earlier phases based on new insights or changes and focuses on the development of architecture. Finally, applying iteration across the whole Architecture Development Method cycle requires the entire cycle to be repeated iteratively. This is often applied to different layers of the architecture. It is used to develop a comprehensive Architecture Landscape.

13.3.2. How to Apply Iteration

The TOGAF Standard indicates that for a typical architecture approach, it is prudent to apply the iteration cycles of the Architecture Development Method. There are four iteration groups related to architecture activities. Each group has a specific purpose. Figure 13-6 illustrates the iteration groups:

Architecture Capability: These iterations of the Architecture Development Method are designed to support the establishment and continued growth of the Architecture Capability. This means mobilizing the organization to proceed with the implementation of architecture activities. A motivator for this may be that the organization in question wants to start working with architecture or that a need for an architecture approach has arisen. During this iteration, the focus is on establishing the architecture approach, defining principles, and determining the scope of the architecture. Developing an architecture vision and establishing architecture governance are also part of this iteration group. The phases of the Architecture Development Method used are the Preliminary Phase and Phase A (Architecture Vision).

Architecture Development: Iterations whose purpose is to develop architecture content. The phases associated with architecture development are phases B–F. Going through these phases of the Architecture Development Methodology creates a cohesive architecture. Typically, to shape the Baseline Architecture, the iteration will focus on Phase B (Business Architecture), Phase C (Information Systems Architectures), and Phase D (Technology Architecture). The stakeholder reviews related to the architecture are often broader in scope during this iteration. As the iterations move toward the Target Architecture, Phase E (Opportunities and Solutions) and Phase F (Migration Planning) in particular will ensure the feasibility of the architecture to be implemented.

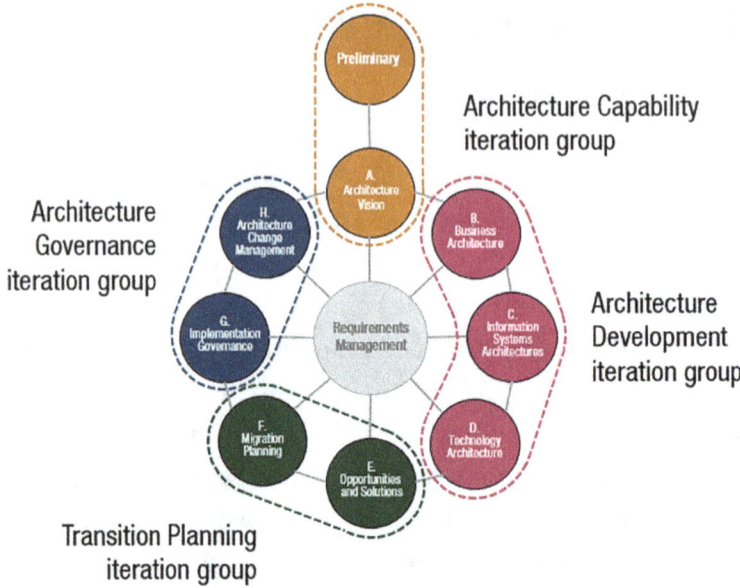

Figure 13-6. Iteration groups of the ADM

Transition Planning: Iterations in this group support the creation of roadmaps for change initiatives related to the architecture.

The purpose of the iterations is to properly plan the transition from the Baseline to the Target Architecture. Phase E (Opportunities and Solutions) and Phase F (Migration Planning) are the Architecture Development Method phases used within this group of iterations.

Architecture Governance: All iterations that take place with the goal of establishing and monitoring architecture governance. The iterations of the Architecture Development Method emphasize Phase G (Implementation Governance) and Phase H (Architecture Change Management).

The four iteration groups and their corresponding Architecture Development Method phases are listed in Table 13-1. The table also shows the possible iteration cycles for each iteration group.

Table 13-1. Activity by iteration cycle

Iteration Group		Architecture Development			Transition Planning		Architecture Governance	
Architecture Development Method Phase		Iteration 1	Iteration 2	Iteration n	Iteration 1	Iteration n	Iteration 1	Iteration n
Preliminary								
Architecture Vision								
Business Architecture	Baseline	X						
	Target		X	X				
Application Architecture	Baseline	X		X				
	Target		X	X				
Data Architecture	Baseline	X		X				
	Target		X	X				
Technology Architecture	Baseline	X		X				
	Target		X	X				
Opportunities and Solutions					X	X		
Migration Planning					X	X		
Implementation Governance							X	X
Change Management							X	X

CHAPTER 13   APPLYING THE ADM

575

CHAPTER 13   APPLYING THE ADM

Establishing the architecture approach, defining the principles, developing an architecture vision, and defining the scope do not need to be repeated for every type of architecture work. For this reason, the corresponding phases of the Architecture Development Method (Preliminary Phase and Phase A (Architecture Vision)) in Table 13-1 are not a regular part of the more common iterations (shown in Figure 13-7) that aim to develop the architecture, plan its transition, and monitor its implementation.

Architecture Development   Transition Planning   Architecture Governance

Figure 13-7. *Iteration cycles Architecture Development, Transition Planning, and Architecture Governance*

The first iteration cycle of the Architecture Development iteration group has architecture development as its goal. This cycle can be initiated in one of two ways. The cycle can be started with a focus on the *Baseline Architecture* or with a focus on the *Target Architecture*.

Baseline First: The first iteration within this cycle is to create a Baseline Architecture (see Figure 13-8). The phases to be completed are the Business Architecture, Information Systems Architectures, and Technology Architecture, focusing on the definition of the baseline. Opportunities, solutions, and migration plans are also considered to drive out the focus for change and test feasibility.

CHAPTER 13   APPLYING THE ADM

The second iteration is aimed at defining the Target Architecture and identifying the gaps between the Baseline and the Target Architecture. This iteration again consists of going through the Business Architecture, Information Systems Architectures, and Technology Architecture phases of the Architecture Development Method. Only this time, it focuses on defining the Target Architecture and analyzing the gaps against the Baseline Architecture.

Opportunities, solutions, and migration plans are included in the feasibility review process. All subsequent iterations are aimed at correcting and refining the Baseline and Target Architecture and associated gaps to achieve an outcome that is beneficial, feasible, and viable.

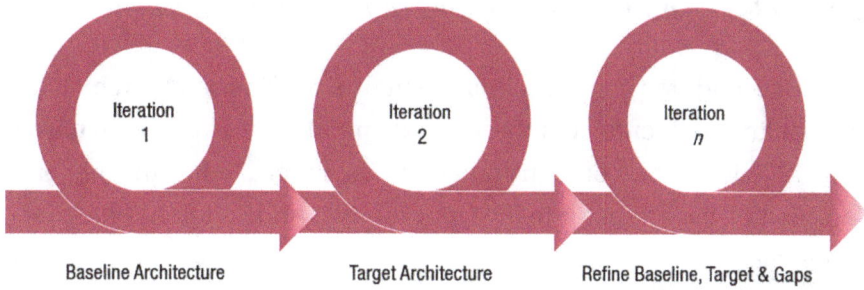

Figure 13-8. Iteration cycles of the Architecture Development activity

Target First: The second way to apply the first iteration cycle is to develop the Target Architecture as the first step. The approach to the cycle is much the same as starting with the Baseline Architecture, except that the focus is on first mapping the desired architecture and then comparing it to the Baseline Architecture. This is a common starting point in many organizations.

The second iteration cycle of the Architecture Development iteration group deals with transition planning to get from the Baseline to the Target Architecture. The goal of the first iteration within this cycle is to define and agree on a set of improvement opportunities, aligned against a provisional Transition Architecture. The initial iteration of Transition Planning (see

Figure 13-9) seeks to gain buy-in to a portfolio of solution opportunities in the Opportunities and Solutions Phase of the Architecture Development Method. This iteration also delivers a provisional migration plan.

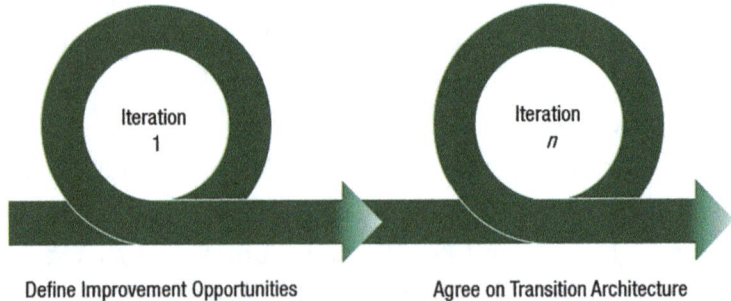

Figure 13-9. Iteration cycles of the Transition Planning activity

The second step in this cycle is to agree on the Transition Architecture and to refine the identified improvement opportunities in order to make them fit. This iteration of Transition Planning seeks to refine the migration plan, feeding back issues into the Opportunities and Solutions Phase for refinement.

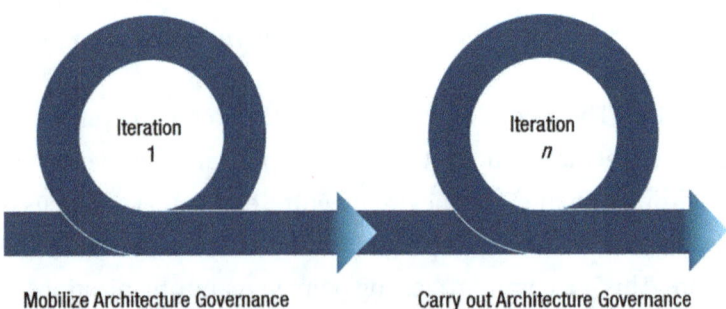

Figure 13-10. Iteration cycles of the Architecture Governance activity

Finally, it is also important to pay attention to Architecture Governance (see Figure 13-10). This is done in the third and final iteration cycle. For this topic, the TOGAF Standard suggests two iterations of the Architecture

Development Method, namely, to *mobilize architecture governance* and change management processes during the first iteration and to *carry out architecture governance* and change control during the second.

The initial Architecture Governance iteration establishes a process for governance of change and also puts in place the appropriate people, processes, and technology to support managed access to and change of the defined architecture.

Subsequent iterations of the Architecture Governance cycle focus on periodic reviews of change initiatives to resolve issues and ensure compliance. Results of a change request may trigger another phase to be revisited, for example, feeding back a new requirement to the Preliminary Phase to improve the Architecture Capability or a new requirement for the architecture into the Architecture Development phases.

Many architects often perceive the Architecture Development Method as a linear process model. However, this interpretation is a misconception. The traditional Architecture Development Method diagram is *not an activity diagram*; rather, it is a logical method that clusters essential activity steps to link actions and information flow, producing specific outputs.

Each time an architect performs activities within the scope of the Architecture Development Method, they contribute to the development of the Enterprise Architecture Landscape through iteration. The phase being performed corresponds to the appropriate domain. Attempting to impose a one-pass linear sequence on the Architecture Development Method can result in convoluted, looping phase diagrams. Instead, consider the steps as a checklist that allows for flexibility and adaptability in the architecture process.

By applying iteration, the Architecture Development Method provides a comprehensive and flexible approach to Enterprise Architecture development that can adapt to changes in scope, context, and requirements. This enhances the likelihood of creating an architecture that evolves with the organization's needs over time.

Further Reading

The TOGAF Standard Reference Cards in the TOGAF Library provide a concise overview of the use of iteration in implementing the Architecture Development Method:

https://publications.opengroup.org/togaf-library

The Series Guide, "A Practitioners' Approach to Developing Enterprise Architecture Following the TOGAF ADM", document number G186, provides additional practical information on applying the Architecture Development Method with an eye toward iteration.

13.4. Architecture Partitioning

Architecture partitioning refers to the practice of dividing an Enterprise Architecture into smaller, manageable parts to improve governance, scalability, and efficiency in the architecture development process.

Organizations are sometimes large and complex, making it difficult to develop and govern architecture as a single, monolithic structure. Architecture partitioning allows breaking the architecture into smaller, more manageable segments, improving focus and clarity.

13.4.1. Why Use Architecture Partitioning

One of the primary reasons for implementing architecture partitioning is the fact that different areas of the architecture (Business, Data, Application, and Technology) may have different stakeholders, policies, and compliance requirements. Partitioning ensures that governance can be effectively applied to specific areas without imposing unnecessary constraints on the entire organization.

In addition, architecture partitioning allows different teams to work in parallel on different architecture segments or projects. In fact, it prevents bottlenecks by ensuring that architecture efforts are distributed rather than centralized in a single group. Not only can multiple teams work on pieces of the architecture, but those pieces can be created in different timeframes. Architecture development does not have to happen all at once; it can evolve incrementally. The Architecture Development Methodology can be applied iteratively to different partitions, allowing for incremental improvements.

In layman's terms, architecture partitioning is the process of dividing the architecture work among multiple teams. Along with dividing the workload, definitions are provided so that each team understands what they are supposed to create in terms of the architecture. This also prevents the creation of architecture that doesn't integrate with other parts of the architecture.

The relationship between architecture partitioning and agile approaches is evident. Both involve breaking down work into manageable units.

Figure 13-11. *Architecture created without an overall picture*

However, it is crucial to maintain a holistic perspective during this process. Without considering the broader context, there's a risk of developing disjointed architecture components that lack integration and coherence. Figure 13-11 illustrates this undesirable outcome.

13.4.2. Supporting Architecture Partitioning

During the Preliminary Phase, the necessary steps need to be taken in order for the architecture work to support architecture partitioning. The following aspects need to be determined in order for the partitioning to work:

- **Organizational Structure:** This step should determine the teams tasked with the creation of (parts of) the architecture.

- **Responsibilities for the Teams:** It is imperative that a definition of the architecture to be created is agreed upon. At the very least, the definition should include the subject matter areas being covered, the level of detail to use, and what the time periods are during which to deliver the architecture.

- **Relationship Between Architectures:** Needless to say that all teams should work toward an integrated architecture. Any overlap between the various parts of the architecture should be addressed so that reusable building blocks remain.

Once the Preliminary Phase is complete, the teams conducting the architecture should be understood. Each team should have a defined scope, and the relationships between teams and architecture should be understood.

13.5. Summary

The TOGAF Standard is renowned for its adaptability, allowing the Architecture Development Method to be tailored to various scenarios and situations. This flexibility makes it an excellent choice when addressing diverse process styles or specialized architectures.

CHAPTER 13 APPLYING THE ADM

Three key scenarios include:

- **Using the TOGAF Standard with Different Architectural Styles:** This scenario demonstrates how the TOGAF Standard can be adapted to various architectural styles.

- **Applying Iteration to the Architecture Development Method:** This scenario explores the concept of iteration and provides an overview of its application within the Architecture Development Method.

- **Architecture Partitioning:** This scenario illustrates how partitions can be employed to structure and simplify architecture development.

When leveraging the TOGAF Standard to support a specific architectural style, architects must consider the unique characteristics of that style. The initial step involves identifying these unique traits, followed by determining how to address them. During phases B, C, and D, architects are expected to select relevant architectural resources – including models, viewpoints, and tools – to accurately describe the architecture domains and ensure that stakeholder concerns are addressed. This approach may introduce new elements, emphasize existing ones, adjust notations, and focus attention on specific stakeholders or their concerns.

The concept of the Architecture Development Method is often misunderstood, particularly regarding its iterative nature. Its graphic is frequently misinterpreted as a linear, waterfall-style process, leading to confusion. In reality, it is a logical method that groups key activities to clarify their relationships and information flow. The classic diagram represents this flow, not a sequential activity model.

CHAPTER 13   APPLYING THE ADM

According to the TOGAF Standard, there are three ways to apply iteration to the Architecture Development Method:

- **Iteration Across the Whole Architecture Development Method Cycle:** This approach involves executing the entire ADM cycle multiple times for different levels of architecture or across varying scope definitions (e.g., organization-level versus project-level architectures).

- **Iteration Within an Architecture Development Method Cycle:** Here, iteration occurs between two or more phases of the ADM cycle, allowing for revisiting previous phases as new information or insights emerge.

- **Iteration to Manage the Architecture Capability:** This type involves executing the Preliminary Phase and Phase A to adjust parts of the Architecture Capability.

Applying iteration cycles within the Architecture Development Method is prudent. There are four iteration groups related to architecture activities:

- **Architecture Capability:** These iterations support the establishment and continued growth of the Architecture Capability.

- **Architecture Development:** Iterations aimed at developing architecture content, typically associated with phases B-F.

- **Transition Planning:** Iterations that support the creation of roadmaps for change initiatives related to the architecture.

- **Architecture Governance:** Iterations focused on establishing and monitoring architecture governance.

The first iteration group, *Architecture Capability*, involves executing the Preliminary Phase and Phase A to adjust parts of the Architecture Capability. The second group, *Architecture Development*, is commonly used for developing architecture content during Phases B through F. The third group, *Transition Planning*, involves iteration between two or more phases with the goal of arriving at an executable Architecture Roadmap or Implementation and Migration Plan. This type of iteration is triggered when implementation details and scope changes necessitate a change or reprioritization of stakeholder requirements. Finally, the fourth group, *Architecture Governance*, supports governance of change activities when progressing toward a Target Architecture.

The key distinction between applying iteration to two or more phases and other forms of iteration is that this approach involves moving backward or looping between distinct phases to refine the architecture as it evolves across the lifecycle.

By applying iteration, the ADM offers a comprehensive and flexible approach to Enterprise Architecture development that can adapt to changes in scope, context, and requirements.

Architecture partitioning is an essential practice in Enterprise Architecture, involving the division of the architecture into smaller, manageable segments.

One primary reason for implementing architecture partitioning is the recognition that different domains within the architecture – such as Business, Data, Application, and Technology – have distinct stakeholders, policies, and compliance requirements. Partitioning allows governance to be effectively tailored to specific areas without imposing unnecessary constraints across the entire organization. Moreover, architecture partitioning facilitates parallel development, enabling various teams to work concurrently on different segments or projects. This approach prevents bottlenecks by distributing architectural efforts, rather than

CHAPTER 13 APPLYING THE ADM

centralizing them within a single group. Additionally, partitioning supports the incremental evolution of the architecture, allowing teams to develop different components at varying paces, ensuring that architecture development is a continuous process rather than a one-time event.

The relationship between architecture partitioning and agile methodologies is evident. Both involve breaking down work into manageable units. However, it is crucial to maintain a holistic perspective during this process. Without considering the broader context, there's a risk of developing disjointed architecture components that lack integration and coherence.

CHAPTER 14

ADM Techniques

 This chapter is part of the GREEN and BLUE reading tours. See Figure 1-1 in Chapter 1, Section 1.3, for alternative reading tours.

A number of topics that were once an integral part of the TOGAF Framework have been split off and given their own place in this chapter. This includes the broader topics of Architecture Principles, Stakeholder Management, and Risk Management. But topics such as architecture patterns, gap analysis, interoperability requirements, Business Transformation Readiness Assessments, and Architecture Alternatives and Trade-Offs have also found their way into this chapter.

The reason for this is that they describe a specific application. One that is invoked, so to speak, at several points in the implementation of the TOGAF Standard.

The unbundling of these topics in turn contributes to the modularity that The Open Group has achieved with the publication of the Standard.

14.1. Technique or No Technique

The Architecture Development Method is often regarded as the core of the TOGAF Standard, providing a structured approach to developing and managing Enterprise Architectures. It consists of a cyclic method that helps organizations align business and IT needs effectively. Various techniques are applied throughout the Architecture Development

CHAPTER 14 ADM TECHNIQUES

Method cycle to ensure efficient decision-making, risk mitigation, and stakeholder engagement. These techniques support different phases of the Architecture Development Method and enhance architectural practice.

When the TOGAF Standard talks about techniques, it does not always mean a *specific way of doing things*. What is really meant by techniques is an *elaboration of a topic* or component *that is frequently used* within the framework or during the implementation of the Architecture Development Method. A recurring topic that requires further explanation and clarification in order to be properly applied.

Some topics are essential aspects of developing an architecture. However, they are not necessarily a specific technique. Take Stakeholder Management or Risk Management, for example. Both are essential and of course should not be forgotten or overlooked. But they are not techniques. Otherwise, one could also argue that Architecture Roadmapping is a technique. Or creating a Baseline Architecture. No, some topics that are listed as techniques are simply not.

Conducting a Business Transformation Readiness Assessment, on the other hand, is. So is performing a gap analysis. The reason for this is that the Business Transformation Readiness Assessment can be seen as a complement to an existing process or activity. In fact, Phases A (Architecture Vision) and E (Opportunities and Solutions) examine an organization's ability to implement change. Conducting a Business Transformation Readiness Assessment serves to complement these activities.

The same applies to the gap analysis. Phases B, C, and D (Business, Data/Application, and Technology Architecture, respectively) establish the Baseline and Target Architectures. To get from the Baseline to the desired Target Architecture, it is necessary to create (candidate) roadmap items. This activity is complemented by a gap analysis.

In short, not all of the topics that the TOGAF Standard refers to as techniques are actually techniques. This does not change the fact that all

CHAPTER 14 ADM TECHNIQUES

of them are essential to the correct execution of architecture activities and the development of an architecture.

That said, the TOGAF Standard distinguishes between eight *techniques*:

- **Architecture Principles:** These are fundamental rules and guidelines that influence architecture development and governance. They ensure consistency and coherence across an organization's architecture.

- **Stakeholder Management:** Engaging stakeholders effectively is crucial for architectural success. The TOGAF Standard provides methods for identifying, analyzing, and managing stakeholders.

- **Architecture Patterns:** Architecture patterns provide standardized solutions to common architectural problems.

- **Gap Analysis:** Gap analysis is a technique used to identify differences between the Baseline and Target Architectures.

- **Interoperability Requirements:** Describes a technique for determining interoperability requirements.

- **Business Transformation Readiness Assessment:** A technique for identifying business transformation issues.

- **Risk Management:** Risk Management ensures that potential issues are identified and mitigated before impacting the architecture.

- **Architecture Alternatives and Trade-Offs:** Identify alternative Target Architectures and perform trade-offs between the alternatives.

591

The eight techniques mentioned above are detailed in the following sections.

14.2. Principles

In the TOGAF Standard, principles play a crucial role in guiding decision-making and ensuring alignment between business objectives and strategies. Principles are often established within different domains and at different levels of an organization. Although the use of principles is not really a technique (rather a necessity), the TOGAF Standard classifies their creation and development as such.

The TOGAF Standard distinguishes between *Enterprise Principles* and *Architecture Principles*, each serving different purposes. These are the two key domains that inform the development and utilization of architecture:

- **Enterprise Principles:** High-level, general guidelines that shape the overall strategy, culture, and decision-making framework of an organization. These principles apply broadly across the entire organization, influencing business operations, IT strategy, and governance. They typically address aspects such as business agility, innovation, customer focus, security, compliance, and sustainability.

- **Architecture Principles:** More specific guidelines that direct the design and development of Enterprise Architecture. They apply to architecture development efforts, ensuring consistency, scalability, and alignment with business goals. These principles provide direction for technology choices, system integration, and IT governance.

The key differences between Enterprise Principles and Architecture Principles are outlined in Table 14-1.

Table 14-1. Differences between Enterprise and Architecture Principles

Aspect	Enterprise Principles	Architecture Principles
Scope	Organization-wide, applies to business and IT	Focused on Enterprise Architecture
Purpose	Guides overall decision-making and strategy	Provides a framework for architecture design and execution
Application	Broad, influences policies and culture	More technical, used in architecture development
Examples	Agility, customer focus, security	Reusability, scalability, standardization

14.2.1 The Added Value of Principles

Principles, in general, provide clarity, consistency, and direction in decision-making. Their value lies in ensuring that business and IT strategies support the same objectives. In other words, *principles guide strategic alignment*. They also allow for standardizing approaches across different teams, reducing conflicts and inefficiencies. Principles enable better decision-making by providing a structured framework to evaluate options and trade-offs. They help establish clear policies and accountability in architecture development. Last but not least, principles ensure a shared understanding of organization and architecture goals and objectives.

Within an Enterprise Architecture approach, principles serve as a foundation for designing flexible and scalable architectures that align with the aforementioned business goals and objectives. They ensure compliance with regulations and industry standards and are able to reduce complexity and duplication by enforcing standardization and interoperability.

14.2.2 The Structure of a Principle

By defining principles clearly and applying them consistently, organizations can create a well-structured, effective Enterprise Architecture that supports both short-term and long-term goals.

Each principle in the TOGAF Standard is typically structured as shown in Table 14-2.

Table 14-2. Structure of a principle

Structure Element	Description
Name	A short, clear title.
Statement	A concise declaration of the principle's intent.
Rationale	The justification for the principle, explaining why it is important.
Implications	The practical impact of the principle, including potential trade-offs and required actions.

There are some essential elements that a principle must incorporate. Table 14-2 lists four structural elements that every principle must contain:

Name: The key is to choose a short title that leaves nothing to the imagination. The best result is achieved by using the core essence of the principle. To ensure the applicability of the principle, it is also wise to avoid using specific product names. Also, the use of words that can be understood in more than one way is out of the question.

Statement: The structure element statement should clearly and concisely convey the basic rule without ambiguity. The intent of the principle is often used for this purpose.

Rationale: With the rationale, the added business value of adhering to the principle is explained. Often, consistency with other principles is also mentioned. If a particular principle takes precedence over another principle, that is a reason to describe it in the rationale. This structure element describes the justification for the principle. It explains why adherence to the principle is important.

Implications: This structure element identifies the requirements for applying the principle. This is usually done by expressing the requirements in terms of resources, time, money, or specific activities required. The impact of a principle after it has been adopted should be clearly articulated, including potential trade-offs and required actions. In effect, the implications answer the question, "How does this affect the business?"

In addition to using the structural elements mentioned above (name, statement, rationale, and implications), it is important to ensure that principles meet a number of quality requirements. Principles are used in decision-making, and for that reason alone it is wise to make them as clear as possible. The following five criteria distinguish a good set of principles [20]:

Understandable: The underlying tenets can be quickly grasped and understood by individuals throughout the organization. The intent of the principle is clear and unambiguous, minimizing violations, whether intentional or unintentional.

Robust: Principles should enable high-quality decisions about architectures and plans and create enforceable policies and standards. Each principle should be sufficiently definitive and precise to support consistent decision-making in complex, potentially controversial situations.

Complete: Every potentially important principle for managing information and technology for the organization is defined – the principles cover every conceivable situation.

Consistent: Strict adherence to one principle may require a loose interpretation of another principle. The set of principles must be expressed in a way that allows for a balance of interpretations. Principles should not be so contradictory that adherence to one principle would violate the spirit of another. Each word in a statement of principles should be carefully chosen to allow for consistent yet flexible interpretation.

Stable: Principles should be enduring, yet able to accommodate change. An amendment process should be established to add, delete, or modify principles after their initial ratification.

Enterprise and Architecture Principles are stored in the Governance Repository section of the Architecture Repository, as shown in Figure 14-1.

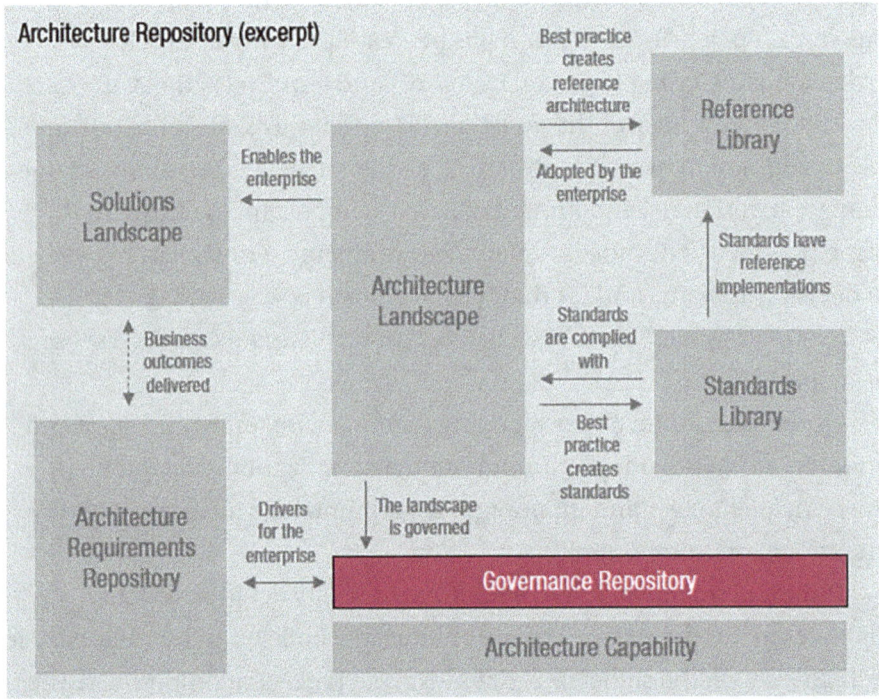

Figure 14-1. Excerpt of the Architecture Repository, showing Enterprise and Architecture Principles stored in the Governance Repository

> The Series Guide "Using the TOGAF Standard in the Digital Enterprise" uses the term Enterprise Architecture Principles [23].
>
> With this, the TOGAF Standard seems to introduce a third variant of the entity principle.
>
> In addition to Enterprise and Architecture Principles – which are very different in form and application – the term *Enterprise Architecture Principles* actually refers to Architecture Principles.
>
> This inconsistency in the Series Guide seems to inadvertently merge Enterprise Principles and Architecture Principles when they should be distinct.

Enterprise and Architecture Principles are different and should not be confused. They should certainly not be considered one and the same. However, it is common for an organization to simply talk about *principles*. Nine times out of ten, these principles (which sometimes resemble requirements) come from IT. Most other departments in an organization are unaware of the existence of principles or do not apply them.

14.2.3 Enterprise Principles

Enterprise Principles are high-level, general guidelines that shape the overall strategy, culture, and decision-making framework of an organization. This type of principles ensures that decisions in an organization are made in harmony. They are an essential part of architecture governance.

Given the primarily strategic nature of Enterprise Principles, they would ideally be drafted by the board of directors or another senior management body within the organization. However, the reality is that they are initially created by the Enterprise Architect and later approved by the board of directors (or the aforementioned senior management). The buy-in of all stakeholders involved in the creation of the Enterprise Principles is necessary to provide the required support.

Enterprise Principles provide a basis for decision-making throughout an enterprise and inform how the organization sets about fulfilling its mission [20].

Because Enterprise Principles (see Table 14-3) are described at a high level, it is quite common for several of these types of principles to exist in an organization. For example, it is not inconceivable for different departments to have their own set of principles. Of course, it is important that all Enterprise Principles are aligned to avoid overlap or contradiction between them.

Table 14-3. Example of an Enterprise Principle

Name	Statement	Rationale	Implications
Comply with laws and regulations.	Information management processes comply with all relevant laws, policies, and regulations.	To comply with laws and regulations, relevant laws and regulations are considered in the development, procurement, and design of systems, products, and services.	• The organization must be mindful to comply with laws, regulations, and external policies regarding the collection, retention, and management of data. • Efficiency, need, and common sense are not the only drivers. Changes in the law and changes in regulations may drive changes in our processes or applications.

The Principles Catalog artifact shown in Table 12-5 in Chapter 12, Section 12.3.2, lists additional Enterprise Principles.

Rooted in business fundamentals, Enterprise Principles play a pivotal role in shaping the foundation for effective architecture governance. After thoroughly understanding the organizational context, it's crucial to define a tailored set of Enterprise Principles that aligns with the organization's unique needs. This ensures a strong, adaptable framework that supports both governance and strategy.

14.2.4 Architecture Principles

Architecture Principles are intended to provide guidance for the architecture process. They also provide direction for the development, maintenance, and use of the Enterprise Architecture. These principles are drawn up by the Enterprise Architect in collaboration with key stakeholders. They are formally approved by the Architecture Board.

Architecture Principles are a set of principles that relate to architecture work [20].

What is often seen in organizations is the existence of one or more sets of principles. These sets then often form a hierarchy. For example, it is quite common to have multiple Architecture Principles for each segment of the architecture.

Figure 14-2 shows how these *segment principles* (derived from segments such as business, data, application, or technology domains) are collectively informed by Enterprise Principles. Architecture Principles build on what has been established at the high level. In this way, the Enterprise Principles define the boundaries for the Architecture Principles.

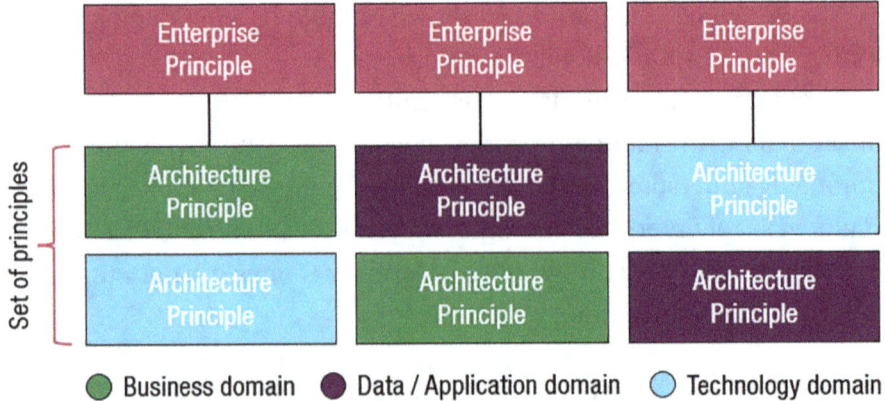

Figure 14-2. Sets of Architecture Principles that refer to specific segments

Table 14-4. Example of an Architecture Principle

Name	Statement	Rationale	Implications
Common vocabulary and definitions	Data is defined consistently throughout the organization, and the definitions are understandable and available to its users.	The data that will be used in the development of applications must have a common definition.	• The organization must establish the initial common vocabulary for the business; the definitions will be used uniformly throughout the organization. • Whenever a new definition is required, the effort will be coordinated and reconciled with the corporate glossary of descriptions [20].

Architecture Principles (see Table 14-4 for an example) are fundamental guidelines that dictate the use and deployment of resources and assets throughout an organization. They represent a consensus among various organizational elements and serve as a foundation for future decision-making. Each Architecture Principle should be explicitly linked to the organization's business goals and key architectural drivers.

14.3. Stakeholder Management

Being able to communicate properly with stakeholders – the key players within the organization – is essential. To identify these key players in an organization, *Stakeholder Management* can be applied. Stakeholder Management is the process of identifying and understanding the needs, interests, and expectations of an organization's various stakeholders and planning and implementing actions to address those interests and improve cooperation and communication with them [20].

The TOGAF Standard provides a structured approach to Stakeholder Management, ensuring that architectural projects align with organizational objectives and gain the necessary support from all involved parties.

The stakeholder concept is fundamental to representing an individual, group, or organization that has an interest in the outcome of an architecture effort. Stakeholders are critical to the success of any Enterprise Architecture initiative because they provide input, have concerns, and influence decision-making throughout the architecture development process. The stakeholder concept is used to *identify* and *analyze* their *concerns* and *interests*. Understanding stakeholder needs and expectations helps define architecture requirements and ensures that the architecture effectively addresses business goals.

Stakeholders are actively involved throughout the architecture development process. Enterprise Architects work with stakeholders to gather information, obtain feedback, and validate architectural decisions. Effective stakeholder engagement results in a shared understanding of business goals and ensures buy-in for the proposed architecture.

The stakeholder concept facilitates effective communication with various groups within the organization. Architects use the concept to document stakeholder relationships, communication channels, and key roles within the architecture effort. It helps manage expectations and ensure that stakeholders are kept informed of progress and results.

Stakeholders also play a critical role in assessing the impact of architecture decisions on various business areas. By understanding their perspectives (also called views; see Chapter 7, Section 7.6), architects can identify potential risks and opportunities associated with architectural changes and align the architecture with the organization's strategy. The primary goal of Enterprise Architecture is to align all the architecture work originating from the architecture domains and segments with an organization's goals. Understanding stakeholder perspectives helps architects prioritize initiatives and allocate resources to projects that deliver the most value to the organization and its stakeholders.

By interacting with stakeholders, architects can elicit and document their needs, which become the basis for developing architectural models and solutions. Stakeholders are an important source of requirements for the architecture. The stakeholder concept helps architects make informed decisions. By considering the interests and concerns of different stakeholders, architects can balance conflicting requirements and ensure that architectural decisions are well accepted and aligned with business priorities.

The stakeholder concept is closely related to governance in Enterprise Architecture. Understanding who the stakeholders are and what their interests are helps architects establish governance mechanisms to address their concerns, maintain transparency, and ensure accountability.

In the context of the TOGAF Standard, a stakeholder is any individual, group, or organization that can affect, be affected by, or perceive themselves to be affected by an Enterprise Architecture initiative. Effective Stakeholder Management involves identifying these parties, understanding their concerns and interests, and engaging with them appropriately to ensure the success of the architectural endeavor.

Early identification of influential or key stakeholders allows their input to shape the architecture, ensuring support and enhancing the quality of the models produced.

Gaining support from powerful stakeholders can help secure necessary resources, increasing the likelihood of project success. Knowing who the stakeholders are ensures that the architecture and related work can be communicated and aligned. It is therefore essential to identify them as early as possible and to understand their needs and expectations. Analysis of the identified stakeholders will help to paint this picture.

Engaging stakeholders early and often ensures that they understand the architecture process and its benefits and can more actively support it when needed. This can be achieved through effective communication. A *Communications Plan* will be of great help in achieving this. Such a plan is often used to ensure effective communication of the necessary information to the relevant stakeholders at the right time.

In order for an architecture team to anticipate reactions from stakeholders, it is important to understand their perspectives. This will help in planning actions to leverage positive responses while addressing negative ones. Identifying conflicting objectives among stakeholders early enables the development of strategies to resolve arising issues.

The TOGAF Standard recommends incorporating stakeholder analysis during Phase A (Architecture Vision) of the Architecture Development Method and updating it throughout each following phase. This form of continuous engagement ensures that new stakeholders are identified and their concerns are addressed as the project progresses through phases such as Phase E (Opportunities and Solutions), Phase F (Migration Planning), and Phase H (Architecture Change Management).

A practical approach to Stakeholder Management is described in the step-by-step guide outlined below.

 Identify Stakeholders: Knowing who the stakeholders are ensures that the architecture and related work can be communicated and aligned.

Analyze and Classify Stakeholder Positions: Stakeholders need to be analyzed to understand their needs and expectations.

Plan Communications: Creating a Communications Plan ensures effective communication of the necessary information to the relevant stakeholders at the right time.

Engage Stakeholders: Explore options for additional, more effective ways and means of communication.

Monitor and Review: Evaluate the effectiveness of stakeholder interactions on a regular basis. If *necessary*, adjust the stakeholder engagement and communication strategies.

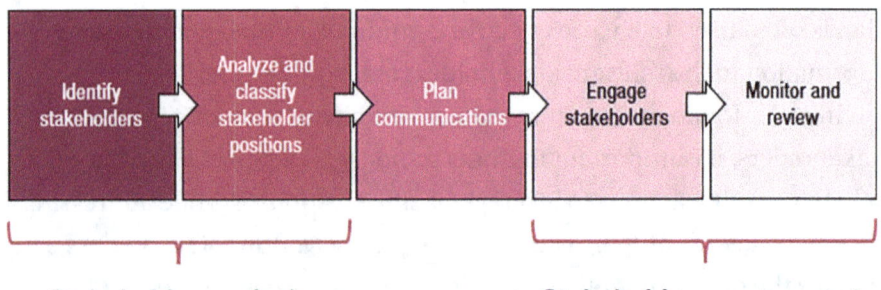

Figure 14-3. *Steps to stakeholder analysis and stakeholder engagement*

The first two steps, *identify stakeholders* and *analyze and classify stakeholder positions,* are part of the overall stakeholder analysis process, as shown in Figure 14-3. The step called *plan communications* is all about creating a Communications Plan. Finally, the last two steps, *engage stakeholders* and *monitor and review,* are part of stakeholder engagement.

14.3.1 Stakeholder Analysis

The Architecture Vision Phase details the process of identifying the stakeholders and determining their requirements. Knowing who the stakeholders are (this includes both organizations and individuals within those organizations) ensures that the architecture and related work can be communicated and aligned. A critical part of communicating with stakeholders is creating an effective Communications Plan.

Having a clear view of the stakeholders and their requirements and concerns is an important prerequisite for creating the Solution Concept Diagram. This diagram is based on the organization's goals and objectives and stakeholder requirements and constraints.

After identifying stakeholders, it is imperative to *analyze* these groups to understand their needs and expectations. This helps to prioritize their requirements and determine the best way to design the architecture to meet their needs. In order to achieve this, a Stakeholder Analysis Model (see Table 14-5) is created.

Table 14-5. Stakeholder Analysis Model

Stakeholder Group	Stakeholder	Ability to Disrupt Change	Current Understanding	Required Understanding	Current Commitment	Required Commitment	Required Support
CxO	Maria Summers (CEO)	H	H	H	M	H	H
	Chris West (CIO)	H	M	H	L	M	H
	James Noble (CFO)	M	M	M	L	M	M
HR Management	Anna Carter (Manager HR)	L	L	M	L	M	M

CHAPTER 14   ADM TECHNIQUES

The Stakeholder Analysis Model includes the stakeholder group, the specific stakeholder, and the stakeholder's ability to disrupt change. The latter is a reference to the level of power a particular stakeholder has within the organization. It illustrates the likelihood that a particular stakeholder can disrupt or actively promote the architecture work.

A stakeholder's current understanding of and commitment to the architecture work are also captured in the model. The level of commitment can be compared to the level of interest mentioned in the Stakeholder Power Grid (see Figure 14-4). Finally, the required levels of understanding, commitment, and support are identified.

A Stakeholder Power Grid is used to determine the power and interest levels of the stakeholders captured in the Stakeholder Analysis Model. The Stakeholder Power Grid allows you to assign a classification to each stakeholder. The grid consists of four quadrants, each representing one of the following four types of classifications:

Key Players: This classification indicates that these are stakeholders with a high level of interest and power within the organization (*high power/high interest*). Therefore, the frequency of communication to this group of stakeholders will be the highest. An example of this stakeholder classification is data and process owners stakeholder group in Table 14-8.

Keep Satisfied: Stakeholders classified as keep satisfied have the same level of power in the organization as the key players, but their level of interest is lower (*high power/low interest*). Therefore, this group can be informed with a slightly lower frequency. Most C-level positions in an organization fall into this classification.

Keep Informed: The level of interest among this group of stakeholders is high, but the power they can exercise within the organization is lower (*low power/high interest*). Nevertheless, it is wise to keep this group informed on a regular basis. For this group, the choice is between informing them by email and in the form of a meeting (physical or digital). An example of a stakeholder in this classification is the program manager.

Minimal Effort: Although this term has a negative connotation, this is far from the intention. It refers to the fact that stakeholders in this group have low levels of both interest and power. However, this group can still contain stakeholders that are very important to the architecture work. The employees group mentioned in Table 14-8 falls into this classification.

Using a Stakeholder Power Grid (see Figure 14-4), the architect can get a good understanding of each stakeholder's role, their level of influence, and their interest in the project. These levels are visualized in Figure 14-4 as *level of power* and *level of interest,* respectively.

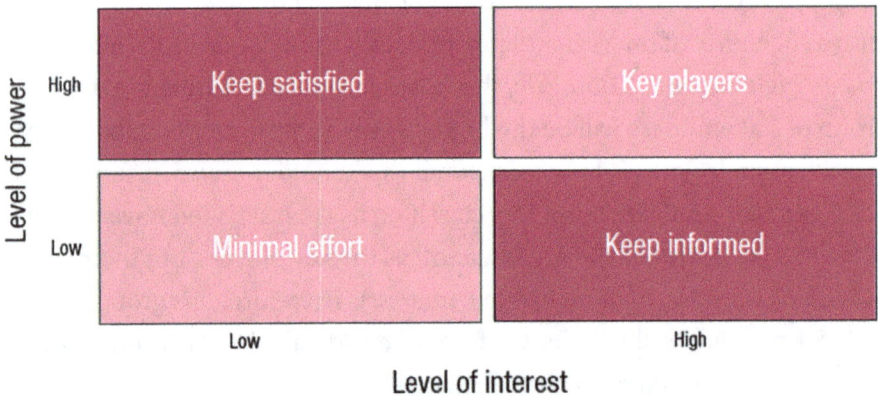

Figure 14-4. Stakeholder Power Grid

The stakeholders that were analyzed at the beginning of step 2 (see Figure 14-3) are classified using the Stakeholder Power Grid. Their power and interest levels, as well as their classification, are recorded in the Stakeholder Analysis Model (see Table 14-6).

Table 14-6. *Stakeholder Analysis Model with Stakeholder Classification*

Stakeholder Group	Stakeholder	...	Power	Interest	Classification
CxO	Maria Summers (CEO)	...	H	L	Keep satisfied
	Chris West (CIO)	...	H	H	Key player
	James Noble (CFO)	...	L	H	Keep informed
HR Management	Anna Carter (Manager HR)	...	L	L	Minimal effort

Each classification comes with a set of requirements regarding the stakeholder's communication needs. For example, it contains information about the appropriate frequency of communication and the most appropriate method of communication. This information forms the basis for the Communications Plan.

14.3.2 Stakeholder Mapping

Another essential Stakeholder Management tool is the creation of a Stakeholder Catalog. This map contains not only the stakeholders and their classification, but also the *key concerns* of the stakeholders. These detail their primary areas of focus. The Stakeholder Catalog (see Table 14-7) also contains references to the various catalogs, matrices, diagrams, and maps that are of interest to a particular (group of) stakeholder(s).

Table 14-7. Stakeholder Catalog

Stakeholder	Key Concerns	Classification	Catalogs, Matrices, Diagrams, and Maps
Maria Summers (CEO)	The high-level drivers, goals, and objectives of the organization and how these are translated into an effective process to advance the business	Keep satisfied	Strategy/Goal Matrix Goal/Objective Matrix
Chris West (CIO)	Alignment of IT strategy with business goals, optimizing IT costs, and supporting business agility through technology	Key player	Goal/Objective Matrix Objective/Initiative Matrix
James Noble (CFO)	Budget efficiency and cost control in organizational projects, long-term financial sustainability of architecture decisions	Keep informed	Application Portfolio Catalog Technology Portfolio Catalog
Anna Carter (Manager HR)	Supporting workforce needs in alignment with organizational transformation, ensuring employee adoption of new processes and technologies	Minimal effort	Business Process Map Organization/Business Process Matrix

A completed Stakeholder Catalog is an important resource that can be used frequently. The stakeholders identified in the Stakeholder Catalog need to be approached and informed on a regular basis. A Communications Plan is used for this purpose.

14.3.3 Communications Plan

Effective stakeholder communication is critical to keeping all stakeholders sufficiently invested in the architecture work. One way to achieve this is to create a Communications Plan. Such a plan ensures effective communication of the necessary information to the relevant stakeholders at the right time. It is therefore one of the critical success factors (CSFs) for a sound implementation of an Enterprise Architecture [11].

The Communications Plan should describe the communication needs of each stakeholder group, such as the information they need, the frequency of communication, and the methods for sharing that information. It should also state the purpose of the communication to the stakeholders. Of course, this will not be the same for each stakeholder group, and this distinction needs to be well-defined.

The Stakeholder Power Grid shown in Figure 14-4 illustrates how stakeholders can be classified. With each classification comes a set of requirements regarding the communication needs of the stakeholders. It is important to include the moments – especially the frequency of these moments – at which stakeholders are informed or updated.

The TOGAF Standard indicates that the scope of a Communications Plan is limited to key stakeholders. In my view, communication does not stop there.

It is precisely by communicating to the entire organization what the expected impact of implementing the Enterprise Architecture will be that everyone becomes involved. It becomes more of everyone, rather than just those directly involved. This strengthens the acceptance of the Enterprise Architecture as well as working with architecture.

There are two components that are essential to include in a Communications Plan.

The first is to include and interpret what everyone's role will be in implementing the Enterprise Architecture. This does not need to be detailed; an overall articulation in terms of roles to be fulfilled is sufficient. If people can see themselves in the roles outlined, with the associated roles and responsibilities, it will be easier to buy into the implementation. Everyone likes to make a meaningful contribution. What is sometimes missing in practice is the interpretation of that meaning. Once the meaning is unclear, people's enthusiasm for making the desired contribution diminishes. So be clear and precise in your descriptions.

The second essential part of the Communications Plan is to schedule consultation moments when the organization as a whole is updated or informed about the progress of the implementation and the method(s) to be used. Transparency in the approach is very important.

Of course, this does not always mean that all employees are informed at the same time; it is certainly possible to create a division into small(er) groups.

Once the Communications Plan has been created, the architect must implement it to foster effective communication and collaboration. From then on, stakeholders are regularly updated on the progress of the implementation, and feedback sessions are offered as opportunities for stakeholders to provide input.

Table 14-8. Communications Plan

Stakeholder Group	Classification	Communication Goals	Communication Method	Communication Frequency
CxO	Keep satisfied	Provide insight into how the high-level drivers, goals, and objectives are translated into architecture work.	Meeting	Every fortnight
Data and process owners	Key players	Ensure the consistent use and governance of the organization's business, data, application, and technology assets.	Meeting	Monthly
Program manager	Keep informed	On-time, on-budget delivery of change initiatives that will lead to the implementation of the architecture.	Meeting	Monthly

(*continued*)

Table 14-8. (*continued*)

Stakeholder Group	Classification	Communication Goals	Communication Method	Communication Frequency
Employees	Minimal effort	Assess the impact of the implementation of the Enterprise Architecture with regard to the day-to-day duties and tasks.	Email message, interspersed with meeting	Monthly
Human resources	Minimal effort	Ensure that the roles and actors who are required to support the implementation of the architecture are available.	Email message	Quarterly

A standard Communications Plan (Table 14-8) often looks static and business-like and does not usually provide the most effective means of communication in practice. It is therefore advisable to explore options for additional means of communication in order to enhance stakeholder engagement.

14.3.4 Stakeholder Engagement

A more effective way of communicating might be to use appealing and sophisticated visuals and designs that can be used to capture and hold the attention of stakeholders. Common forms of this are presentations and infographics, but posters or specially decorated rooms where the information is highlighted can also help. Encourage stakeholders to provide input and express concerns to foster a collaborative environment.

Assess stakeholder engagement by regularly evaluating the effectiveness of stakeholder interactions. If necessary, adjust the stakeholder engagement and communication strategies. Also update the Stakeholder Analysis Model on a regular basis. When new stakeholders emerge (e.g., new employees are joining the organization), it is advisable to revise stakeholder information and the accompanying engagement plans to reflect any changes in stakeholder positions or project dynamics.

Effective Stakeholder Management, as outlined in the TOGAF Standard, is essential for the success of Enterprise Architecture initiatives. By systematically identifying, analyzing, and engaging stakeholders, organizations can ensure that architectural projects are aligned with business objectives, gain necessary support, and address concerns proactively. Implementing the step-by-step plan detailed above will facilitate structured Stakeholder Management, contributing to the overall success of the architecture development process.

Further Reading

For more information on Stakeholder Management, refer to *The TOGAF Standard, 10th Edition*, document number C220.

14.4. Architecture Patterns

Chapter 7, Section 7.5.2, discussed building blocks. They are considered the backbone of the architecture and serve as essential, reusable components in architectural solutions. Building blocks form a modular foundation that is designed to be generic and adaptable, ensuring that they can be seamlessly integrated with other building blocks. Their inherent reusability allows architects to combine them in different configurations, promoting efficiency and scalability in the development of architectural solutions.

Architecture patterns, on the other hand, put building blocks into context. They use building blocks to create a reusable solution to a problem.

Building blocks are what you use; patterns can tell you how you use them, when, why, and what trade-offs you have to make in doing so [20].

Architecture patterns offer architects a way to use combinations of Architecture and Solution Building Blocks that have proven to deliver an effective solution in the past.

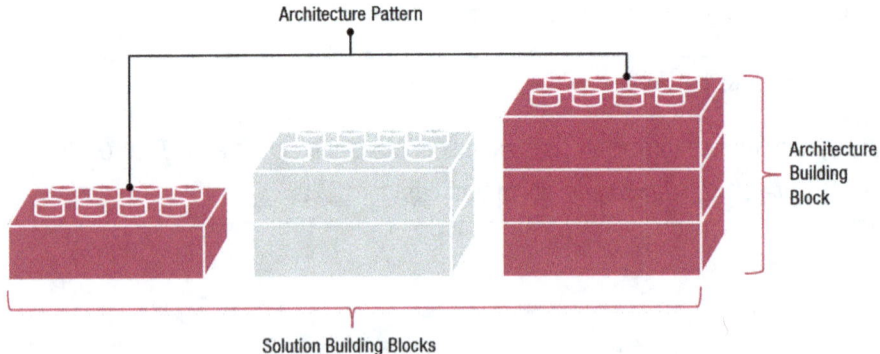

Figure 14-5. Architecture patterns and building blocks

As shown in Figure 14-5, architecture patterns are composed of Architecture and Solution Building Blocks. The figure illustrates a set of *Solution Building Blocks* (horizontal) and *Architecture Building Blocks* (vertical). An architecture pattern can be a *combination* of one or more Solution Building Blocks and one or more Architecture Building Blocks.

In everyday practice, when the word *pattern* is used, it is still common to think of software design patterns. This is because many people in the software field use the term architecture to refer to software. For this reason, an architecture pattern is often confused with a design pattern.

The TOGAF Standard has not yet formally incorporated the use of architecture patterns into the Architecture Development Method.

Architecture patterns have not (as yet) been integrated into the TOGAF Standard [20].

However, it can be stated that during the execution of phases B, C, and D – the moment when reusable building blocks are identified – architecture patterns can also be defined.

14.5. Gap Analysis

A gap analysis is a process that identifies the difference between an organization's Baseline and Target Architectures in terms of its capabilities, processes, applications, and technology. The purpose of a gap analysis is to identify areas where improvements are needed to achieve the desired future state.

The basic premise is to highlight a shortfall between the Baseline Architecture and the Target Architecture, that is, items that have been deliberately omitted, accidentally left out, or not yet defined [20].

CHAPTER 14 ADM TECHNIQUES

There are a few basic steps to perform a gap analysis:

- Identify Baseline and Target Architectures.
- Compare the two states to highlight gaps.
- Categorize gaps as missing functionalities, technology gaps, or process inefficiencies.
- Create work packages of grouped activities to fill the gaps.
- Prioritize work packages based on business impact and feasibility.

The results of the gap analysis are used to inform subsequent phases of the Architecture Development Method, such as phases B, C, and D, where the detailed architecture is developed based on the architectural vision.

Remember the fictional hospital described in Chapter 5 (Section 5.1.1)?

This particular hospital concluded earlier on that it lacked a required capability, Inflow Support, based on the Business Capability Heat Map in Figure 12-9 (Chapter 12, Section 12.5.2).

To automate the missing capability later, a new application (HR Recruitment Application B) would have to be implemented.
Table 12-57, the Consolidated Gaps, Solutions, and Dependencies Matrix, lists the implementation of the new application as a possible solution for automating the capability.

Figure 14-6 shows one way to identify and compare the Baseline and Target Architectures by highlighting the differences between them. The figure shows the identified capabilities and applications that exist either in both the Baseline and the Target Architectures, only in the Baseline Architecture, or only in the Target Architecture.

CHAPTER 14   ADM TECHNIQUES

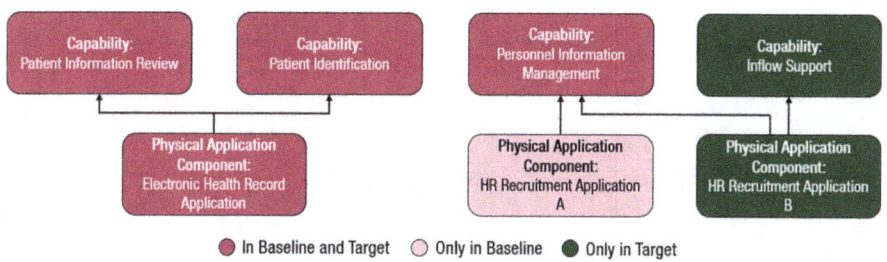

Figure 14-6. Identification of the Baseline and Target Architectures

The figure illustrates that HR Recruitment Application A exists in the Baseline Architecture, but is replaced by HR Recruitment Application B in the Target Architecture. After the new application is implemented, it enables the automation of the Inflow Support capability, which does not exist in the Baseline Architecture. Due to the implementation of the new application, the Personnel Information Management capability is also automated by HR Recruitment Application B.

Figure 14-7 shows the phases of the Architecture Development Method in which the gap analysis steps take place. Most of the work is done in phases B, C, and D. This is where the gaps between the Baseline and Target Architectures are identified. Then, in Phase E (Opportunities and Solutions), the real analysis takes place. The necessary changes to the architecture are identified, and the activities required to close the gaps are grouped into work packages.

The figure shows three architecture states at the top row: Baseline, Transition, and Target. Right below that are two gaps visible. Gap A illustrates the gap between the Baseline and Transition Architectures, and Gap B illustrates the gap between the Transition and Target Architectures. The bottom row shows two work packages. The one on the left contains the activities necessary to close Gap A, and the other one contains activities to close Gap B.

CHAPTER 14 ADM TECHNIQUES

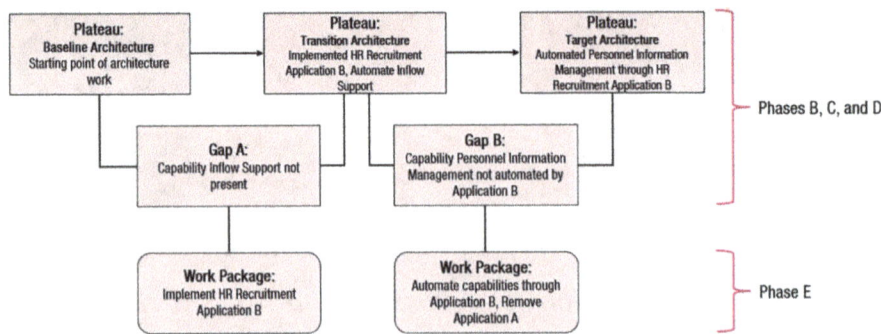

Figure 14-7. *Gap analysis during phases B–E of the ADM*

As shown in Figure 14-7, the gaps between the Baseline and Transition Architectures (Gap A) on the one hand and the Transition and Target Architectures (Gap B) on the other are identified. Gap A shows that the capability Inflow Support is not yet present. A work package with activities is created to close the gap. In this case, the work package contains the activity to implement HR Recruitment Application B, which will eventually automate the capability Inflow Support.

Gap B, which details the gap between the Transition and Target Architectures, shows that the Personnel Information Management capability is not yet automated by HR Recruitment Application B. Again, a work package is created containing activities to automate the Personnel Information Management and Inflow Support capabilities through HR Recruitment Application B and to remove HR Recruitment Application A from the architecture.

It is recommended that the actions listed above be described in a Gap Analysis Matrix (see Table 14-9).

Set Up the Matrix: Begin by drawing a grid. Label the rows with the Architecture Building Blocks from the Baseline Architecture. Label the columns with the Architecture Building Blocks from the Target Architecture.

Add New and Eliminated Sections: Extend the matrix by adding an extra row at the bottom labeled *New* and an extra column on the right labeled *Eliminated*. This will help track additions and removals.

Identify Common Building Blocks: For each Architecture Building Block that exists in both the Baseline and Target Architectures, mark the intersecting cell in the matrix with *Included*. This shows which components remain unchanged.

Review Missing Building Blocks: Look for Architecture Building Blocks that are present in the Baseline Architecture but absent in the Target Architecture. These will align with the *Eliminated* column. Assess whether removing these components is appropriate and document the reasoning.

Document New Building Blocks: Identify Architecture Building Blocks that appear in the Target Architecture but not in the Baseline. Place these in the *New* row of the matrix. Consider how these additions align with the architectural goals and document any necessary implementation steps.

If an Architecture Building Block was correctly eliminated, mark it as such in the appropriate *Eliminated* cell. If it was not, this means that an accidental omission in the Target Architecture has been uncovered that must be addressed by reinstating the Architecture Building Block in the next iteration of the architecture design. Mark it as such in the appropriate *Eliminated* cell.

Table 14-9. *Example Gap Analysis Matrix*

Baseline Architecture ▼	Target Architecture				
	Patient Information Review	Patient Identification	Personnel Information Management	Inflow Support	Eliminated
Patient Information Review	Included				
Patient Identification		Included			
Personnel Information Management			Included		HR Recruitment Application A
New			Gap: HR Recruitment Application B	Gap: HR Recruitment Application B	

When the exercise is completed, anything under *Eliminated* or *New* is a gap that should either be explained as correctly eliminated or marked as to be addressed by performing activities to reinstate or develop/procure the building block. Work packages are used to group these activities. The work packages find their way to the Architecture Roadmap, which is described in Chapter 12, Section 12.8.3.

14.6. Interoperability Requirements

The idea behind the interoperability technique is to compare a set of entities and determine what form and degree of interoperability is required between them. This involves determining the degree of interoperability needed to share information and services.

For example, interoperability may be required between two stakeholders or between a stakeholder and an application. Cross-mapping between the two entities provides a picture of what can exchange information with whom and in what manner. The cross-mapping technique is also used to determine the requirements for the desired interoperability.

A definition of interoperability is "the ability to share information and services". Defining the degree to which the information and services are to be shared is a very useful architectural requirement, especially in a complex organization and/or extended enterprise [20].

The TOGAF Standard recommends using the cross-mappings just described to determine the *interoperability requirements*. An example is shown in Table 14-10.

Table 14-10. Interoperability Requirements between stakeholders

Stakeholders ▶ ▼	A	B	C	D	E
A		2	3	2	3
B	2		3	2	3
C	3	3		2	2
D	2	2	2		3

CHAPTER 14   ADM TECHNIQUES

Table 14-10 shows a cross-mapping between stakeholders (A–E) and lists the associated degree of interoperability. Since this particular table contains the entity *stakeholder*, it is created from a Business Architecture perspective. Therefore, it is developed during the corresponding phase (Phase B) of the Architecture Development Method. The content of a cross-mapping is related to the phase of the Architecture Development Method in which it is created.

Table 14-10 hints at the fact that the TOGAF Standard distinguishes four degrees (levels) of interoperability. These levels are indicated by numbers:

- **Degree 1 (Unstructured Data Exchange):** Involves the exchange of human-interpretable unstructured data, such as the free text found in operational estimates, analyses, and papers

- **Degree 2 (Structured Data Exchange):** Involves the exchange of human-interpretable structured data intended for manual and/or automated handling, but requires manual compilation, receipt, and/or message dispatch

- **Degree 3 (Seamless Sharing of Data):** Involves the automated exchange of data between systems based on a common exchange model

- **Degree 4 (Seamless Sharing of Information):** Is an extension of Degree 3 to the universal interpretation of information through data processing based on cooperating applications

Each of the aforementioned degrees can be further specified by assigning more specific descriptions to the degree. In this case, an alphabetic letter is added to the number of the degree. This is illustrated in Figure 14-8.

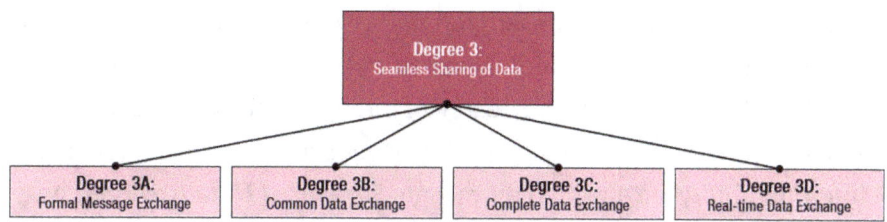

Figure 14-8. *Detailed degrees of interoperability*

Further detailing of the degrees can be used, for example, to express the level of standardization or compatibility of interoperability.

These degrees are very useful for specifying the way that information has to be exchanged between the various systems and provide critical direction to the projects implementing the systems.

In addition to cross-mapping between entities of the same type, it is also possible to compare different entities with each other. Table 14-11 shows an example of such a comparison between stakeholders on the one hand and applications on the other hand.

Table 14-11. *Cross-mapping of Interoperability Requirements between Stakeholder and Application Entities*

Applications Stakeholders ▼	CRM Application	Recruitment Application	Financial Application
Business Unit Managers	4D	3C	2A
HR Employees	3B	4D	–
Financial Controllers	3C	–	4D

The example in Table 14-11 shows that three applications were identified (CRM Application, HR Recruitment Application, Financial Application) as well as three stakeholder groups (Business Unit Managers, HR Employees, and Financial Controllers).

CHAPTER 14   ADM TECHNIQUES

The cells in the table represent the degree of interoperability between each application and stakeholder. For example, the cell in *row 1, column 3*, of Table 14-11 represents the degree of interoperability between the *Financial Application* and *Business Unit Managers*. In this case, the degree of interoperability is 2A, which represents *Structured Data Exchange* as the base degree. It also contains the letter B, indicating the detail level of *Common Data Exchange*. This means that the Financial Application can exchange data with Business Unit Managers in a structured format, but there may be some limitations or restrictions on the types of data that can be exchanged.

During the execution of the Architecture Development Method, the issue of interoperability arises several times. Figure 14-9 illustrates the integration of interoperability requirements into the Architecture Development Method. The Preliminary Phase has been removed for better readability.

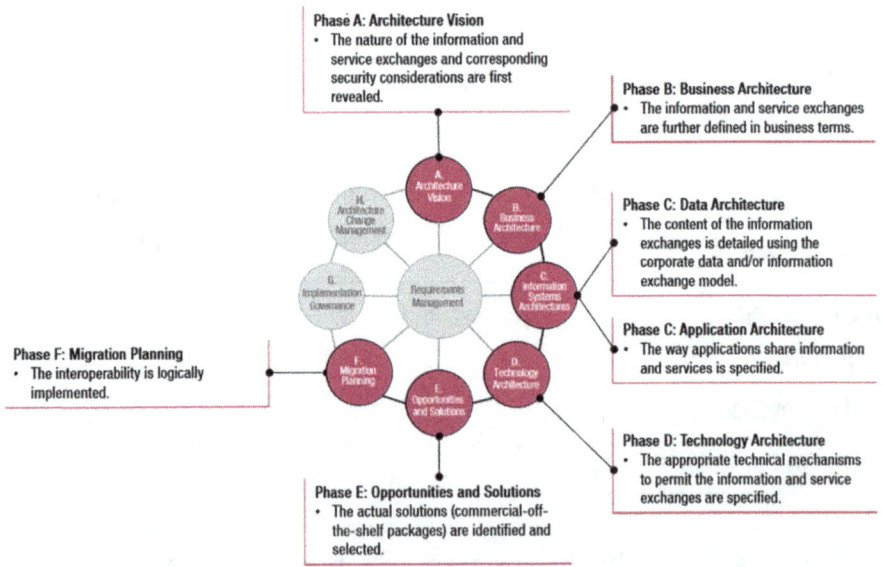

Figure 14-9. Integration of interoperability requirements in the ADM

One major advantage of applying the interoperability technique is *enhanced integration and communication between systems*. Interoperability ensures that different entities (such as people, systems, applications, and services) can effectively exchange and use information, reducing data silos and integration complexity.

Further Reading

For more information about interoperability requirements and how to apply them, refer to *The TOGAF Standard, 10th Edition*, document number C220.

The following website contains additional information about interoperability requirements and how to apply them:

https://guides.visual-paradigm.com/togaf-adm-top-10-techniques-part-6-interoperability-requirements/

14.7. Business Transformation Readiness Assessment

In the dynamic landscape of Enterprise Architecture, implementing an innovative architecture vision is not just a technical task. It is a profound organizational change that requires careful consideration of the human element. The TOGAF Standard recognizes the critical need to assess and address the organization's readiness for business transformation.

In the early phases of the Architecture Development Method, particularly Phase A (Architecture Vision) and the subsequent architecture definition phases (phases B–D), organizations conceptualize and design innovative solutions that often involve significant changes. These changes are not limited to technological shifts as they have a profound impact on the workforce, culture, and existing skills and competencies within the organization.

For example, the fact that a recruitment application will lead to a more effective way of managing candidate information and potentially reduce the number of staff required requires a keen understanding of the human resource implications. A change-averse culture and a workforce with narrow skill sets can potentially hinder the implementation of even the most well-designed architecture.

To ensure successful implementation of the architecture transformation in phases E and F, organizations must conduct a comprehensive readiness assessment. This is a collaborative effort between the organization's staff, particularly human resources, lines of business, and (IT) planners.

The steps in conducting a Business Transformation Readiness Assessment consist of identifying and defining the readiness factors that will significantly impact the organization during the transformation. These factors may include cultural readiness, workforce skills, leadership alignment, and change management capabilities.

Maturity models can be used to represent the readiness factors. They provide a structured framework for assessing the organization's Baseline and Target Architecture with respect to each readiness factor. The maturity model format described in Chapter 15 can be used as a template.

To assess the organization's readiness, a readiness factor rating must be determined. This assessment involves analyzing the current maturity level and identifying gaps that need to be addressed for a successful transformation.

Of course, the risks associated with each readiness factor also need to be identified and strategies developed to mitigate these risks. This includes understanding potential obstacles and challenges that could impede the transformation process. The Risk Management technique described in Section 14.8 can be used to accomplish this.

Finally, the results of the readiness assessment must be incorporated into the Implementation and Migration Plan created in Phase E and finalized in Phase F. This ensures that the organization's readiness for change is considered and addressed in the overall transformation strategy.

CHAPTER 14 ADM TECHNIQUES

A Business Transformation Readiness Assessment table typically includes various factors that determine an organization's preparedness for a significant change. Table 14-12 shows an example of a Business Transformation Readiness Assessment.

Table 14-12. Business Transformation Readiness Assessment

Readiness Factor	Assessment Criteria	Score (1–5)
Leadership commitment	Level of commitment from top leadership	4
Employee engagement	Willingness and enthusiasm of employees for change	3
Organizational culture	Openness to new ideas and adaptability	5
Skills and competencies	Availability of required skills and competencies for the transformation	3
Risk Management	Identification and mitigation of potential risks	4
Stakeholder alignment	Alignment of stakeholders' interests and expectations	5

In the example presented in Table 14-12, each readiness factor is assessed on a scale of 1–5, with 5 being the highest readiness. The criteria can be customized based on the specific needs and nature of a business transformation.

Each readiness factor can be assigned a rating to score the factor based on its urgency, readiness status, and the degree of difficulty to remediate any issues (shown in Table 14-13). The urgency indicates how quickly action is needed, the readiness status provides an overall assessment of the current state (which refers to the score provided in Table 14-12), and the remediation difficulty represents the effort required to address any identified issues.

CHAPTER 14 ADM TECHNIQUES

Table 14-13. *Business Transformation Readiness Assessment using a Readiness Factor Rating*

Readiness Factor	Urgency	Readiness Status	Remediation Difficulty
Leadership commitment	High	Good	Easy
Employee engagement	High	Acceptable	Moderate
Organizational culture	Moderate	Excellent	Moderate
Skills and competencies	Moderate	Good	Moderate
Risk Management	Moderate	Acceptable	Moderate
Stakeholder alignment	High	Excellent	Easy

The *readiness status* in Table 14-13 is based on the number that was assigned in the Business Transformation Readiness Assessment table (Table 14-12). The scores of 1–5 represent *poor, fair, acceptable,* good, and *excellent,* respectively.

Table 14-13 provides a quick overview for management, helping to understand the critical areas that require immediate attention and the overall readiness of the organization for the transformation initiative. The readiness factors, as part of an overall Implementation and Migration Plan, will have to be continuously monitored (Phase G) and rapid corrective actions taken through the Governance Framework to ensure that the defined architectures can be implemented.

Identify Key Readiness Factors: Begin by pinpointing the critical factors that influence the organization's ability to undergo transformation. These may include cultural readiness, workforce skills, leadership alignment, and change management capabilities. Understanding these elements is essential for a successful transformation journey.

Use Maturity Models for Assessment: Maturity models can be used to evaluate each readiness factor systematically. These models offer a structured framework to assess the organization's Baseline and Target Architecture concerning each factor.

Determine Readiness Factor Ratings: Analyze the current maturity levels of each readiness factor to identify gaps that need addressing. This evaluation helps in understanding the areas requiring improvement to facilitate a smooth transformation process.

Identify and Mitigate Associated Risks: Recognize potential risks linked to each readiness factor. Develop strategies to mitigate these risks by understanding possible roadblocks and challenges that could hinder the transformation. Proactive Risk Management ensures preparedness for unforeseen obstacles.

Integrate Findings Into Implementation and Migration Plan: Incorporate the insights gained from the readiness assessment into the Implementation and Migration Plan. This ensures that organizational preparedness is a core component of the overall transformation plan.

In the ever-evolving landscape of Enterprise Architecture, success hinges not only on the brilliance of solutions but also on the organization's ability to embrace and adapt to change. The TOGAF Standard's emphasis on performing a Business Transformation Readiness Assessment acknowledges the significance of the human element in the success of architectural endeavors.

By proactively identifying, evaluating, and mitigating the risks associated with organizational readiness, organizations can navigate the complexities of change with agility. A well-executed readiness assessment ensures that the human factors align harmoniously with the architectural vision, propelling the organization toward a future of innovation and resilience.

Further Reading

For more information about the Business Transformation Readiness Assessment, refer to *The TOGAF Standard, 10th Edition*, document number C220.

14.8. Risk Management

Risk Management is the second-to-last technique in the Architecture Development Method. Since it is addressed in several phases (Phase A and phases E-H), it can be seen as an integral part of architecture development. Applying Risk Management techniques ensures that risks are identified, assessed, and mitigated as part of the architecture development process.

According to the TOGAF Standard, Risk Management is specifically addressed in the following phases of the Architecture Development Method (see also Figure 14-10):

- **Phase A (Architecture Vision):** During this phase, an initial risk assessment is performed to identify high-level risks associated with the architecture work. This phase identifies significant risks that could affect the feasibility or success of the architecture initiative. The risks identified during Phase A are captured in the Statement of Architecture Work document.

CHAPTER 14 ADM TECHNIQUES

- **Phase E (Opportunities and Solutions):** After the Baseline and Target Architectures have been created in phases B–D, the gaps discovered through the use of gap analysis allow for the discovery of risks related to the implementation of the Target Architecture. These implementation-related risks must be evaluated for potential architecture solutions. Phase E also evaluates risk mitigation strategies and ensures that risk considerations influence the selection of implementation approaches.

- **Phase F (Migration Planning):** The Consolidated Gaps, Solutions, and Dependencies Matrix (created in Phase E) serves as input for a detailed risk assessment related to migration and transition plans. Phase F uses Risk Management to identify risks standing in the way of a successful implementation and integration. It also ensures that mitigation plans and fallback strategies are in place.

- **Phase G (Implementation Governance):** The governance phase of the Architecture Development Method manages risks that arise during the implementation of the architecture. Managing risks from this phase ensures compliance with the architecture specification and governance standards. Phase G addresses emerging risks through governance oversight.

- **Phase H (Architecture Change Management):** Continuous risk monitoring is performed during Phase H with respect to changes to the architecture. The Architecture Change Management Phase identifies risks associated with new requirements, technologies, or business changes. It implements risk response mechanisms as part of change management.

There are a few phases of the Architecture Development Method in which Risk Management is *not* explicitly addressed:

- **Preliminary Phase:** This phase focuses on architecture governance and Architecture Capability setup, but does not perform a risk assessment.
- **Phases B, C, and D:** Do not formally include Risk Management, although risks may be informally identified within the architectures. These risks, derived from the gap analysis performed during the development of the Baseline, Transition, and Target Architectures, are recorded in the Consolidated Gaps, Solutions, and Dependencies Matrix during Phase E.

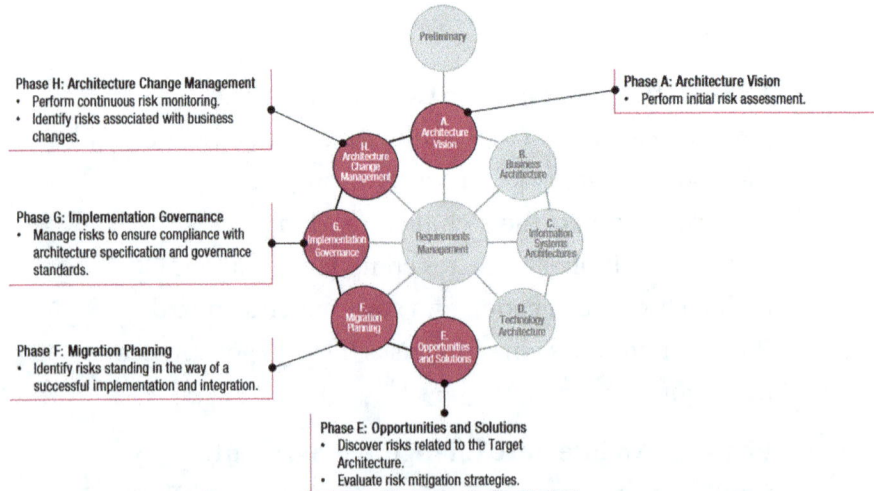

Figure 14-10. Risk Management in specific phases of the ADM

CHAPTER 14 ADM TECHNIQUES

There is a difference between the technique called *Risk Management* and *Enterprise Risk Management*. The technique described in this chapter is a simplified version of the comprehensive Enterprise Risk Management. The latter is described in great detail in the Series Guide "Integrating Risk and Security within a TOGAF® Enterprise Architecture". A link to this document is provided in the *Further Reading* information box at the end of this section.

Risk Management in the TOGAF Standard primarily focuses on architecture project risk. This is only one type of risk. The scope of Enterprise Risk Management, as presented in the "Integrating Risk and Security within a TOGAF® Enterprise Architecture" Series Guide as part of the Enterprise Security Architecture, is much broader. It includes business, system, information, project, privacy, compliance, and organizational change risk, among other categories, too [24].

When a risk is identified in the course of architecture work, the first thing to consider is whether it is an *initial (preliminary) risk* or a *residual risk*. The former indicates a new, previously unaddressed risk, while the latter concerns a previously encountered – and possibly (partially) mitigated – risk. The approach is different for each type of risk.

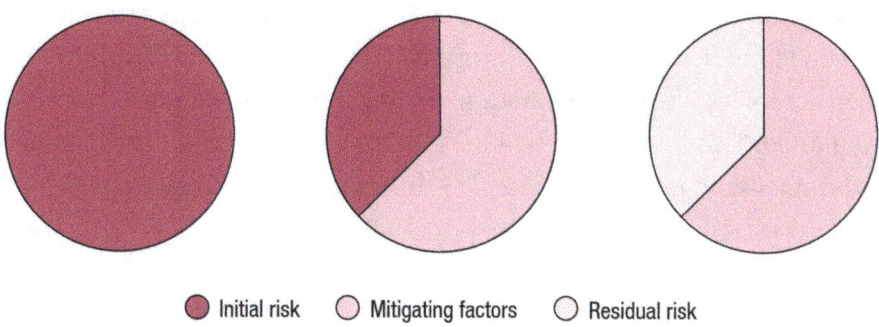

Figure 14-11. Difference between initial and residual risks and mitigating factors

For an initial risk – a categorization given to a risk prior to identifying and implementing mitigating actions – the Risk Management technique should be applied. Existing or residual risks have usually already been accepted and mitigating actions implemented. Figure 14-11 illustrates the difference between initial risk, residual risk, and mitigating factors.

The technique of applying Risk Management consists of a number of steps. First, risks are typically classified as time, cost, and scope, but can also include contractual, technology, and complexity risks. In addition, risks can be classified by architecture domains or segments. The Risk Management technique involves identifying the Baseline and Target Architectures and the work packages required to arrive at the Target Architecture. The consequences of *not* achieving the Target Architecture can lead to the discovery of risks.

These risks are classified by impact and frequency according to scales used in the organization. When impact and frequency are combined, they create a preliminary risk score that can be visualized in a Risk Heat Map (see Figure 14-12 in Section 14.8.6).

After the risks are identified and classified, they need to be mitigated (or at least as much of the risk as possible). This means reducing the risk to an acceptable level. Once the initial risk has been mitigated, the remaining risk is referred to as *residual risk*. The key consideration is that the mitigation effort actually reduces the impact to the organization and does not simply move the risk to another similarly high quadrant.

One final step of the Risk Management technique involves tracking and updating risk assessments over time. Once the residual risks have been accepted, the implementation of mitigating actions must be carefully monitored. This is to ensure that the organization is dealing with residual risks rather than initial risks. Phase G (Implementation Governance) is where risk monitoring takes place.

Classify the Risks: First, identify the types of risks involved. Common categories include time, cost, and scope. But don't stop there. Also consider contractual commitments, technological challenges, and the complexity of the project. In addition, consider risks specific to different parts of the architecture.

Identify the Risks: Next, review the Baseline and Target Architectures. Identify the steps required to reach the Target Architecture. By analyzing these steps, potential risks that could occur if the Target Architecture is not achieved can be identified.

Assess the Risks: Once the risks have been identified, assess how often they could occur and how much impact they could have. Use the organization's standard scales to make this assessment. Combining these factors provides a preliminary risk score that helps prioritize which risks need more attention.

Mitigate the Risks and Assess What's Left: Plan and implement actions to reduce the significant risks to acceptable levels. After these mitigation efforts, assess the remainder of the risks, known as residual risks. Ensure that the mitigation strategies effectively reduce the impact on the organization and don't just shift the risk elsewhere.

Monitor the Risks: Finally, keep track of the risks over time. Once the residual risks are accepted, closely monitor the implementation of the mitigation actions. This ongoing oversight ensures that the residual risks are managed effectively, rather than facing the initial risks all over again. This continuous monitoring happens during Phase G (Implementation Governance).

Each of the steps to Risk Management is further detailed in the sections below.

Further Reading

The Risk Management technique is further described in *The TOGAF Standard, 10th Edition*, document number C220.

More information about Enterprise Risk Management is available through the Series Guide "Integrating Risk and Security within a TOGAF® Enterprise Architecture", document number G152.

14.8.1 Risk Classification

Effective Risk Management is critical for organizations seeking to protect their objectives and operations. A structured approach to risk classification enables rapid and appropriate mitigation strategies.

A common method is to categorize risks based on their potential impact on the organization. This approach ensures that risks with significant impact are escalated to the appropriate governance levels, facilitating timely and effective responses.

Risks are often neatly categorized. This segmentation helps identify and effectively address potential pitfalls. Primary categories include

- **Time (Schedule) Risks:** These pertain to delays that can derail project timelines.
- **Cost (Budget) Risks:** These involve expenditures that exceed the allocated budget.
- **Scope Risks:** These arise when there's a deviation from the project's defined objectives.

Given the multifaceted nature of risks, organizations should also consider the following risk categories when classifying risks:

- **Contractual Risks:** These relate to potential issues arising from contractual terms and obligations. Misunderstandings or ambiguities in contracts can lead to disputes or unmet expectations.

- **Technology Risks:** As technology rapidly evolves, existing systems may become obsolete or incompatible. Keeping abreast of technological advances is critical to mitigating these risks.

- **Scope and Complexity Risks:** Projects with large or complex scope can be challenging to execute and manage. Clear definition and understanding of project boundaries is critical.

- **Environmental (Corporate) Risks:** Internal cultural or structural challenges within an organization can impede progress. Aligning organizational structures and cultures with project goals is essential.

- **People Risks:** Staffing issues, such as skills shortages or turnover, can affect project delivery. Investing in training and retention strategies helps mitigate these risks.

- **Client Acceptance Risks:** Deliverables may be met with resistance from customers or stakeholders. Engaging clients throughout the project lifecycle ensures that their needs and expectations are addressed.

Another effective strategy is to classify risks by architecture domain or segment:

- **Business Risks:** Threats that could derail business processes and objectives. Think of them as hurdles that could derail strategic goals.

- **Information Risks:** Data integrity and security concerns. It's all about making sure information is accurate and protected.

- **Application Risks:** These are potential problems within software applications. Whether it's bugs, integration problems, or performance issues, they fall into this category.

- **Technology Risks:** This domain covers vulnerabilities in the hardware and supporting technologies. It's about keeping the backbone of the IT infrastructure robust and resilient.

The Implementation Factor Catalog (see Section 12.8.2) may be supplemented by any of the classifications listed above.

CHAPTER 14 ADM TECHNIQUES

Table 14-14. Implementation Factor Catalog Supplemented by a Risk Classification

Risk Classification	Factor Category	Factor	Description	Deduction
Environmental (corporate) risk People risk	Effect (impact)	Change in Applications	Replace HR Recruitment Application A with Application B to streamline storage of candidate information.	Need for personnel training to use new application. New application has major personnel savings and should be given priority.
Scope and complexity risk	Dependency	Consolidation of Application Services	Multiple information streams will be consolidated.	New application will automate most application services.
Business risk	Action	Introduction of New HR Recruitment Application		

The first row in the Implementation Factor Catalog (Table 14-14) is about replacing HR Recruitment Application A with Application B. The deduction of this factor shows that it is expected that there will be a *need for personnel training* to adequately handle the new application. In addition, the implementation of the new application is likely to result in a

641

reduction in headcount (*major personnel savings*). Because the first factor in the catalog involves *internal cultural or structural challenges* and *staffing issues that (may) affect project delivery*, it is classified as an *environmental (corporate) risk* and a *people risk*.

Organizations may have unique categorizations tailored to their specific contexts. It's advisable for the Architecture Capability of the organization to adopt or extend these classifications to maintain consistency and relevance.

14.8.2 Risk Identification

During the execution of phases B, C, and D of the Architecture Development Method, the Baseline and Target Architectures are formed. The gaps between these architecture states will undoubtedly allow one or more risks to surface. Performing Business Transformation Readiness Assessments does the same. Identifying these risks is critical because it allows organizations to develop strategies to address them throughout the transformation process.

In Phase E (Opportunities and Solutions), these risks are captured in the Implementation Factor Catalog. This is where they are further detailed. The catalog is then used during Phase F (Migration Planning) to proactively address the risks.

Capability Maturity Models (see also Chapter 15) are particularly effective for assessing specific factors related to architecture delivery. They help establish baseline and target states and outline the actions required to move to the desired state. Failure to achieve the target state may reveal potential risks that require attention.

Documenting how to approach these risks is typically done in a Risk Management Plan. Standard project management methodologies such as the Project Management Body of Knowledge (PMBOK) and PRINCE2 provide templates.

A Risk Management Plan contains at least three basic components:

- **Document Information:** Things like the project name, date, and author of the document, including the version number.

- **Approvals:** A list of the dates, names, and signatures of the people who approved the approach described in the document.

- **Approach:** A detailed description of how risks will be managed. It also includes a summary of the tools that will be used and an outline of the information that will be recorded. Perhaps more importantly, it defines the metrics that will be used to classify risks. Risk tolerance is also described in the document.

The aforementioned project management methodologies typically include procedures for contingency planning, tracking and assessing risk levels, responding to changes in risk factors, and processes for documenting, reporting, and communicating risks to stakeholders.

Further Reading

Additional information about project management methodologies can be found at these URLs:

https://www.pmi.org/standards/pmbok
https://www.axelos.com/certifications/propath/prince2-project-management

---

14.8.3 Initial Risk Assessment

The initial risk assessment takes place during phases E and F of the Architecture Development Method. In the Opportunities and Solutions Phase (Phase E), the Implementation Factor Catalog and the Consolidated Gaps, Solutions, and Dependencies Matrix are used to capture risks. Phase F proactively addresses these risks by creating a Risk Management Plan. This plan covers the potential risks identified during migration planning with proposed mitigation strategies.

Once risks have been identified, the first step is to determine whether they are *initial* or *residual*. The approach to these types of risks differs. Section 14.8 describes these approaches.

There are no hard and fast rules with respect to measuring impact (effect) and frequency (likelihood). The guidelines described in the TOGAF Standard are based upon existing Risk Management best practices [20].

Risks are usually classified with respect to *impact* (effect) and *frequency* (likelihood) in accordance with scales used within the organization. In most organizations, it is common to assign a value between 1 and 5 to both impact and frequency. The sample criteria shown in Tables 14-15 and 14-16 can be used to classify the impact and frequency of risks. Of course, these criteria can be adapted to the organization.

CHAPTER 14 ADM TECHNIQUES

Table 14-15. Criteria to determine Risk Impact Level

Impact (Effect)	Impact Level	Description
Catastrophic	5	Critical financial loss that could lead to the organization's bankruptcy.
Significant	4	Causes severe financial loss in more than one line of business, resulting in lost productivity and no return on (IT) investment.
Moderate	3	Results in a small financial loss to a line of business and a reduced return on (IT) investment.
Low	2	Infers a very minor impact on the ability of a line of business to provide services and/or products, not accompanied by any type of financial loss.
Negligible	1	Assumes minimal impact on business unit's ability to deliver services and/or products.

Table 14-16. Criteria to determine Risk Frequency Level

Frequency (Likelihood)	Frequency Level	Description
Frequent	5	Likely to occur very often and/or continuously.
Probable	4	Occurs several times during a transformation cycle.
Occasional	3	Occurs sporadically.
Seldom	2	Possible, but unlikely to occur more than once in the course of a transformation cycle.
Unlikely	1	Not likely to occur in the course of a transformation cycle.

CHAPTER 14 ADM TECHNIQUES

The two factors (impact and frequency) can be combined to provide a score and classification of the actual risk. This method is described in more detail in Section 14.8.6.1. Table 14-17 shows the possible risk classifications.

Table 14-17. Criteria to determine Risk Classification

Classification	Indication	Description
Critical risk	C	The transformation effort is likely to fail, with serious consequences.
High risk	H	Significant failure of parts of the transformation effort, resulting in failure to achieve certain goals.
Moderate risk	M	Noticeable failure of parts of the transformation effort, jeopardizing the success of certain goals.
Low risk	L	Certain goals will not be fully achieved.

Something very important to mention here is that the risk being measured is a risk to the organization. It is not a risk to the architecture or the work around it. The risks that emerge from the initial risk assessment are risks that the organization faces if the architecture described at this point is implemented.

When the impact and frequency values are plotted against each other as is shown in Table 14-18, it is possible to see which risk classification applies to each intersecting cell. This risk classification scheme forms the basis for the Risk Heat Map (used for plotting and visualizing risks), which is discussed in more detail in Section 14.8.6.

Table 14-18. Risk Classification Scheme

Impact ▼	Frequency				
	Frequent	Probable	Occasional	Seldom	Unlikely
Catastrophic	C	C	H	M	L
Significant	C	H	M	M	L
Moderate	H	M	M	L	L
Low	M	M	L	L	L
Negligible	L	L	L	L	L

For the initial risk assessment to work, everyone in the organization must agree on the definitions of impact levels and frequency. Once this agreement is reached, the values can be used to visualize the risks encountered in a Risk Heat Map. The use of this technique is explained in more detail in Section 14.8.6.

14.8.4 Risk Mitigation and Residual Risk Assessment

Risk mitigation is a critical aspect of strategic planning for any organization. It involves identifying potential risks, developing plans to address them, and executing actions to reduce those risks to an acceptable level. Mitigation strategies can range from simply monitoring or accepting a risk to comprehensive contingency plans, such as implementing full redundancy as part of a business continuity plan. Each approach has its own scope, cost, and time implications. Given the significant impact of risk assessments, it's important to conduct them in a pragmatic yet systematic manner. Prioritizing frequent, high-impact risks ensures that each risk is addressed effectively and efficiently.

CHAPTER 14 ADM TECHNIQUES

During Phase E (Opportunities and Solutions), the Implementation Factor Catalog (see Section 12.8.2) is created. This catalog is used as input to the Risk Management technique. When the architecture activities reach Phase F (Migration Planning), this is where Risk Management is incorporated into the Implementation and Migration Plan.

As described in the previous sections, a risk is first identified as an initial or preliminary risk. The classification is then extended by categorizing the risk (e.g., time, cost, scope). Next, the impact and frequency of the risk are determined. A final step in the risk identification process is to define the mitigating factors that will either eliminate the risk or reduce the risk level. If the risk is not fully mitigated, what remains of the risk is called the residual risk. It is important to re-evaluate the impact and frequency level of this residual risk to see if the mitigation effort has really made an acceptable difference. Table 14-19 shows an example of how this information can be recorded.

Table 14-19. Risk Identification and Mitigation Assessment Matrix

#	Risk	Preliminary Risk			Mitigation	Residual Risk		
		Imp.	Freq.	Risk Level		Imp.	Freq.	Risk Level
1	Reduction in headcount	3	4	12	Automate capabilities	2	2	4
2	Unable to use application	2	3	6	Provide training for staff	2	1	2

It is important to note whether the mitigation efforts have actually reduced the impact of the risk on the organization. The risk should not simply have been moved to a different quadrant of similar importance.

CHAPTER 14 ADM TECHNIQUES

For example, if mitigating factors reduce a risk from catastrophic to critical, it is still an extremely high risk. In this case, the mitigation effort must be reconsidered. Mitigating factors should have a noticeable effect on the risks.

Part of the Risk Management technique described in this chapter is to perform a transformation risk assessment by creating a Risk Identification and Mitigation Assessment Matrix. Table 14-19 illustrates this deliverable.

The Risk Identification and Mitigation Assessment Matrix contains the *risk ID* along with a *description* of the risk. This description is based on the *Deduction column* from the Implementation Factor Catalog.

The Preliminary Risk columns contain the *impact* and *frequency values* of the initial risk. Together, they result in a risk score (*impact* x *frequency*), which is noted in the *Risk Level* column.

The Mitigation column of the Risk Identification and Mitigation Assessment Matrix contains the *mitigation factors* that will be applied to the risk, followed by the impact, frequency, and risk levels after these factors are implemented.

List All Potential Risks: Start by identifying every possible risk that could impact the architecture work. Assign a unique ID and provide a clear description for each risk.

Determine Initial Impact and Frequency: For each risk, assess

- **Impact:** How severe would the consequences be if this risk occurred?

- **Frequency:** How often is this risk likely to happen?

Multiply these two values to get an initial risk score.

Record Initial Risk Level: Based on the initial risk score, classify the risk into a category.

Plan Mitigation Strategies: For each risk, decide on actions to reduce its impact or likelihood. Document these mitigation measures.

Reassess Impact and Frequency After Mitigation: After implementing mitigation strategies, evaluate

- **New Impact:** What would be the consequences now if the risk occurred?

- **New Frequency:** How likely is the risk to happen after mitigation?

Multiply these new values to get a post-mitigation risk score.

Record Post-mitigation Risk Level: Classify the risk again based on the new score to see how the mitigation efforts have changed its severity.

14.8.5 Risk Monitoring

Residual risks must gain approval through the Governance Framework, potentially extending to corporate governance for business acceptance. Once accepted, it's crucial to meticulously monitor the execution of mitigating actions, ensuring the organization addresses residual rather than initial risks. The Risk Identification and Mitigation Assessment

Matrix serves as a governance artifact. It is maintained and updated during Phase G (Implementation Governance), where risk monitoring occurs. This phase can spotlight critical risks that remain unmitigated, possibly necessitating another full or partial Architecture Development Method cycle.

Later on, during Phase H, risk monitoring is performed with respect to *changes to the architecture*. The Architecture Change Management Phase identifies risks associated with new requirements, technologies, or business changes. It implements risk response mechanisms as part of change management.

14.8.6 Visualizing Risks

After conducting the initial risk assessment as described in Section 14.8.3, it may be a good idea to visualize the various risks identified and compare them to the organization's risk appetite or tolerance. Although the TOGAF Standard does not formally describe how risk can be applied in visualizations, there are a number of ways to approach this.

A Risk Heat Map (see Figure 14-12) is a visualization tool that helps organize, define, and quickly communicate key risks. It is an essential tool in any Risk Management toolbox. Such a heat map is often used to focus management's attention on the most important threats and opportunities and to lay the groundwork for risk responses. The heat map is based on the *risk classification scheme* (Table 14-18) that was detailed in Section 14.8.3.

Risk heat mapping is a powerful technique that adds *risk visualization* to the Risk Management technique already described in the TOGAF Standard. The Risk Heat Map shows *risk frequency* (likelihood) on the horizontal axis (x) and *risk impact* (effect) on the vertical axis (y). Together, these axes can help analyze a risk and decide what actions to take to mitigate it.

CHAPTER 14 ADM TECHNIQUES

The heat map is a two-dimensional representation of data in which values are typically represented by colors (often red, green, and yellow). It allows the architect to effectively and visually communicate which risks are of most and least concern to the organization, depending on its risk appetite or tolerance. Typically, green represents low risk, yellow represents medium risk, and orange/red represents high/critical risk. This information can guide risk mitigation and minimization strategies. In the risk assessment process, visualizing risks using a heat map provides a concise, big-picture view of the entire risk landscape, opening the door for discussion as decisions are made about the likelihood and impact of risks within the organization.

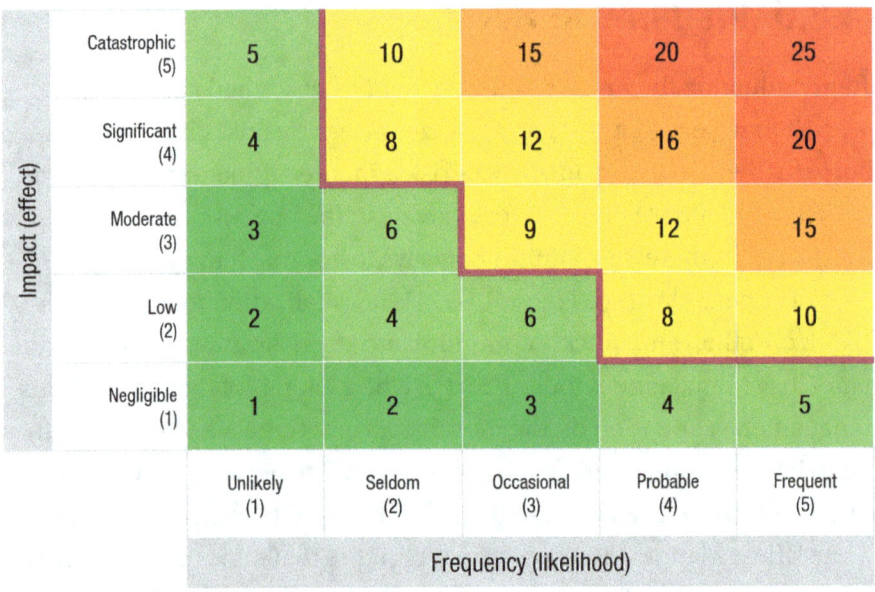

Figure 14-12. Risk Heat Map

The use of a Risk Heat Map, as shown in Figure 14-12, is very effective in communicating which risks have the highest significance when considering their potential impact and the likelihood of that impact.

Psychologically, people looking at the heat map are naturally drawn to the upper-right quadrant (high impact and high likelihood), while items in other quadrants receive less attention.

Calculating Risk

To use a Risk Heat Map, it's important for the organization to establish a common language for discussing risk. Section 14.8.3 outlined that terms such as *impact* or *effect* and *frequency* or *likelihood* need to be defined and used throughout the organization and in the design of the heat map. This allows everyone to be on the same page when discussing risk. Using a Risk Heat Map also requires a common understanding of the organization's *risk appetite* or *tolerance*. Risk tolerance refers to the highest level of risk that is still considered acceptable. In Figure 14-12, the risk tolerance is visualized by the maroon-colored line. All the risk values below the line are considered to be acceptable and are therefore given a green color.

Risks are generally assigned a score for risk frequency (the likelihood that the risk will occur) and the impact or effect the risk will have on the organization. Unfortunately, there is no scientific approach to assigning these scores. Common sense is the best way to go here.

The formula for calculating risk is as follows:

Risk = Impact (effect) × Frequency (likelihood)

Once each risk has been calculated, it can be plotted on the Risk Heat Map as shown in Figure 14-13.

CHAPTER 14   ADM TECHNIQUES

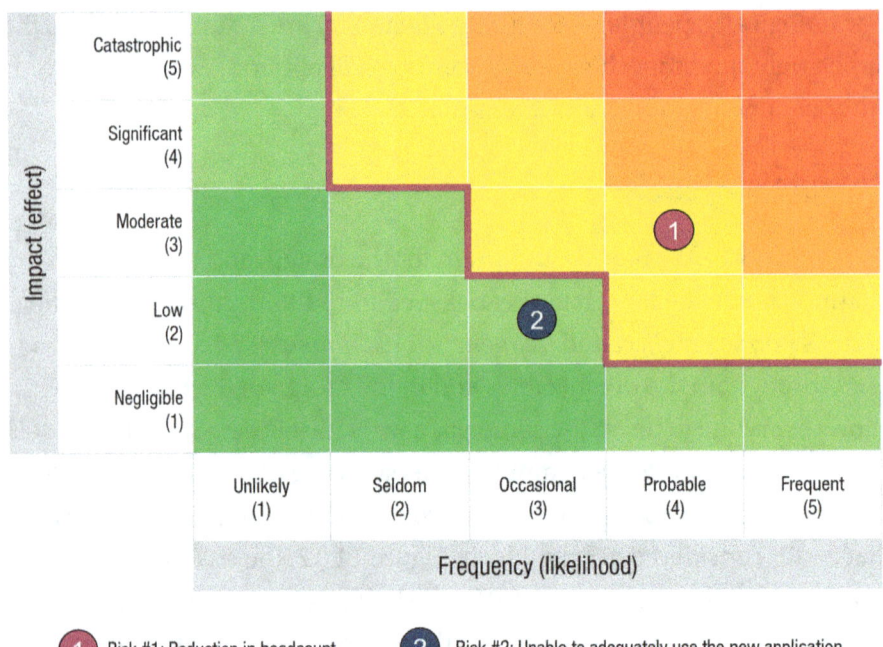

Figure 14-13. *Risk Heat Map with mapped risks*

Figure 14-13 shows the Risk Heat Map described earlier with two risks. Each risk is represented by a number in a circle corresponding to the risk. For the first risk (Risk #1), the impact is estimated to be *moderate* (3) and the likelihood of the risk occurring is estimated to be *probable* (4). Using the formula above, the risk score is 12, meaning *medium* risk classification. The second risk (Risk #2) has a *low* impact (2) and a frequency (likelihood) of *occasional* (3). Therefore, the risk score for the second risk is 6, classifying it as a *low* risk.

Scoring each risk in this way and plotting them in the heat map provides a good picture of which risks need the most attention. Implementation plans can be adjusted accordingly.

CHAPTER 14   ADM TECHNIQUES

Adding Risks to Diagrams

Although the TOGAF Standard does not formally describe how risk can be applied in visualizations, the ArchiMate standard (another standard from The Open Group) does. As mentioned in Chapter 1, Section 1.3, this book is not about the ArchiMate standard, nor is it about modeling ethics. However, to give a sense of how risks can be incorporated into diagrams, the following example is given.

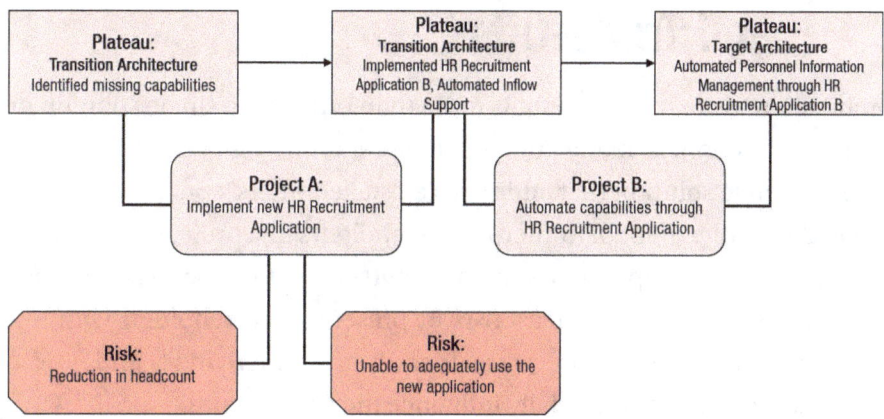

Figure 14-14. *Plateau Planning Diagram with added risks*

In Chapter 12, Section 12.8.2, a Plateau Planning Diagram was shown. This diagram illustrated the use of projects instead of gaps and work packages. Figure 14-14 shows the addition to this diagram of risks derived from the Implementation Factor Catalog regarding the implementation of HR Recruitment Application B.

There is, of course, much more to be said about modeling risk in diagrams. However, this book deals with the TOGAF Standard, not the ArchiMate standard. Therefore, the example above is for illustrative purposes only.

Further Reading

For more information about visualizing risks, refer to the URL below:

https://bizzdesign.com/blog/the-value-of-enterprise-architecture-in-managing-risk-compliance-and-security/

14.9. Architecture Alternatives and Trade-Offs

Architecture development entails more than pursuing a single solution or solution direction. In many architectural scenarios, it is even uncommon to find a single solution that addresses all stakeholders' concerns. Given the involvement of numerous stakeholders in a project or in the development and implementation of architecture, it is essential to propose not just one solution but at least two. By presenting stakeholders with multiple alternatives, architects can identify any underlying agendas or principles that could potentially influence the architecture.

The TOGAF Standard provides a structured approach to explore various alternatives and engage stakeholders in meaningful discussions. This technique is called Architecture Alternatives and Trade-Offs and is illustrated in Figure 14-15.

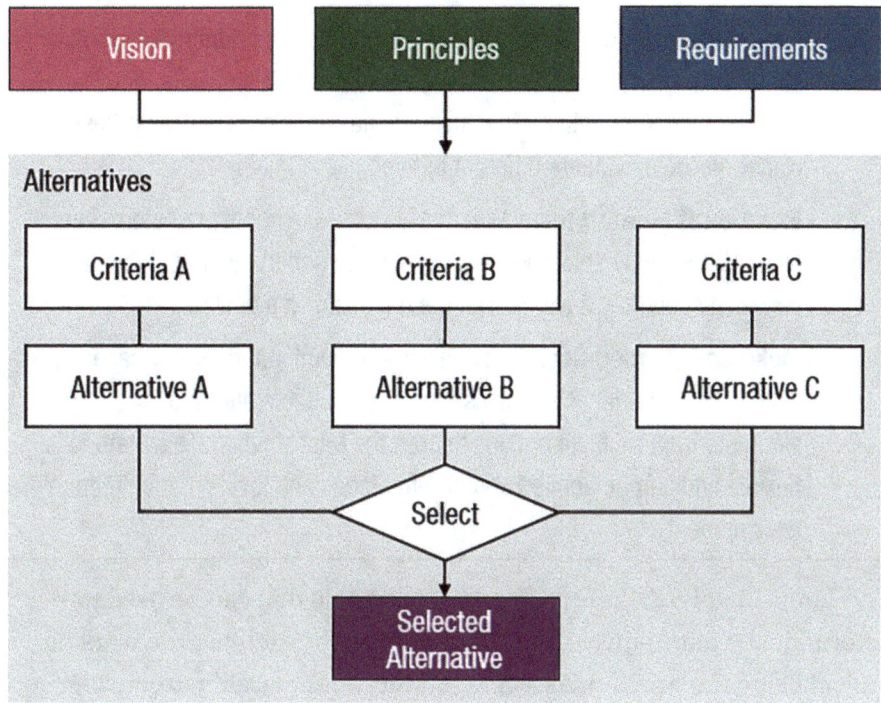

Figure 14-15. Architecture Alternatives and Trade-Offs method

The approach begins by leveraging the organization's vision, principles, requirements, and other pertinent information to establish a set of criteria suitable for assessing different alternatives. Then, based on these criteria, different alternatives are formulated, leading to a deeper understanding of each option. Finally, the process involves selecting the most appropriate alternative or combining elements from multiple options to develop a proposed solution. This approach involves conducting thorough analysis to support informed decision-making. This method is versatile and can be applied at various stages and levels of architectural development.

Gather Information: Begin by gathering all the important information about the organization – its vision, guiding principles, specific needs, and any other relevant details. This foundational knowledge will help to set clear criteria to evaluate different options.

Explore Different Options: With these criteria in hand, brainstorm various alternatives. This exploration will deepen the understanding of each possibility and how it aligns with the organization's goals.

Make an Informed Decision: Analyze each option thoroughly based on the established criteria. Then, choose the most suitable alternative or combine elements from multiple options to craft the best solution. This method is flexible and can be applied at different stages and levels of architecture development.

Unfortunately, there is no fixed set of criteria that can be used to determine the alternatives. Much depends on the architecture work or project being undertaken. As a general rule, it is advisable to consider such things as Architecture Principles, requirements, the architecture vision, and stakeholder concerns when developing the alternative scenarios.

Of course, each alternative has advantages and disadvantages. These need to be discussed with the stakeholders involved and agreement reached on the final chosen alternative.

Selecting the right alternative isn't just a checkbox exercise. It involves a nuanced process influenced by a tapestry of inputs. Architecture Principles, stakeholder concerns, requirements, and the overarching vision all play pivotal roles in shaping the criteria used for evaluation.

Each alternative brings its own set of advantages and challenges. Engaging stakeholders in open discussions ensures that these pros and cons are thoroughly examined and understood. Sometimes, to paint a clearer picture, it might be necessary to introduce additional viewpoints or perspectives. This helps stakeholders delve deeper into the alternatives, uncovering dependencies, assessing risks, and navigating uncertainties.

The types of alternatives to consider depend on a number of criteria:

- **Flexibility:** How adaptable is the alternative to changing business needs?
- **Time and Cost:** What are the implications in terms of implementation timeframes and associated costs, including transitional phases and stabilization periods?
- **Benefit Realization:** Over what period can the reaping of the benefits of this alternative be expected?
- **Architectural Alignment:** Does the alternative adhere to the established architecture styles or guidelines?
- **Delivery Method:** Are existing solutions being reused, developed anew, or are off-the-shelf products being purchased?
- **Business Impact:** How minimal is the disruption to existing business capabilities during implementation?
- **Risk Profile:** What risks are associated with this alternative, and what mitigation strategies are in place?

In my other book, *Getting Started with Enterprise Architecture* [9], I gave an example of a fictional organization called Lemon-A-de. This organization spent the early days of its existence developing a strategy (see Figure 14-16).

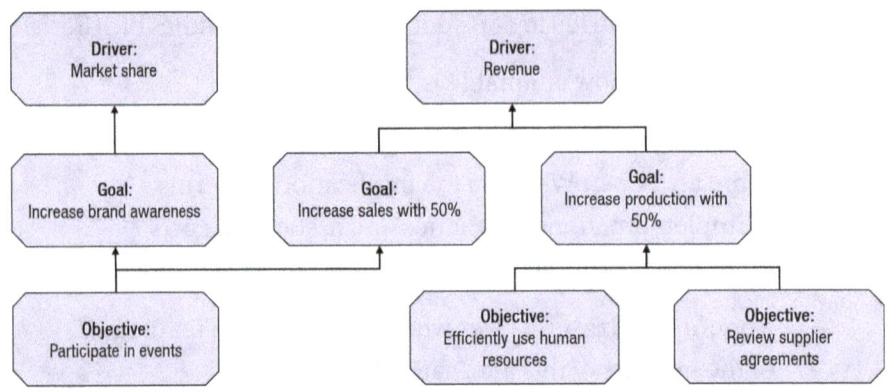

Figure 14-16. Lemon-A-de's strategy

The organization's architect was asked to visualize the impact of the strategy. To do this, this person looked at the goals the organization had in mind and translated them into the activities (initiatives) the company needed to develop to achieve them.

The architect's research showed that it was not immediately possible to achieve all the goals at once, given all kinds of organizational constraints. So the architect presented two scenarios to the CEO.

The Architecture Alternatives and Trade-Offs method was used here (see Figure 14-17).

CHAPTER 14   ADM TECHNIQUES

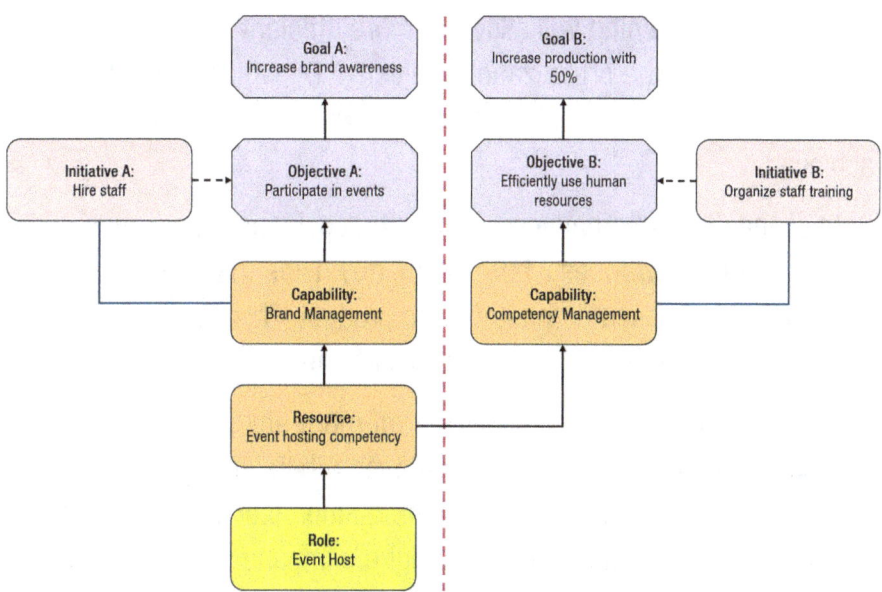

Figure 14-17. Architecture Alternatives and Trade-Offs diagram

Figure 14-17 shows that achieving the goal of *increasing brand awareness* (Goal A) is considered achievable if the organization commits to *hiring additional staff* (Initiative A). This will allow the organization to *plan and host several events* (Objective A), which will ultimately lead to an increase in brand awareness. An additional business role must also be defined and filled to make this possible.

The second scenario, which will achieve the goal of *increasing production by 50%* (Goal B), consists of the activity of *organizing staff training* (Initiative B).

Participation in training will allow for more efficient deployment of personnel within the organization (Objective B), which may allow for more effective work. In all likelihood, this will lead to the goal being achieved.

Both scenarios – alternatives – come at a cost. The first scenario will increase brand awareness, but has the downside that this awareness may lead to increased demand for the product. This demand may not be able to be met immediately due to staff inefficiencies.

Scenario 2, which focuses on using staff more efficiently by providing appropriate training, will lead to increased production. The downside is that if brand awareness lags, this production stock will not be sold within the desired timeframe, leaving the company with an inventory.

The Architecture Alternatives and Trade-Offs method provides insight into the advantages and disadvantages of all the scenarios or solutions mentioned and tries to satisfy all the wishes and requirements of the parties involved. Often, the application of this method leads to a phased roadmap, in which, in this example, both scenarios are eventually realized.

Presenting multiple, well-considered alternatives gives the organization confidence that scenarios have been considered from a variety of perspectives. It also demonstrates that all stakeholder concerns have been considered.

Further Reading

For more information about the Architecture Alternatives and Trade-Offs method, refer to *The TOGAF Standard, 10th Edition*, document number C220.

14.10. Summary

The Architecture Development Method is often regarded as the core of the TOGAF Standard. It consists of a cyclic method that helps organizations align business and IT needs effectively. Various techniques are applied throughout the Architecture Development Method cycle to ensure efficient decision-making, risk mitigation, and stakeholder engagement. These techniques support different phases of the Architecture Development Method and enhance architectural practice.

When the TOGAF Standard talks about techniques, it does not always mean a *specific way of doing things*. What is really meant by techniques is an elaboration of a topic or component that is frequently used within the framework or during the implementation of the Architecture Development Method. A recurring topic that requires further explanation and clarification in order to be properly applied.

Some topics are essential aspects of developing an architecture. However, they are not necessarily a specific technique. Stakeholder Management and Risk Management both are essential, but they are not techniques. However, the TOGAF Standard still refers to them as if they are.

The framework distinguishes between eight techniques:

- **Architecture Principles:** Fundamental rules and guidelines that influence architecture development and governance.

- **Stakeholder Management:** A method for identifying, analyzing, and managing stakeholders.

- **Architecture Patterns:** Provide standardized solutions to common architectural problems.

- **Gap Analysis:** Identifies differences between the Baseline and Target Architectures.

- **Interoperability Requirements:** Determine interoperability requirements.

- **Business Transformation Readiness Assessment:** Describes a technique for identifying business transformation issues.

- **Risk Management:** Ensures that potential issues are identified and mitigated before impacting the architecture.

- **Architecture Alternatives and Trade-Offs:** Describes a technique to identify alternative Target Architectures and perform trade-offs between the alternatives.

Principles play a crucial role in guiding decision-making and ensuring alignment between business objectives and strategies. Principles are often established within different domains and at different levels of an organization.

The TOGAF Standard distinguishes between Enterprise Principles and Architecture Principles:

- **Enterprise Principles:** High-level, general guidelines that shape the overall strategy, culture, and decision-making framework of an organization.

- **Architecture Principles:** More specific guidelines that direct the design and development of Enterprise Architecture.

Within an Enterprise Architecture approach, principles serve as a foundation for designing flexible and scalable architectures that align with the aforementioned business goals and objectives. They ensure compliance with regulations and industry standards and are able to reduce complexity and duplication by enforcing standardization and interoperability.

Stakeholder Management is the process of identifying and understanding the needs, interests, and expectations of an organization's various stakeholders and planning and implementing actions to address those interests and improve cooperation and communication with them. The stakeholder concept is closely related to governance in Enterprise Architecture. Understanding who the stakeholders are and what their interests are helps architects establish governance mechanisms to address their concerns, maintain transparency, and ensure accountability.

The TOGAF Standard recommends incorporating stakeholder analysis during Phase A (Architecture Vision) of the Architecture Development Method and updating it throughout each following phase. This form of continuous engagement ensures that new stakeholders are identified and their concerns are addressed as the project progresses through phases such as Phase E (Opportunities and Solutions), Phase F (Migration Planning), and Phase H (Architecture Change Management).

The "Architecture Patterns" section is a different story. Although this topic is included as such in the TOGAF Standard, it states that this technique has not yet been incorporated into the Architecture Development Method [20]. In fact, the TOGAF Standard is indicating that it is not useful enough. Why it is included in the framework remains a mystery. It is possible that a future version of the TOGAF Standard will provide more information on this subject.

A *gap analysis* is a process that identifies the difference between an organization's Baseline and Target Architectures in terms of its capabilities, processes, applications, and technology. The purpose of a gap analysis is to identify areas where improvements are needed to achieve the desired future state. The results of the gap analysis are used to inform subsequent phases of the Architecture Development Method, such as phases B, C, and D, where the detailed architecture is developed based on the architectural vision.

CHAPTER 14   ADM TECHNIQUES

During Phase E (Opportunities and Solutions) and Phase F (Migration Planning) of the Architecture Development Method, five artifacts can be created to help fine-tune the migration planning:

- **Phase E (Opportunities and Solutions):**
  - **Implementation Factor Catalog:** This catalog documents factors impacting the architecture Implementation and Migration Plan.
  - **Consolidated Gaps, Solutions, and Dependencies Matrix:** Can be used to group the gaps identified in the gap analysis results and assess potential solutions and dependencies to one or more gaps.
  - **Architecture Definition Increments Table:** Used to plan a series of Transition Architectures outlining the status of the Enterprise Architecture at specified times.
- **Phase F (Migration Planning):**
  - **Transition Architecture State Evolution Table:** This table shows the proposed state of the architectures at various levels using the defined taxonomy.
  - **Business Value and Risk Matrix:** Effective assessment of business value can be achieved by creating a matrix based on a value index dimension and a risk index dimension.

The idea behind the *interoperability technique* is to compare a set of entities and determine what form and degree of interoperability is required between them.

One major advantage of applying the interoperability technique is enhanced integration and communication between systems. Interoperability ensures that different entities (such as people, systems, applications, and services) can effectively exchange and use information, reducing data silos and integration complexity.

In the early phases of the Architecture Development Method, particularly Phase A (Architecture Vision) and the subsequent architecture definition phases (phases B–D), organizations conceptualize and design innovative solutions that often involve significant changes. These changes are not limited to technological shifts as they have a profound impact on the workforce, culture, and existing skills and competencies within the organization.

To ensure successful implementation of the architecture transformation in phases E and F, organizations must conduct a comprehensive *Business Transformation Readiness Assessment*. This is a collaborative effort between the organization's staff, particularly human resources, lines of business, and (IT) planners. A Business Transformation Readiness Assessment table typically includes various factors that determine an organization's preparedness for a significant change.

Risk Management is an integral part of architecture development. As such, it is addressed in several phases (Phase A and phases E–H) of the Architecture Development Method. Applying this technique ensures that risks are identified, assessed, and mitigated as part of the architecture development process.

There are two types of risks:

- **Initial (Preliminary) Risk:** A new, previously unaddressed risk.

- **Residual Risk:** A previously encountered – and possibly (partially) mitigated – risk.

For an *initial risk* – a categorization given to a risk prior to identifying and implementing mitigating actions – the Risk Management technique should be applied. Existing or *residual risks* have usually already been accepted and mitigating actions implemented.

A common method is to categorize risks based on their potential impact on the organization. This approach ensures that risks with significant impact are escalated to the appropriate governance levels, facilitating timely and effective responses.

The initial risk assessment takes place during phases E and F of the Architecture Development Method. Risks are usually classified with respect to *impact* (effect) and *frequency* (likelihood) in accordance with scales used within the organization. The two factors (impact and frequency) can be combined to provide a score and classification of the actual risk.

Risk mitigation is a critical aspect of strategic planning for any organization. It involves identifying potential risks, developing plans to address them, and executing actions to reduce those risks to an acceptable level. If the risk is not fully mitigated, what remains of the risk is called the residual risk. It is important to re-evaluate the impact and frequency level of this residual risk to see if the mitigation effort has really made an acceptable difference.

After conducting the initial risk assessment, risks can be visualized and compared to the organization's risk appetite or tolerance. Although the TOGAF Standard does not formally describe how risk can be applied in visualizations, there are a number of ways to approach this. For example, creating a Risk Heat Map allows the visualization of risks.

Architecture development entails more than pursuing a single solution or solution direction. In many architectural scenarios, it is even uncommon to find a single solution that addresses all stakeholders' concerns. The TOGAF Standard provides a structured approach to explore various alternatives and engage stakeholders in meaningful discussions. This technique is called *Architecture Alternatives and Trade-Offs*. The approach leverages the organization's vision, principles, requirements, and other pertinent information to establish a set of criteria suitable for assessing different alternatives. Based on these criteria, different alternatives are formulated. The process involves selecting the most appropriate alternative or combining elements from multiple options to develop a proposed solution.

CHAPTER 15

Architecture Maturity Models

 This chapter is part of the GOLD reading tour. See Figure 1-1 in Chapter 1, Section 1.3, for alternative reading tours.

An Enterprise Architecture maturity model serves as a framework to assess how well an organization's architecture aligns with its business objectives. By outlining an ideal Enterprise Architecture implementation – typically conforming to a specific architecture framework – a maturity model framework provides a benchmark to evaluate the effectiveness of architectural processes. This assessment not only highlights the current maturity level but also offers a structured pathway for enhancement, guiding organizations toward improved alignment between their IT infrastructure and business goals.

In the "Architecture Maturity Models" Series Guide, the TOGAF Standard refers to two applicable methods for measuring architecture maturity. These are the Architecture Capability Maturity Model (ACMM) and Capability Maturity Model Integration (CMMI). Organizations can also develop their own organization-specific model to fit their unique circumstances.

CHAPTER 15   ARCHITECTURE MATURITY MODELS

15.1. Determining Architecture Maturity

In the Preliminary Phase of the Architecture Development Method, a maturity model is often used to measure and provide insight into the maturity of the internal Architecture Capability. The "Architecture Maturity Models" Series Guide discusses two common frameworks for measuring maturity. One is the Architecture Capability Maturity Model framework. It has a structure in which the core components of an architecture process are represented and measured, often in relation to the organization's IT. The purpose of using such a framework is to identify weaknesses and provide a step-by-step approach to improving them.

The second is the Capability Maturity Model Integration framework. This framework places more emphasis on maturing the Architecture Capability in order to achieve organizational goals. This makes this type of maturity framework more appropriate for use in an Enterprise Architecture environment.

Enterprise Architecture is considered a strategic business management tool, not a technical instrument [9].

Enterprise Architecture is a strategic discipline that focuses on translating an organization's strategy into its execution and impact.

15.1.1. Architecture Capability Maturity Model

An Architecture Capability Maturity Model consists of three parts:
- The model itself
- The elements (characteristics) of the architecture processes, grouped into different maturity levels
- A scoring mechanism

CHAPTER 15   ARCHITECTURE MATURITY MODELS

The first two parts explain the maturity levels and the corresponding architecture elements for each level. The third section is designed to help determine the current level of maturity. This level can then be reported on.

The Architecture Capability Maturity Model (see Table 15-1) considers the following nine architecture *elements* when assessing architecture maturity.

Table 15-1. Elements of ACMMs

Element	Description
Architecture process	Refers to the extent to which architecture operations are systematically defined, meticulously documented, and consistently followed.
Architecture development	Encompasses the methods and tools used to create architecture solutions. Critical evaluation includes assessing how well they function.
Business linkage	Refers to the degree of alignment between IT investments and core business objectives.
Senior management involvement	How effectively leadership involvement is aiding the development and maintenance of architecture initiatives.
Business unit participation	The degree to which different business units and departments are involved in the architecture process.
Architecture communication	How effectively architecture processes are communicated throughout the organization.
Architecture governance	Involves the establishment of mechanisms and structures to oversee and guide architecture efforts.
IT security	Addresses the organization's ability to effectively meet cybersecurity requirements.
IT investment and acquisition strategy	Focuses on the judicious allocation of resources to existing systems and the acquisition of new capabilities.

Architecture Process: The architecture process refers to the extent to which architecture operations are systematically defined, meticulously documented, and consistently followed. In organizations where architectural practices are well integrated, these processes are characterized by rationality, repeatability, and clarity, thereby facilitating efficient execution and alignment with organizational goals.

Architecture Development: Architecture development encompasses the methods and tools used to create architecture solutions. Critical evaluation includes assessing the effectiveness of these methods and tools, determining the standards they are intended to meet, and verifying adherence to established best practices. Such scrutiny ensures that the architecture framework remains robust, adaptable, and aligned with industry benchmarks.

Business Linkage: Business linkage refers to the degree of alignment between IT investments and core business objectives. A well-structured architecture approach ensures that the IT infrastructure not only supports but also enhances business strategies, delivering measurable return on investment and driving business growth.

Senior Management Involvement: Senior management involvement is critical to the development and maintenance of architecture initiatives. Effective leadership provides the necessary sponsorship and resources to protect architecture projects from potential derailment due to resource constraints or lack of strategic direction.

Business Unit Participation: The involvement of various business units and departments in the architecture process indicates the depth of integration within the organization. A high level of participation indicates a collaborative environment where architecture practices are embraced throughout the organization, promoting coherence and optimal performance.

Architecture Communication: Effective communication of architecture processes within the organization ensures that all stakeholders have a clear understanding of their roles and overall goals. Transparent dissemination of information promotes alignment, facilitates informed decision-making, and increases the overall effectiveness of architecture efforts.

Architecture Governance: Architecture governance involves the establishment of mechanisms and structures to oversee and guide architecture efforts. Robust governance frameworks assign clear roles and responsibilities, ensure compliance and accountability, and foster innovation within architecture projects.

IT Security: IT security, in the context of architecture, addresses the organization's ability to effectively meet cybersecurity requirements. Mature architecture practices incorporate comprehensive security policies and cost-effective measures to safeguard sensitive data and protect IT assets from potential threats.

IT Investment and Acquisition Strategy: IT investment and acquisition strategy focuses on the judicious allocation of resources to existing systems and the acquisition of new capabilities. A strategic and economical approach to the use of IT resources ensures that investments are aligned with organizational goals and contribute to sustainable value creation.

In addition to the architecture process characteristics, an Architecture Capability Maturity Model has *six* maturity levels against which the Architecture Capability is measured (see Table 15-2). In practice, the lowest level (Level 0) is often omitted. The reason for this is that Level 0 indicates that nothing is being done with respect to architecture at all. In this case, there is no need to complete a maturity model.

CHAPTER 15 ARCHITECTURE MATURITY MODELS

Table 15-2. Maturity Levels of ACCMs

Level	Name	Description
0	None	Indicates a complete lack of architecture development.
1	Initial	Elements have been identified as worth developing but not yet have any processes defined.
2	Under development	Elements have begun being issued processes, tools, and methods for the purpose of architecture optimization.
3	Defined	Elements are in active development, having been given a standardized and content-complete set of architecture practices.
4	Managed	Elements have been developed to the extent that performance metrics have come into focus.
5	Measured	Element have been optimized as fully as the architecture team could envision, making it an example for other elements.

None: Indicates a complete lack of architecture development.

Initial: At this preliminary stage, the architecture element has been recognized as a candidate for further development, although no formalized processes have been established. Any changes or experimentation to date has been done in an unstructured and isolated manner, without broader organizational visibility or buy-in.

Under Development: At this stage, initial methodologies, tools, and frameworks are being implemented to support the development of the architectural element. Preliminary documentation may exist, but activities remain informal and efforts have not yet achieved consistency or alignment with organization-wide standards.

Defined: The architectural element is undergoing structured development within a comprehensive and standardized architectural

CHAPTER 15   ARCHITECTURE MATURITY MODELS

framework. Practices are now formally adopted, and the element has a clear alignment with strategic business imperatives. A dedicated architecture team is actively involved in its evolution.

Managed: The architecture element has reached a level of maturity where key performance indicators are monitored and operational effectiveness is actively reviewed. Collaboration between the architecture team and executive stakeholders is focused on identifying opportunities for refinement and improving return on investment.

Measured: In this final phase, the architecture element is considered a mature and optimized asset within the architecture portfolio. It serves as a reference model for similar initiatives. Emphasis is placed on leveraging data analytics to inform continuous improvement cycles and guard against performance plateau.

A fully completed maturity model based on the Architecture Capability Maturity Model framework is shown in Table 15-3.

Table 15-3. *Architecture Capability Maturity Model Framework*

	Initial	Under Development	Defined	Managed	Measured
Architecture process	Processes are ad hoc and localized. Some architecture processes are defined. Success depends on individual efforts.	The basic architecture process is clearly documented. The architecture process has developed clear roles and responsibilities.	The architecture is well-defined and communicated with operating unit IT responsibilities. The process is largely followed.	Architecture process is part of the culture. Quality metrics associated with the architecture process are captured.	Concerted efforts to optimize and continuously improve architecture process.
Architecture development	Architecture processes, documentation, and standards are established by a variety of ad hoc means and are localized or informal.	IT vision, principles, business linkages, Baseline Architecture, and Target Architecture are identified. Standards exist, but not necessarily linked to Target Architecture.	Gap analysis and migration plan are completed. Fully developed TRM and Standards Profile. IT goals and methods are identified.	Architecture documentation is updated on a regular basis to reflect the updated architecture. Architectures are defined by appropriate standards.	A standards and waivers process is used to improve architecture development process.

Business linkage	Minimal or implicit linkage to business strategies or business drivers.	Explicit linkage to business strategies.	Architecture is integrated with capital planning and investment control.	Capital planning and investment control are adjusted based on the feedback received and lessons learned. Periodic re-examination of business drivers.	Architecture process metrics are used to optimize and drive business linkages. Business involved in the continuous architecture process improvements.
Senior management involvement	Limited management team awareness or involvement in the architecture process.	Management awareness of architecture effort.	Senior management team aware of and supportive of the organization-wide architecture process. Management actively supports architectural standards.	Senior management team directly involved in the architecture review process.	Senior management involvement in optimizing process improvements in architecture development and governance.

(continued)

Table 15-3. (continued)

	Initial	Under Development	Defined	Managed	Measured
Business unit participation	Limited operating unit acceptance of the architecture process.	Responsibilities are assigned and work is underway.	Most elements of operating unit show acceptance of or are actively participating in the architecture process.	The entire operating unit accepts and actively participates in the architecture process.	Feedback on architecture process from all operating unit elements is used to drive architecture process improvements.
Architecture communication	The latest version of the operating unit's architecture documentation is online. Little communication exists about the architecture process and its improvements.	The architecture web pages are updated periodically and are used to document architecture deliverables.	Architecture documents updated regularly on the architecture web page.	Architecture documents are updated regularly and frequently reviewed for the latest architecture developments and standards.	Architecture documents are used by every decision-maker in the organization for every IT-related business decision.

Architecture governance	No explicit governance of architectural standards.	Governance of a few architectural standards and some adherence to existing Standards Profile.	Explicit documented governance of majority of IT investments.	Explicit governance of all IT investments. Formal processes for managing variances feed back into the architecture.	Explicit governance of all IT investments. A standards and waivers process is used to make governance process improvements.
IT security	IT security considerations are ad hoc and localized.	IT Security Architecture has defined clear roles and responsibilities.	IT Security Architecture Standards Profile is fully developed and is integrated with architecture.	Performance metrics associated with IT Security Architecture are captured.	Feedback from IT Security Architecture metrics is used to drive architecture process improvements.

(continued)

Table 15-3. (continued)

	Initial	Under Development	Defined	Managed	Measured
IT investment and acquisition strategy	Little or no involvement of strategic planning and acquisition personnel in the architecture process. Little or no adherence to existing standards.	Little or no formal governance of IT investment and acquisition strategy. Operating unit demonstrates some adherence to existing Standards Profile.	IT acquisition strategy exists and includes compliance measures to IT Architecture. Cost benefits are considered in identifying projects.	All planned IT acquisitions and purchases are guided and governed by the architecture.	No unplanned IT investment or acquisition activity.

CHAPTER 15   ARCHITECTURE MATURITY MODELS

Using the Architecture Capability Maturity Model framework provides a picture of how the IT Architecture fits with the rest of the organization. It clarifies where the organization stands in terms of applying the architecture to day-to-day operations and processes.

The Architecture Capability Maturity Model framework focuses primarily on IT architecture, which makes it less applicable to an Enterprise Architecture environment.

Something else that stands out about the Architecture Capability Maturity Model framework is the element called IT security. Judging by its name and content, it is focused solely on security in relation to IT. This contradicts what the TOGAF Standard said earlier about security being a cross-cutting concern (see Chapter 11, Section 11.2).

All in all, the Capability Maturity Model Integration framework described in the next section is a much better fit for determining and measuring Enterprise Architecture maturity.

15.1.2. Capability Maturity Model Integration

The use of a Capability Maturity Model Integration framework is preferred when it comes to defining and measuring Enterprise Architecture maturity because of its more strategic approach to viewing Enterprise Architecture. As mentioned in the previous section, the Architecture Capability Maturity Model framework puts a lot of emphasis on the IT side. Enterprise Architecture is a strategic discipline, and it is very different from IT Architecture. For this reason, it is more appropriate to use a Capability Maturity Model Integration framework to measure and provide insight into Enterprise Architecture maturity.

CHAPTER 15 ARCHITECTURE MATURITY MODELS

A maturity model based on Capability Maturity Model Integration (CMMI) defines five *elements* of Enterprise Architecture maturity. These are shown in Table 15-4.

Table 15-4. Elements of CMMI Models

Element	Description
Strategy and vision	A clear strategy and vision for architecture development that is closely aligned with the organization's goals is required.
Architecture governance	Well-trained, experienced, and competent people are responsible for developing the architecture.
Architecture method and process	The organization has standardized processes and methods for developing the architecture.
Architecture deliverables	The appropriate tools and technologies are available to design, implement, and manage the architecture.
Business alignment	There are clear controls and governance for architecture development, including measurable performance indicators and quality standards.

Strategy and Vision: The organization has a clear strategy and vision for architecture development that is closely aligned with the organization's goals. There are clear priorities and guidelines for architecture development.

Stakeholder involvement in the architecture is critical to the success of an Enterprise Architecture implementation. Stakeholders include not only architects, but also administrative roles such as senior (business and IT) management. Project and portfolio management play an important role in the implementation of an Enterprise Architecture as well. Ideally, there is support at all levels of the organization for the use of the architecture and the adoption of its results.

The Enterprise Architecture is evaluated on a regular basis and suggestions for improvement or further development are identified. In this way, the effectiveness of the Enterprise Architecture is continuously improved.

Architecture Governance: The organization has well-trained, experienced, and competent people responsible for developing the architecture. There are clear roles and responsibilities for the architecture function. These are clearly defined so that the entire organization understands what can and cannot be expected of the Architecture Capability. This prevents discussions and disagreements about the architecture.

Senior management training is organized so that the management level of the organization has sufficient substantive knowledge of Enterprise Architecture. The architecture discipline is also introduced during the onboarding process of new employees. Awareness programs are used to make employees aware of Enterprise Architecture.

Opportunities for collaboration with other similar organizations are actively sought. Experiences and ideas are exchanged to improve and develop the Enterprise Architecture.

Architecture Method and Process: The organization has standardized processes and methods for developing the architecture.

The architecture process, like any other process, must be maintained. This ensures the effectiveness and efficiency of the Enterprise Architecture. Captured metrics can be used proactively to identify and make improvements to the architecture process, framework, or products.

However, developing and implementing an Enterprise Architecture is not an end in itself. Going back to the earlier definition of Enterprise Architecture (see Chapter 4), the discipline ensures that an organization can implement the necessary initiatives to achieve its intended goals and objectives. Not surprisingly, this is the purpose of Enterprise Architecture.

In practice, of course, the use and application of architecture can vary. For example, it can be used simply as an information conduit (architecture in support of solution delivery) or as a means to drive individual projects (architecture in support of projects).

The ultimate goal of Enterprise Architecture is to be a tool that guides the entire organization in achieving business goals (architecture in support of strategy). The uses of the Architecture Capability are described in more detail in Chapter 9.

Architecture Deliverables: The organization has the appropriate tools and technologies to develop, implement, and manage the architecture.

Architecture is not just about creating a set of architecture deliverables (such as Architecture Principles, requirements, diagrams, and other models). Of course, architecture artifacts are very important to have, but they also need to be used and maintained. Maintaining architecture artifacts means updating them and, if necessary, removing artifacts that are no longer applicable. Active maintenance of architecture artifacts ensures that the architecture is and remains current and functional.

Working with architecture can be supported by the use of architecture tools. These tools must be well suited to the task for which they are used. The integrated use of tools, preferably with repository support, maximizes efficiency and effectiveness.

Business Alignment: The organization has clear controls and governance for architecture development, including measurable performance indicators and quality standards.

The ideal implementation of Enterprise Architecture is characterized by its use in support of strategy. Simply stating that projects must conform to the architecture is generally not enough. Ultimately, it is up to the organization to decide how to use an Enterprise Architecture (and that may mean using it to manage the portfolio, some individual projects, or even occasionally to achieve specific solutions).

The evolution of the Enterprise Architecture and its alignment with the organization requires regular coordination with various stakeholders. Stakeholders such as senior management, information and project managers, and subject matter experts should be involved. For the architecture process to run smoothly, recurring and regular consultation with stakeholders is essential.

It is necessary to create and increase awareness of the Architecture Capability. This introduces the organization to the use and application of architecture concepts and processes.

A maturity model based on Capability Maturity Model Integration defines not only its *elements*, but also five *levels* of Enterprise Architecture maturity against which the Architecture Capability is measured. These are shown in Table 15-5.

Table 15-5. *Maturity Levels of CMMI Models*

Level	Name	Description
1	Ad hoc	Frameworks and standards have not yet been established, and the architecture process is generally conducted in an informal manner.
2	Repeatable	The architecture process is formalized, and there is a vision for the architecture.
3	Defined	The architecture framework to be used has been determined, and there is a formally defined process for adhering to the architecture process.
4	Managed	Measurement data is actively used to improve and refine the architecture process.
5	Optimal	The architecture process is mature; the organization has drivers, goals, and objectives aligned with the application of Enterprise Architecture.

Ad Hoc: Frameworks and standards have not yet been established, and the architecture process is generally conducted in an informal manner. The organization is aware of the benefits that Enterprise Architecture can provide, but has not yet defined a process to track and monitor the evolution of the architecture process. The organization relies on the knowledge and skills of individual employees. A business strategy exists, but has not yet been translated into implementation.

Repeatable: The architecture process is formalized, and there is a vision for the architecture. The architecture process is repeatable, and templates and standards are being developed. The need for adherence to architectural frameworks is recognized by senior management, and efforts are made to formalize it through process capture. Metrics are also being captured to evaluate the process. The organization's strategy is beginning to be translated into concrete goals and objectives. The words *enterprise* and *architecture* are used with some regularity in conversations at the strategic level.

Defined: The architecture framework to be used has been determined, and there is a formally defined process for adhering to the architecture process. The architecture framework is now formalized, and standards are widely used within the organization. There is a roadmap for the evolution of the architecture, and related activities are performed in accordance with the roadmap. Metrics are maintained and monitored for the purpose of evolving the architecture process.

Managed: Measurement data is actively used to improve and refine the architecture process. Data is analyzed and used to drive architecture performance. The Enterprise Architecture is invariably used to inform the organization's strategy. Adherence to the architecture framework is a standard activity within the organization.

CHAPTER 15 ARCHITECTURE MATURITY MODELS

Optimal: The architecture process is mature; the organization has drivers, goals, and objectives aligned with the application of Enterprise Architecture. Continuous improvements are made to the architecture process, and projects and other initiatives can no longer proceed without Enterprise Architecture input. Work is also done outside the boundaries of one's own organization to improve and refine the architecture together with similar organizations. Experiences are shared and suggestions for improving processes are discussed.

A fully completed maturity model based on the Capability Maturity Model Integration framework is shown in Table 15-6.

Table 15-6. *Capability Maturity Model Integration Framework*

	Ad Hoc	Repeatable	Defined	Managed	Optimal
Strategy and vision	Architecture activities are not formally initiated and happen ad hoc.	The organization has started drafting a vision for EA.	EA deployment is clearly defined, including governance roles and responsibilities.	The EA is evaluated and adjustments are identified to improve the EA.	Action plans are proactively implemented to increase EA effectiveness based on measured data.
Architecture governance	The need to define processes and standards is recognized.	The need for governance has been identified.	Architecture governance is defined (a consultative body has been created).	The communication process is revised to improve EA activity.	The organization collaborates with similar organizations exchanging ideas to improve their EA.
Architecture method and process	Architecture processes are ad hoc and inconsistent.	The architecture method is beginning to be reused to capture crucial EA information.	Templates are used so that capture of information is consistent.	EA is used to guide organizational development.	Business influences technology and technology influences business.

Architecture deliverables	The documentation of business drivers, goals and objectives, architecture standards, etc. is not formally defined.	Drivers, goals, and objectives have been identified.	Classification of existing standards is consistent.	Documentation of drivers, goals, and objectives has become a standard activity.	Captured business and technology information is used to proactively identify technology that will improve business operations.
Business alignment	The organization recognizes that staff need to become more familiar with working with architecture.	The organization has started raising awareness and understanding of EA concepts and processes.	The organization starts to operate as a team, using the defined architecture and standards.	Staff throughout the organization have a good understanding of the Architecture Principles.	Departments work together as contributors to the architecture and its processes.

15.2. Maturing the Architecture Capability

After defining the desired framework for measuring the maturity of the Architecture Capability in the organization, the task is to further professionalize and mature the capability. The maturing of the capability takes place along the axis of the aforementioned elements of the framework.

This example is based on the Capability Maturity Model Integration framework. The framework has the following elements:

- Strategy and vision
- Architecture governance
- Architecture method and process
- Architecture deliverables
- Business alignment

The CMMI framework also suggests five maturity levels for each of the elements (see Table 15-5). Together with the identified current maturity level, it is possible to identify the actions or activities required for each element to move from a lower level to a higher level. Figure 15-1 shows an excerpt from Table 15-6 illustrating the *Strategy and vision* element for the Ad hoc, Repeatable, and Defined maturity levels.

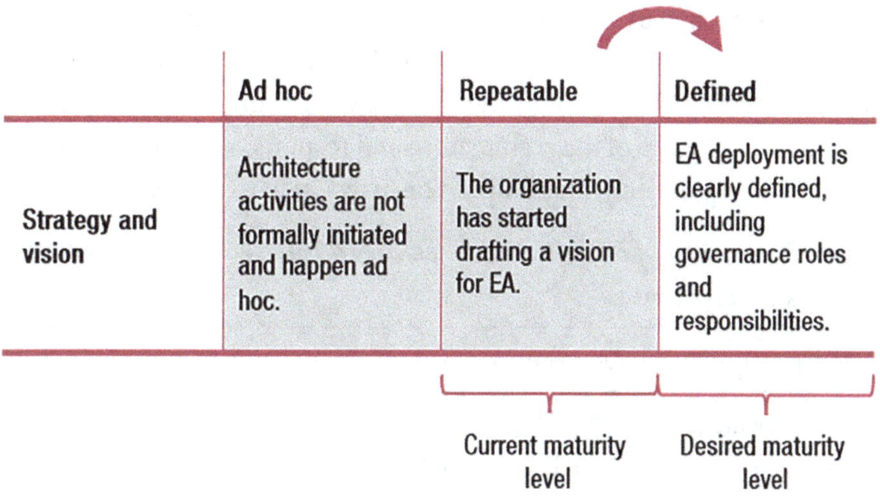

CHAPTER 15 ARCHITECTURE MATURITY MODELS

	Ad hoc	Repeatable	Defined
Strategy and vision	Architecture activities are not formally initiated and happen ad hoc.	The organization has started drafting a vision for EA.	EA deployment is clearly defined, including governance roles and responsibilities.

Current maturity level → Desired maturity level

Figure 15-1. *Excerpt from the CMMI framework for the Strategy and vision element*

To move from the second level (Repeatable) to the next level (Defined), the vision for the use and application of the architecture must be supplemented with the necessary governance, roles, and responsibilities. This ensures that the architecture work has a sponsor from the executive level of the organization.

Questions that can be asked at this point are as follows:

- Has ownership of the architecture been assigned at the executive level?
- Can an owner be identified for the architecture processes?
- Is ownership of the architecture invested at the top management level?
- Is architecture also the responsibility of business management?
- Is the architecture owner held accountable for the contribution of the architecture to the business goals?

CHAPTER 15 ARCHITECTURE MATURITY MODELS

To answer these (and other) questions, it is important (among other things) to clarify the necessary responsibilities and accountability. One of the core components of doing architecture work properly is being able to establish ownership of the architecture and responsibility for its content. Table 15-7 lists the activities that can be invested in to achieve this.

Table 15-7. *List of activities to raise the maturity level of the Strategy and vision element*

Activity	Description
Obtain mandate for architecture	Ask senior management to express their commitment to architecture and explicitly assign responsibility for architecture.
Create responsibility matrix	Create a responsibility (RACI) matrix that maps architecture-related tasks to different functions within the organization.
Establish architecture board	Establish an architecture board to formally approve architecture products and provide an escalation platform.
Appoint architecture process owner	Delegate ownership of the architecture process. The process owner is responsible for the effectiveness and efficiency of the architecture processes.
Assign final responsibility for architecture	Ensure that senior management is effectively involved in architecture, especially on the business side.

After the above activities have been successfully implemented, the maturity level for the *Strategy and vision* element can be reviewed and reassessed. If the assessment is positive, the maturity level can be increased.

A second example of increasing maturity is to improve *the Architecture method and process* element. In Figure 15-2, this element is shown with the first three maturity levels.

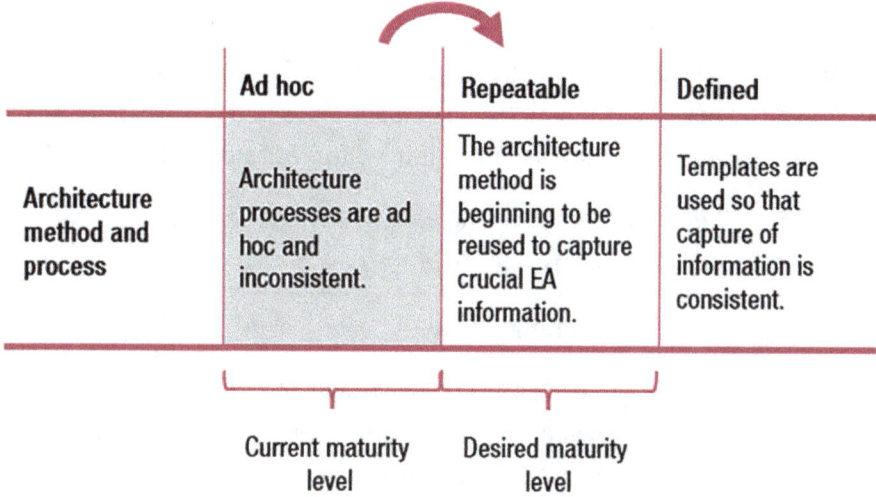

Figure 15-2. Excerpt from the CMMI framework for the Architecture method and process element

To move from the first level (Ad hoc) to the next level (Repeatable), it is necessary to examine how the architecture is created and applied in the organization. An indication that the maturity of the *Architecture method and process* element is at the *Ad hoc* level is that little or nothing has been done to formalize the architecture processes and their application.

Questions that can be asked at this point are as follows:

- Is an architecture only created when there is a client for it?

- Is an action plan created before an architecture is designed?

- Is the architecture process described?

- Is the architecture process known throughout the organization?
- Is there a regular review of whether the architecture process is still adequate?

After answering the above questions, the activities listed in Table 15-8 can be performed. Carrying out the listed activities will ensure that a higher level of maturity can also be achieved for this element.

Table 15-8. List of activities to raise the maturity level of the Architecture method and process element

Activity	Description
Plan architecture	Determine which architecture products are needed. Confirm this with stakeholders. Then create an action plan for creating the architecture.
Describe architecture processes	Describe the architecture processes, such as creating and maintaining the architecture, formalizing the architecture products, and the role of the architecture in projects.
Establish architecture product management process	Establish procedures for handling changes to the architecture. The goal is to actively maintain and update the architecture as a whole.

Once the activities listed in Table 15-8 have been performed and the processes and procedures are in place, it is possible to reassess whether the architecture is now at a higher level of maturity. If so, the maturity model can be updated with a higher rating.

CHAPTER 15 ARCHITECTURE MATURITY MODELS

  Identify the Current Maturity Level: Begin by assessing where each element currently stands within the five maturity levels defined by the CMMI framework. This provides a clear starting point for improvement.

Determine the Next Desired Level: Decide on the next maturity level to achieve for each element. This sets a clear goal for the improvement efforts.

Ask the Right Questions: To understand what changes are needed, ask targeted questions about each element. This helps identify specific activities required to progress to the next level.

Implement the Necessary Activities: Based on the insights gained, carry out the identified activities. These actions are designed to address gaps and enhance the maturity of each element.

Review and Reassess: After implementing the activities, evaluate the changes to determine if the maturity level has improved. If the assessment is positive, acknowledge the advancement to the higher maturity level.

The listed actions and activities that can be undertaken to increase maturity for each element are just a few examples [25]. It is advisable to determine the actions in consultation with senior management and other key stakeholders.

15.3. Alignment with Competencies

The previous sections illustrated how a maturity framework and model can be used to determine the Architecture Capability maturity. To take this maturity to the next level, and thus properly implement the further professionalization of the capability in an organization, appropriate competencies are required.

Chapter 10, Section 10.4, describes the skills and competencies for each core architecture role. Table 15-9 shows how these competencies map to the elements of the Capability Maturity Model Integration framework.

Table 15-9. CMMI elements mapped to competencies

Element	Competency	Description
Strategy and vision	Strategic thinking and vision	Ability to conceptualize the long-term view of enterprise systems and their alignment with organizational goals.
	Strategic decision-making	Making informed, strategic decisions under uncertainty or complex scenarios.
Architecture governance	Leadership and influence	Architects need to have the capability to guide and inspire teams toward shared goals.
	Governance and compliance	Focuses on enforcing standards, policies, and guidelines that ensure architectural consistency.
Architecture method and process	Communication and collaboration	Ability to communicate complex ideas clearly to both technical and non-technical audiences.
	Adaptability and resilience	The ability to maintain composure and thus effectiveness in high-pressure situations.
	Analytical thinking and problem solving	Ability to break down complex challenges into manageable components.
Architecture deliverables	Systems thinking	Understanding the interconnectedness of components within an enterprise ecosystem is an absolute must.
Business alignment	Change leadership	Driving organization-wide change initiatives with resilience and adaptability.
	Business and technology alignment	Architects must be able to bridge the gap between business strategy and technology implementation.

CHAPTER 15   ARCHITECTURE MATURITY MODELS

Table 15-9 shows, between the lines, which architectural roles are required to fulfill each of the elements in the maturity model. These roles, along with the elements and competencies, are shown in Table 15-10.

Table 15-10. CMMI elements, competencies, and core architecture roles

Element	Competency	Core Architecture Role
Strategy and vision	Strategic thinking and vision	Enterprise Architect, Segment Architect
	Strategic decision-making	Enterprise Architect, Segment Architect
Architecture governance	Leadership and influence	Enterprise Architect
	Governance and compliance	Enterprise Architect, Segment Architect, Solution Architect
Architecture method and process	Communication and collaboration	Enterprise Architect, Segment Architect, Solution Architect
	Adaptability and resilience	Enterprise Architect, Segment Architect, Solution Architect
	Analytical thinking and problem solving	Enterprise Architect, Segment Architect, Solution Architect
Architecture deliverables	Systems thinking	Segment Architect, Solution Architect
Business alignment	Change leadership	Enterprise Architect, Segment Architect
	Business and technology alignment	Enterprise Architect, Segment Architect

CHAPTER 15 ARCHITECTURE MATURITY MODELS

The tables showing the mapping between the elements of the CMMI framework on the one hand and the required competencies and core architecture roles on the other show that several roles are needed to take the Architecture Capability within an organization to the next level. It is clear that having only one architecture role in an organization is not enough to significantly increase maturity.

It is also clear from the mapping that the role of the Enterprise Architect has a significant stake in further professionalizing the Architecture Capability.

The TOGAF Standard does not (yet) have its own maturity model. It is not inconceivable that one will be included in a later edition of the framework. Hopefully, a maturity model will be chosen that focuses on the more strategic nature of the discipline and less on the IT domain.

A maturity model is a powerful tool. Its use enables an organization to identify opportunities for growth and development. By translating the activities associated with the maturity elements into actionable initiatives, it becomes possible to increase the level of architecture maturity within the organization.

Further Reading

The TOGAF Standard Series Guide "Architecture Maturity Models", document number G203, offers additional information about the use of maturity models.

An interesting article on architecture maturity models and how to read them can be found at the following link:

https://www.ardoq.com/knowledge-hub/enterprise-architecture-maturity-model

CHAPTER 15   ARCHITECTURE MATURITY MODELS

15.4. Summary

Maturity models are commonly used to measure and provide insight into the maturity of the internal Architecture Capability. The TOGAF Standard highlights two well-known maturity frameworks.

The *Architecture Capability Maturity Model* framework has a structure in which the *core components of an architecture process* are represented and measured, often in relation to the organization's IT. The purpose of using such a model is to identify weaknesses and provide a step-by-step approach to improving them.

The Architecture Capability Maturity Model considers the following nine architecture elements when assessing architecture maturity:

- Architecture process
- Architecture development
- Business linkage
- Senior management involvement
- Operating unit participation
- Architecture communication
- Architecture governance
- IT security
- IT investment and acquisition strategy

The *Capability Maturity Model Integration* framework, on the other hand, places more emphasis on maturing the Architecture Capability in order to *achieve organizational goals*. This makes this type of maturity model more appropriate for use in an Enterprise Architecture environment.

A maturity model based on Capability Maturity Model Integration defines five elements of Enterprise Architecture maturity:

- Strategy and vision
- Architecture governance
- Architecture method and process
- Architecture deliverables
- Business alignment

Appropriate competencies are needed to properly implement the further professionalization of the Architecture Capability in an organization.

In Chapter 10, Section 10.4, architecture roles were mapped to competencies. This chapter maps these competencies (and their corresponding architecture roles) to the elements of the Capability Maturity Model Integration framework.

In order to take the Architecture Capability within an organization to the next level, several architecture roles are needed in an organization. The role of the Enterprise Architect has a significant stake in further professionalizing the Architecture Capability.

A maturity model is a powerful tool. Its use enables an organization to identify opportunities for growth and development. By translating the activities associated with the maturity elements into actionable initiatives, it becomes possible to increase the level of architecture maturity within the organization.

CHAPTER 16

Alignment with Other Frameworks

 This chapter is part of the BLUE reading tour. See Figure 1-1 in Chapter 1, Section 1.3, for alternative reading tours.

The TOGAF Standard is the most commonly used framework for developing Enterprise Architectures. Needless to say that organizations also use other types of frameworks and methods to steer the governance of planning and execution, information systems and operation, and management and measurement. In some cases, an organization might even use industry-specific frameworks. It is therefore important to align these existing frameworks and methods with the TOGAF Standard to ensure a solid governance and related outcomes.

16.1. Mapping Existing Frameworks and Methods

In order to align all existing frameworks and methods in an organization, try to answer the following questions:

- Which framework takes precedence?

- What is the organizational level of commitment on using this framework?
- In what way does the TOGAF Standard fit in?

The main idea behind answering the aforementioned questions is to determine in what way the organization looks at planning, execution, and governance functions. It is useful to get a clear picture of how committed the organization is in using the existing frameworks.

Build the Framework Catalog: Start by compiling a list of all the frameworks the organization uses. These include those for planning and execution, information systems governance and operations, and management and measurement. Don't forget to add industry-specific frameworks that provide insight into business processes and capabilities.

Organize Frameworks by Key Models: Next, group these frameworks according to models such as risk, accounting, and planning. This will help to understand how each framework contributes to different aspects of the organization's operations.

Determine What's Needed from Each Framework: For each framework, identify the specific elements, methods, or techniques that are essential for effectively implementing the recommendations from the Architecture Capability. This ensures that the focus is on what's really needed to achieve the target state.

Assess Organizational Commitment: It's important to recognize how committed the organization is to each framework. Some frameworks may be fully integrated into operations, while others may only be partially adopted. Focus the analysis on the frameworks to which the organization is truly committed, as these will be the most effective in driving meaningful change.

CHAPTER 16   ALIGNMENT WITH OTHER FRAMEWORKS

In addition to mapping existing frameworks in the organization to the TOGAF Standard, it is also possible to reverse the approach. This means looking at how other frameworks and methods can connect to the TOGAF Standard. An example of such a connection is given in the next section.

16.2. Enterprise Architecture Implementation Wheel

The Enterprise Architecture Implementation Wheel method is a great example of both the tailorability of the TOGAF Standard and how well the framework allows for alignment with other frameworks and methods.

The Enterprise Architecture Implementation Wheel method is detailed in my other book, *Getting Started with Enterprise Architecture* [9].

The book describes a practical and pragmatic approach to developing and implementing a fundamental Enterprise Architecture.

The Enterprise Architecture Implementation Wheel is a practical and pragmatic approach to developing a fundamental Enterprise Architecture. It is based on the TOGAF Standard. The method differs from the framework in that it adapts TOGAF's structure and flow by reorganizing its content into four main stages: Document, Define, Execute, and Control. Each stage is broken down into one or more steps, each of which includes key focus areas and actionable artifacts. This offers organizations a streamlined entry point.

The straightforward description of the steps within the method ensures an accessible approach. The simplicity with which actions and outcomes are described and explained makes the method applicable to almost any situation and scenario.

CHAPTER 16 ALIGNMENT WITH OTHER FRAMEWORKS

The method is visually represented by a wheel consisting of a core and three rings (see Figure 16-1).

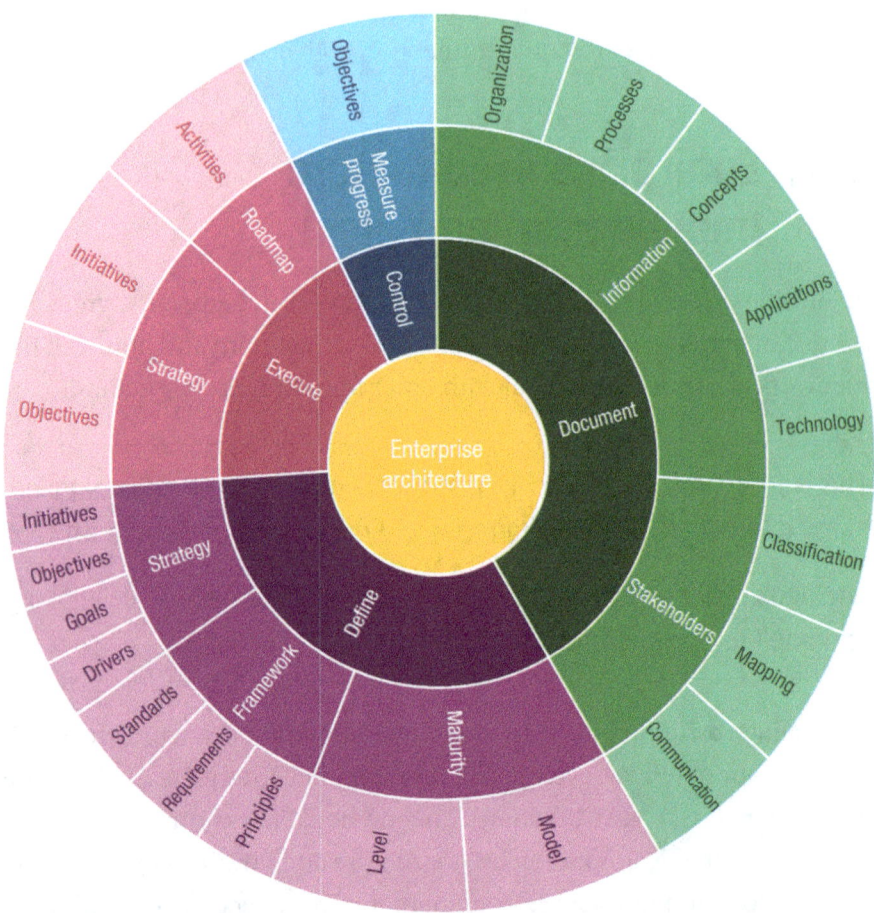

Figure 16-1. *The Enterprise Architecture Implementation Wheel*

The *first ring* around the center is the ring that defines the *four main stages* of the Implementation Wheel. Each stage consists of one or more steps.

The steps form the *second ring* around the center circle and provide valuable information on how to approach each stage. They also define a number of *key focus areas* that are relevant to these steps.

The key focus areas are presented in the *third and final ring* around the center circle. Each of the key focus areas details the various *architectural deliverables* that can be created to complete the previously mentioned steps.

The Enterprise Architecture Implementation Wheel should preferably be read clockwise, from the inside out. However, an organization may have already gone through an extensive inventory phase and documented everything it does, has, and uses. The organization may want to focus first on translating its strategy into implementation. To accommodate these wishes, the method also supports an iterative approach. This allows for a starting point of choice.

For organizations new to practicing architecture, the Enterprise Architecture Implementation Wheel argues that the approach described in the TOGAF Standard will be even more effective when the sequence of phases (and steps within those phases) is slightly changed. To this end, the method defines four stages:

Document (Stage one): This first stage involves gathering information about the organization and identifying and communicating with key stakeholders. This stage consists of two steps. The first step is to *capture and document the entities of interest* to the organization. The entities are in all architecture domains (Business Architecture, Information Architecture, Application Architecture, and Technology Architecture). The second step of the first stage is to *classify and map the key stakeholders* and *create a Communications Plan*.

Define (Stage two): The *Define* stage focuses on determining the organization's maturity level, principles, and guidelines and outlining its strategy. The stage has three steps. First, it determines the *current and desired maturity* of the Architecture Capability. Second, it describes what is involved in *establishing basic principles, requirements,* and *standards*[1]. Finally, it focuses on defining the organization's strategy (using the available information). Guidance is provided on how to *determine the drivers, goals,* and *objectives* that the organization is pursuing. It also shows how to identify related *initiatives*.

Execute (Stage three): This stage is about executing the strategy and providing a plan to do so. It consists of two steps. The first step explains how to achieve an *execution of the strategy* defined in the previous stage. This takes into account the previously established goals and objectives. The second step deals with the *creation of a roadmap* based on the objectives to be achieved and the related initiatives and activities.

Control (Stage four): The final stage, *Control*, is concerned with *measuring the progress* of the implementation of a basic Enterprise Architecture. It also explains how to visualize the progress of the realization of the organizational goals and objectives. The initiatives and activities to be carried out are monitored and made measurable. To this end, it discusses a number of measurement tools and techniques that can be used to visualize progress.

The Enterprise Architecture Implementation Wheel offers a practical and pragmatic approach to implementing a fundamental Enterprise Architecture, rather than a complete one. The method can be thought of as a simplified, tailored version of the TOGAF Standard. It is developed to help organizations get started with architecture, so that they can build on it over time.

---

[1] The word "standard" refers to particular Solution Building Blocks that further detail specific products or services used to meet requirements. For example, if a principle requires the "use of standard products and services" and a requirement indicates the "use of cloud-based services for commoditized services", a "standard" in this context may refer to "Microsoft Azure web services" or "AWS web services".

16.3.3. Mapping to the Architecture Development Method

All the aforementioned stages, steps, and key focus areas defined in the Enterprise Architecture Implementation Wheel method can be mapped to phases of the Architecture Development Method. Table 16-1 shows the mapping of the first stage (Document) to the ADM.

Table 16-1. Mapping of Stage one (Document) to the ADM

Stage	Step	Key Focus Area	ADM Phase
Document	Information	Organization	Phase B
		Processes	Phase B
		Concepts	Phase C
		Applications	Phase C
	Stakeholders	Classification	Phase A
		Mapping	Phase A
		Communication	Phase A

According to the TOGAF Standard, Phase B of the Architecture Development Method is about developing the Baseline and Target Business Architectures that describe how the organization needs to operate to achieve the business goals and respond to the strategic drivers. This aligns with what the Enterprise Architecture Implementation Wheel describes as gaining insight into the organizational structure and its processes.

Phase C directs the organization to, among other things, capture its application portfolio within the Architecture Repository. The same goes for the technology products that are being used. They, too, need to be captured in diagrams and catalogs (during Phase D). Classifying the products during inventory is a welcome addition. The Implementation Wheel supports this approach by emphasizing the importance of its inventory stage.

Finally, Phase A is all about identifying stakeholders and their concerns, as well as knowing how to communicate the architecture with them. To do so correctly, creating a stakeholder mapping based on their classification is essential. The Enterprise Architecture Implementation Wheel addresses all three of these important topics – stakeholder classification, mapping, and communication – during the first stage *(Document)*.

Table 16-2. Mapping of Stage two (Define) to the ADM

Stage	Step	Key Focus Area	ADM Phase
Define	Maturity	Model	Phase A
		Level	Phase A
	Framework	Principles	Preliminary Phase
		Requirements	Phases B, C, D
		Standards	Phases B, C, D
	Strategy	Drivers	Preliminary Phase
		Goals	Preliminary Phase
		Objectives	Preliminary Phase
		Initiatives	Preliminary Phase

During the execution of Phase A, a maturity assessment should take place. The Enterprise Architecture Implementation Wheel describes the same need in its second stage *(Define)*, but supplements it by providing guidance regarding the use of a maturity model (see Table 16-2).

The TOGAF Standard mentions defining Architecture Principles as being part of the Preliminary Phase. According to the framework, defining a set of Architecture Principles that is appropriate for the organization should be done once the organizational context is understood. The Implementation Wheel therefore suggests performing an inventory of the organization and its context prior to defining the principles.

CHAPTER 16   ALIGNMENT WITH OTHER FRAMEWORKS

With regard to strategy, the Preliminary Phase states that strategic and tactical business objectives and aspirations need to be met. The Enterprise Architecture needs to reflect this requirement and enable the architecture discipline to operate within the organization. Using the Implementation Wheel, this requirement is met by following the guidance provided by Business Architecture. The BIZBOK Guide (to which the TOGAF Standard refers) provides useful resources for developing and defining strategic drivers, goals, objectives, and initiatives. The Implementation Wheel adopts this methodology and details how these elements can be implemented.

Table 16-3. Mapping of Stage three (Execute) to the ADM

Stage	Step	Key Focus Area	ADM Phase
Execute	Strategy	Objectives	Phase F
		Initiatives	Phase F
	Roadmap	Activities	Phases E, F

The TOGAF Standard implies in Phase F that a business value needs to be assigned to each work package. This allows for focused governance during a later stage. The Enterprise Architecture Implementation Wheel hooks into this during stage three (*Execute*) by detailing the objectives that originated from the organization's strategy into initiatives or work packages (see Table 16-3).

Phase E of the framework describes the creation of an Architecture Roadmap, which is finalized in Phase F. The Architecture Roadmap contains initiatives based on the candidate roadmap items identified during earlier phases. The Implementation Wheel suggests linking the initiatives to their corresponding goals and objectives defined in the second stage (*Define*). This allows for a visually understandable governance in the fourth stage (*Control*).

CHAPTER 16   ALIGNMENT WITH OTHER FRAMEWORKS

Table 16-4. Mapping of Stage four (Control) to the ADM

Stage	Step	Key Focus Area	ADM Phase
Control	Measure progress	Objectives	Phases G, H

The objective of Phase G is to ensure conformance with the Target Architecture. Priorities are defined and the necessary transitions to reach the Target Architecture are realized. Stage four (*Control*) of the Implementation Wheel provides tools and techniques for measuring and monitoring the progress of these transitions. This aligns with Phase H of the Architecture Development Method (see Table 16-4).

The fourth stage of the Enterprise Architecture Implementation Wheel also enables the architect to prioritize initiatives based on their business value.

The description of the Enterprise Architecture Implementation Wheel clearly shows how easy it is to hook into the TOGAF Standard. The framework described in this book offers tremendous flexibility and adaptability so that it can truly be applied in all kinds of different situations and scenarios.

Further Reading

General information about the alignment with other frameworks can be found in the "The TOGAF® Leader's Guide to Establishing and Evolving an EA Capability" Series Guide, document number G168.

The Enterprise Architecture Implementation Wheel is described in more detail in the book *Getting Started with Enterprise Architecture* (ISBN 9781484298572) and at the following URL:

https://eawheel.com/wheel/

CHAPTER 16 ALIGNMENT WITH OTHER FRAMEWORKS

16.4. Summary

When it comes to developing Enterprise Architectures, the TOGAF Standard stands out as the most commonly used framework. It is often the go-to choice for organizations. However, these organizations also employ other frameworks. Therefore, aligning these existing frameworks with the TOGAF Standard becomes crucial.

To successfully achieve this alignment, it's essential to determine which framework takes precedence, understand the organizational commitment to existing frameworks, and identify how the TOGAF Standard can integrate seamlessly. Addressing these considerations helps clarify how the organization approaches planning, execution, and governance functions. This understanding reveals how the TOGAF Standard can complement existing frameworks, providing a solid foundation for developing and implementing an Enterprise Architecture.

Beyond mapping existing frameworks to the TOGAF Standard, it's also possible to reverse the approach – examining how other frameworks can connect to the TOGAF Standard.

The Enterprise Architecture Implementation Wheel exemplifies both the tailorability of the TOGAF Standard and its capability for alignment with other frameworks and methods. The Enterprise Architecture Implementation Wheel largely follows the TOGAF Standard but distinguishes itself by adapting the sequence of steps (or phases) in architecture development. All stages, steps, and key focus areas defined in the Enterprise Architecture Implementation Wheel can be easily mapped to phases of the Architecture Development Method.

The description of the Enterprise Architecture Implementation Wheel demonstrates how effortlessly it integrates with the TOGAF Standard. The framework described in this book offers remarkable flexibility and adaptability, making it applicable across various situations and scenarios.

CHAPTER 17

Tailoring the TOGAF Standard

 This chapter is part of the GREEN, BLUE, and GOLD reading tours. See Figure 1-1 in Chapter 1, Section 1.3, for alternative reading tours.

The TOGAF Standard is a framework with a high degree of flexibility and adaptability. The framework itself states in one of the first chapters that it should be adapted to the needs and requirements of the organization in which it is used. Several factors can play a role in determining the extent to which the framework needs to be adapted. These may include industry regulations, best practices, and compliance requirements, project scope and complexity, alignment with the organization's current maturity level, and specific modeling languages, tools, and techniques.

CHAPTER 17 TAILORING THE TOGAF STANDARD

17.1. Designed for Customization

Adapting the TOGAF Standard to one's own organization, to one's own needs and desires, and to the situation in which the organization finds itself is of great importance. There is a reason why adaptability is one of the first points mentioned in the framework.

> The TOGAF Framework is designed to require interpretation or customization [18].

The framework can be modified in many areas. The most important parts and aspects are described below.

The adaptability of the Enterprise Metamodel and the Content Framework has already been discussed in detail in Chapter 5, Section 5.2, and in Chapter 6, Section 6.4, respectively. However, it does not hurt to review all the ways in which the framework can be customized.

The TOGAF Standard can be customized in the following ways:

- Tailoring for organizational context
- Tailoring the Architecture Development Method
- Tailoring the Content Framework and deliverables
- Tailoring governance and compliance
- Tailoring tools and techniques

Tailoring for Organizational Context: The framework's entities can be modified to better fit the organization's strategic goals and operating model. This allows for better alignment with business strategy.

The TOGAF Standard's Enterprise Metamodel (see Chapter 5) can also be adapted to specific industry regulations, best practices, and compliance requirements. This type of tailoring focuses on industry-specific adaptation.

CHAPTER 17 TAILORING THE TOGAF STANDARD

A third way to tailor the framework to the organizational context is based on the size and structure of the organization. The TOGAF Standard can be easily customized for small, medium, or large organizations, taking into account centralization and decentralization.

A final way to customize the architecture framework is to tailor the governance and processes based on the organization's experience with architecture practices. This final approach aims to tailor the framework to an organization's culture and maturity level.

The essential scaffolding of the TOGAF Framework is the concepts. Everything else in the TOGAF Framework is either an example or a starter set to get you moving. If you do not like the example, then you can take advantage of the modular structure of the TOGAF Framework and substitute it [18].

Tailoring the Architecture Development Method: Depending on the scope of the project, some phases may be modified or even omitted. It is by no means necessary to go through all the phases of the Architecture Development Method for the sake of going through them. For example, if a project's scope is small enough that it doesn't require a new set of Architecture Principles, then it's possible to skip the corresponding Preliminary Phase steps. The same is true for a situation that does not require the creation of an architecture team. There is no need to go through all the necessary steps if a team already exists. Skipping or even reordering phases is entirely possible, as is combining phases where efficiencies can be gained (e.g., combining phases B, C, and D for a high-level architecture definition).

The Architecture Development Method also allows for integration with iterative or agile methodologies such as SAFe. It is even possible to include additional governance checkpoints to align with corporate governance structures. The Enterprise Architecture Implementation Wheel method described in Chapter 16, Section 16.2, is a great example of the flexibility and adaptability of the Architecture Development Method.

Tailoring the Content Framework and Deliverables: As discussed in Chapters 5 and 6, the Enterprise Metamodel and Content Framework can be easily aligned with the organization's proprietary data models. Architecture deliverables can be simplified (or expanded) to better match the organization's level of maturity in working with architecture. These base models can be aligned with ITIL, COBIT, the Zachman Framework, ArchiMate, or agile frameworks.

For example, certain ITIL processes can be added to the Content Framework as business functions or even capabilities. The same is true for parts of the COBIT framework. The TOGAF Standard can also be seamlessly integrated with ArchiMate, as was demonstrated in Chapter 5, Section 5.2, when the addition of the stakeholder to the Enterprise Metamodel entity was proposed.

Tailoring Governance and Compliance: If the organization uses specific policies, the governance structure of the TOGAF Standard can be modified to include these policies. Like the Risk Management or Stakeholder Management technique, these can be added to or implemented within the governance structure of the framework. The techniques provided can be tailored to the needs of the organization.

Tailoring Tools and Techniques: Finally, the TOGAF Standard allows for customization at the level of adoption of specific modeling languages, tools, and techniques. It is possible to use preferred tools such as ArchiMate, UML, BPMN, or even custom solutions to visualize the architecture artifacts that may be generated during the execution of the phases of the Architecture Development Method. In addition to using preferred modeling languages, the TOGAF Standard allows integration with Enterprise Architecture tools such as Bizzdesign or LeanIX.

Align with Organizational Goals: Start by aligning the framework's core components with the organization's strategic goals and operating model. This ensures that the Enterprise Architecture is not just a theoretical model, but a practical tool that moves the business forward.

Adapt to Industry Standards: Modify the Enterprise Metamodel to incorporate specific industry regulations, best practices, and compliance requirements. Tailoring the framework in this way helps the architecture remain relevant and compliant within the specific industry context.

Scale to the Size of the Organization: Tailor the framework to the size and structure of the organization, whether it's a small startup or a large organization. In large organizations, it may make sense to implement the architecture in its full breadth. Smaller organizations may be better off with only parts of the architecture. Therefore, only those parts of the TOGAF Standard that add value to the organization should be used. This flexibility ensures that the framework remains practical and effective regardless of the size or complexity of the organization.

Reflect the Maturity of the Organization: Adapt governance structures and processes based on the organization's experience and maturity with architecture practices. This approach aligns the framework with the organization's culture and readiness, facilitating smoother implementation and adoption.

Adjust the Architecture Development Method: Modify the phases of the Architecture Development Method to fit the scope of the project. It's acceptable to skip or combine phases as appropriate. This adaptability eliminates unnecessary work and enables a more efficient architecture development process.

Customize the Content Framework and Deliverables: Align the Content Framework with the organization's data models and existing entities. Simplify or extend deliverables based on the organization's maturity level. Customizing the content ensures that the architecture deliverables are relevant and manageable for the organization.

Change Governance and Compliance Structures: Incorporate the organization's specific policies into the architecture governance structure. Use techniques such as Risk Management and Stakeholder Management. This will ensure that the architecture framework supports and enhances the organization's governance and compliance efforts.

Select Appropriate Tools and Techniques: Select modeling languages and tools that match the organization's preferences, such as ArchiMate, UML, BPMN, or custom solutions. Integrate with Enterprise Architecture tools. Using familiar tools facilitates better visualization and management of architecture artifacts, leading to more effective implementation.

As described in the introduction to this book, many practitioners end up abandoning the framework. The main reason given for this is that the framework does not describe *how* it can be implemented. It focuses too much on describing *what* can be done. Another common argument against the use of the TOGAF Standard is that it lacks real-world examples that show how it can be used in practice.

As far as real-world examples are concerned, this can be directly refuted. The following section demonstrates the adaptability of the TOGAF Standard in a real-world situation.

17.2. Tailoring the TOGAF Standard for the Healthcare Industry

The steps outlined above for adapting the framework are fairly generic and therefore applicable to a wide range of sectors. To illustrate the high degree of adaptability of the TOGAF Standard, consider the need to adapt the framework to better meet the specific needs of the healthcare industry. The sector has unique challenges, including regulatory compliance, interoperability, patient data security, and rapidly evolving technology. Tailoring the TOGAF Standard for healthcare will require adjustments in several areas.

When developing an industry-specific metamodel, architects may choose not to include entities and relationships from the Enterprise Metamodel that are not relevant and/or add additional entities and relationships [2].

The various components of the TOGAF Standard that can be adapted were applied to the case study described below. For each part, the steps taken are described.

Tailoring for Healthcare-Specific Needs: To meet industry-specific needs, the framework can be aligned with standards such as HIPAA (US), GDPR (Europe), HL7 FHIR (interoperability standards), and ISO 13485 (medical devices)[1] to meet regulatory compliance requirements.

The Foundation Enterprise Metamodel can be augmented with industry-specific entities and a patient-centric focus. This leads to the modification of artifacts to incorporate *patient experience, electronic health records* (EHRs), and *digital health solutions*.

---

[1] ISO 13485 is the internationally recognized standard for quality management systems in the design and manufacture of medical devices. It outlines specific requirements that help organizations ensure their medical devices meet both customer and regulatory demands for safety and efficacy (https://www.iso.org/standard/59752.html).

CHAPTER 17   TAILORING THE TOGAF STANDARD

Tailoring the framework to healthcare-specific needs ensures that the Enterprise Architecture supports both clinical workflow and hospital administration by achieving clinical and operational alignment.

Last but not least, the TOGAF Standard encourages its user to implement or even strengthen *information security architectures* to meet healthcare-specific threats and vulnerabilities.

Tailoring the Architecture Development Method: The healthcare industry requires rapid adaptation to new technologies (like telemedicine, AI diagnostics). Therefore, implementing an agile approach within the Architecture Development Method cycle allows the industry to respond quickly. It is quite possible to introduce what is called phase prioritization. This can be achieved by tailoring the Architecture Development Method to:

- **Emphasize Business Architecture** (Phase B) to map patient journeys, hospital workflows, and public health initiatives.

- **Strengthen Technology Architecture** (Phase D) to focus on cloud adoption, data lakes for research, and AI-driven diagnostics.

- **Enhance migration planning** (Phase F) to address legacy system modernization and EHR system upgrades.

Tailoring the Content Framework and Deliverables: The introduction of *FHIR-based data models*[2] for interoperability enables the development of a customized Content Framework for healthcare. This allows the use of tailored deliverables by leveraging *clinical workflows* and *decision support systems* in Business Architecture models. Some of the key architecture deliverables to be customized are:

---

[2] The Fast Healthcare Interoperability Resources (FHIR) standard is a set of rules and specifications for the secure exchange of electronic healthcare data.

CHAPTER 17   TAILORING THE TOGAF STANDARD

- **Capability Models:** These models enable the definition of hospital capabilities such as *telemedicine, remote patient monitoring*, and *AI diagnostics* (see

- **Interoperability Maps:** Ensure that EHR systems, lab systems, pharmacy management, and imaging systems (PACS/RIS) can communicate seamlessly.

- **Risk and Compliance Models:** Allow for addressing risks associated with data breaches, AI misdiagnosis, and critical care system downtime.

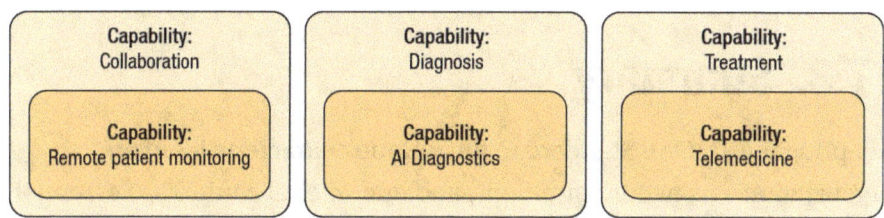

Figure 17-1. Tailored industry-specific Business Capability Diagram

Tailoring Governance and Compliance: The TOGAF Standard can also be adapted for healthcare-specific governance by establishing an Architecture Board that includes clinical representatives, IT security, and compliance officers. There are known situations where medical informatics and health IT experts have been included in and on the Architecture Board.

In addition, tailoring the governance and compliance structures of the TOGAF Standard allows an industry to consider implementing regulatory decision gates. These could include regulatory checkpoints such as HIPAA risk assessments or country-specific regulatory approvals for AI-driven diagnostics.

Tailoring Tools and Techniques: Use industry-specific Enterprise Architecture tools. For example, implement ArchiMate models with healthcare extensions and/or use FHIR-based interoperability testing tools. Adopt data-driven decision-making by using AI-driven analytics in architecture modeling. Finally, introduce predictive modeling to optimize patient flow and resource allocation.

The example scenario described above demonstrates a fully tailored version of the TOGAF Standard. By customizing the framework for the healthcare industry, organizations can improve patient outcomes, ensure regulatory compliance, and enhance interoperability. The focus should be on security, interoperability, AI adoption, and cloud-based healthcare ecosystems. The TOGAF Standard's flexibility and adaptability makes everything possible.

17.3. Summary

Adapting the TOGAF Standard to the unique characteristics of an organization – its needs, ambitions, and operational context – is essential for achieving its intended value. Adaptability, prominently emphasized within the framework, serves as a foundational principle precisely for this reason.

The TOGAF Standard offers extensive opportunities for customization:

- **Tailoring for Organizational Context:** The framework's core components may be adapted to better align with the organization's strategic objectives and operational realities. This ensures a closer connection between Enterprise Architecture efforts and overarching business imperatives.
- **Tailoring the Architecture Development Method:** The phases of the Architecture Development Method can be selectively modified or omitted in accordance with

the specific scope and requirements of an initiative. Rigid adherence to all phases is neither prescribed nor necessary where it does not add value.

- **Tailoring the Content Framework and Deliverables:** The Enterprise Metamodel and Content Framework are designed to integrate seamlessly with proprietary data models. Furthermore, the complexity or simplicity of architecture deliverables can be adjusted to correspond with the organization's maturity level in Enterprise Architecture practices.

- **Tailoring Governance and Compliance Structures:** Where existing policies and regulatory frameworks are in place, the governance model outlined in the TOGAF Standard can be extended to incorporate these organizational-specific requirements.

- **Tailoring Tools and Techniques:** The selection and application of modeling languages, tools, and techniques may be customized, allowing organizations to adopt methods best suited to their operational environment and technological landscape.

The TOGAF Standard can provide a tailored solution for virtually any situation, scenario, organization, or industry by utilizing the customization options mentioned above.

CHAPTER 18

Agile Architecture

 This chapter is part of the BLUE reading tour. See Figure 1-1 in Chapter 1, Section 1.3, for alternative reading tours.

Agile architecture is not an oxymoron. It is a conscious approach to designing and developing architecture in a way that embraces change, fosters collaboration, and delivers value incrementally. In the context of the TOGAF Standard, agile architecture represents a harmonious blend of structured methodology and adaptive practices.

The TOGAF Standard provides a robust framework that can be tailored to support agile principles. By incorporating iterative cycles, continuous stakeholder engagement, and incremental delivery, the Architecture Development Method can be adapted to facilitate agility within Enterprise Architecture initiatives. This adaptation ensures that architecture development is responsive to evolving business needs while maintaining alignment with strategic goals.

This chapter examines how the TOGAF Standard accommodates agile architecture. It explores the integration of agile practices within the Architecture Development Method phases and discusses the concept of just enough architecture. Through this exploration, it is demonstrated that the TOGAF Standard is not only compatible with agile methodologies, but also enhances their effectiveness by providing a structured approach to managing complexity and change.

CHAPTER 18 AGILE ARCHITECTURE

18.1. What Is Agile Architecture

The term *agility* is often thrown around in different contexts. But what does it really mean? While interpretations may vary, certain characteristics consistently emerge when discussing agility.

At its core, agility is about *adaptability*. It's a mindset that *anticipates change* and adapts accordingly. It involves short, iterative cycles and frequent reassessment of priorities to ensure that organizations remain responsive and relevant in dynamic environments.

Agility emphasizes the *delivery of tangible value*. Activities are driven by their potential to have the greatest impact. This approach involves continually evaluating and reprioritizing tasks to focus on high-value outcomes, often minimizing intermediate products and extensive documentation. Agility encourages a culture of experimentation. Rather than exhaustive theoretical analysis, it encourages trying ideas, learning from results, and iterating quickly. This *fail-fast* approach fosters innovation and continuous learning.

Agile environments thrive on the autonomy of skilled, multidisciplinary teams. These teams are empowered to make decisions, which promotes a sense of ownership and accountability for their results.

Agility is a journey, not a destination. It embodies an ongoing commitment to refining processes, improving performance, and striving for excellence. This relentless pursuit of improvement ensures that organizations evolve and adapt over time.

Agile means to move/change quickly and easily, often to provide value-generating outcomes [23].

Today, when the word *agile* is mentioned, many people immediately think of the *Agile Software Development Process*, which is rooted in the "Manifesto for Agile Software Development" [26]. While these principles are invaluable, it's important to recognize that organizational agility encompasses a broader spectrum than just agile methodologies.

CHAPTER 18   AGILE ARCHITECTURE

18.1.1. The Basic Concepts Explained

The approach to developing and ultimately implementing an agile architecture requires an understanding of a few concepts. So let's start with the basics. Figure 18-1 shows a conceptual visualization of an architecture. This architecture is divided into three levels of granularity (Strategic, Segment, and Capability Architectures). One of the levels (the Capability Architecture) is broken down into three smaller pieces.

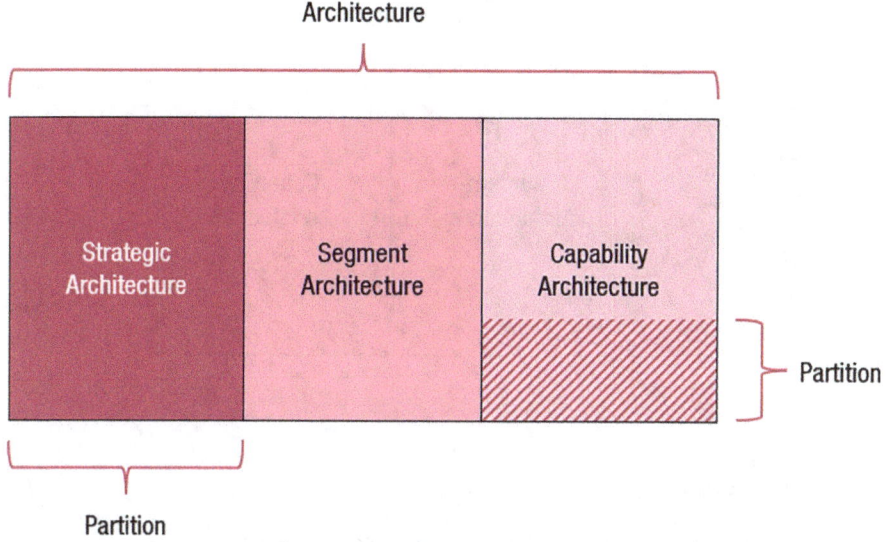

Figure 18-1. *Architecture divided into partitions*

Splitting an architecture or level of architecture granularity into one or more smaller parts is called *architecture partitioning*. This concept is explained in Chapter 13, Section 13.4. An architecture partition is a logical scope of work that can be independently architected and managed. In layman's terms, an architecture partition refers to *a smaller piece of the overall architecture*.

CHAPTER 18   AGILE ARCHITECTURE

Architecture partitions can be developed and implemented using *agile sprints*. A sprint is a short, time-bound iteration that delivers working functionality. In agile environments, sprints are often used to deliver features or increments within a partitioned architecture. As shown in Figure 18-2, each architecture partition can be mapped to one or more agile sprints that focus on building solutions within that partition.

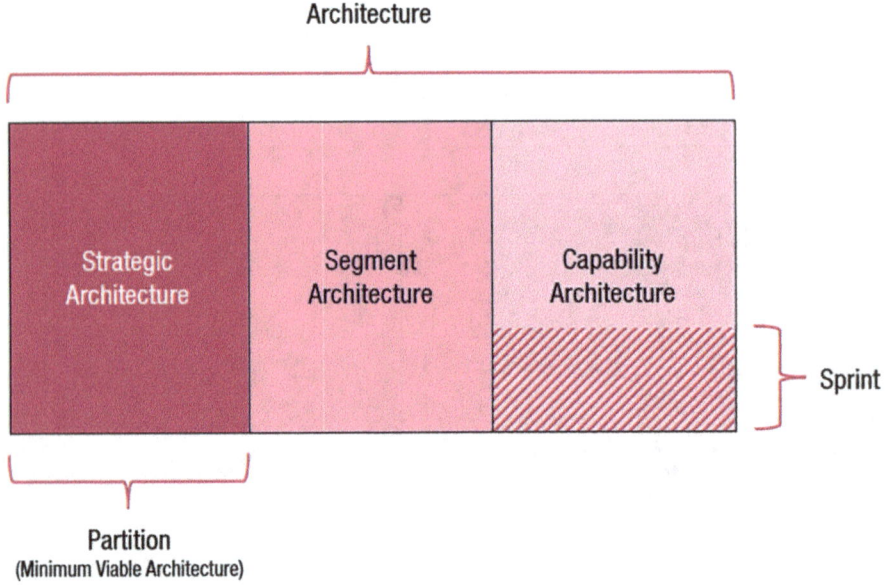

Figure 18-2. Architecture showing partitions and sprints

An architecture partition contains at least the minimum architecture that is realizable and adds business value. In agile terms, this is called a *Minimum Viable Architecture*.

A Minimum Viable Architecture defines the minimum architecture that is realizable and add business value [23].

CHAPTER 18 AGILE ARCHITECTURE

In essence, an agile approach in the context of architecture means doing work in manageable chunks. Where architecture using a non-agile approach is often seen as a long process, an agile approach is about being able to deliver pieces of the whole quickly and in short cycles. Each piece is developed and delivered in one or more sprints (short, time-bound iterations that deliver working functionality). An important point to note here is that the work does not have to be done sequentially, as is common in a waterfall approach. An agile approach supports doing work in *parallel*. But with the understanding that the big picture must not be lost.

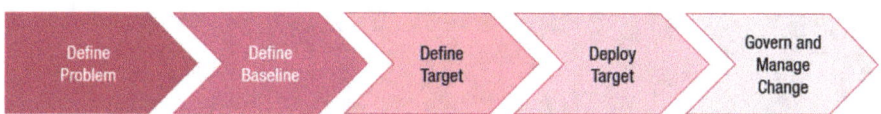

Figure 18-3. *Regular approach to architecture development*

The regular approach to architecture development, shown in Figure 18-3, consists of five components:

- Define Problem
- Define Baseline
- Define Target
- Deploy Target
- Govern and Manage Change

The five components listed above are directly related to the phases of the Architecture Development Method. For example, the first component, *Define Problem*, is addressed in Phase A (Architecture Vision). The *Define Baseline* and *Define Target* components are touched on during phases B–D (Business Architecture, Information Systems Architectures, and Technology Architecture). The *Deploy Target* section relates to Phase E

CHAPTER 18 AGILE ARCHITECTURE

(Opportunities and Solutions), Phase F (Migration Planning), and Phase G (Implementation Governance). Finally, Phase H (Architecture Change Management) can be seen in the *Govern and Manage Change* section.

As mentioned above, an agile way of working does not use a *sequential* approach. Instead, its strength lies in the ability to perform short-cycle tasks (sprints) in *parallel*. During each of the components described above, one or more sprints take place. Each sprint accomplishes a *portion of the total work* to be done for a component. Sprints can run in parallel. This is visualized in Figure 18-4.

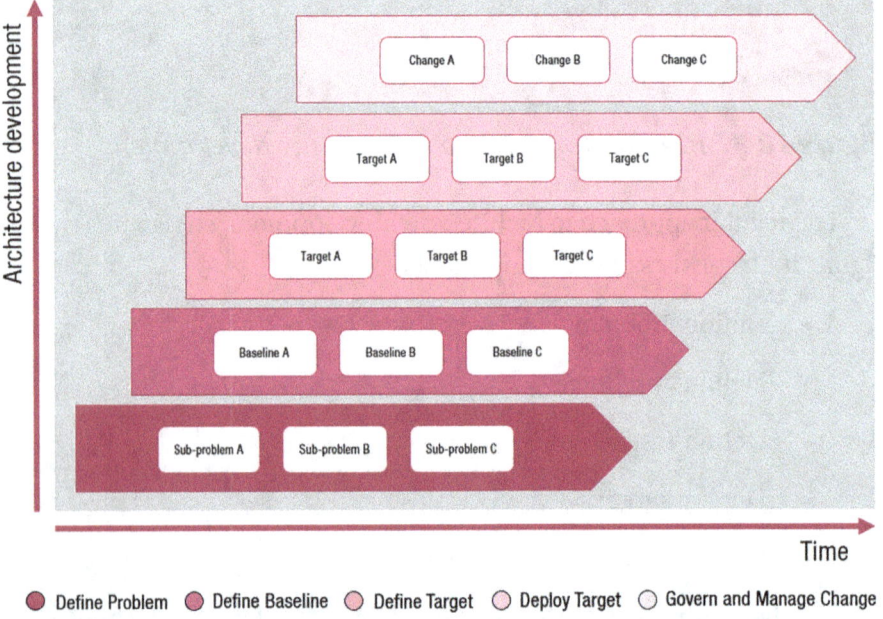

Figure 18-4. Components of a regular approach that includes sprints

It is by no means necessary to complete the problem definition before beginning other activities. It is sufficient to define enough of the problem to provide a context for other activities. As these other activities are initiated, work can continue to refine and extend the problem definition.

CHAPTER 18 AGILE ARCHITECTURE

There are of course obvious dependencies. It is not possible to implement any part of the target until the parts of the baseline and target that describe that particular part of the target have been sufficiently defined.

> It's like creating a painting. If you don't know what you want the painting to look like, you cannot start painting.

The TOGAF Standard embraces agile working through the use of architecture partitioning. Architecture partitioning is the process of dividing architecture work into smaller chunks across multiple teams. Along with partitioning the workload, definitions are provided so that each team understands what they need to create in terms of the architecture. The architecture is developed and implemented by performing sprints.

18.2. Three Levels of Architecture Granularity

Remember the breakdown of the architecture levels that were described in Chapter 7, Section 7.2, that existed in the Architecture Landscape? The TOGAF Standard recognizes a model that identifies three levels of detail that can be used to partition the architecture development (see also Figure 18-5):

- **Strategic Architectures:** Provide a long-term summary view of the entire organization. These architectures provide an organizing framework for operational and change activities and enable executive-level direction setting.
- **Segment Architectures:** Provide more detailed operating models for areas within an organization. Segment Architectures can be used at the program or

portfolio level to organize and operationally align more detailed change activities. Another term often used for Segment Architectures is Domain Architectures.

- **Capability Architectures:** Show in more detail how the organization can support a particular unit of capability. Capability Architectures are used to provide an overview of current capability, target capability, and capability increments and allow individual work packages and projects to be grouped into managed portfolios and programs.

Referring back to Figure 18-3 in Section 18.1, the *problem* illustrated in the figure is defined at the strategic and segment levels. At the segment and capability levels, a *target* is defined, and at the capability level a target is *deployed* by agile teams supported by the architect and the Capability Architecture definition.

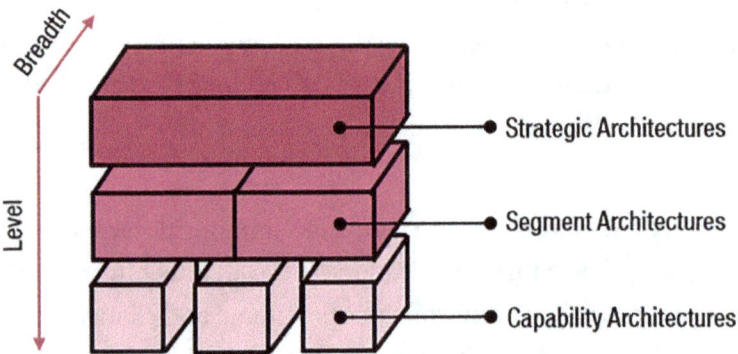

Figure 18-5. Three levels of architecture granularity

Strategic Architectures: Serve as the compass that guides an organization's long-term vision. They provide a broad, high-level perspective that contextualizes different segments and capabilities within the organization. This overarching view is essential for orchestrating comprehensive planning and design efforts, thereby mitigating the risk of unanticipated outcomes.

In agile organizations, Strategic Architecture is not a static blueprint, but rather a *dynamic iteration*. It is supported by Phase A (Architecture Vision) of the Architecture Development Method, which sets the strategic direction to inform decision-making processes. This strategic iteration can be further elaborated in Phase B to provide a more detailed view of the organizational landscape. It also helps to identify the necessary architecture and solution delivery capabilities required to achieve the organization's objectives.

Segment Architectures: Typically operate at the portfolio, program, or product level. These segments are not arbitrary; they are aligned with the structural contours of the organization, providing clarity and focus.

In the realm of Segment Architecture, the principle of *just enough* is paramount. It's not about exhaustive detail, but about identifying key features and both functional and non-functional requirements. This approach doesn't require going through every phase and step of the Architecture Development Method. Instead, it emphasizes the minimum necessary to ensure effective results.

Multiple iterations can be delivered simultaneously by different agile teams, fostering a dynamic and responsive development environment. Involving delivery teams in defining Segment Architectures creates a common understanding of the solution, setting the stage for cohesive and efficient implementation.

At its core, Segment Architecture is about strategic precision, providing just enough detail to guide development without stifling agility. It's a testament to the power of focused, collaborative design to achieve business goals.

CHAPTER 18 AGILE ARCHITECTURE

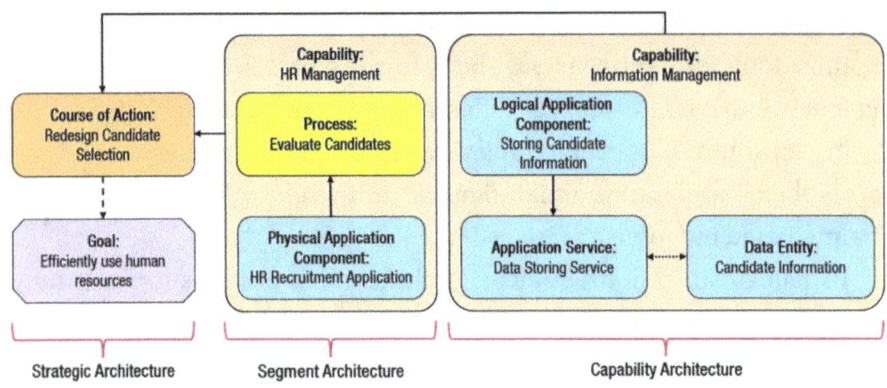

Figure 18-6. Strategic, Segment, and Capability Architectures

Capability Architectures: Serve as the essential link between high-level strategy and hands-on implementation. Think of them as the detailed blueprints that guide delivery teams as they bring business capabilities to life. They delve into the specifics of business capabilities and break them down into manageable increments. Each increment is meticulously detailed, ensuring that developers have clear, actionable guidance. Whether a capability is aligned with a single delivery sprint or spans multiple sprints, the focus remains on delivering tangible value.

It's worth noting that while sprints can occur at various levels, they are most commonly aligned with the delivery of capabilities or their increments. This alignment ensures that each sprint contributes meaningfully to the overarching business goals.

In addition, Capability Architectures don't exist in isolation. They build on Segment Architectures, adding layers of detail to ensure that solutions are not only well designed, but also implementable. This layered approach ensures that each architectural component is seamlessly aligned, facilitating a cohesive and effective implementation strategy.

Figure 18-6 shows that the Capability Architecture to be developed, which consists of three components, is part of the business capability Information Management. Partitioning the work associated with this architecture results in the three components being developed separately. Each sprint focuses on developing a specific part of the total. An example of this is shown in Figure 18-7.

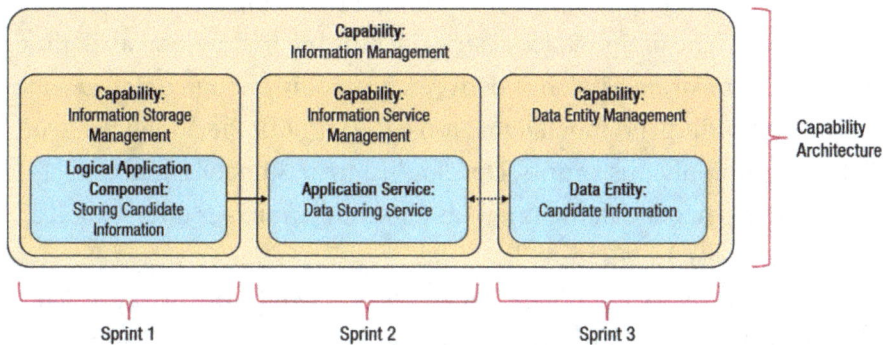

Figure 18-7. *Development of a partitioned architecture using three sprints*

The Capability Architecture from Figure 18-6 is shown in Figure 18-7 in a zoomed view. Clearly shown is the implementation of the development of a *logical application component,* an *application service,* and a *data entity.* Each component is implemented during a separate sprint.

Mapping architecture partitions to agile sprints is described in the next section.

18.3. Architecture Partitions and Agile Sprints

The idea behind an agile approach is not to use the waterfall principle and wait to do work until all previous issues have been completed. The development of Segment Architectures can therefore begin as soon as all relevant areas of interest are known. Even if not everything has been worked out in detail, the architecture work can begin. A similar approach applies to Capability Architectures. The development of these architectures does not have to wait for the completion of the Segment Architectures. As illustrated in Figure 18-4 in Section 18.1.1, much of the architecture work and development can be done in parallel.

Of course, it is important not to lose sight of the big picture. The relationship to the organizational strategy must be maintained at all times. The advice is to strike a balance between providing enough detail at each level and providing too much detail too early without the added value of feedback from previous sprints. Developing an architecture that reflects exactly what is needed at the time – and nothing more – is often called a *Minimum Viable Architecture*.

It is important to identify the segments that need to be involved. This will largely determine the scope of the Segment Architecture. Once this scope is clear, it can be further refined into Capability Architectures. These can be implemented using an agile approach. The incrementally delivered Capability Architectures can be used to iteratively construct the final outcome (product, service, or solution). These iterations can be performed in parallel, depending on the degree of interdependency between them. Capability Architectures corresponding to a Segment Architecture can also be executed in parallel.

The end result of the Capability Architecture is the solution specification that is built and deployed on demand by agile teams according to the architecture guidelines, metrics, and compliance considerations.

The TOGAF Standard recognizes the need to recursively partition the Enterprise Architecture into more granular levels (Strategic, Segment, and Capability Architectures) that can be architected using an agile approach.

Partitioning the architecture work is key to agile delivery and involves defining segmentation based on the Strategic and Segment Architecture concepts discussed earlier. These smaller pieces, covering a specific area of the organization, can then be more easily specified and implemented using an agile approach. Partitions are considered to be the agile equivalent of a *Minimum Viable Architecture*.

CHAPTER 18   AGILE ARCHITECTURE

Let's say that an average architecture project is being developed. The architecture work consists of the Baseline Architecture, a Transition Architecture, and a Target Architecture (see Figure 18-8).

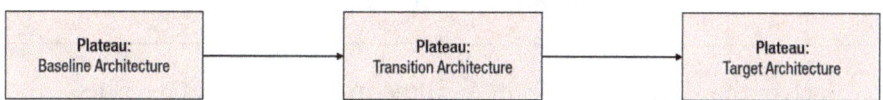

Figure 18-8. *Architecture project showing Baseline, Transition, and Target Architecture plateaus*

Using architecture partitioning (see Chapter 13, Section 13.4), the work packages or projects containing the activities necessary to create the Transition Architecture could be split (partitioned) into smaller pieces, creating multiple Transition Architectures, as demonstrated in Figure 18-9.

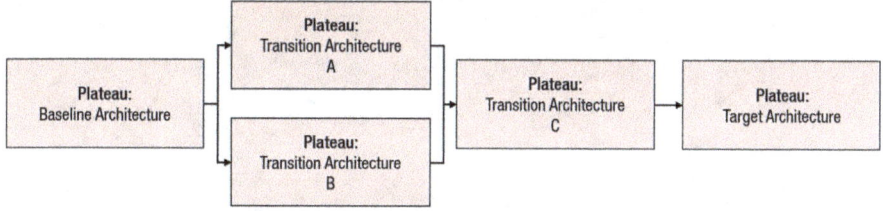

Figure 18-9. *Architecture project showing Baseline Architecture, partitioned Transition Architectures, and Target Architecture plateaus*

Figure 18-9 shows the application of partitioning to the architecture development design shown earlier. Here, the Transition Architecture (still referred to as a single plateau in Figure 18-8) is partitioned into several Transition Architectures (A, B, and C).

737

CHAPTER 18   AGILE ARCHITECTURE

A Transition Architecture is a formal description of one state of the architecture at an architecturally significant point in time, usually including a number of capability increments [23].

Each of the shown Transition Architectures (A, B, and C) has its own architecture to be developed and implemented. Figures 18-7 and 18-9 are combined in Figure 18-10, which shows the architecture to be implemented per sprint as part of each individual Transition Architecture.

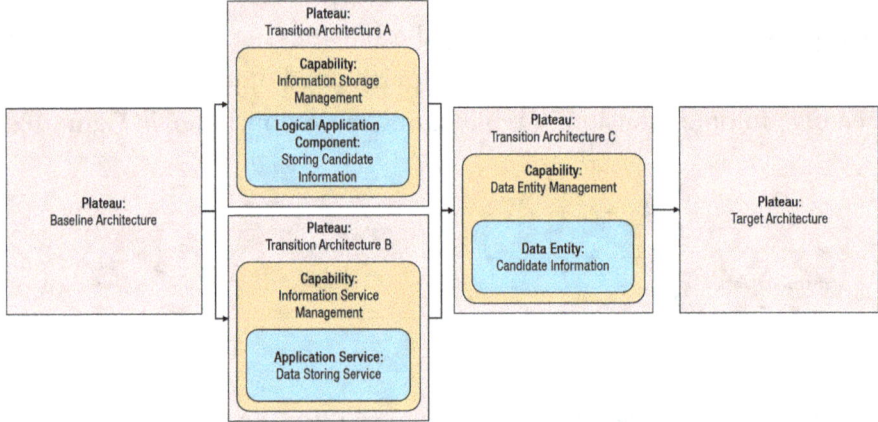

Figure 18-10. Architecture project showing Transition Architectures and sprints

Splitting or partitioning the work associated with the Transition Architecture into multiple Transition Architectures creates smaller pieces of deliverable architecture. These pieces (partitions) can be developed and implemented using one or more agile sprints.

Now, combining the individual sprints from Figure 18-10 and the basic concepts from Figure 18-2 (Section 18.1), it is possible to see how the sprints relate to these basic concepts. This is shown in Figure 18-11.

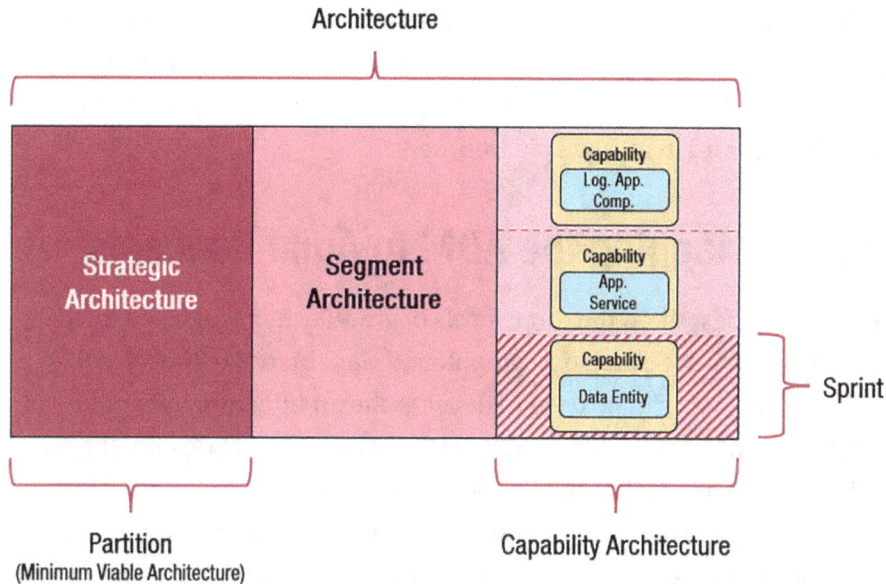

Figure 18-11. Architecture showing partitions and detailed sprints

The TOGAF Standard fully supports an agile way of working. The Architecture Development Method can be easily adapted to support agile concepts, and the phases of the Architecture Development Method can be mapped to them. It is also quite possible to indicate for each phase of the Architecture Development Method what needs to be done in relation to the execution of agile sprints.

18.4. Developing an Agile Architecture

The TOGAF Standard supports an agile approach through the use of the Architecture Development Method. The three levels of granularity and associated phases of the Architecture Development Method map well to the concepts of an agile approach. Due to its iterative nature, this approach can be used very effectively to interpret the relationship to the agile concepts.

The phases of the Architecture Development Method can also be aligned with an agile approach. Each phase separately supports the agile approach, making it possible to apply the architecture approach of the TOGAF Standard in an agile environment.

18.4.1. Mapping the ADM to Agile Concepts

Section 18.2 described the three levels of granularity recognized by the TOGAF Standard (Strategic, Segment, and Capability Architectures). These three levels can be used to develop the architecture in parallel using different iterations. This approach is equivalent to developing an architecture using techniques such as SAFe and Scrum.

Table 18-1 shows how the three levels of architecture granularity, combined with the corresponding phases of the Architecture Development Method, can be mapped to different agile concepts.

Table 18-1. Architecture granularity levels and ADM Phases mapped to agile concepts

Architecture Granularity Level	ADM Phase	SAFe/Agile/Scrum Layer
Strategic Architecture	Preliminary Phase Phase A: Architecture Vision	SAFe – Strategic Themes – Enterprise Portfolio Delivery
Segment Architecture	Phase B: Business Architecture Phase C: Information Systems Architectures Phase D: Technology Architecture	Agile/SAFe – Product/Portfolio Backlog – Change Plans – Epics

(continued)

Table 18-1. (*continued*)

Architecture Granularity Level	ADM Phase	SAFe/Agile/Scrum Layer
Capability Architecture	Phase E: Opportunities and Solutions	Agile/SAFe – Product/Solution Backlog – Capabilities to be delivered
	Phase F: Migration Planning	
	Phase G: Implementation Governance	Agile/SAFe – Product/Program Backlog – Capability Increments (Features)
	Phase H: Architecture Change Management	Agile/SAFe – Program Team Backlog – Deliverable User Stories
		Scrum – Sprint Team Backlog – Deliverable User Stories

At the *strategic level*, the focus is more on the Preliminary Phase (if changes to the architecture capabilities are required) and Phase A to provide the basis for defining the cross-organizational and strategic change time horizon view. This results in a set of high-level strategic plans known as *courses of action*.

Agile techniques typically address this with concepts such as high-level strategic themes and the highest level of an organization's product portfolio backlog. At this level, cross-functional teams (business, technical, design, implementation, and operations) must be involved to develop an Enterprise Architecture that meets both the business goals and objectives of the organization and is also potentially deliverable.

At the *segment level*, the focus is on *partitioning the courses of action* among the relevant organizational units so that the work of implementing the change can be organized effectively and efficiently. If the information gained from performing Phase A is not sufficient for this activity, the focus may be on exploring phases B, C, and D in more detail.

The work can be approached by factoring the work into self-organizing teams at different levels (in accordance with the chosen organizational unit structure), along with a high-level iteration through phases C and D that provides more detailed information for product or solution delivery, going deeper into smaller organizational areas (segments). The outputs of this iteration are the Epics, which reflect large or long-running user stories, and the segment-based initial portfolio and/or backlog. The output from this level can be used to test and experiment with new products (if necessary), providing descriptions for prototypes to test ideas in the relevant segment market.

At the *capability level*, phases B, C, and D identify a more specific solution-oriented architecture specification, including the Architecture Building Blocks, covering both the functional and non-functional aspects of the solution to be implemented. These architecture specifications are then further developed in phases E and F as the basis for the Solution Building Blocks and their integration into the desired solutions, services, and/or products.

This approach to rapid, continuous implementation at the lowest level of capability creates a Transition Architecture of deliverable units, similar to the sprints at the lowest level of the backlog in agile-style approaches. This smallest level of capability is often referred to as the *Minimum Viable Product* in agile approaches.

A Minimum Viable Product is an output that satisfies a minimum set of functional and non-functional requirements and can be realized when implemented in a live operational environment [23].

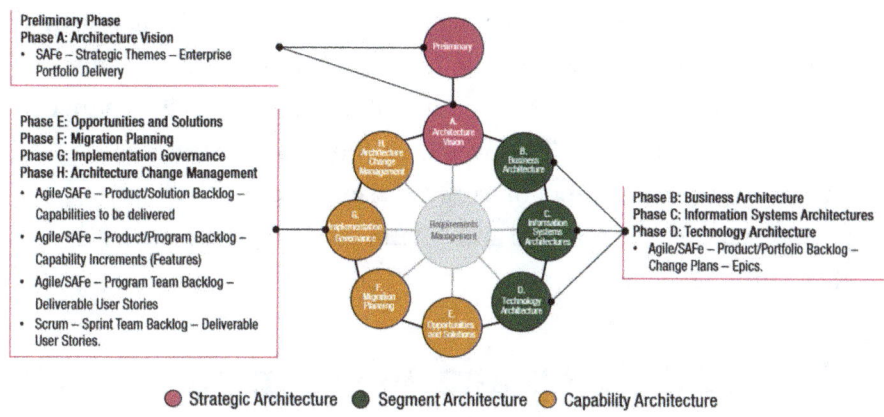

Figure 18-12. *The ADM mapped to agile concepts*

The backlogs shown in Figure 18-12 are not simple lists of activities. They are refined to reflect the Scrum concept of a sprint backlog. This refinement ensures that each sprint is not only deliverable, but also ready for implementation within weeks, or a month at most. Such sprints transition specifications from the phases of the Architecture Development Method (specifically, phases A–E) into tangible implementations. The focus is on fostering integrated teams and environments that ensure that design, build, implementation, and operations processes interact seamlessly. This approach facilitates continuous integration and implementation, efficiently delivering each unit of the Minimum Viable Product.

The culmination of this process is observed in Phase G (Implementation Governance). This phase is critical in confirming that the agreed contracts have delivered the expected capabilities in accordance with the contractual agreements. It also ensures that all necessary information required to operate and modify the product, solution, or service is meticulously created, stored, and made accessible. This meticulous approach supports faster and higher-quality changes in the future.

Phase H (Architecture Change Management) plays a critical role in confirming that the anticipated benefits of a capability enhancement have been realized. This phase is about evaluating whether the changes implemented have tangibly improved operational and business performance. It's the moment of truth to assess whether the value propositions have been realized and whether the end users are truly satisfied with the results.

18.4.2. ADM Phases and Agile Sprints

Going through the phases of the Architecture Development Method ensures that an agile architecture can be created. It is important to develop a *Minimum Viable Architecture* during each iteration.

Each Minimum Viable Architecture created by the Enterprise Architecture team can contain some or all Architecture Development Method phases. Creating partitions (which are the equivalent of Minimum Viable Architectures) of the architecture is already part of Architecture Development Method iteration. Figure 18-13 lists the Architecture Development Method phases and describes how they can be adapted in a sprinting environment. The Preliminary Phase has been intentionally omitted.

CHAPTER 18   AGILE ARCHITECTURE

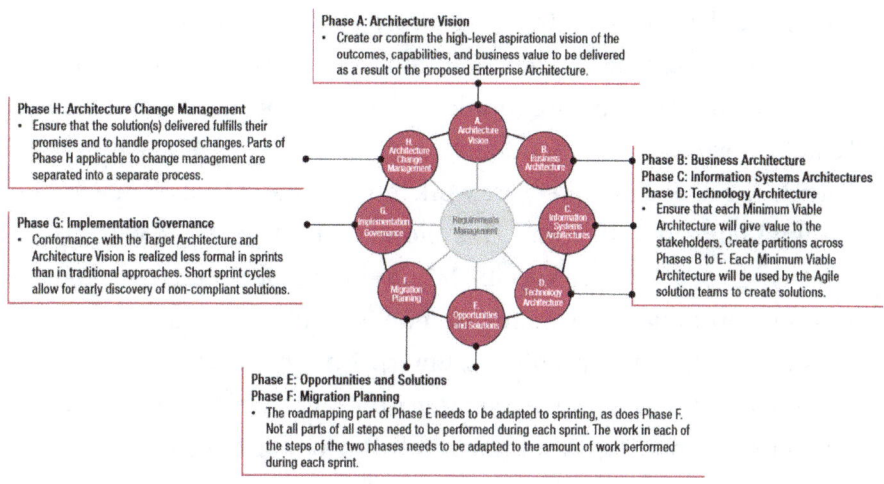

Figure 18-13. *ADM Phases and sprints*

Phase A (Architecture Vision): Start by outlining a high-level vision of what the organization aims to achieve with the proposed Enterprise Architecture. This includes identifying the desired outcomes, capabilities, and the business value expected from the initiative. Recognize that this vision is not set in stone. As the business and technology environments evolve, and as feedback is received from stakeholders, the vision may need to adapt. However, these changes are typically evolutionary, not revolutionary. This means that there may be a gradual development or change in the vision over a period of time. It does not mean that there will be a sudden, complete, or radical change in vision.

Develop the vision in a way that allows for future growth and refinement. It should serve as a stable foundation upon which the Enterprise Architecture can be built and expanded over time. Ensure that the vision offers clear direction to the teams involved. It should guide their efforts within the defined scope and timeframe, helping them align their work with the overarching goals of the Enterprise Architecture.

Phase B (Business Architecture), Phase C (Information Systems Architectures), Phase D (Technology Architecture): Instead of attempting to design the entire architecture at once, break it down into smaller, manageable parts called Minimum Viable Architectures. Each Minimum Viable Architecture should deliver specific value to stakeholders and can be developed independently.

Divide the overall architecture work into partitions that align with the Minimum Viable Architectures. This allows multiple teams to work in parallel, each focusing on a specific Minimum Viable Architecture. Such partitioning promotes agility and faster delivery of solutions.

Ensure that Business Architects, Enterprise Architects, and solution teams collaborate closely when creating Minimum Viable Architectures and their corresponding Minimum Viable Business Designs.

Table 18-2. *Steps to adapt Phase E (Opportunities and Solutions) for Agile Sprints*

Step	Description
Understand stakeholder readiness	At the start, assess how ready stakeholders are for change. If the same stakeholders are involved throughout, this might only need to be done once.
Identify business constraints	Before each sprint, check for any new business constraints that could affect implementation. Also, review constraints at the end of a series of sprints to understand inter-team dependencies.
Review gap analysis	Regularly revisit the gap analysis to ensure it is still accurate and relevant. Refer to Chapter 14, Section 14.5.
Manage requirements across functions	In each sprint, review and manage requirements that span multiple business areas.

(continued)

Table 18-2. (continued)

Step	Description
Ensure interoperability	Confirm that the solutions developed in the sprint can work together seamlessly. Refer to Chapter 14, Section 14.6.
Refine dependencies	Identify and validate dependencies between agile teams to ensure smooth collaboration.
Assess readiness and risks	Determine if the organization is ready for the Minimum Viable Product and identify any risks. Refer to Chapter 14, Section 14.8.
Develop implementation strategy	Create a strategy for implementing the architecture that aligns with the sprint's goals.
Define work packages	List the major tasks (work packages) needed for the sprint, which will feed into team backlogs. Refer to Chapter 12, Section 12.8.3.
Plan Transition Architectures	Decide how the sprint's outputs will be transitioned to all users or a select group.
Create roadmap and migration plan	Collaborate with product owners and agile teams to define the backlog and plan for implementation. Refer to Chapter 12, Section 12.8.3.

This collaboration ensures a common understanding and alignment with business objectives.

The TOGAF Standard doesn't mandate a strict sequence for phases B–E. Feel free to iterate across these phases as needed, adapting to the project's and organization's specific requirements. This flexibility supports continuous improvement and responsiveness to change.

Phase E (Opportunities and Solutions), Phase F (Migration Planning): Not every activity needs to happen in every sprint. Therefore, the work can be tailored to match the sprint's scope and focus. When multiple solution teams are involved, coordination becomes more complex, requiring careful planning.

Phase E (Opportunities and Solutions) can be adapted for sprints by following the steps detailed in Table 18-2.

Possible adaptations for Phase F (Migration Planning) are listed in Table 18-3.

Table 18-3. Steps to adapt Phase F (Migration Planning) for Agile Sprints

Step	Description
Confirm management framework	Ensure that the management structures are in place to support the migration plan. This is typically done before starting the agile journey.
Assign business value	For each sprint, assign a business value to the work packages to prioritize efforts (see Chapter 12, Section 12.9.2).
Estimate resources and timelines	During sprint planning, estimate the resources needed, timelines, and delivery methods.
Prioritize projects	Use cost/benefit analysis and risk assessment to prioritize migration projects for each sprint.
Update roadmap and Architecture Definition Document	After each sprint, update the overall roadmap and the Architecture Definition Document (see Chapter 12, Section 12.2) to reflect progress.
Complete migration plan	Finalize the plan for implementing the next sprint's work, effectively updating the backlog.
Document lessons learned	At the end of each sprint, document what was learned to improve future sprints.

Phase G (Implementation Governance): When it comes to Phase G – with its objective of making sure everything aligns with the Target Architecture and the architecture vision – things work a little differently in a sprint-based approach. First, the entire compliance process becomes much less formal. And second, the short sprint cycles are a game changer. They allow non-compliant solutions to be identified early on.

Phase G (Implementation Governance) can be further adapted by following the suggested steps in Table 18-4.

Table 18-4. Steps to adapt Phase G (Implementation Governance) for Agile Sprints

Step	Description
Confirm scope and priorities	This is usually done at the sprint start, when the sprint goal and sprint scope are defined.
Identify deployment resources and skills	This is usually done at the beginning of each sprint for each implementation.
Guide development of solutions deployment	This is an informal process for guiding development teams through implementation.
Perform architecture compliance reviews	The difference with the traditional approach is how the compliance review is performed. This is an ongoing process, as opposed to the traditional gate/review process at a specific point in time.
Implement business and IT operations	After each sprint with an approved demo, a solution increment can be deployed.
Perform post-implementation review and close the implementation	Because reviews in sprints are an ongoing process, rather than end-user test-style reviews, this step can be the formal summary of all previous reviews of portions of the implementation.

Phase H (Architecture Change Management): The main objective of Phase H is to ensure that the solution(s) delivered fulfills its promises and to handle proposed changes. It is suggested that the parts of Phase H applicable to change management are separated into a separate process:

- **Establish a Value Realization Process:** Define a process to ensure that the business value produced during the sprint is received.
- **Deploy Monitoring Tools:** Deploy the process and associated tools to measure the business value.

Phase H (Architecture Change Management) can be further adapted by following the suggested steps in Table 18-5. These steps are applicable specifically to change management and are separated into a separate process. This process runs in parallel with agile sprints.

Table 18-5. Steps to adapt Phase H (Architecture Change Management) for Agile Sprints

Step	Description
Manage risks	Collect risks from demos and retrospectives and create mitigation activities.
Provide analysis for architecture change management	Analyze business value monitoring results. Evaluate other incoming change requests and move them into the business change backlog.
Develop change requests to meet performance goals	Define change requests to address poor business value creation and risk mitigation.

(continued)

Table 18-5. (continued)

Step	Description
Manage governance process	Part of the sprint work. The business team governs the Enterprise Architecture team, and the Enterprise Architecture team governs the solution team, and vice versa.
Activate the change implementation process	This is part of the normal planning work for a sprint.

Sprints are a great way to create Enterprise Architecture deliverables. It all starts with *sprint zero*, which lays the groundwork. After that, each sprint runs like clockwork, delivering a Minimum Viable Architecture each time.

At each *sprint demo*, the Enterprise Architecture team presents what they've built. This gives stakeholders a clear picture of where things stand and what's coming. And at every *retrospective*, the team steps back and reviews two things: the quality of the architecture itself and how well the process is working. It's about making sure that both the work and the way it's done are getting better and better. Measuring speed – such as tracking how requirements were allocated and how accurate effort estimates were – can really help.

When it's time to plan the next sprint, the team picks the highest-value items from the business change backlog. It's all about delivering maximum impact with each sprint. And if the workload demands it, multiple Enterprise Architecture teams can work in parallel to move even faster.

After several sprints, the architecture starts to come together, and implementation can begin. Depending on the approach, solution teams may even start building while some parts of the architecture are still being finalized. That's why decoupled design is so important. It allows teams to move forward without getting stuck waiting for each other.

When it's time to deliver solutions, agile is a great fit. Whatever method is chosen, make sure that the governance approach adapts to support it. The best path depends on where the organization is in its maturity journey.

Start with Sprint Zero: Begin with an initial preparation sprint, often called sprint zero. This sprint sets the foundation for the work ahead. Once completed, its outcomes are handled just like the results of any other sprint.

Deliver a Minimum Viable Architecture: Each sprint should focus on producing a Minimum Viable Architecture – a usable and understandable version of the architecture that can evolve over time.

Demonstrate Progress to Stakeholders: At the end of every sprint, the Enterprise Architecture team should present the results to stakeholders. This ensures that everyone is aligned and understands what is being developed.

CHAPTER 18 AGILE ARCHITECTURE

Inspect and Adapt: After the sprint demo, hold a retrospective. Review both the quality of the Enterprise Architecture deliverables and how the team performed during the sprint. Look at key indicators like team velocity, accuracy of effort estimates, and how well requirements were allocated. Use the findings to improve future sprints.

Plan the Next Sprint: When planning each new sprint, select the most valuable and impactful items from the business change backlog. If needed, multiple Enterprise Architecture sprint teams can work in parallel to accelerate progress.

Transition to Implementation: After completing a series of sprints, the architecture is ready for implementation. In many cases, solution teams can start building before the full architecture is finished, as long as the necessary parts are defined and independent. This is where a decoupled design strategy shows its value.

Support and Govern: Throughout the implementation phase, the Enterprise Architecture team, together with the business, provides support and governance to ensure the solutions align with the architectural vision.

Adapt the Delivery Approach: Solutions can be built using agile methods or other approaches, depending on what best fits the organization's maturity and way of working. The governance approach should adapt accordingly.

Sprint planning isn't just about tasks and schedules. It's also a great time for Enterprise Architecture to sharpen desired outcomes, define measurable success, and lock down both functional and non-functional requirements. Throughout each sprint, architects should stay close to the business and solution teams to provide support and guidance. A successful iteration depends on collaboration, clarity, and course correction.

> **Further Reading**
>
> More information on using an agile approach in combination with the TOGAF Standard can be found in the Series Guides "Enabling Enterprise Agility", document number G20F, and "Applying the TOGAF® ADM using Agile Sprints", document number G210.
>
> Additional information about modeling agile architecture can be found at the Portfolio of Digital Open Standards, in the document "Agile Architecture Modeling Using the ArchiMate® Language", document number G20E.

18.5. Summary

At its core, agility is about one thing: adaptability. It's a mindset built around expecting change, embracing it, and pivoting when necessary. In practice, it means working in short, focused cycles, constantly re-evaluating priorities, and ensuring that the work being done remains relevant and valuable in a world that won't sit still.

Agility puts results first. It means doing the things that create real value, not just staying busy. Rather than getting bogged down in endless plans and documentation, agility encourages organizations to focus on the results that matter most and to change course quickly when necessary. It's a culture of action, experimentation, and rapid learning. Less theory, more doing.

Building an agile architecture follows the same philosophy. And to do it well, a few key ideas come into play.

The TOGAF Standard introduces a simple but powerful model for managing architecture development at different levels of detail:

- **Strategic Architectures** set the big picture.
  They provide a high-level view across the whole organization, helping leadership steer the ship and shape future operations.

- **Segment Architectures** zoom in a little closer. They organize and align change efforts within specific areas, programs, or domains.
- **Capability Architectures** go deep. They describe how to deliver a particular capability, breaking it down into manageable pieces that can be grouped into programs and projects.

A smart approach is to divide the work into smaller, logical chunks. This is called architecture partitioning. Think of it as breaking the big picture into bite-sized chunks that can be tackled individually. Each partition can be developed independently and deliver real business value on its own. They can be developed in sprints. These are short, time-bound cycles designed to deliver working solutions quickly. In agile terms, the minimum amount of architecture needed to deliver value is called *Minimum Viable Architecture*. It's about doing just enough to move forward, not trying to design everything up front. Simply put, agile architecture is about making progress in manageable chunks. Traditional architecture can sometimes feel like a long, slow process. Agile turns that on its head.

In an agile way of working, tasks don't have to be done one after the other. The goal is to make parallel progress while respecting obvious dependencies. It's a balancing act. Some pieces need to be built before others, but the focus is always on moving fast and delivering value early.

The TOGAF Standard fully supports an agile way of working. The Architecture Development Method is flexible and aligns well with agile principles. In fact, each phase of the Architecture Development Method can easily be aligned with agile sprints and iterative delivery:

- **Strategic Level:** Focus on the *Preliminary Phase* and *Phase A* to set the high-level direction and outline strategic plans (also known as courses of action).

- **Segment Level:** Organize the strategic plans into actionable parts across different organizational units. Depending on the need, *phases B, C, and D* might be explored more deeply here.

- **Capability Level:** Dig into *phases B, C, and D* to create detailed Solution Architectures. These get developed further through *phases E and F* into actual Solution Building Blocks.

All of this leads naturally to *Phase G (Implementation Governance)*, where sprint-based development allows for early identification of non-compliant solutions and rapid course correction when needed.

Finally, *Phase H (Architecture Change Management)* ensures that the delivered solutions do their job: deliver real, measurable value. In an agile environment, change management activities can run in parallel with sprints, keeping momentum high and ensuring that the architecture continues to evolve in line with real-world needs.

CHAPTER 19

Digital Transformation

 This chapter is part of the BLUE reading tour. See Figure 1-1 in Chapter 1, Section 1.3, for alternative reading tours.

The TOGAF Standard structures digital transformation through its Architecture Development Method, which provides a phased approach to defining, planning, implementing, and governing organization-wide change.

The transformation begins with a strategic vision (Phase A (Architecture Vision)) that aligns business goals with digital capabilities, followed by the Business Architecture (Phase B) that optimizes processes for digital readiness. The Information Systems Architectures and Technology Architecture phases (C and D) ensure that data, applications, and infrastructure are designed to support the new digital ecosystem.

A key aspect of digital transformation under the TOGAF Standard is its emphasis on governance and iterative improvement. The framework ensures that transformation initiatives are continuously evaluated through opportunities and solutions (Phase E), migration planning (Phase F), and implementation governance (Phase G).

Additionally, the TOGAF Standard supports ongoing innovation and adaptability through architecture change management (Phase H), allowing organizations to refine their digital strategies as new technologies and business needs emerge.

CHAPTER 19 DIGITAL TRANSFORMATION

19.1. What Is Digital Transformation?

Digital transformation is the process of integrating digital technologies into all aspects of a business or organization to improve efficiency, enhance customer experiences, and drive innovation. It goes beyond simply adopting new software or automating tasks – it involves a fundamental shift in how an organization operates, delivers value, and adapts to market changes. It is primarily based on a major change in the business and operating models. Successful digital transformation requires a combination of strategic planning, cultural change, and technological advancements to create a more agile, data-driven, and customer-focused organization.

At its core, digital transformation entails rethinking business models, processes, and technology infrastructure to leverage the power of modern digital tools. It includes several key components:

- **Technology Adoption:** Implementing cloud computing, artificial intelligence (AI), automation, and data analytics to enhance decision-making and operational efficiency

- **Process Optimization:** Re-engineering workflows through automation, digital collaboration tools, and AI-driven insights to streamline operations

- **Customer Experience Enhancement:** Leveraging digital platforms, self-service portals, and personalized experiences to better engage customers and meet their evolving expectations

- **Cultural Change:** Encouraging a digital-first mindset among employees by fostering innovation, upskilling the workforce, and embracing agile methodologies

- **Data-Driven Decision-Making:** Utilizing big data, real-time analytics, and business intelligence to gain insights, predict trends, and drive strategic growth

Ultimately, digital transformation is an ongoing journey, not a one-time project. It requires organizations to continuously adapt, innovate, and integrate new digital capabilities to remain competitive in an ever-evolving digital landscape.

Digital transformation, as defined by the TOGAF Standard, is the strategic and systematic evolution of an organization's capabilities, processes, and technologies to enhance business outcomes through digital innovation. It is not merely about implementing new technology but ensuring that the architecture aligns with business objectives, governance, and operational efficiency.

19.2. Digital Transformation and Enterprise Architecture

Digital transformation encompasses the comprehensive integration of digital technologies into every aspect of an organization. This integration fundamentally alters operational models and the manner in which value is delivered to customers. It is not just the digitization of existing processes. Rather, it requires a profound cultural shift that encourages organizations to challenge traditional norms, experiment with emerging technologies, and adapt to continually evolving market expectations.

Enterprise Architecture serves as a critical discipline in this context, providing a structured framework to align technology, processes, and human resources with overarching business objectives. It functions as a blueprint designed to optimize the organization's capabilities and ensure coherence across its operational landscape.

CHAPTER 19   DIGITAL TRANSFORMATION

Both digital transformation and Enterprise Architecture must be regarded as strategic initiatives. Each represents a business-driven approach aimed at redefining how an organization thinks, operates, and responds to change. Achieving successful transformation requires the reorganization of internal – and, where necessary, external – information environments to support new modes of operation.

It is also essential to distinguish clearly between commonly conflated terms such as *digitization, digitalization,* and *digital transformation,* as their interchangeable use often contributes to widespread confusion. The distinctions are as follows (see also Figure 19-1):

- **Digitization:** Refers to the process of converting physical assets into digital formats, typically through scanning or data conversion

- **Digitalization:** Involves the application of digital technologies to modify existing business processes, thereby improving workflows and operational efficiency

- **Digital Transformation:** Signifies a broader shift wherein the fundamental business model and value propositions are reimagined, moving from primarily physical to predominantly digital engagement and delivery mechanisms

CHAPTER 19   DIGITAL TRANSFORMATION

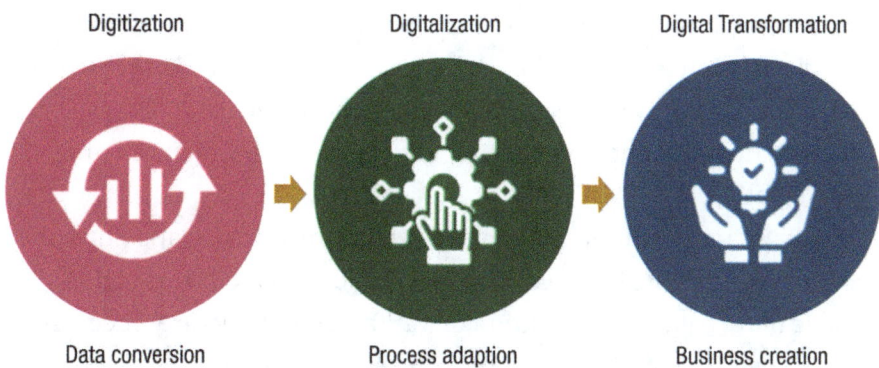

Figure 19-1. *Focus areas of digitization, digitalization, and digital transformation*

When organizations think of *digital*, they often associate it with information technology. However, true transformation must come from within the business itself. Digital transformation involves a fundamental shift in an organization's business model from a traditional approach to one driven primarily by digital capabilities.

In today's dynamic environment, digital transformation must be approached as an integrated, *business-driven initiative*. Isolated IT, data, service, or information strategies need to be carefully revised to align with overarching business objectives. The focus must extend beyond individual components, emphasizing a coordinated effort to maximize the value derived from information assets.

The notion of building a transformation strategy independent of the underlying business structures is analogous to building a house without a solid foundation. True digital transformation encompasses the delivery of innovative services, the improvement of organizational efficiency, and the enhancement of decision-making capabilities. It represents a complex synergy of people, processes, and technology that collectively drive the organization to higher performance and greater strategic advantage.

CHAPTER 19 DIGITAL TRANSFORMATION

The disciplines of Enterprise Architecture and digital transformation are intricately intertwined, with each playing a pivotal role in the success of the other. Despite ongoing debates about the sequencing of these initiatives, the most effective strategy involves the continuous alignment and integration of both disciplines.

Digital transformation initiatives often inform and reshape Enterprise Architecture. Insights from emerging technologies and the art of the possible may require the evolution of the Target Architecture. Successful digital transformation initiatives often begin with specific use cases or projects, which can then be scaled with the structured guidance of an Enterprise Architecture Roadmap.

Enterprise Architecture, in turn, provides the foundational framework and strategic blueprint essential to the success of digital transformation initiatives. By clearly defining the Target Architecture and the organization's operating model, Enterprise Architecture ensures that digital transformation efforts are cohesive rather than fragmented. It also identifies the Baseline Architecture and critical gaps, providing essential input for setting transformation priorities and guiding investment decisions.

A hybrid approach that thoughtfully integrates elements of both Enterprise Architecture and digital transformation is the most advantageous strategy. This balanced methodology enables organizations to realize immediate value from digital transformation while establishing a robust, long-term architectural foundation. Finding the optimal balance between these disciplines is essential to achieving sustainable success in the evolving digital landscape.

CHAPTER 19   DIGITAL TRANSFORMATION

Further Reading

Recommended blogs on digital transformation can be found at the following URLs:

https://eawheel.com/blog/2024/09/digital-transformation/

https://www.linkedin.com/pulse/what-comes-first-enterprise-architecture-digital-omar-gawad-34nff

19.3. A Structured Approach to Digital Transformation

Digital transformation is a complex process that requires careful planning and execution. The TOGAF Standard provides a structured approach to digital transformation, following the Architecture Development Method. It consists of iterative phases to guide organizations through their transformation journey (see Figure 19-2).

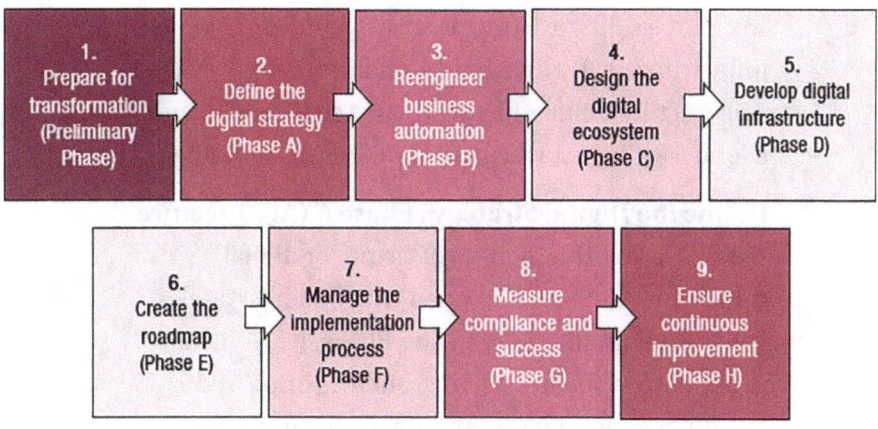

Figure 19-2. Nine steps for performing a digital transformation using the ADM

1. **Prepare for Transformation (Preliminary Phase):** Lay the foundation for transformation by defining the principles, governance, and capabilities that indicate the need for digital transformation. Identify the key stakeholders in the organization who have an interest in the digital transformation effort. Common stakeholder groups include the CxO, business executives, and other architects. Setting the boundaries of the transformation effort is another important step. This is usually done by defining the digital transformation principles (e.g., customer centricity, data-driven decision-making, cloud-first technology). Third, assess the maturity and readiness of the organization for such a transformation. Determine where the organization stands in terms of digital maturity. This includes conducting a digital maturity assessment by evaluating customer engagement strategies, workforce skills, existing processes, and technology infrastructure. Establishing an Enterprise Architecture team and governance structure will go a long way in ensuring the success of the initiative.

2. **Define the Digital Strategy (Phase A (Architecture Vision)):** Create a high-level vision for digital transformation that is aligned with business goals. During the Architecture Vision Phase, it is critical to define key business drivers such as improved customer experience, operational efficiency, and/or compliance. These drivers are an essential starting point for the digital strategy. Once the drivers have been defined, they must be linked

to organizational goals. Therefore, it is necessary to establish digital transformation goals and key performance indicators (KPIs). Be sure to formulate the goals and KPIs in a SMART[1] manner. Next, identify transformation enablers such as AI, process automation, cloud computing, and/or IoT. These enablers should become part of the high-level Target Architecture that outlines the business, data, application, and technology requirements. At this stage, it is essential to gain stakeholder buy-in and approval for the transformation vision.

3. **Re-engineer Business Automation (Phase B (Business Architecture)):** Define how digital transformation impacts business capabilities, and processes, by conducting a gap analysis between current and target business models. Try to redesign core business processes with automation and digital tools in mind. Also, define new roles and responsibilities to align with the digital initiatives of the organization. Integrating business services with customer-centric platforms (e.g., self-service portals and/or AI-driven interactions) is an absolute necessity. Last but not least, develop a business transformation roadmap with prioritized initiatives. This is where the Work Package Portfolio Map (see Chapter 12, Section 12.8.3) comes in.

---

[1] Specific, Measurable, Achievable, Realistic, and Timebound (SMART).

4. **Design the Digital Ecosystem (Phase C (Information Systems Architectures)):** Enable the digital approach by integrating data, applications, and IT services:

    - **For the Data Architecture:** Define data governance policies to ensure security and compliance (GDPR, HIPAA). Establish a centralized data strategy (e.g., data lakes, real-time analytics) and implement AI-driven data insights for decision-making.

    - **For the Application Architecture:** Identify the applications needed for transformation. This means taking another hard look at the current application landscape and perhaps planning to phase out certain existing applications and replace them with modernized versions. Also, enable cloud-based and API-driven integration across systems. The Application Portfolio Catalog (see Chapter 12, Section 12.6.4) is essential for defining legacy modernization plans.

5. **Develop Digital Infrastructure (Phase D (Technology Architecture)):** Build the foundational IT infrastructure for digital transformation by choosing the right cloud strategy (public, private, hybrid) and adopting microservices and containerization for scalability. Strengthen cybersecurity measures (zero-trust security, AI-driven threat detection) and ensure high system availability and disaster recovery mechanisms. From an agile perspective, the organization could also adopt DevOps and CI/CD practices for faster deployment cycles.

6. **Create the Roadmap (Phase E (Opportunities and Solutions))**: Identify transformation projects and prioritize their execution by evaluating and prioritizing initiatives based on business value and feasibility. Use the Business Value and Risk Matrix as described in Chapter 12, Section 12.9.2. This artifact allows you to define the quick-win projects with the highest business value and lowest risk. The resulting roadmap should include both short-term and long-term initiatives.

7. **Manage the Implementation Process (Phase F (Migration Planning))**: Develop a structured plan for moving from the Baseline Architecture to the Target Architecture. This can be done by defining incremental deployment phases (e.g., Transition Architectures consisting of pilots or phased rollouts). Identifying risks and creating mitigation strategies is also critical to this phase and will help establish a change management plan to ensure smooth adoption. Finally, encourage a digital-first mindset by training employees to embrace digital tools and processes.

8. **Measure Compliance and Success (Phase G (Implementation Governance))**: Monitor and govern the execution of digital initiatives by defining governance frameworks to track project execution and compliance. Implement feedback loops to adjust digital strategies based on performance. It can be helpful to use Enterprise Architecture performance dashboards to track (real-time)

progress (an example is shown in Figure 12-54 in Chapter 12, Section 12.10.2.). This phase is about addressing security, legal, and compliance risks.

9. **Ensure Continuous Improvement (Phase H (Architecture Change Management)):** Adapt to changing business needs and emerging technologies. A simple but effective way to enable the organization to do this is to monitor new market trends and innovations. It is imperative to conduct regular Enterprise Architecture reviews to assess the impact of digital initiatives. Ensure that business and IT teams remain aligned with the organization's evolving goals and objectives. Implement a continuous innovation strategy to sustain digital growth.

I recently helped a hospital with its digital transformation.

The reason for this hospital's transformation had to do with government policy. A new theme, the right care in the right place, was imposed on all hospitals by the government. Hospitals also had to save more on the cost of providing care. This meant that, on the one hand, hospitals had to focus on supporting and providing the right care in the best place from the patient's perspective. On the other hand, costs had to come down. The answer to both problems lay in providing care in the home environment.

To meet these demands, the hospital had to drastically change its business model – and the operating model that flowed from it. Much more effort had to be put into digitalization and automation. On paper, the approach seemed straightforward. It looked like a simple series

CHAPTER 19 DIGITAL TRANSFORMATION

of steps to follow. In practice, however, the biggest challenge was getting the people in the hospital – the employees – on board.

You're probably familiar with the phrase "every change meets resistance", and the proposed change to transition to a more digital hospital was no different.

Digital transformation begins with a change in the organization's business model. The hospital realized that in order to provide services such as telemedicine, remote patient monitoring, and self-medication, it would need to change its business and operating model. Of course, the success of the transformation depended largely on the willingness of employees to actively participate. To this end, many conversations were held to identify employee reluctance and to properly articulate their concerns. This was essential, both to ensure that people felt heard and to gain insight into what the digital transformation needed to consider in order to be successful.

Over a period of one-and-a-half to two years, I was able to implement the transformation initiative using a phased approach (see Figure 19-3).

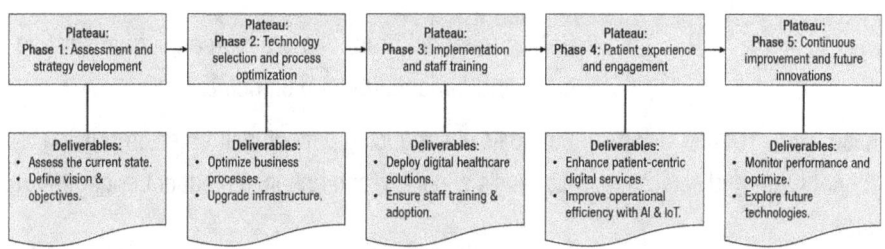

Figure 19-3. Five phases of realizing digital transformation

CHAPTER 19 DIGITAL TRANSFORMATION

The core driver for digital transformation came from a change in the hospital's strategy, due to government-imposed policies. The hospital needed to focus more on telehealth, remote patient monitoring, and self-medication. This alignment with strategy required a much more digital approach to existing capabilities, business processes, and workflow systems.

The first phase of the transformation included a readiness assessment of the current situation and the development of a strategy to implement the digital transformation (see Table 19-1).

Table 19-1. Phase 1: Assessment and Strategy Development (0–3 Months)

ADM Phase	Activity	Description
Preliminary Phase	Assess transformation readiness.	Evaluate existing IT infrastructure (EHR systems, telemedicine, patient management tools).
		Identify inefficiencies in patient care, billing, and administration.
		Conduct a cybersecurity risk assessment to ensure compliance with HIPAA or relevant regulations.
		Gather feedback from doctors, nurses, administrative staff, and patients about pain points.
Phase A: Architecture Vision	Define vision and objectives.	Set goals like reducing patient wait times, improving diagnosis accuracy, and enhancing patient engagement.
		Define KPIs (e.g., reduction in paperwork, increased telehealth adoption, faster billing cycles).
		Get leadership and key stakeholders (hospital board, department heads) aligned on the transformation vision.

Going through the steps of the first phase provided a clear picture of the application landscape and technology in use. The first phase also led to the mapping of processes in use and the identification of pain points. At the same time, it established the goals the hospital wanted to pursue with the transformation.

The second phase introduced new capabilities (see Figure 19-4), more automated processes, and the use of AI. This not only increased the efficiency of healthcare processes, but also enabled modern technologies to streamline processes.

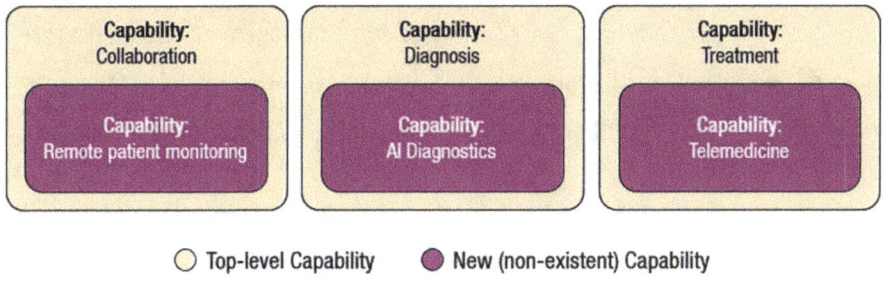

Figure 19-4. Introduction of new capabilities

The second phase was also used to determine the desired technology that could contribute to the hospital's stated goals. Selecting the most appropriate technology was part of this phase. In addition, this phase addressed the previously identified pain points in existing capabilities and business processes. By implementing the new technology, the existing operations could be significantly improved (see Table 19-2).

CHAPTER 19   DIGITAL TRANSFORMATION

Table 19-2. Phase 2: Technology Selection and Process Optimization (3-6 Months)

ADM Phase	Activity	Description
Phase B: Business Architecture Phase C: Information Systems Architectures	Optimize business processes.	Automate patient scheduling and billing to reduce manual paperwork. Introduce AI-based chatbots for answering patient queries and booking appointments. Implement digital HR tools for better workforce management and training.
Phase D: Technology Architecture	Upgrade infrastructure.	Implement or enhance electronic health record (EHR) systems for seamless data sharing. Adopt AI-driven diagnostic tools for faster and more accurate disease detection. Upgrade hospital network security to protect sensitive patient data.

Phase 3 went beyond automating existing processes and making them more efficient. This phase provided opportunities to achieve previously set goals. The use of appropriate remote care and support for devices that facilitate home care contributed positively to the achievement of goals. Training internal staff in the new way of working also had a positive impact on the overall operation of the hospital (see Table 19-3).

Table 19-3. *Phase 3: Implementation and Staff Training (6–12 Months)*

ADM Phase	Activity	Description
Phase E: Opportunities and Solutions	Ensure staff training and adoption.	Conduct digital literacy programs for doctors, nurses, and administrative staff.
		Organize workshops on cybersecurity best practices to prevent data breaches.
		Encourage change management by addressing resistance and demonstrating benefits.
Phase F: Migration Planning	Deploy digital healthcare solutions.	Launch telemedicine services to enable remote consultations.
		Implement wearable health monitoring devices to track patient vitals remotely.
		Integrate AI-powered Clinical Decision Support Systems (CDSS) for doctors.

The third phase brought new opportunities to better serve patients by eliminating the need for them to come in for each visit. The focus on providing the right care in the right place led to a huge increase in patient satisfaction.

Phase 4 provided patients with information services that were available 24 hours a day. Virtual assistants using AI technology were made available with increasing frequency. The virtual assistants handled consultations for patients who did not require immediate acute care. As an added benefit, physicians were able to step back and focus more on patients who needed longer-term or acute care (see Figure 19-5).

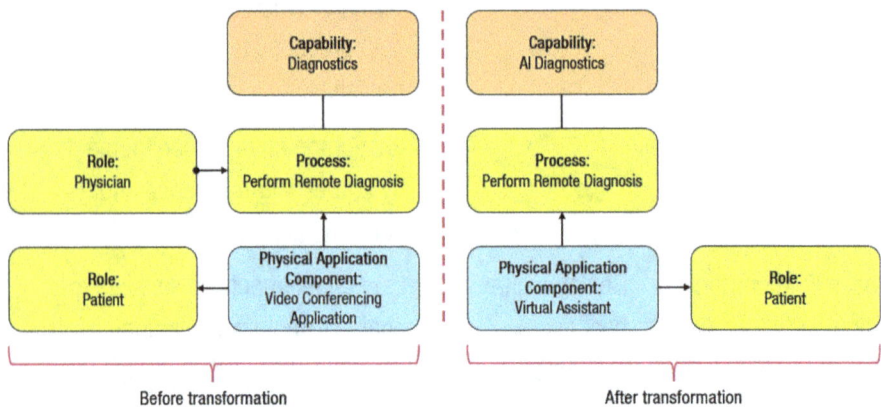

Figure 19-5. Deployment of new capability eliminates need for physician role

A second benefit of using virtual assistants was that care could be provided regardless of location or time. This was in line with government policy (the right care in the right place). Table 19-4 lists the activities performed during phase 4.

CHAPTER 19 DIGITAL TRANSFORMATION

Table 19-4. *Phase 4: Patient Experience and Engagement (12–18 Months)*

ADM Phase	Activity	Description
Phase G: Implementation Governance	Enhance patient-centric digital services.	Introduce patient portals for easy access to medical records and test results.
		Enable AI-driven virtual assistants for answering patient queries 24/7.
		Implement mobile health apps to help patients track medications and appointments.
	Improve operational efficiency with AI and IoT.	Use IoT-based smart beds to monitor patient vitals in real time.
		Deploy AI-powered predictive analytics for managing hospital inventory and predicting patient inflow.
		Implement Robotic Process Automation (RPA) for faster insurance claim processing.

In Phase 4, internal processes were augmented with far-reaching transformation techniques, including the use of AI to manage the hospital's inventory and predict patient flow. This allowed the hospital to better prepare for peak periods (think flu season) and ensure that sufficient resources were available during these busy times.

775

CHAPTER 19   DIGITAL TRANSFORMATION

The final phase focused on further optimizing the adjustments that had been made. Targets and associated KPIs were closely monitored so that adjustments could be made in a timely manner (see Table 19-5).

Table 19-5. Phase 5: Continuous Improvement and Future Innovations (18+ Months)

ADM Phase	Activity	Description
Phase H: Architecture Change Management	Monitor performance and optimize.	Track KPIs (e.g., reduced patient wait times, increased telehealth usage, improved diagnosis accuracy).
		Gather feedback from healthcare professionals and patients for continuous improvement.
		Regularly update software and security protocols.
	Explore future technologies.	Implement blockchain for medical records to enhance security and interoperability.
		Explore AI-driven drug discovery and personalized treatment plans.
		Consider Augmented Reality (AR) and Virtual Reality (VR) for surgical training and patient rehabilitation.

Phase 5 has ensured that the hospital no longer views the use of technology as an obstacle. It is more receptive to using new technologies to further streamline daily processes and operations.

By undertaking a digital transformation, the hospital has been able to reduce its healthcare costs by 30%. The hospital owes this to the fact that a large proportion of people requiring non-acute care have been moved to home care. The adjustments made as part of the digital transformation – think telehealth solutions, remote patient monitoring, and self-medication – have ensured that the hospital has been able to achieve its goals.

Ultimately, the implementation steps described above have ensured that the hospital has not only embraced digital transformation, but also convinced staff that it contributes to improved patient care, operational efficiency, and data security.

By following the TOGAF Standard and its Architecture Development Method, organizations can systematically execute a digital transformation while ensuring business–IT alignment, risk mitigation, and scalability.

19.4. Summary

Digital transformation is a complex process. It requires careful planning and execution. By leveraging the TOGAF Standard, a structured approach to digital transformation following the Architecture Development Method can be developed.

The approach to performing a digital transformation consists of the following steps:

- **Outline the Digital Vision:** Identify the key reasons for pursuing digital transformation. Link these drivers to the organization's goals. Set specific, measurable goals and determine how to track progress. Consider technologies that can support the goals. Create a high-level plan that shows how the Business, Data, Application, and Technology Architectures will evolve. Ensure that all stakeholders are aligned with this vision.

- **Get Ready for Change:** Start by identifying the key people in the organization who care about digital transformation. Set clear boundaries for what the transformation will encompass. Define principles and assess how prepared the organization is for this change. Establish a team and governance structure to guide the transformation.

- **Rethink Business Automation:** Examine current business processes and compare them to the Target Architecture. Look for gaps and opportunities to redesign processes using digital tools and automation. Define new roles and responsibilities that align with digital goals. Integrate services and develop a roadmap that prioritizes initiatives to guide business transformation.

- **Build the Digital Framework:** Establish policies that ensure security and compliance. Identify which applications are needed for transformation. Review the current landscape and plan for modernization.

- **Deploy the Right Technology:** Choose a strategy that fits the organization's needs.

- **Plan the Journey:** Identify the projects that will drive the transformation and evaluate them based on their business value and feasibility. Develop a roadmap that includes short- and long-term projects.

- **Manage the Transition:** Create a structured plan for moving from the current state to the desired future state. Implement changes in phases and identify potential risks. Develop mitigation strategies. Establish a change management plan to ensure adoption of new processes and technologies. Provide training programs to help employees adapt to new digital tools and workflows.

- **Monitor Progress and Compliance:** Establish governance frameworks to oversee the execution of digital initiatives and ensure compliance with relevant standards. Implement feedback mechanisms to adjust strategies. Use performance dashboards to track progress and address any security, legal, or compliance risks that arise.

- **Continue to Improve:** Stay abreast of new market trends and emerging technologies. Conduct regular Enterprise Architecture reviews. Evaluate the impact of digital initiatives. Ensure that business and IT teams remain aligned with evolving business goals. Implement a continuous innovation strategy to sustain and enhance digital growth.

The TOGAF Standard and its Architecture Development Method provide a robust approach for organizations to systematically execute a digital transformation initiative.

CHAPTER 20

Putting the TOGAF Standard into Practice

 This chapter is part of the GREEN and GOLD reading tours. See Figure 1-1 in Chapter 1, Section 1.3, for alternative reading tours.

The TOGAF Standard describes in great detail the complete approach to an architecture implementation. All related aspects, such as adapting the Enterprise Metamodel and Content Framework, setting up the Architecture Repository, establishing architecture governance, and applying various techniques, are also richly described in the framework.

Applying all of this information in day-to-day practice still presents a challenge or two. To demonstrate that this need not be a showstopper, this chapter describes a case scenario in which all aspects of an architecture implementation are examined and executed using the Architecture Development Method. The case follows Jim, an Enterprise Architect for an organization called Academic Diagnostic Medicine.

CHAPTER 20   PUTTING THE TOGAF STANDARD INTO PRACTICE

20.1. A Case Scenario

Academic Diagnostic Medicine is a modern healthcare institution committed to delivering high-quality, patient-centered care while ensuring operational excellence and financial sustainability. In response to growing demands for value-based healthcare, the hospital has defined a strategic agenda for 2028-2030 that includes *improving patient satisfaction, optimizing internal workflows,* and *ensuring long-term financial health.*

To effectively realize this strategy, the hospital is engaging Jim, the Enterprise Architect it employs, to guide the transformation. Jim's role is to translate strategic drivers into executable capabilities, align business and IT domains, and ensure the change is planned and implemented in a structured, traceable, and value-driven manner.

Academic Diagnostic Medicine has been used to working with architecture for several years. As a result, the healthcare organization has a fairly high level of architecture maturity, as evidenced by the various strategic meetings Jim has had with the board. The TOGAF Standard has been the preferred standard since the introduction of working with architecture. Architecture work is done in accordance with the framework. Academic Diagnostic Medicine faces a number of challenges, and the board has decided that following and executing its newly defined strategy will allow them to realize their business goals.

20.1.1 Scoping the Work

March 2027-April 2027

During two strategy meetings with the board of directors, Jim learned about the plans that had been drawn up. Now, he has a clear understanding of what the upcoming period will entail. He visualizes the strategy using a Driver/Goal/Objective Diagram (Figure 20-1).

CHAPTER 20 PUTTING THE TOGAF STANDARD INTO PRACTICE

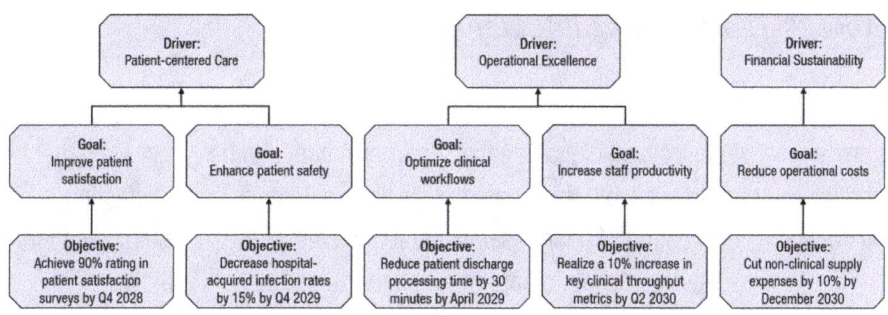

Figure 20-1. *Strategy of Academic Diagnostic Medicine*

The visualization suggests that implementing three major architectural projects can achieve the strategy. These projects appear to focus on the drivers and aim to improve patient satisfaction and safety, optimize clinical workflows and staff productivity, and reduce operational costs.

During a follow-up meeting with the board two weeks later, Jim explains the diagram he created earlier. He shares his vision for a phased, project-based approach, planning to divide the strategy into three projects.

The *improve patient satisfaction* project will focus on improving patient satisfaction and safety. The *improve clinical workflows* project will address improving workflow processes and will indirectly improve staff productivity. The third project will focus on *reducing operating costs*. The board recognizes the strategy and formally approves the diagram. The board also provides a set of requirements for the work. Jim records these in a Requirements Catalog (see Table 20-1).

783

Table 20-1. Requirements Catalog

Project	Requirement	Goal
Improve patient satisfaction	A centralized digital feedback platform needs to be made available that enables real-time collection and routing of patient satisfaction and safety-related feedback to appropriate departments.	Improve patient satisfaction. Enhance patient safety.
Optimize clinical workflows	Clinical workflows must be digitized and integrated into the hospital's EHR system, allowing automated role-based task assignments.	Optimize clinical workflows. Increase staff productivity.
Reduce operational costs	A cost-tracking dashboard must be implemented that links departmental spending to patient outcomes, enabling quarterly reviews of cost-effectiveness.	Reduce operational costs.

Jim directly links the requirements to the goals of Academic Diagnostic Medicine. This makes the requirements for achieving the goals clear to everyone at all times.

During the implementation of the strategy, Jim anticipates that a number of risks will arise that need to be identified. Jim is therefore considering adding the risk entity to the Enterprise Metamodel used by Academic Diagnostic Medicine. The inclusion of this new entity will allow Jim to define and capture risks in a Risk Catalog. From this artifact, they can be visualized as needed.

20.1.2 Envisioning the Architecture

May 2027–June 2027

The board's approval of the strategy visualization is the green light for Jim to begin his first inventory. He starts talking to different people in the organization. The reason for this is to get as complete a picture as possible of the work to be done. To keep track, Jim creates an initial Stakeholder Catalog (see Table 20-2). This artifact initially contains the names, roles, and classifications of the stakeholders with whom he will interact.

Table 20-2. Stakeholder Catalog

Stakeholder	Classification
Maria Summers (CEO)	Keep satisfied
Chris West (CIO)	Key player
James Noble (CFO)	Keep satisfied
Anna Carter (Manager HR)	Minimal effort
Susan Delgado (CNO[1])	Key player
Benjamin Lee (CMIO[2])	Key player
Elena Petrova (Medical Director of Quality and Safety)	Key player
Kevin O'Brien (Patient Experience Manager)	Key player
Hannah Duarte (Charge Nurse/Unit Manager)	Keep informed

---

[1] Chief Nursing Officer.

[2] Chief Medical Information Officer.

Jim talks to everyone directly involved. In addition to the board of directors, this includes various executives, clinical staff, and various operational teams. Jim wants to understand how the strategy is perceived at the operational level. Gradually, a picture emerges of the best way to communicate with the various stakeholders. Every time Jim finds a way that works well, he records it in the Communications Plan. He can use this plan throughout the execution of the work to communicate with all stakeholders. The different conversations reveal the stakeholders' concerns and constraints (see Table 20-3).

Table 20-3. Strategy concerns and constraints

Goal	Concern	Constraint
Improve patient satisfaction		Low digital literacy for older patients
Enhance patient safety	Will feedback loops criticize care rather than improve it?	
Optimize clinical workflows	Will workflow changes require system re-training?	
Increase staff productivity		Resistance to process standardization
Reduce operational costs	How fast will cost savings materialize?	

The concerns and constraints are all things he needs to consider. Jim adds the concerns to the Stakeholder Catalog that he created earlier.

During one of Jim's weekly board meetings, he provides feedback on the status of the process. He indicates that he now has an understanding of the stakeholders involved and their concerns and constraints. All of this information, along with the requirements established by the

CHAPTER 20 PUTTING THE TOGAF STANDARD INTO PRACTICE

board, is captured in a Statement of Architecture Work. This document contains a high-level overview of the work, in addition to the components mentioned above.

20.1.3 Starting the Development

July 2027

To get a good idea of what exactly needs to be done, Jim turns to the existing content of the Architecture Repository. This helps him determine what the current situation looks like. A quick look at the documented material from previous projects provides building blocks in the form of a Business Capability Map (Table 20-4) and a Capability/Business Process Matrix (Table 20-5) that can be reused.

Table 20-4. Business Capability Map

Capability (Level 1)	Capability (Level 2)	Description
Patient Engagement	Patient Feedback Management	Ability to collect, analyze, and respond to feedback
Clinical Services	Clinical Workflow Optimization	Ability to manage patient care workflows efficiently
	Workflow Coordination	Ability to coordinate patient care workflows
Operations	Resource Scheduling	Ability to align staff and resource schedules
Financial Management	Cost Reduction Management	Ability to manage cost effectively

The Capability/Business Process Matrix is a tailored artifact that he created in the past. It contains an overview of the existing processes in the organization that are related to the capabilities.

CHAPTER 20 PUTTING THE TOGAF STANDARD INTO PRACTICE

Table 20-5. Capability/Business Process Matrix

Capability (Level 2)	Business Process
Patient Feedback Management	Feedback Collection
Clinical Workflow Optimization	Optimized Discharge Planning
	Surgical Workflow Coordination
Resource Scheduling	Staff Roster Optimization
	Equipment Usage Scheduling
Cost Reduction Management	Procurement

Jim uses the space provided for this task to get a good sense of Academic Diagnostic Medicine's strategy. He creates a Strategy/Capability Diagram (see Figure 20-2) by combining the Driver/Goal/Objective Diagram with the Business Capability Map. This new artifact ensures that he has a talking point when discussing the impact of the strategy within the organization.

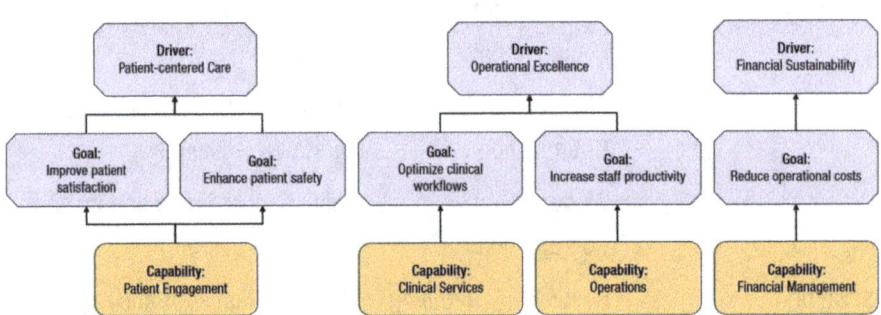

Figure 20-2. Strategy/Capability Diagram

The relationship between strategy and capabilities makes it easier for many stakeholders to understand where the strategy impacts the organization and what it means for them. This allows Jim to show at a glance where the challenges lie in the context of the strategy.

CHAPTER 20 PUTTING THE TOGAF STANDARD INTO PRACTICE

All the information gathered so far is added to the Statement of Architecture Work document. Jim has this document formally approved by the board and then shares it with the relevant stakeholders. They also sign off on the document.

August 2027–October 2027

With all the available and retrieved information in hand, Jim gets to work. His first goal is to establish the Baseline Business Architecture for each of the identified capabilities. The baseline consists of a number of pre-existing building blocks. These include capabilities, business processes, and the associated organizational structure. By iteratively going through the architecture process, Jim is able to establish the Baseline Architecture for each of the identified capabilities. One of the tools he uses is a Business Capability Map. Using this artifact, he will be able to easily identify where pain points (such as missing capabilities) exist in the current situation. The missing capabilities are shown in the Business Capability Map in Table 20-6.

Table 20-6. Business Capability Map listing missing capabilities

Capability (Level 1)	Missing Capability (Level 2)	Description
Patient Engagement	Feedback Analysis	Ability to analyze patient satisfaction feedback
Financial Management	Supply Chain Management	Ability to manage and rationalize the supply chain

The lack of Level 2 capabilities for Patient Engagement and Financial Management indicates that there are no processes in place to analyze feedback and streamline the supply chain. The missing processes are shown in Table 20-7.

Table 20-7. Capability/Business Process Matrix listing missing processes

Capability (Level 2)	Missing Process
Patient Feedback Management	Feedback trend analysis
	Infection prevention audit
Cost Reduction Management	Supply chain rationalization

Jim's mapping of existing roles in the organization, using an Organization/Business Roles Catalog (see Table 20-8), shows that there are currently no appropriate roles to perform the new processes. Jim makes a note to himself to bring this issue to the attention of HR. Perhaps one or more vacancies will need to be advertised to fill the current staffing gap.

Table 20-8. Organization/Business Roles Catalog

Organization Unit	Business Role	Status of Role
Board of Directors	Chief Executive Officer	Existing
Finance	Chief Financial Officer	Existing
	Financial Analyst	Non-existent
Operations	Chief Operations Officer	Existing
IT Department	Chief Information Officer	Existing
Quality and Safety	Medical Director	Existing
	Patient Safety Officer	Non-existent
	Patient Experience Manager	Existing
Clinical Operations	Chief Nursing Officer	Existing
	Chief Medical Information Officer	Existing
	Scheduling Coordinator	Existing
	Charge Nurse/Unit Manager	Existing

Once the Baseline Business Architectures have been established, Jim can use the established strategy to determine what the Target Architectures should look like. Again, the building blocks from the Architecture Repository come in handy. Jim recognizes the importance of aligning the newly created business capabilities with the goals of Academic Diagnostic Medicine. To convince the board of the need, he uses a Capability Heat Map (Figure 20-3), which clearly shows which capabilities are working properly and which are not. The artifact also shows where new capabilities are needed.

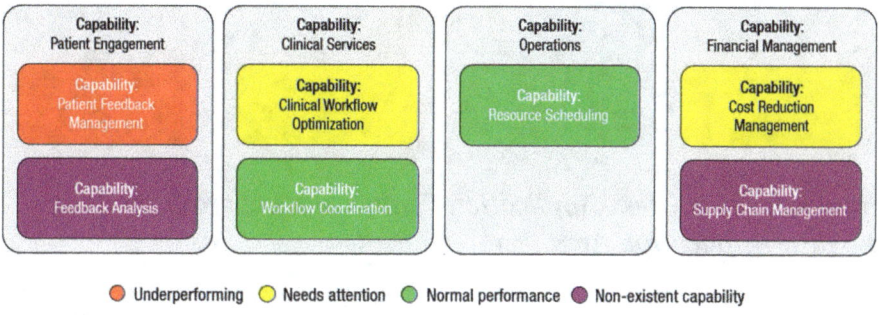

Figure 20-3. Capability Heat Map

The Patient Feedback Management capability, in particular, requires attention. It is underperforming and must be addressed. The Clinical Workflow Management and Cost Reduction Management capabilities will also need to be examined more closely and improved where possible. This can be achieved by automating the processes associated with these capabilities. Currently, the Clinical Workflow Management capability uses manually performed tasks by a specific business role. The other capability seems to focus exclusively on order processing. Much less attention is paid to mapping the entire supply chain. To address this, two new capabilities, Feedback Analysis and Supply Chain Management, will have to be created. These new capabilities will support Patient Engagement and Financial Management, respectively.

After mapping the capabilities, Jim focuses on the Baseline and Target Architectures for the Application and Technology segments. He uses a Process/Application Realization Diagram (see Figure 20-4) to illustrate the current connection between the application landscape and business processes.

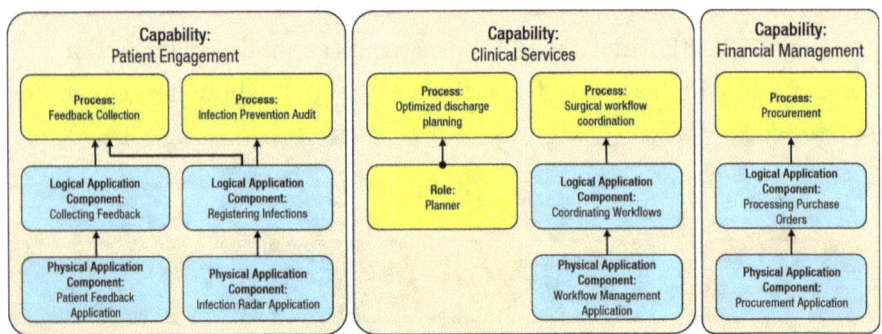

Figure 20-4. *Process/Application Realization Diagram showing Baseline Application Architecture*

As the diagram shows, there is currently no functionality to analyze incoming feedback. Incoming feedback is simply stored in a system. No analysis is performed on it. The board of directors has indicated the need for a central digital feedback platform that can process and route patient satisfaction and safety-related feedback in real time to the relevant departments.

Within the Clinical Services capability, discharge planning is currently managed manually by the Planner business role. One of the board's requirements indicates that this process should be automated.

For the Financial Management capability, the current method of processing purchase orders is inadequate. A cost-tracking dashboard must be made available to provide insight into the costs incurred by each department. To this end, the supply chain will also need to be examined.

CHAPTER 20 PUTTING THE TOGAF STANDARD INTO PRACTICE

Based on the stated requirements, Jim creates a visualization of the possible Target Application Architecture (see Figure 20-5). He fleshes out the requirements by creating an application function that analyzes incoming feedback. Jim also replaces the manual process within the Clinical Services capability with an automated variant that provides real-time mapping of departmental expenditures. The Supply Chain Management Application and the Managing Supply Chain application function address this.

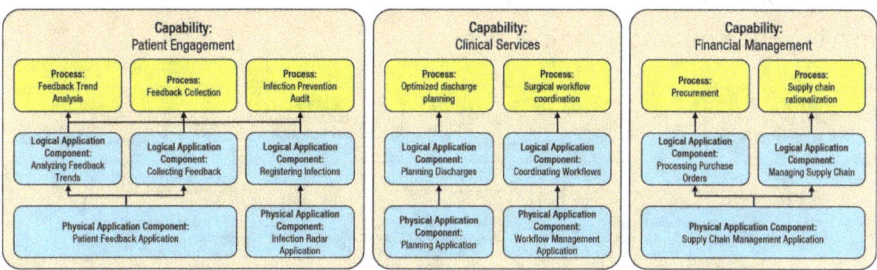

Figure 20-5. Process/Application Realization Diagram showing Target Application Architecture

Jim was recently informed by the COO that no changes will be made to the existing technical infrastructure. Academic Diagnostic Medicine invested heavily in a cloud-based setup last year and migrated all its services to it. There is no budget to make adjustments. Therefore, Jim will have to adapt the architecture to the infrastructure and services currently available.

Jim draws up an initial version of the Architecture Definition Document. In it, the main objectives and scope of the architecture are clearly defined. He makes sure the document contains detailed descriptions of the Baseline and Target Architectures.

20.1.4 Setting the Direction

November 2027

Of course, the transition from the Baseline Architecture to the Target Architecture will not happen all at once. Jim assumes that two Transition Architectures will be needed as intermediate phases. He creates a Plateau Planning Diagram, shown in Figure 20-6, in which the Baseline, Transition, and Target Architectures are visible. He also uses the diagram to visualize the gaps between the architectures. These gaps will need to be closed by carrying out a number of initiatives.

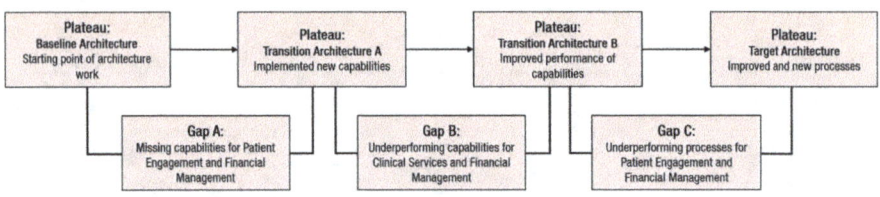

Figure 20-6. *Plateau Planning Diagram*

Closing the gaps requires multiple work packages. Work packages consolidate well into projects, which is an advantage. With the help of a project leader, the required activities can be fleshed out. To explain these activities to the project leader, Jim compiles a Work Package Portfolio Map (see Table 20-9). In this artifact, he lists all the work packages, the gaps they close, the associated initiatives (activities), and the goals to which the work packages are linked.

Table 20-9. Work Package Portfolio Map

Work Package Name	Gap	Activity	Associated with Goal(s)
Implement new capability for Patient Engagement	A	Implement new capability Feedback Analysis.	Improve patient satisfaction.
Implement new capability for Financial Management	A	Implement new capability Supply Chain Management.	Reduce operational costs.
Improve capability for Clinical Services	B	Improve capability Clinical Workflow Optimization.	Optimize clinical workflows.
Improve capability for Financial Management	B	Improve capability Cost Reduction Management.	Reduce operational costs.
Implement and improve processes for Patient Engagement	C	Implement process Feedback trend analysis.	Improve patient satisfaction.
	C	Improve process Feedback collection.	
	C	Implement process Infection prevention audit.	
Improve processes for Financial Management	C	Improve process Supply chain rationalization.	Reduce operational costs.

Jim begins to build a first draft of the Architecture Roadmap based on the gaps and corresponding initiatives captured in work packages. These work packages form the basis of the draft roadmap and are a welcome addition. Jim understands that the roadmap does not need to be finalized at this stage of the architecture development process. It will evolve as the architecture work progresses. New business capabilities and processes are added to the roadmap in the form of work packages. Jim pays close attention to aligning these components with organizational goals and objectives. Jim takes extra care to map the work packages to the timeframes mentioned in the corresponding objectives.

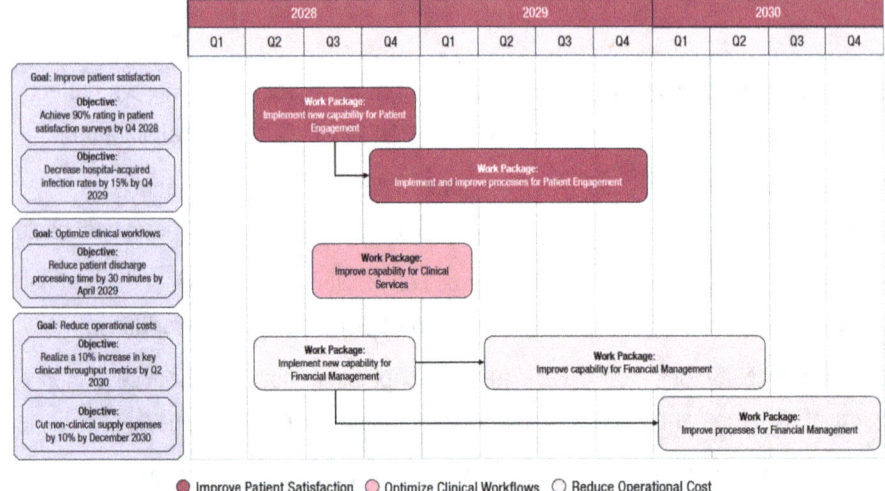

Figure 20-7. Draft Architecture Roadmap containing work packages

Figure 20-7 illustrates the relationship between the work packages and the goals and objectives. The colors indicate which work packages belong to which project.

Armed with the roadmap, Jim visits the HR department. He discusses how the absence of certain roles complicates the realization of the strategy. These roles include the newly created positions of Patient Safety Officer in the Quality and Safety Department and Financial Analyst in the Finance Department. After consulting with an HR business partner, Jim determines that filling these roles is necessary. The HR business partner takes the initiative to contact the board of directors and coordinate posting the needed vacancies. Jim will be informed of the outcome.

CHAPTER 20 PUTTING THE TOGAF STANDARD INTO PRACTICE

December 2027

One of Jim's next steps is to determine the impact of the architecture work on the existing one. He meets with the project manager, who is already aware of the impending work. After a brief discussion, they conclude that there is currently no reason to believe the new architecture will interfere with ongoing projects, or vice versa.

Over the next week, Jim records all the information he has gathered into a rough draft of the Architecture Definition Document. He also uses this time to schedule appointments with various stakeholders. He wants to explain the upcoming work so that everyone involved has a clear understanding of what is to come.

During the group sessions, Jim discusses the current state of affairs and the direction the organization has chosen. He explains that the recorded Target Architecture will be realized using Plateau Planning, a phased approach. Several business stakeholders, executives, process owners, and subject matter experts are present during the work sessions. Their insights are invaluable in shaping an architecture that reflects the organization's needs and goals.

Jim reviews the requirements provided by the board in front of the stakeholders. After the session, all stakeholders agree that these should be the criteria for the architectural work.

Jim determines the necessary building blocks, or capabilities, based on the information available to him. Then, he performs a final cross-check to ensure that the architecture aligns with the organization's overarching business goals. This alignment is vital for the architecture's effectiveness and relevance. The reasoning behind every decision made during this process is documented. This practice maintains a clear link between architectural choices and organizational objectives.

The Baseline, Transition, and Target Architectures are described in detail and added to the Architecture Definition Document. The same goes for the gaps between the architecture states. The Driver/

Goal/Objective Diagrams, Business Capability Map, and Capability/Business Process Matrix are also included in the Architecture Definition Document. The document also includes the Plateau Planning Diagram and the Process/Application Realization Diagrams for the Baseline and Target Architectures. A draft version of the Architecture Roadmap is also incorporated. These diagrams, map, and matrix effectively communicate key aspects of the architecture. Visual representations simplify complex information, making the document more accessible to stakeholders.

After finishing the first draft of the document, Jim shares the Architecture Definition Document with key stakeholders to gather their insights and feedback. He then uses this feedback to make the necessary adjustments to the document. After two short iterations, everyone agrees that the written plan is what needs to happen. Any concerns raised by stakeholders have been addressed, and everyone feels that this approach will align with organizational objectives.

20.1.5 Getting Ready for Implementation

January 2028–February 2028

The next step in the architecture process is creating an Implementation and Migration Plan. To do so, Jim must first identify any existing factors that could impact the execution of the architectural work. Only three factors that will potentially affect the work emerged during the preparation of the Implementation Factor Catalog (see Table 20-10).

Table 20-10. Implementation Factor Catalog

Factor Category	Factor	Description	Deduction
Effect (impact)	Change in business process	Automation of discharge planning	New application will automate work done by hand.
Action	Introduction of new Planning Application	Implementation of new application	Planner role needs to be trained to use the new application.
Assumption	Introduction of Feedback Analysis application function	Implementation of new application functionality	Feedback analysis • may result in both negative and positive feedback • needs to be able to distinguish between both types of feedback

Implementing the new Planning Application will replace the Planner role, which is currently responsible for manually executing the process, with an automated variant. This means that those filling the role will need to be trained to use the new application. Additionally, the new application's functionality will provide greater visibility into feedback received from patients. This may include criticism, in addition to positive feedback. The Patient Feedback Application must be able to distinguish between these two types of feedback. Jim realizes that the three risks that emerged from the deduction of the Implementation Factor Catalog may affect the implementation of the Target Architecture. He makes a note to himself to properly identify and clearly visualize the risks. There is an opportunity to do this when determining the business value of the projects to be implemented. Fortunately, he included risk as an entity in the Enterprise Metamodel at an earlier stage.

To adequately flesh out the work to be performed and establish a sound architecture, Jim reviews the earlier noted concerns and constraints.

One of the identified constraints is the low digital literacy of older patients. Fortunately, little to nothing is changing for patients at this time. The changes are more on the back end of the process. The Target Architecture will change how incoming feedback is processed and analyzed. This will address the aforementioned constraint.

A second constraint is the possible resistance to process standardization. This mainly concerns optimizing the clinical workflow process. Since the current manual process will be automated and a new application will be introduced for this purpose, it is crucial to provide thorough training to users of the new application. Once users feel confident using the application, their resistance to standardizing the process will diminish, if not disappear.

The risk analysis should address the concern that the feedback process will generate more criticism than valuable information. Although there is little change on the patient side, negative feedback is still a possibility. If the analyses show this, Academic Diagnostic Medicine will have to take action.

The concern about whether staff training is needed is also valid. However, this can easily be overcome by providing training for the Planner role, which will work with the newly implemented application. Other than that, no customization is required.

Finally, the concern about how quickly cost savings will be visible cannot be addressed at this time. In any case, cost savings are not the primary objective. First, the board of directors wants insight into costs per department. Cost reduction is not planned until the end of 2030.

After ensuring that the concerns and constraints have been adequately addressed, Jim prepares an initial draft of the Implementation and Migration Plan, which is presented in Table 20-11. He divides the plan into three phases corresponding to the two Transition Architectures and the Target Architecture.

Table 20-11. Implementation and Migration Plan

Phase	Timeframe	Work Packages	Key Milestones	Dependencies
1	Q2 2028 to Q4 2028	Implement new capability for Patient Engagement.	Implemented new capability	–
		Implement new capability for Financial Management.	Implemented new capability	–
2	Q3 2028 to Q2 2030	Improve capability for Clinical Services.	Improved performance of capability	Staff training complete
		Improve capability for Financial Management.	Improved performance of capability	–
3	Q4 2028 to Q4 2030	Implement and improve processes for Patient Engagement.	Improved and new process	Implemented new capability for Patient Engagement
		Improve processes for Financial Management.	Improved process	Improved capability for Financial Management

Jim uses the Plateau Planning Diagram (Figure 20-6) to define the phases. The Work Package Portfolio Map (Table 20-9), which he created earlier, comes in handy. In it, the work packages are neatly grouped and described, including the gaps that will be closed by executing the activities. The goals that the work packages will accomplish are also listed.

CHAPTER 20 PUTTING THE TOGAF STANDARD INTO PRACTICE

Since the work packages and their associated activities relate to the gaps, it is easy to identify the work packages' dependencies. For example, Jim emphasizes the importance of carefully implementing the two new capabilities for Patient Engagement and Financial Management (Feedback Analysis and Supply Chain Management, respectively). These new capabilities are prerequisites for later steps. In the Implementation and Migration Plan, Jim takes this into account.

During an earlier meeting with the project leader, it was revealed that there are currently no ongoing projects that would prevent the implementation of the architecture. Nevertheless, to avoid any surprises, Jim contacts the project leader again. During their conversation, they confirm again that there are no overlapping projects that could limit the implementation of the new architecture.

Now that everything seems to be in order, Jim conducts a Business Transformation Readiness Assessment (see Table 20-12). This assessment helps Jim determine if there are any roadblocks to the organization's readiness for implementation.

Table 20-12. Business Transformation Readiness Assessment

Readiness Factor	Assessment Criteria	Score (1–5)
Leadership commitment	Level of commitment from top leadership	5
Employee engagement	Willingness and enthusiasm of employees for change	3
Organizational culture	Openness to new ideas and adaptability	5
Skills and competencies	Availability of required skills and competencies for the transformation	3
Risk Management	Identification and mitigation of potential risks	4
Stakeholder alignment	Alignment of stakeholders' interests and expectations	5

The assessment reveals that both employee engagement and the necessary skills and competencies require attention. This is likely related to the automation of tasks that are currently performed manually. Specifically, the *optimized discharge planning* process within the Clinical Services capability will be automated. After the implementation of the architecture, specialized skills will be required to run these tasks through the automated process. Training the relevant personnel is one possible solution.

Jim schedules another meeting with the relevant stakeholders and presents the migration plan. A few questions are asked about the order of the planned activities, which Jim can easily answer using the Plateau Planning Diagram and the Work Package Portfolio Map to explain how the order was established. The stakeholders unanimously approve the Implementation and Migration Plan.

Later, Jim runs into the HR business partner he spoke with a few weeks ago. She was just on her way to see him with good news. The positions for Patient Safety Officer and Financial Analyst have been filled.

March 2028–May 2028

Jim asks the project manager overseeing the execution of the work to incorporate relevant parts of the Implementation and Migration Plan into the appropriate project management framework. Jim also creates a Business Value and Risk Matrix to illustrate the business value of the projects to be implemented (see Figure 20-8).

CHAPTER 20   PUTTING THE TOGAF STANDARD INTO PRACTICE

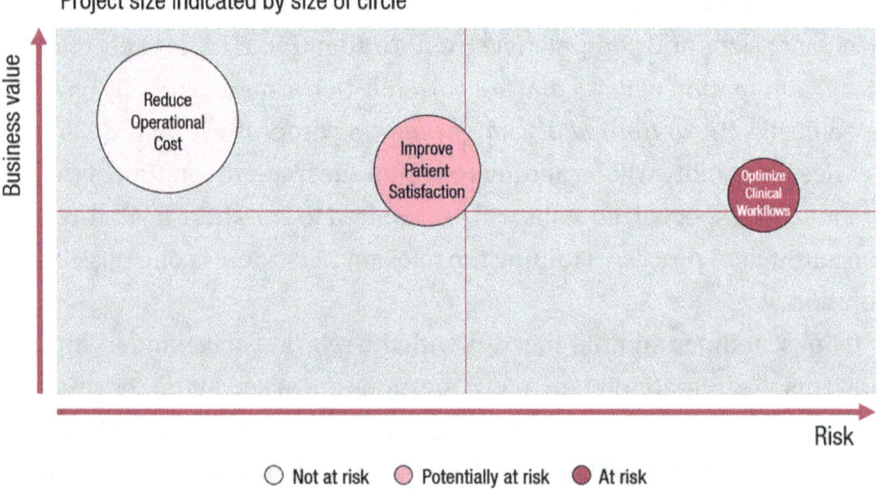

Figure 20-8. Business Value and Risk Matrix

The matrix shows the three previously defined projects: reduce operational costs, improve patient satisfaction, and optimize clinical workflows. Jim captures the risks he noticed earlier when creating the Implementation Factor Catalog in a Risk Catalog. Then, he plots the risks on the projects. The project with the highest risk is the *optimize clinical workflows* project. This is because it involves automating tasks that are currently performed manually. Those responsible for executing these tasks will need to reinterpret the Planner role. They will also need to undergo training to use the Planning Application correctly. The *improve patient satisfaction* project has a significantly lower risk because the only question is whether making feedback more available will lead to more criticism.

Jim combines the previously created Risk Catalog with the Process/Application Diagram and uses a Risk/Process/Application Realization Diagram to visualize the risks, as shown in Figure 20-9.

CHAPTER 20   PUTTING THE TOGAF STANDARD INTO PRACTICE

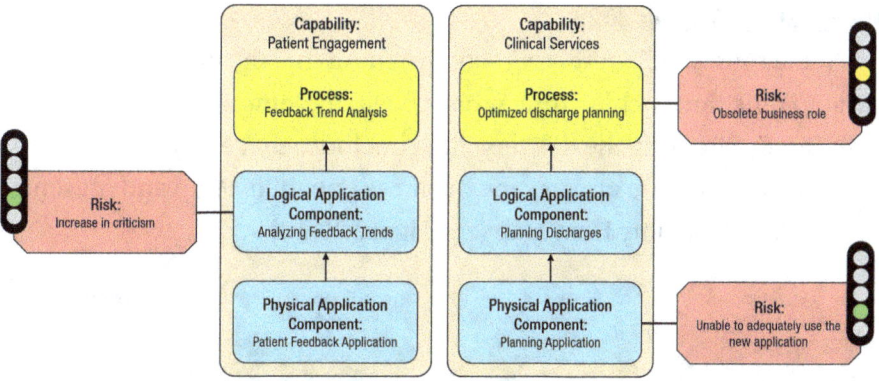

Figure 20-9. Risk/Process/Application Realization Diagram

The figure shows the two capabilities for which one or more risks have been identified. These risks have been assigned to the corresponding building blocks of the planned architecture. Each risk has also been assigned a score, which has been visualized using a traffic light. The five color levels in the graphic represent the five risk levels used in the Risk Management technique. Jim draws up a set of risk mitigation measures and submits them to the relevant stakeholders. Along with the latest versions of the Architecture Roadmap and the Implementation and Migration Plan, Jim obtains formal approval for the work to be performed.

20.1.6 Overseeing the Architecture Work

June 2028–February 2029

The work is then carried out according to the terms of the Architecture Contract. Jim supervises the execution of activities and the implementation of the overall architecture. He regularly performs random checks to ensure that the agreements are still being met. He shares the Architecture Compliance Report with the involved stakeholders so they are well-informed.

March 2029–August 2029

Some time later, Jim receives confirmation that the new capabilities, part of Transition Architecture A, have been implemented and improved. After briefly checking the delivery, Jim concludes that everything went according to plan. He updates the Business Capability Map and presents an updated Capability Heat Map (see Figure 20-10).

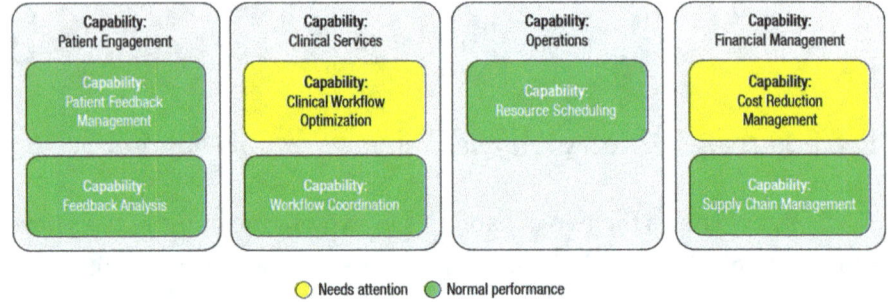

Figure 20-10. Updated Capability Heat Map

The Capability Heat Map shows the addition of Feedback Analysis and Supply Chain Management capabilities. Both capabilities are operating satisfactorily. Furthermore, implementing the former capability has positively impacted Patient Feedback Management.

20.1.7 Maintaining Control

September 2029–July 2030

Jim regularly meets with the most involved stakeholders and asks them about the state of affairs using questionnaires. This gives him a good picture of how the architecture is being implemented. This also enables him to make adjustments where necessary. Based on the performance reviews he conducts, Jim effectively manages the architecture change management process.

Around January 2030, Jim receives a message from the board of directors. They inquire about the effectiveness of his work and ask him to explain it in person. To prepare for the upcoming consultation, Jim uses the Driver/Goal/Objective Diagram with the heat mapping technique (see Figure 20-11). This technique allows him to easily visualize where progress has been made and where attention is still needed in the coming period.

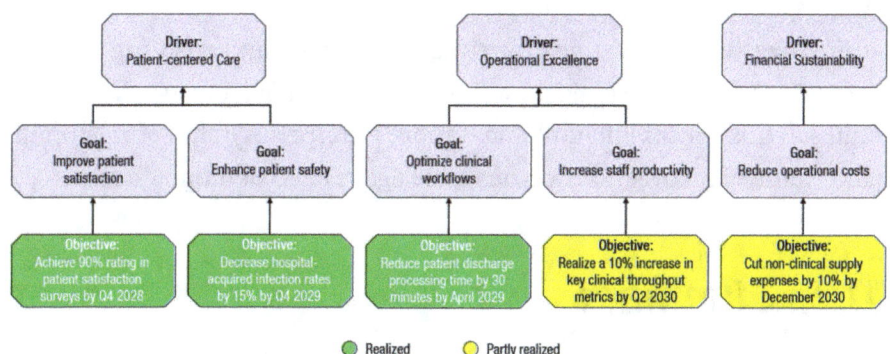

Figure 20-11. Driver/Goal/Objective Diagram Heat Map

The heat map of the Driver/Goal/Objective Diagram shows that not all objectives have been achieved yet. In particular, increasing key clinical throughput and reducing non-clinical supply expenses have not yet been fully realized. This is not surprising, since the year 2030 has not yet ended. Nevertheless, Jim is conducting an investigation to determine if any unforeseen issues have arisen.

The investigation quickly revealed that the learning curve for the new Planning Application used in Clinical Operations was steeper than initially assumed. This delayed the adoption of the new way of working. Consequently, the goal of a 10% increase in throughput has not yet been achieved. However, it is expected to be achieved by the end of Q2 2030.

Regarding the reduction of operating costs, the survey shows that the targeted 10% reduction has not yet been achieved. Initially, however, the board focused on identifying where expenses were occurring. As a result, more attention was given to recording expenditures than to reducing costs. Jim concludes that this objective will be achieved in a timely manner.

CHAPTER 20   PUTTING THE TOGAF STANDARD INTO PRACTICE

August 2030

Several months pass once again, and the architecture is approaching its final form. Significant efforts are currently being made to enhance the processes employed in capabilities. This will be the final phase that achieves the Target Architecture.

Jim invested a lot of time preparing for the work and ensuring that Academic Diagnostic Medicine would succeed in realizing its strategy. Thanks to this successful implementation, Enterprise Architecture's ability to add value to an organization has once again been demonstrated.

20.2. Summary

In the case scenario described in this chapter, Enterprise Architect Jim demonstrated how the TOGAF Standard can be used to develop and implement an architecture through a tailored Architecture Development Method.

He covered all phases of the Architecture Development Method while developing and implementing the architecture. Some phases were iterative and thus performed repeatedly. At other times, Jim briefly transitioned to subsequent phases, only to return almost immediately to the current phase. The TOGAF Standard's adaptability allows for this flexibility, making it possible to develop and implement architecture in a way that suits different scenarios, organizations, and industries.

CHAPTER 20   PUTTING THE TOGAF STANDARD INTO PRACTICE

March 2027–April 2027 (Scoping the Work)
Preliminary Phase

Table 20-13. Steps performed during the Preliminary Phase

Steps Performed	Description of Step
Identify key drivers and organizational context	Pinpointing the critical factors that influence the architecture, such as key drivers and the organizational context.
Assess Enterprise Architecture maturity	Evaluating where the organization stands in its architectural development.
Choose and tailor the architecture framework	Identifying and tailoring the TOGAF Standard, Architecture Development Method, and the Content Framework.

During the Preliminary Phase (see Table 20-13), Jim used the Driver/Goal/Objective Diagram to visualize the key drivers and organizational context. He reviewed the diagram with the most important stakeholders at the time, the board of directors, and coordinated with them.

Since Academic Diagnostic Medicine has been working according to the TOGAF Standard for years, Jim saw no need to conduct an architecture maturity measurement. The TOGAF Standard has been considered the preferred framework for working with architecture since its introduction. Later in the project, Jim adapted the framework by adding the risk entity to the Enterprise Metamodel. He also incorporated the Risk Catalog as an artifact in the Content Framework.

May 2027-June 2027 (Envisioning the Architecture)
Phase A: Architecture Vision

Table 20-14. Steps performed during Phase A

Steps Performed	Description of Step
Establish the architecture project	Obtain formal project recognition, secure endorsement from corporate leadership, and gain support and commitment from relevant line management.
Identify stakeholders and determine requirements	Identify stakeholders and requirements to allow for the architecture and related work to be communicated and aligned.
Create a high-level architecture vision	Develop a high-level description of the architecture solution that will meet business requirements and outline how it aligns with the overall business strategy.
Develop Statement of Architecture Work	Develop Statement of Architecture Work and secure formal approval.

During Phase A (see Table 20-14), Jim defined and outlined the projects to be implemented. Three project initiatives emerged based on the strategy. Jim sought and received formal board approval for the project format.

He identified the key stakeholders he would encounter during the development and implementation of the projects. In doing so, he considered the requirements established by the board of directors and the concerns and constraints raised by the stakeholders.

Based on all the input he collected, Jim determined which capabilities would be the subject of the work and visualized this using a Strategy/Capability Diagram. This diagram shows the high-level architecture vision. Jim documented the high-level work to be performed in a Statement of Architecture Work. This document includes stakeholder concerns and constraints, as well as the requirements set by the board of directors. The board of directors and the stakeholders formally approved the document.

July 2027–October 2027 (Starting the Development)
Phase B: Business Architecture
Phase C: Information Systems Architectures
Phase D: Technology Architecture

Table 20-15. *Steps performed during Phases B, C, D*

Steps Performed	Description of Step
Select reference models, viewpoints, and tools	Choose relevant models and viewpoints from the Architecture Repository to match the business goals and stakeholder interests.
Develop Baseline and Target Business Architecture description	Describe the current and future Business Architecture to a degree that supports the architecture vision, building on the Architecture Repository's resources as needed.
Perform gap analysis	Confirm internal consistency, identify any gaps between baseline and target, and validate that models align with business objectives.
Define candidate roadmap components	Using the baseline, target, and gap analysis, build an initial roadmap to guide future actions and prioritization across the remaining architecture phases.
Conduct formal stakeholder review	Revisit the architecture's initial motivation and make any necessary refinements to ensure it's fit to support other architecture domains.
Create the Architecture Definition Document	Complete all relevant sections in the Architecture Definition Document, adding models, reports, and graphics as needed to capture a clear architectural view.

Jim used existing material from the Architecture Repository. He extracted the previously defined building blocks, the Business Capability Map and the Capability/Business Process Matrix, for reuse.

Using these building blocks, he developed the Baseline and Target Architectures. Jim performed a gap analysis to define the candidate building blocks for the Architecture Roadmap.

Throughout the process, Jim coordinated with the stakeholders involved through formal stakeholder reviews. After establishing the Baseline and Target Architectures, Jim prepared an Architecture Definition Document. In the document, he outlined the detailed designs of the Baseline and Target Architectures (see Table 20-15).

November 2027–December 2027 (Setting the Direction)
Phase E: Opportunities and Solutions

Table 20-16. Steps performed during Phase E

Steps Performed	Description of Step
Determine/confirm key corporate change attributes	Align the architecture with an organization's business culture. This involves creating an Implementation Factor Catalog and assessing the transition capabilities of organizational units.
Consolidate and reconcile interoperability requirements	Identify and address constraints that may hinder interoperability.
Identify and group major work packages	Assess and address gaps in business capabilities in order to enable the organization to effectively plan and implement necessary changes aligned with strategic objectives.
Identify Transition Architecture(s)	Identify Transition Architecture(s) to provide structured, value-driven steps toward achieving the Target Architecture, tailored to the organization's implementation strategy and capacity for change.
Create Architecture Roadmap and Implementation and Migration Plan	Create a structured approach to transition from Baseline to Target Architectures through a detailed Architecture Roadmap and Implementation and Migration Plan.

CHAPTER 20 PUTTING THE TOGAF STANDARD INTO PRACTICE

Jim used an Implementation Factor Catalog to identify potential risks in the architecture work (see Table 20-16). He also used the catalog to identify areas of the organization with limited change capacity. The architecture to be implemented has been divided into several phases. Each phase represents a Transition Architecture. Jim indicated which activities (initiatives) were associated with each Transition Architecture. These activities were consolidated into work packages and used as plannable components in the Architecture Roadmap. Based on the roadmap, Jim drew up an Implementation and Migration Plan.

January 2028–May 2028 (Getting Ready for Implementation)
Phase F: Migration Planning

Table 20-17. Steps performed during Phase F

Steps Performed	Description of Step
Confirm management framework interactions	Align the Implementation and Migration Plan with an organization's management frameworks.
Assign business value to work packages	Define and measure business value for work packages, identify corresponding projects, assess associated risks, and estimate business value.
Prioritize migration projects through cost/benefit assessment and risk validation	Determine the sequence of projects based on business priorities, technical dependencies, resource availability, and risk factors.
Confirm Architecture Roadmap	Update the Architecture Roadmap, incorporating any necessary Transition Architectures.
Complete Implementation and Migration Plan	Develop a detailed timeline with clear milestones for each transition stage.

Since Jim wanted a project manager to coordinate the work, aligning the architectural approach with the project management framework was useful. The TOGAF Standard can easily be linked to and aligned with other frameworks.

Jim assigned a business value to each work package. He used the Business Value and Risk Matrix he created earlier to do so. He also managed to visualize the risks using a Risk/Process/Application Realization Diagram. The architecture developed thus far has been checked and verified with the stakeholders (see Table 20-17).

June 2028–August 2029 (Overseeing the Architecture Work)
Phase G: Implementation Governance

Table 20-18. Steps performed during Phase G

Steps Performed	Description of Step
Confirm scope and priorities	Identify and address gaps within the existing architecture. Review the Work Package Portfolio Map to determine which initiatives lead to optimal value extraction.
Perform architecture compliance reviews	Regularly assess the implementation activities to ensure that they are consistent with the architectural standards and principles.
Perform post-implementation review	Examine the end result of the architecture work. If there are deviations, submit change requests to correct them.

Jim took on a leading role during the execution of the work. As an Enterprise Architect, he did not need to operate the controls himself. From his coordinating position, he oversaw the implementation of the work.

The Work Package Portfolio Map, which had been drawn up previously, provided Jim with the tools necessary to prioritize the activities (initiatives), ensuring that the work was carried out in the correct order. Throughout the implementation process, Jim conducted random compliance reviews to ensure everything was on track. He reviewed the results with the relevant stakeholders repeatedly.

After each successful implementation of a Transition Architecture, he conducted a post-implementation review (see Table 20-18).

September 2029–July 2030 (Maintaining Control)
Phase H: Architecture Change Management

Table 20-19. Steps performed during Phase H

Steps Performed	Description of Step
Deploy monitoring tools	Deploy tools to monitor the business and technology changes that could impact the architecture, as well as the maturity of the Architecture Capability.
Manage risks	Identify, classify, and mitigate risk(s) before business transformation.
Provide analysis for architecture change management	Analyze information from monitoring tools, stakeholder feedback, and performance reviews.

Jim used a heat map to track the strategy, which allowed him to provide the board of directors with progress updates at any time. Based on performance reviews, stakeholder feedback, and additional monitoring tools, such as dashboards displaying the progress of the work, he could accurately update the heat map. Any new risks could quickly be translated into changes that could be implemented during the next Architecture Development Method iteration (see Table 20-19).

Jim customized the Architecture Development Method for his work. Not all of the method's steps were necessary. Therefore, he only used the parts that added value to the development and implementation of the architecture they had devised. He used the TOGAF Standard as intended.

CHAPTER 21

The TOGAF Standard on a Page

 This chapter is part of the GOLD reading tour. See Figure 1-1 in Chapter 1, Section 1.3, for alternative reading tours.

The TOGAF Standard comprises several key components (or concepts), which are detailed in previous chapters of this book. These include the Governance Framework, Architecture Repository, Architecture Development Method, Enterprise Metamodel, and Content Framework. These components are closely linked and form the foundation of the TOGAF Standard together. These components enable the structured development of architecture while maintaining mutual dependencies. Visualizing these components reveals how structured and well-thought-out the framework is.

21.1 Connecting the Concepts

Bringing together the basic concepts mentioned in the introduction and relating them to each other in a visualization clarifies the connection between these concepts. Figure 21-1 illustrates the relationships between the most important components and aspects of the framework[1].

1. **Enterprise Metamodel Is Created During Execution of the ADM:** During the Preliminary Phase, a tailored Enterprise Metamodel is developed. This model contains all entities relevant to the organization.

2. **Enterprise Metamodel Specifies Entities to Be Used As Building Blocks:** The entities identified in the tailored metamodel serve as the foundation for the creation of building blocks.

3. **Enterprise Metamodel Realizes the Content Framework:** The metamodel also serves as the foundation for the Content Framework, which also needs to be adapted. The Content Framework contains a collection of models representing the architecture. The Content Framework's categorization mechanism can be used to structure a representation of the Enterprise Metamodel, which provides a detailed view of the entities used within an organization.

4. **Content Framework Complements the ADM:** When applying the Architecture Development Method, it may be necessary to add new artifacts to define the architecture more clearly. This could lead to changes (additions or deletions) in certain parts of the Content Framework.

---

[1] Figure 21-1 can be downloaded in a large format from https://eawheel.com/books/media/

5. **ADM Develops Content for Architecture Domains:** The architecture is developed for the four core architecture domains during the execution of phases B, C, and D of the Architecture Development Method.

6. **Architecture Domains Need Corresponding Architecture Roles:** To properly develop architecture, it is best to create a specialist role for each domain. This role is the Segment Architect.

7. **Governance Framework Governs Architecture Roles:** The various architecture roles within an organization are governed by a central Architecture Governance Framework.

8. **Governance Framework Enables the Use of the ADM:** The Architecture Repository is part of the Governance Framework. This repository contains the processes necessary for implementing architecture activities.

9. **Governance Framework Enables the Architecture Repository:** An important aspect of architecture work is creating artifacts and deliverables, which are stored in the Architecture Repository. The Governance Framework oversees management of the repository.

10. **Architecture Repository Stores Artifacts and Building Blocks:** The developed artifacts and reusable building blocks are stored in the Architecture Repository.

11. **Architecture Repository Contains the Enterprise Metamodel:** The metamodel that was previously created and tailored is also stored in the Architecture Repository.

12. **Artifacts Are Created Out of Building Blocks:** Artifacts use reusable building blocks.

13. **Building Blocks Are Used to Create Catalogs, Diagrams, Maps, and Matrices:** Catalogs, diagrams, maps, and matrices are created using reusable building blocks from the Architecture Repository.

CHAPTER 21 THE TOGAF STANDARD ON A PAGE

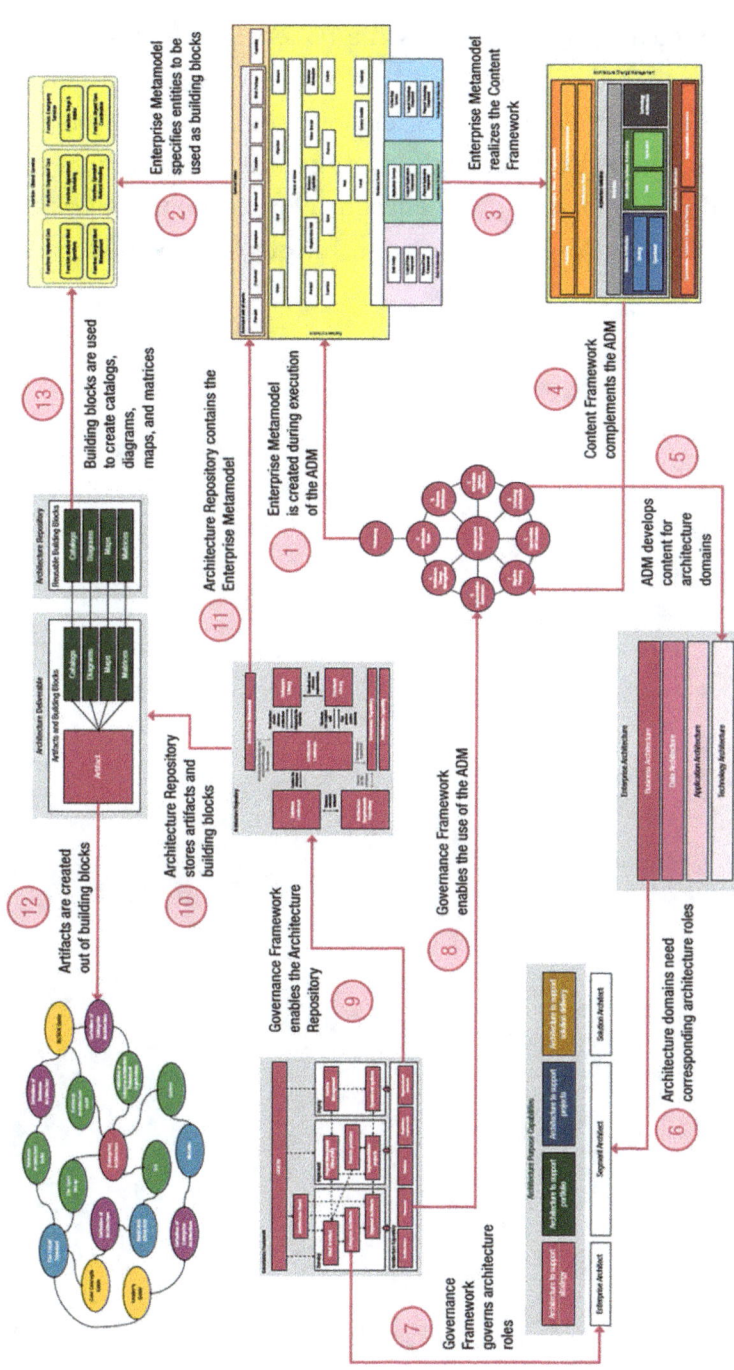

Figure 21-1. The TOGAF Standard on a page

CHAPTER 21 THE TOGAF STANDARD ON A PAGE

As discussed in previous chapters, the TOGAF Standard is much more complex than shown in Figure 21-1. Nevertheless, the visualization provides a good overview of the interrelationships between several essential concepts of the framework. In any case, it clearly shows that the TOGAF Standard is not a random collection of components but rather the result of careful design and consideration, making it the world's leading Enterprise Architecture framework.

CHAPTER 22

Closing Remarks

 This chapter is part of the PINK reading tour. See Figure 1-1 in Chapter 1, Section 1.3, for alternative reading tours.

So this is it. This marks the end of my attempt to translate the *world's leading architecture framework* into a *practical approach*. One that sheds new light on applying the TOGAF Standard in a more accessible way.

I strongly believe that the TOGAF Standard is still highly relevant today. It covers all the necessary aspects for creating a sound Enterprise Architecture. The framework is supplemented with various methods and techniques that provide the right tools for an effective approach. Yet, many practitioners do not apply it as intended. I believe this is because they do not fully know *how* to apply it. This is partly due to the amount of information in the framework and how that information is explained. Sometimes, it may seem vague or even slightly academic. This causes many architects to abandon the framework, leaving room for myths to emerge.

Many of the myths that surround the framework stem from a lack of understanding of how the framework's theory is applied. This leads to further misunderstandings in practice.

CHAPTER 22 CLOSING REMARKS

The seven most common myths are listed below:

- The TOGAF Standard is only for large organizations.
- The TOGAF Standard is a rigid, prescriptive framework.
- The TOGAF Standard is only about IT Architecture.
- The TOGAF Standard replaces other frameworks.
- The TOGAF Standard is outdated and no longer relevant.
- The TOGAF Standard is just about documentation.
- The Architecture Development Method must be followed step by step.

As I bring this book to a close, I feel compelled to dispel these myths.

The TOGAF Standard Is Only for Large Organizations: In practice, the TOGAF Standard is used fairly regularly in large organizations and less frequently in smaller ones. However, it is a misconception that the TOGAF Standard is only applicable to large organizations. In fact, the use of any framework is independent of an organization's size. Every organization benefits from using an architectural framework.

Granted, not every organization will use a framework to the same extent because there is not always a need to do so. Architecture frameworks – and the TOGAF Standard is no different – are scalable and adaptable to any situation or organization size. They provide insight, overview, and guidance for a structured approach to work, which is beneficial to any organization, large or small. Every organization should want its goals translated into the impact they will have on operations. A framework such as the TOGAF Standard can be tremendously helpful in this regard.

CHAPTER 22 CLOSING REMARKS

The TOGAF Standard Is a Rigid, Prescriptive Framework: Contrary to popular belief, the TOGAF Standard is a highly flexible framework. Rather than following a strict set of rules, organizations are free to tailor it to their specific needs. Chapter 17 is entirely devoted to tailoring the framework.

The TOGAF Standard is very complete in that it contains a lot of information on how to approach different situations and scenarios. This wealth of information is sometimes seen as something that must be followed exactly. However, this is not the case.

The essential scaffolding of the TOGAF Framework is the concepts. Everything else in the TOGAF Framework is either an example or a starter set to get you moving. If you do not like the example, then you can take advantage of the modular structure of the TOGAF Framework and substitute it [18].

The framework simply describes every component, entity, or concept encountered in an architecture practice. This does not mean that architecture must adhere to everything written in the TOGAF Standard. The framework acts as a blueprint that needs to be customized. Then, it can be applied as best suits the organization.

The TOGAF Standard Is Only About IT Architecture: Given the scope of Enterprise Architecture, it's clear that IT Architecture (also referred to as Technology Architecture in the TOGAF Standard) is part of the bigger picture. However, Technology Architecture is only *one of the four core domains* of Enterprise Architecture. The other three are Business Architecture, Data Architecture, and Application Architecture (see Figure 22-1).

CHAPTER 22 CLOSING REMARKS

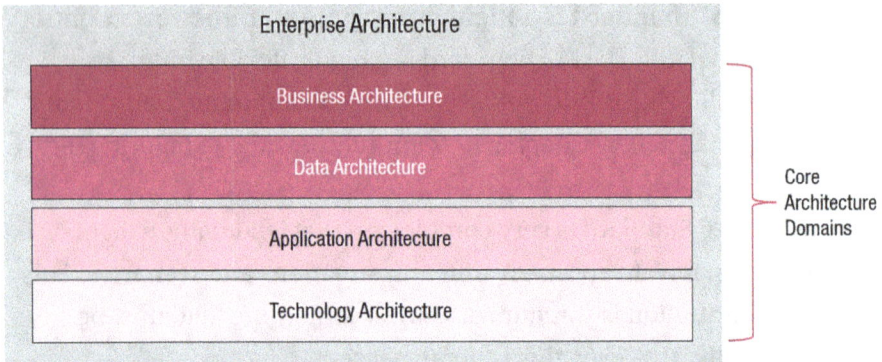

Figure 22-1. Four core architecture domains

The TOGAF Standard is an Enterprise Architecture framework because it covers *all four domains*. It provides a holistic view of an organization.

The framework references the BIZBOK Guide, among others, for its Business Architecture concepts and techniques. This is because the BIZBOK Guide describes the content of this particular architecture domain in much more detail than the TOGAF Standard does. Rather than duplicating the information, The Open Group decided to refer to this other source.

The TOGAF Standard Replaces Other Frameworks: The TOGAF Standard *is not in competition* with other frameworks, such as Zachman, SAFe, ITIL, or COBIT. In fact, it is sometimes beneficial not to source all the architecture-related information from a single framework. The TOGAF Standard acknowledges this.

Because the TOGAF Standard is a generic framework and intended to be used in a wide variety of environments [...], it may be used either in its own right, with the generic deliverables that it describes, or else these deliverables may be replaced or extended by a more specific set, defined in any other framework that the architect considers relevant [2].

The TOGAF Standard provides generic deliverables and suggests a generic approach to architectural practice. However, if there are components from other frameworks that better suit the organization, they should certainly be used.

The TOGAF Standard Is Outdated and No Longer Relevant: The TOGAF Standard is neither outdated nor irrelevant. Over its many years of existence, the framework has continued to evolve. It now reflects modern architectural trends, such as digital transformation, cloud computing, and agile methodologies.

A new and updated edition of the TOGAF Standard is released every few years. Each new edition addresses the shortcomings of previous versions and adds new content, trends, or techniques. The Core Content remains largely the same because it doesn't change much over the years.

The TOGAF Standard Is Just About Documentation: Many practitioners still believe that architectural work requires producing extensive documentation. This seems to be based on the idea that the TOGAF Standard deliberately captures everything related to the profession and its work. However, this assumption is unwarranted because it is not stated anywhere in the framework.

Of course, the TOGAF Standard recognizes the need to document important matters, such as agreements made at the beginning of architectural work. There are four architecture documents that need to be created. Each document has a set structure that must be reflected in the final product. However, this does not mean that every aspect of architectural work must be endlessly documented.

One source of confusion may be that architects think every diagram, matrix, or catalog must be included in the Architecture Repository, which is another misconception. The Architecture Repository should contain Architecture and Solution Building Blocks, not elaborate solution diagrams, matrices, or catalogs.

CHAPTER 22 CLOSING REMARKS

The TOGAF Standard certainly does not suggest endless documentation. This is one of the myths that arise from a lack of understanding of the framework.

The Architecture Development Method Must Be Followed Step by Step: The Architecture Development Method graphic is still frequently misinterpreted as a linear, waterfall-style process. In reality, however, it is a *logical method* that *groups key activities* to clarify their relationships and information flow. The well-known graphic is a stylized representation showing these information flows. It is *not a representation of an activity sequence*.

The Architecture Development Method should not be understood as a process model [18].

Like the full framework, the Architecture Development Method allows for the same amount of tailoring and supports four different types of iteration, making it suitable for any situation. It is *not limited* to being followed step by step.

These myths exist because the TOGAF Standard is often misinterpreted. This is usually caused by *not carefully reading* the framework's content. Another reason is that practitioners continue to use *outdated information*. Often, they rely on what they learned years ago when they first took their TOGAF exam. Other times, they simply accept information they've heard somewhere else as truth without giving it any further thought.

In this book, I have addressed and refuted the aforementioned myths and arguments against using the TOGAF Standard. To this end, I have provided numerous examples of the TOGAF Standard's practical application. These examples are easy to understand and are drawn from daily practice. They demonstrate the framework's added value.

I have presented a practical approach that clarifies the methodology and encourages more practitioners to use and tailor the framework. After all, this is precisely how the TOGAF Standard is intended to be used.

CHAPTER 22 CLOSING REMARKS

The idea is to apply the *relevant parts* to your situation or organization in a way that solves architectural challenges.

This book covers the most important parts of the framework and includes additional topics that I felt were necessary to provide a comprehensive overview of how to use the framework. I included additional aspects that needed more context and explanation. The primary purpose of this book, however, is to help you master the most important parts of the world's leading architecture framework.

In the introduction of the book, I mentioned my primary reasons for writing it:

- To **explain and clarify** the methodology of the TOGAF Standard
- To present a **practical approach** to using the framework
- To show you how to **correctly apply and tailor** the framework
- And, by doing so, to **refute the unfounded criticism**

Throughout the various chapters of the book, I have explained and clarified the most fundamental concepts of the framework in detail. I provided actionable steps for approaching various scenarios and situations and explained how to implement these concepts. I also demonstrated how to tailor the framework by showing how to adjust the Enterprise Metamodel and the Content Framework and how to apply a tailored Architecture Development Method.

I have demonstrated how the TOGAF Standard can be utilized to implement Enterprise Architecture, using compelling examples.

I hope to have enabled you to use the world's leading architecture framework as it was intended: in a *practical* and *pragmatic* way that can be *tailored to the situation.*

Your feedback is invaluable. Leaving a review on websites like Amazon not only supports the work but also helps future readers decide if this book is right for them.

APPENDIX A

List of Copyrighted Figures

Name	Description	Original Image Location
Figure 3-2	Structure of the TOGAF Standard	https://pubs.opengroup.org/togaf-standard/introduction/chap02.html
Figure 3-4	The TOGAF Library	https://pubs.opengroup.org/togaf-standard/introduction/chap02.html
Figure 6-1	Content Framework by ADM phase	https://pubs.opengroup.org/togaf-standard/architecture-content/chap01.html
Figure 6-2	The Extended Content Framework	https://pubs.opengroup.org/togaf-standard/architecture-content/chap01.html
Figure 7-1	The Architecture Repository (including all excerpts)	https://pubs.opengroup.org/togaf-standard/architecture-content/chap07.html

(*continued*)

APPENDIX A   LIST OF COPYRIGHTED FIGURES

Name	Description	Original Image Location
Figure 7-3	The Enterprise Continuum	https://pubs.opengroup.org/togaf-standard/architecture-content/chap06.html
Figure 7-6	The four architecture types of the Architecture Continuum (including all variations)	https://pubs.opengroup.org/togaf-standard/architecture-content/chap06.html
Figure 7-8	The four solution types of the Solutions Continuum (including all variations)	https://pubs.opengroup.org/togaf-standard/architecture-content/chap06.html
Figure 7-13	Artifacts associated with the Foundation Enterprise Metamodel	https://pubs.opengroup.org/togaf-standard/architecture-content/chap03.html
Figure 7-22	Minimum Architecture Repository components that must be represented in an architecture tool	https://pubs.opengroup.org/togaf-standard/architecture-content/chap07.html
Figure 8-1	Architecture Governance Framework	https://pubs.opengroup.org/togaf-standard/ea-capability-and-governance/chap03.html

(*continued*)

APPENDIX A   LIST OF COPYRIGHTED FIGURES

Name	Description	Original Image Location
Figure 10-1	Core architecture roles and their relationships	`https://pubs.opengroup.org/togaf-standard/architecture-roles-and-skills/chap04.html`
Figure 10-7	Differences between Segment and Solution Architectures	`https://pubs.opengroup.org/togaf-standard/architecture-skills-framework/chap06.html`
Figure 12-1	The Architecture Development Method (including all amended instances of the ADM)	`https://pubs.opengroup.org/togaf-standard/adm/chap01.html`
Figure 14-4	Stakeholder Power Grid	`https://pubs.opengroup.org/togaf-standard/adm-techniques/chap03.html`
Figure 14-15	Architecture Alternatives and Trade-Offs method	`https://pubs.opengroup.org/togaf-standard/adm-techniques/chap09.html`

References

[1] The Open Group. "Directory of Certified People." https://togaf-cert.opengroup.org/certified-individuals

[2] The Open Group. *The TOGAF® Standard, 10th Edition: Introduction and Core Concepts.* 's-Hertogenbosch: Van Haren Publishing, 2022.

[3] The Open Group. *The TOGAF® Standard, 10th Edition: Leader's Guide.* 's-Hertogenbosch: Van Haren Publishing, 2022.

[4] Zachman, J. A. "A Framework for Information Systems Architecture." https://www.academia.edu/110050217/A_Framework_for_Information_Systems_Architecture_Abstract_of_Tutorial_

[5] Business Architecture Guild®. *A Guide to the Business Architecture Body of Knowledge®*, vol. 13.0. 2024.

[6] O'Higgins, D. "Impacts of Business Architecture in the Context of Digital Transformation." https://arxiv.org/abs/2307.11895. 2023.

[7] ISO (International Organization for Standardization). "Software, systems and enterprise – Architecture description." https://www.iso.org/standard/74393.html

REFERENCES

[8] Greenslade, C. "The Open Group Architecture Framework: The Continuing Story." https://www.opengroup.org/architecture/0307bos/presents/greenslade_togaf.pdf

[9] Jager, E. *Getting Started with Enterprise Architecture.* Apress, 2023.

[10] Gartner. "Gartner Clarifies the Definition of the Term 'Enterprise Architecture.'" https://www.gartner.com/en/documents/740712

[11] The Open Group. *The TOGAF® Standard, 10th Edition: Content, Capability, and Governance.* 's-Hertogenbosch: Van Haren Publishing, 2022.

[12] Ziemann, J. *Fundamentals of Enterprise Architecture Management.* Cham: Springer Nature, 2022.

[13] Homann, U. "A Business-Oriented Foundation for Service Orientation." 2006.

[14] The Open Group. *The TOGAF® Standard, 10th Edition: Business Architecture.* 's-Hertogenbosch: Van Haren Publishing, 2022.

[15] The Open Group. "How to Use the ArchiMate® Modeling Language to Support the TOGAF® Standard," G21E. 2022.

[16] The Open Group. "TOGAF® Series Guide: An Approach to Selecting Building Blocks," G248. 2024.

[17] Gartner. "Gartner Magic Quadrant for Enterprise Architecture Tools." https://www.gartner.com/en/documents/5937407

REFERENCES

[18] The Open Group. *The TOGAF® Standard, 10th Edition: ADM Practitioner's Guide.* 's-Hertogenbosch: Van Haren Publishing, 2022.

[19] The Open Group. "TOGAF® Series Guide: Architecture Roles and Skills," G249. 2024.

[20] The Open Group. *The TOGAF® Standard, 10th Edition: Architecture Development Method.* 's-Hertogenbosch: Van Haren Publishing, 2022.

[21] The Open Group. *Hospital Reference Architecture Guide: The Complete and Expanded English Translation of the Dutch ZiRA.* The Open Group, 2023.

[22] The Open Group. "TOGAF® Series Guide: Information Mapping," G190. 2022.

[23] The Open Group. *The TOGAF® Standard, 10th Edition: Enterprise Agility and Digital Transformation.* Van Haren Publishing, 2022.

[24] The Open Group. "TOGAF® Series Guide: Integrating Risk and Security within a TOGAF® Enterprise Architecture," G152. 2022.

[25] Van den Berg, M. and M. Van Steenbergen. *DYA Stap voor stap naar professionele enterprise-architectuur.* The Hague: Sdu Uitgevers bv, 2004.

[26] Beck, K. "Manifesto for Agile Software Development." https://agilemanifesto.org/

Index

A

Academic Diagnostic Medicine, 781–784, 791, 793, 800, 808, 809
Actor/Role Matrix, 368, 387, 388
ADM, *see* Architecture Development Method (ADM)
ADM phase, Content Framework
   architecture change management, 119
   architecture principles, visions and requirements, 118
   architecture realization, 119
   business architecture, 118
   information systems architectures, 119
   technology architecture, 119
ADM techniques, 591, 663, 664
   architecture alternatives and trade-offs, 591, 656, 657, 668
      architect, 660
      diagram, 660–662
      informed decision, 658
      Lemon-A-de's strategy, 659, 660
      options, 658
      phased roadmap, 662
      rule, 658
      selection, 658
      stakeholders, 658
      types, 659
   architecture patterns, 591, 616, 617, 663, 665
   architecture roadmapping, 590
   business transformation readiness assessment, 590, 591, 628, 667
      human factors, 632
      implementation and migration plan, 628, 631
      maturity models, 628
      readiness factor rating, 628–631
      readiness factor risks, 628
      readiness factors, 629–631
      risks, 631
   gap analysis, 590, 591 (*see also* Gap analysis)
   interoperability requirements, 591, 666
      advantages, 627
      cells, 626

INDEX

ADM techniques (*cont.*)
    cross-mapping, 623, 625
    degrees/levels, 624, 625
    idea, 623
    integration, 626
    stakeholders, 623
    website, 627
  principles, 591, 664
    added value, 593
    architecture principles, 591–593, 597, 599–601, 663, 664
    enterprise architecture, 664
    enterprise principles, 592, 593, 597–599, 664
    structure, 594–596
  risk management (*see* Risk management)
  stakeholder analysis, 665
  stakeholder management, 590, 663 (*see also* Stakeholder management)
Advantages of TOGAF Standard
  adaptability and tailoring, 47, 48
  common language and communication, 49
  comprehensive coverage of architectural domains, 46, 47
  enhanced business and IT alignment, 48
  facilitation of digital transformation, 48
  scalability and resource optimization, 49, 50
  support for process innovation, 49
Agile architecture
  ADM (*see* Architecture Development Method (ADM))
  agile approach, 729, 735
  architecture partitioning, 727, 731, 737
  architecture plateaus, 737
  capability architectures, 732, 734–736, 755
  conceptual visualization/levels, 727
  development, 736
  enterprise architecture, 736
  governance approach, 752, 753
  levels, 727, 731, 732
    capability level, 742, 756
    mapping, 740
    segment level, 741, 756
    strategic level, 741, 755
  minimum viable architecture, 728, 737
  minimum viable product, 742
  modeling, 754
  organizational strategy, 736
  partitioned architecture, 735
  partitioning, 736
  partitions, 728, 738, 739
  regular approach, 729, 730

INDEX

segment architectures, 731, 733–736, 755
sequential approach, 730
splitting/partitioning, 738
sprint demo, 751
sprint planning, 753
sprints, 728, 738, 739, 751, 752
strategic architectures, 731, 733, 734, 754
transition architectures, 737, 738, 742
waterfall approach, 729
Agile environments, 726, 728
Agile techniques, 42, 741
Agility, 726, 754
Application and User Location Diagram, 437, 440
Application Architect, 256
Application Architecture, 47, 93, 257–260, 263
Application Architecture Entities
  application service, 97
  logical application component, 97, 98
  overview, 96
  physical application component, 98
Application Communication Diagram, 436–438
Application/Data Matrix, 411, 414
Application/Function Matrix, 428, 435, 465, 483
Application Interaction Matrix, 428, 435, 436
Application Migration Diagram, 439, 440
Application/Organization Matrix, 433, 434
Application Portfolio Catalog, 429–433
Application service, 97–99, 158, 162, 367, 428
Application/Technology Matrix, 455
Application Use-Case Diagram, 428, 437
AR, *see* Augmented Reality (AR)
ArchiMate modeling language, 104, 108
ArchiMate models, 716, 718, 721
Architects, 286, 602
Architectural assets, 37, 39, 128, 136, 138, 172
Architectural element, 674
Architectural frameworks, 15
Architectural styles, 563, 564
Architecture, 268
  communication, 673, 678, 699
  frameworks, 1, 52
  landscape, 130, 131
  layers, 287, 293
  metamodels, 129
  method and process, 683
  principles, 314, 316, 320, 325, 330, 331, 535, 536, 708
  process, 672
  roles, 279
  sponsorship, 209
  work, 4

INDEX

Architecture Alternatives and Trade-Offs approach, 38, 353, 403, 422, 446, 464
Architecture artifacts, 39, 154–157, 163, 179
  catalogs, 151
  definition, 150
  diagrams and maps, 151–153
  matrices, 153, 154
  relationship of, 150
Architecture Board, 191, 193, 205, 214, 226
  Architecture Capability, 206
  Architecture Governance Framework, 206
  architecture roles, 204
  composition and size, 213–215
  creation, 209
  define engagement model and workflows, 218, 219
  diverse leadership
    CEO, 208
    CEO leadership, 206, 207
    CIO/CTO leadership, 207–209
    IT Architects, 205
    organizations, 226
    organization's Enterprise Architecture governance, 204
    responsibilities, 211–213
    senior management, organization, 205
    setup, 209, 210
  steps, setup, 226–227
    celebrate successes and promote value, 223, 227
    define the purpose and scope, 216, 217
    develop the board's operating model, 218
    ensure ongoing communication and alignment, 222
    establish key artifacts and tools, 220, 221
    identify key stakeholders, 217
    measure and optimize performance, 223
    pilot the Architecture Board, 221
    roll out the Architecture Board organization-wide, 221, 222
  strategic decisions, 205
Architecture Building Blocks, 131, 170, 185, 186
  addressing required capability, 171
  Architecture Repository
    Architecture Landscape, 172
    Reference library, 173
    Standards Library, 173
    storage locations, 172
  capability entities, 170

INDEX

capability implementation, 170
collecting architecture
 requirements, 171
definition, 171
functionality and
 capabilities, 170
fundamental functionality and
 attributes, 171
high-level conceptual
 components, 171
HR-related activities
 management, 172
what *and* why, 170
Architecture Capability, 135, 205,
 215, 218, 224, 259, 572, 698,
 702, 705
 Architecture Development
  Method, 230
 architecture function, 112
 Architecture Maturity
  Assessment, 239
 architecture roles, 241
 architecture to support
  portfolio, 237, 244
 architecture to support projects,
  238, 244
 architecture to support solution
  delivery, 238, 239, 244
 architecture to support strategy,
  237, 244
 Enterprise Architecture
  function within an
  organization, 229
 framework, 39

iteration group, 571
organizations, 229, 236
organization's ability, 229
organization's ability to practice
 architecture, 230
organization-specific, 112
position, 240, 241
Architecture Capability creation
 architecture change
  management
  change management,
   235, 244
 architecture vision
  define constraints, 233, 242
  define scope, 233, 242
  define the project, 232, 242
  identify business goals and
   drivers, 232, 242
  identify stakeholder
   concerns and business
   requirements, 232, 242
  perform architecture
   maturity assessment,
   233, 242
 business architecture
  agree on common language,
   233, 242
  decide on the framework,
   234, 243
  define performance metrics,
   234, 243
  define the process, 234, 243
  Governance Framework,
   234, 243

843

INDEX

Architecture Capability
creation (*cont.*)
identify views and
viewpoints, 234, 243
data, application, and
technology architectures
define requirements,
architecture deliverables,
235, 243
define technology, 235, 243
specify and govern the
structure, architecture
repository, 234, 243
migration planning and
implementation
governance
govern organizational
change, 235, 244
opportunities and solutions
determine organizational
change, 235, 243
requirements
management phase
manage requirements,
236, 244
Architecture Capability Maturity
Model (ACMM), 288, 312,
313, 670
framework, 676, 681
elements, 671, 699
architecture
communication, 673
architecture
development, 672
architecture governance, 673
architecture process, 672
business linkage, 672
business unit
participation, 672
IT investment and
acquisition strategy, 673
IT security, 673
senior management
involvement, 672
maturity levels, 674
defined, 674
initial, 674
managed, 675
measured, 675
none, 674
under development, 674
Architecture change management,
122, 528, 529, 744, 750, 756
activate process to implement
change, 530, 537, 538
deploy monitoring tools,
529, 532
develop change requirements
to meet performance
targets, 530, 534, 535
driver/goal/objective
diagram, 807
establish value realization
process, 530–531
heat map, 815
manage governance process,
530, 536
manage risks, 529, 532

844

INDEX

new planning application, 807
operating costs, 807
performance reviews, 806
provide analysis for architecture change management, 529, 533
steps, 815
target architecture, 808
Architecture Compliance Report, 529, 531, 805
Architecture Content Framework, 24, 38, 45
Architecture Continuum, 138, 184
  architecture types, 141, 142
  Common Systems Architectures, 143
  definition, 141
  Foundation Architectures, 141–143
  Industry Architectures, 143, 144
  Organization-Specific Architectures, 144, 145
Architecture Contracts, 301, 304, 305, 530, 805
Architecture Definition Document, 154, 301, 303, 304, 351, 358, 359, 401, 409, 410, 427, 444, 451, 473, 499, 506, 507, 551
Architecture Definition Increments Table, 472, 473, 483, 484
Architecture deliverables, 150, 684, 716
  architecture artifacts, 154–157, 163

Architecture Definition Document, 154
  building blocks (*see* Building blocks)
  contents, 155
Architecture Development Method (ADM), 4, 234, 236, 295, 572, 573, 589, 672, 715–717, 720, 729, 739, 819
  activity by iteration cycle, 575
  agile concepts, 740, 743
  Agile sprints
    adapt phase E, 747, 748
    adapt phase F, 748
    adapt phase G, 749
    adapt phase H, 750
  architectural styles, 561
  architecture alternatives and trade-offs method, 571
  Architecture capability, 230, 231
  Architecture capability iteration group, 571, 572
  architecture development and management, 111
  architecture development iteration group, 571, 573, 576, 577
  architecture governance iteration, 579
  architecture governance iteration group, 568, 574
  customization, 815
  cycle, 566, 567
  cyclic method, 663

845

INDEX

Architecture Development
  Method (ADM) (*cont.*)
 definition, 16
 development
  Academic Diagnostic
    Medicine, 793
  baseline and target
    architectures, 792
  business architecture,
    789, 791
  business capability map,
    787, 789, 811
  capability/business process
    matrix, 787–789, 811
  capability heat map, 791
  clinical services, 792
  clinical workflow
    management, 791
  documents, 789, 812
  financial management, 792
  incoming feedback, 792
  organization/business roles
    catalog, 790
  patient feedback
    management, 791
  steps, 811
  strategy and capabilities, 788
  strategy/capability
    diagram, 788
  target application
    architecture, 793
 document creation, 300–305
 early version, 26

Enterprise Architecture
  development, 579
Enterprise Architecture
  Landscape, 566, 579
fixed components, 300
focus areas, 231
goals and related initiatives, 569
graphic, 565, 584
important role, 20
iteration across the whole ADM
  cycle, 566–567, 572
iteration cycles
  Architecture Development
    activity, 577
  Architecture Governance
    activity, 578, 579
  Baseline Architecture, 576
  preliminary phase, 571, 572
  Target Architecture, 577
  Transition Planning
    activity, 578
iteration groups, 568, 571, 572
iteration to manage the
  Architecture
  Capability, 571–572
iteration within an ADM
  cycle, 567
linear process model, 579
mapping, 27, 739
minimum viable architecture,
  744, 746
phases, 117, 231, 296, 297, 619,
  620, 666, 740, 744, 745, 808

INDEX

architecture change management, 299, 553 (*see also* Architecture change management)
architecture transformation, 628
architecture vision, 298, 549, 627 (*see also* Architecture vision)
business architecture, 298, 745 (*see also* Business architecture)
implementation governance, 299, 554 (*see also* Implementation governance)
information systems architectures, 298, 303, 745 (*see also* Information systems architectures)
migration planning, 298, 553 (*see also* Migration planning)
opportunities and solutions, 298, 551 (*see also* Opportunities and solutions)
preliminary phase, 297, 548 (*see also* Preliminary phase, ADM)
requirements management, 111, 112, 299, 555 (*see also* Requirements management)
technology architecture, 298, 550, 745 (*see also* Technology architecture)
scenarios, 562
sprint backlog, 743
sprints, 744, 745
summary of steps, 557
TAFIM framework, 20, 28, 30
Technical Corrigendum, 25
techniques (*see* ADM techniques)
TOGAF 7, 21
TOGAF 8, 23
transition planning iteration group, 568, 573
2025 TOGAF Standard, 26
Architecture Development Methodology, 171, 581
Architecture domains, 280, 286, 819, 826
application architecture, 281, 292
architecture descriptions, 285
architecture initiative's goal, 285
vs. architecture layers, 287, 289, 293
BIZBOK Guide, 285
business architecture, 280, 282, 283, 292, 293
categorization, 288
classification, 282
cross-cutting concerns, 288, 289
data architecture, 280, 292

847

INDEX

Architecture domains (*cont.*)
  enterprise architecture, 285, 286, 293
  information concepts, 283, 284
  information systems architectures, 282, 283
  integration, 285
  interoperability model, 283
  organizations, 286
  roles and specializations, 289, 290, 292
    chief/enterprise architect, 290, 294
    segment architect, 291, 294
    solution architect, 291, 292, 294
  security architecture, 293
  technology architecture, 281, 293
Architecture governance, 224, 574, 578, 673, 683
  Architecture Board, 191, 195
  Architecture Repository, 192
  business management tool, 189
  CIO/CTO, 192, 193
  control/manage an organization's architecture/Architecture Capability, 190
  development capability, 191
  Enterprise Architects, 190
  framework, 191, 306, 311, 349, 514, 530, 819
  shared responsibility among business leaders, 190
  TOGAF Standard, 190
Architecture governance implementation
  Architecture Board, 193
  automate governance processes, 202, 203
  communications plan, 193
  define governance roles and responsibilities, 195–197
  develop governance policies and standards, 198
  development, communications plan, 202
  ensure continuous improvement, 203
  establish key performance indicators (KPIs), 200, 201
  establish the architecture governance framework, 194, 195
  example RACI matrix, 196, 197
  governance framework, 193
  implement an architecture review process, 199, 200
  integrate governance into the ADM, 201, 202
  RACI responsibilities, 197
  set up an architecture compliance process, 198, 199
  steps, 194

848

INDEX

Architecture Maturity Assessment, 233, 239
Architecture partitioning, 727, 731, 737, 755
   vs. agile approaches, 581, 587
   architecture development methodology, 581
   enterprise architecture, 580, 586
   governance, 586
   multiple teams work, 581
   organizational structure, 583
   organizations, 580
   primary reasons, 580, 586
   relationship between architectures, 583
   responsibilities, teams, 583
Architecture project
   progress phases
   phase A (Architecture Vision), 123
   phase B, c, d, 123
   phase F, 124
   phase H, 123
Architecture Purpose
   Capabilities, 237
Architecture Repository, 192, 819, 820
   Architecture and Solutions Continua, 138, 139
   Architecture deliverables (*see* Architecture deliverables)
   architectures and solutions collections, 183
   architecture tool, 179
   categories, 150
   central hubs, 128, 129
   centralized hub, 138
   continua (*see* Three continuum)
   database construction, 180
   definition, 136
   intangible concept, 187
   key components (*see* Key components of Architecture Repository)
   word processor/presentation software, 179
Architecture Repository Reference Library, 403, 408
Architecture Requirements Repository, 132–134
Architecture tools
   Enterprise Architecture management functionality, 183
   features, 180
   functionality considerations, 181, 182
   Gartner's view, 180
   minimum architecture repository components, 182
   selection, 182
   selection factors, 180, 187
   usage requirements, 182
Architecture vision, 324, 745
   artifacts, 336
   Business Capability Diagram, 346

849

INDEX

Architecture vision (*cont.*)
  Business Capability Map, 345, 346
  Business Model Diagram, 343–345
  in healthcare services, 340
  primary and support activities, 341, 342
  Solution Concept Diagram, 343
  Stakeholder Catalog, 336, 338
  support activities, 341
  Value Chain Diagram, 339, 340
  Value Stream Map, 346, 347
 assess readiness for business transformation, 329
 board meetings, 786
 create high-level architecture vision, 325, 331, 332
 communications plan, 786
 define Target Architecture value propositions and KPIs, 325, 333
 develop Statement of Architecture Work, 334–335
 elaborate on business goals and Architecture Principles, 325, 330
 establish architecture project, 324–326
 identify business transformation risks and mitigation activities, 325, 333–334
 identify stakeholders and determine requirements, 325, 327
 project initiatives, 810
 RACI matrix, 324
 reference materials, 324
 stakeholder catalog, 785
 stakeholders, 810
 stakeholders' concerns and constraints, 786
 steps, 325, 810
 strategy, 786
 strategy/capability diagram, 810
 strategy visualization, 784
Artifacts, 299, 300
Aspects of Content Framework
 agile and digital transformation alignment, 116
 architecture artifacts, 116
 core content model, 115
 extended content model, 115
 modular structure, 115
 outcome-focused, 117
 tailoring guidance, 116
 views and viewpoints, 116
Assignment, 248
Assumption, 69, 70
Augmented Reality (AR), 776

INDEX

B

Bandwidth Management and Digital Product Management, 35
Baseline Architecture, 70
BBIB, *see* Building Block Information Base (BBIB)
Benefits Catalog, 477, 478
BIZBOK Guide, 60, 135, 399, 709
Bizzdesign, 716
BPMN, 716, 718
Building Block Information Base (BBIB), 19
Building blocks
   architecture artifact making components, 185
   Architecture Building Blocks (*see* Architecture Building Blocks)
   ADM, 132, 134, 155, 185
   building block's design, 166
   Business Capabilities Catalog, 167
   core principles, 167
   definition, 164–166
   key references, 165, 185
   LEGO bricks, 164
   multiple Solution, 166
   potentially reusable component, 165
   reusability and modularity, 164
   smooth integration, 167
   specification process, 168, 169
   types, 170

Business Alignment, 684
Business Architect, 256
Business Architecture, 255, 257, 258, 260, 263, 347, 348
   artifacts, 366–369
   Actor/Role Matrix, 387, 388
   Business Capabilities Catalog, 377, 378
   Business Capability Map, 379
   Business Event Diagram, 396
   Business Footprint Diagram, 388, 389
   Business Glossary Catalog, 385, 386
   Business Interaction Matrix, 386, 387
   Business Service/Function Catalog, 372, 373
   Business Service/Information Catàlog, 389, 390
   Business Service/Information Diagram, 389
   Business Use-Case Diagram, 394
   Capability/Organization Matrix, 383
   Contract/Measure Catalog, 376, 377
   Driver/Goal/Objective Catalog, 370, 371
   Functional Decomposition Diagram, 390, 391

INDEX

Business Architecture (*cont.*)
 Goal/Objective/Business Service Diagram, 393
 Information Map, 396, 397
 Location Catalog, 374, 375
 Organization/Actor Catalog, 369, 370
 Organization Decomposition Diagram, 394, 395
 Organization Map, 383, 384
 Process/Event/Control/Product Catalog, 375, 376
 Process Flow Diagram, 395, 396
 Product Lifecycle Catalog, 392
 Product Lifecycle Diagram, 391, 392
 Role Catalog, 371, 372
 Strategy/Capability Matrix, 382
 Value Stream/Capability Matrix, 381
 Value Stream Catalog, 379, 380
 Value Stream Map, 380, 381
 conduct formal stakeholder review, 351, 356
 craft Target Business Architecture, 348
 create Architecture Definition Document, 351, 358, 359
 define candidate roadmap components, 350, 354
 develop Baseline and Target Business Architecture description, 350, 352–353
 finalize Business Architecture, 351, 357
 identify potential components for Architecture Roadmap, 348
 methods and modeling techniques, 360
  activity models, 364
  Business Capability Heat Mapping, 361, 362
  Business Capability Map, 360
  Business Scenarios, 365
  Information Maps, 363
  logical data model, 364
  Organization Map, 362, 363
  use-case models, 364
  value streams, 362
 non-architectural inputs, 349
 organizational model, 349
 organizational processes, 348
 perform gap analysis, 350, 353
 reference category, information concept, 398
 resolve impacts across Architecture Landscape, 351, 355, 356
 section, 121, 122

INDEX

select reference models, viewpoints and tools, 350–352
steps, 350
transactional category, information concept, 397
Business Architecture Body of Knowledge (BIZBOK), 24, 60, 62, 87–89, 135, 283–285, 366, 399, 709, 826
Business Architecture Entities
  business capability, 80
  concept, 74
  course of action, 79, 80
  driver, 77
  goal, 77
  measure, 78
  objectives, 77
  overview, 74
  value stream (*see* Value streams)
Business automation, 778
Business capabilities, 34, 80
  Catalog, 367, 377–379
  Diagram, 346
  entity, 73
  Heat Mapping technique, 381
  Map, 153, 167, 336, 345, 346, 360, 361, 367, 379
Business constraints, 746
Business-driven approach, 760
Business Event Diagram, 369, 396
Business Footprint Diagram, 368, 388, 389, 393

Business Glossary Catalog, 368, 385, 386
Business Interaction Matrix, 368, 386, 387, 465, 466, 483
Business Leader, 197
Business linkage, 672
Business Model Diagram, 343–345
Business Scenarios, 19
Business Scenarios technique, 365
Business service/function catalog, 367, 372, 373
Business service/function matrix, 373, 374
Business service/information diagram, 368, 389, 390
Business transformation readiness assessment, 313, 325, 329, 461, 468, 469
Business transformation risks, 326, 333
Business unit participation, 672
Business Use-Case Diagram, 369, 394
Business Value and Risk Matrix, 496, 500, 508, 511–513, 517

C

Candidate Evaluation Service, 173
Capability, 72, 73
  architectures, 130, 732, 734, 735, 741, 755
  requirements, 133

INDEX

Capability-based planning, 72, 381, 382, 390
Capability Maturity Model Integration (CMMI), 288, 312, 313, 670
   activities, strategy and vision element, 694
   architecture method and process element, 693
   competencies, 697
   competencies map to the elements, 695, 696
   core architecture roles, 697
   elements, 682, 690, 697, 699
      architecture deliverables, 684
      architecture governance, 683
      architecture method and process, 683
      business alignment, 684
      strategy and vision, 682, 683, 690–692
   framework, 687, 688, 690
   maturity levels, 685
      Ad Hoc, 686
      defined, 686
      managed, 686
      optimal, 687
      repeatable, 686
Capability/Organization Matrix, 368, 383
Catalogs, 151, 184
Categorization framework
  architecture descriptions, 112

CDSS, *see* Clinical Decision Support Systems (CDSS)
CEO Leadership, 206, 207
Chief Architect, 191, 250–252, 259, 277
Chief/Enterprise Architect, 279
Chief Information Officer (CIO), 192, 370, 388
Chief Technology Officer (CTO), 192
CIO, *see* Chief Information Officer (CIO)
CIO/CTO leadership, 207–209
Clinical Decision Support Systems (CDSS), 773
CMMI, *see* Capability Maturity Model Integration (CMMI)
Collaboration, 253
Common Systems Architectures, 141–143, 145, 148, 184
Common Systems Solutions, 148–149, 184
Communications Plan, 325, 328, 349
Competencies, 266
   adaptability and resilience, 270
   analytical thinking, 271
   business and technology alignment, 269
   change leadership, 269
   communication and collaboration, 270
   governance and compliance, 270

INDEX

leadership and influence, 269
problem solving, 271
required architecture
   competencies per core
   architecture role, 272
vs. skills, 266, 267, 275, 276
strategic decision-making, 269
strategic thinking and
   vision, 268
systems thinking, 271
Conceptual Data Diagram, 414, 415
Consolidated Gaps, Solutions and
   Dependencies
   Matrix, 481–483
Constraints, 68
Content framework, 105, 718
   advantages, 114
   aims of, 126
   comprehensive checklist of
      architecture
      deliverables, 111
   content of, 113
   definition, 111
   vs. earlier versions, 117, 126
   and Enterprise Metamodel, 113
   one-to-one translation, 125
   purpose of, 113, 114
   reference materials, 123
   valuable architecture artifacts
      creation, 126
Continuum, 137
Contract/Measure Catalog, 367, 376, 377
Core architecture roles, 251

Chief Architect, 250–252
Enterprise Architect, 251–253
   enterprise, 249, 250
   primary architecture roles, 251
   relationships, 250
   segment, 249, 250
   segment architect, 253, 254
   solution, 249, 250
   solution architect, 254, 255
Course of action, 79, 80
Critical success factors (CSFs), 78, 611
Cross-mapping, 153
CSFs, *see* Critical success
   factors (CSFs)
CTO, *see* Chief Technology
   Officer (CTO)

D

Data and Application
   Architectures, 398, 441, 442, 459
Data Architect, 256
Data Architecture, 257–259, 263, 280, 407, 410, 418
Data Architecture Entities
   data entities, 93, 94
   logical data components, 94–95
   overview, 92
   physical data component, 95
   types, 93
Database Management
   System (DBMS), 100

855

INDEX

Data Dissemination Diagram, 415
Data-driven decision-making, 16, 538, 722, 759
Data entities, 93, 94
Data Entity/Business Function Matrix, 412, 413
Data Entity/Data Component Catalog, 411, 412
Data Lifecycle Diagram, 418
Data Migration Catalog, 416, 417
Data Security Catalog, 415, 416
DBMS, *see* Database Management System (DBMS)
Definitions, 53
Diagrams and maps, 151–153, 184
Digital framework, 778
Digitalization, 760, 761
Digital services, 35
Digital technologies, 759
Digital transformation, 34, 35, 760, 777
    ADM
        compliance and success (phase G), 767
        continuous improvement (phase H), 768
        digital ecosystem (phase C), 766
        digital infrastructure (phase D), 766
        digital strategy (phase A), 764, 765
        implementation process (phase F), 767
        phases, 763
        preliminary phase, 764
        re-engineer business automation (phase B), 765
        roadmap (phase E), 767
    business-driven initiative, 761
    core driver, 770
    definition, 758, 759
    digital technologies, 759
    disciplines, 762
    and Enterprise Architecture, 760
    focus areas, 761
    hospital, 768, 776
    hybrid approach, 762
    implementation steps, 777
    initiatives, 762
    key components, 758, 759
    organization's business model, 761, 769
    phases, 769
        assessment and strategy development (phase 1), 770, 771
        continuous improvement and future innovations (phase 5), 776
        implementation and staff training (phase 3), 772, 773
        new capabilities, 771
        patient experience and engagement (phase 4), 775

technology selection and
process optimization
(phase 2), 772
virtual assistants, 774
structured approach, 777–779
transformation strategy, 761
Digital transformation
key aspect, 757
phases, 757
Digital vision, 777
Digitization, 760
Domain, 256
Draft Architecture Roadmap,
406, 448, 473, 502
Driver/Goal/Objective Catalog,
367, 370, 371
Driver/Goal/Objective
diagram, 309
Drivers, 77
Dutch Healthcare Reference
Architecture (ZiRA), 146

E

EA, *see* Enterprise
Architecture (EA)
EHR, *see* Electronic health
record (EHR)
Electronic health record (EHR),
143, 178, 772
Enterprise Architects, 4, 190, 191,
248, 251–253, 259, 263, 264,
277, 294, 782, 808

Enterprise Architecture (EA), 1, 3,
30, 238, 261, 310, 313, 318,
319, 326, 420, 463, 464, 491,
493, 497, 499, 516, 530, 531,
534, 592, 593, 602, 603, 611,
612, 614, 627, 631, 664, 670,
681, 683, 706, 709, 711, 736,
741, 744, 751, 753, 759, 760,
762, 823
ADM, 561
architecture partitioning,
580, 586
BIZBOK Guide, 60, 62
confusions, 57
Core Concepts Guide, 58
definition, 54, 56, 61
definitions collection, 58, 59
digital transformation
initiatives, 30
evolution, 15, 685
Gartner's definition, 58, 59, 62
history, 13
identifying and resolving
inconsistencies, 56
implementation, 684
individual definitions
sources, 59
ISO/IEC/IEEE 42010:2022
standard, 54, 62
vs. IT Architecture, 57
Leader's Guide, 62
maturity model, 669
organizational strategy, 60, 686

INDEX

Enterprise Architecture (EA) (*cont.*)
   organization's activities, holistic views, 56
   organization's information systems, 14
   primary object, 59
   purpose, 56
   representations of, 60
   significance of, 16
   strategic and business-centric, 16
   strategic business management tool, 57
   strategic management disciplines, 15
   TOGAF Standard, 55, 61
   word system, 54
   Zachman Framework, 14, 15
Enterprise Architecture Implementation Wheel method, 704, 715
   Architecture Principles, 708
   Baseline and Target Business Architectures, 707
   BIZBOK Guide, 709
   Business Architecture, 709
   control stage, 706
   define stage, 705
   description, 710, 711
   developing fundamental Enterprise Architecture, 703
   document stage, 705
   execute stage, 706
   mapping to the ADM
      control stage, 710
      define stage, 708
      document stage, 707
      execute stage, 709
   organizations, 705
   practical/pragmatic approach to implement fundamental Enterprise Architecture, 706
   Preliminary Phase, 709
   stages, 703
   TOGAF Standard, 711
Enterprise Continuum, 137–139, 184
   conceptual framework, 139
   descriptions/specifications of solutions, 140
   four classification models, 140
   overall Enterprise Architecture, 140
   scale, 140
Enterprise Manageability Diagram, 437–438
Enterprise Metamodel, 125, 317, 318, 337, 338, 535, 536, 717, 818
   adding new/removing entities, 103
   Application Architecture, 96–98
   Architecture Development Method, 101, 102, 104, 108

INDEX

Business Architecture (*see* Business Architecture Entities)
   common entities and relationships, 63
   Content Framework, 63, 64, 103, 105–108
   core components of, 107
   Data Architecture area (*see* Data Architecture Entities)
   Foundation Enterprise Metamodel, 63
   general entities (*see* General entities)
   interrelated sections, 64, 107
   model entities, 65
   organization-specific metamodel, 103
   tailorability, 109
   tailorability of TOGAF Standard, 103, 104
   Technology Architecture area (*see* Technology Architecture Entities)
   TOGAF Standard, 102
Environments and Locations Diagram, 456
Evolution of today's leading architecture frameworks, 15
Extended Content Framework
   Architecture Definition, 121
   Architecture Principles development, 120
   Architecture Realization, 122
   Architecture Repository, 122
   Architecture Requirements, 121
   Business Architecture section, 122
   definition, 120
   entities capabilities and work packages, 122
   Motivation section, 121
   preliminary, 120

F

FEAPO, *see* Federation of Enterprise Architecture Professional Organizations (FEAPO)
Federation of Enterprise Architecture Professional Organizations (FEAPO), 59
FHIR-based data models, 720
Foundation Architecture, 19, 20, 141–143, 145
Foundation Enterprise Metamodel, 65
Foundation Solutions, 148
Functional Decomposition Diagram, 368, 372, 390, 391
Fundamental Content, 44, 50, 58
Fundamental Content of TOGAF Standard
   ADM techniques, 37, 38
   architecture content framework, 38

INDEX

Fundamental Content of TOGAF Standard (cont.)
  comprehensive guide and toolbox, 39
  EA capability and governance, 39
  introduction and core concepts, 37
  pivotal documents, 50

G

Gap analysis, 665
  ADM phases, 619, 620
  architecture building block, 621
  baseline and transition architectures, 619, 620
  baseline architecture, 618, 619
  business capability heat map, 618
  definition, 617
  eliminate/new cell, 622
  matrix, 620, 622
  results, 618
  steps, 618
  target architecture, 618, 619
  transition and target architectures, 619, 620
  work packages, 622
Gap entity, 70, 71, 486
General Entities, 108
  assumptions, 69, 70
  capability, 72, 73
  constraint, 68
  gap, 70, 71
  overview, 66
  principle entity, 67, 68
  requirement, 68, 69
  work package, 71, 72
Goal, 77
Goal/Objective/Business Service Diagram, 368, 393
Governance and compliance competency, 270
Governance Framework, 191, 193, 195, 262
Governance Repository, 135, 136, 596

H

Healthcare industry, Tailoring TOGAF Standard
  challenges, 719
  tailored industry-specific business capability diagram, 721
  Tailoring for Healthcare-Specific Needs, 719, 720
  tailoring governance and compliance, 721
  tailoring the architecture development method
  emphasize business architecture, 720
  enhance migration planning, 720

INDEX

strengthen technology architecture, 720
technologies, 720
Tailoring the Content Framework and Deliverables
  capability models, 721
  FHIR-based data models, 720
  interoperability maps, 721
  risk and compliance models, 721
  tailoring tools and techniques, 721
Healthcare-specific needs, 719, 720
Human Resources Management capability, 171, 509

I

Implementation Factor Catalog, 461, 463, 464, 466, 471, 478–481
Implementation governance, 513, 743, 749, 756
  architecture compliance report, 805
  capability heat map, 806
  establish architecture governance, 514
  confirm scope and priorities, 514–516
  guide development of solutions deployment, 514, 517, 518
  identify deployment resources and skills, 514, 516
  implement business and IT operations, 515, 520
  perform architecture compliance reviews, 515, 519
  perform post-implementation review, 521
  steps, 514
performance monitoring, 522
  measure achievement of goals and objectives, 526–528
  measure completion of work packages, 523–525
  steps, 814
  work package portfolio map, 814
Implementation Planning, 29
Industry Architectures, 143, 144
Industry Solutions, 149
Industry-specific Enterprise Architecture tools, 721
Information Maps, 363, 369, 396–397, 411, 414

INDEX

Information systems
architectures, 398
Application Architecture, 399
artifacts, Application
Architecture, 427–429
Application and User
Location Diagram, 437
Application Communication
Diagram, 436
Application/Function
Matrix, 435
Application Interaction
Matrix, 435–436
Application Migration
Diagram, 439–440
Application/Organization
Matrix, 433–434
Application Portfolio
Catalog, 429–433
Application Use-Case
Diagram, 437
Enterprise Manageability
Diagram, 437–438
Interface Catalog, 432–433
Process/Application
Realization
Diagram, 438–439
Role/Application Matrix, 434
Software Distribution
Diagram, 440
Software Engineering
Diagram, 439
artifacts, Data Architecture,
410, 411
Application/Data
Matrix, 414
Conceptual Data Diagram,
414, 415
Data Dissemination
Diagram, 415
Data Entity/Business
Function Matrix, 412, 413
Data Entity/Data
Component Catalog,
411, 412
Data Lifecycle
Diagram, 418
Data Migration Catalog,
416, 417
Data Security Catalog,
415, 416
to craft Target Data
Architecture, 400
Data Architecture, 399
develop Application
Architecture
conduct formal stakeholder
review, 420, 425
define candidate roadmap
components, 419, 423
develop Baseline and Target
Application Architecture
description, 419, 421
finalize, 426
perform gap analysis, 419
resolve impacts across
Architecture Landscape,
420, 424

862

INDEX

select reference models,
viewpoints and tools,
419, 421
steps, 419
update, 427
develop Data Architecture
conduct formal stakeholder
review, 401, 407
define candidate roadmap
components, 401, 405, 406
develop Baseline and Target
Data Architecture
description, 400, 403
finalize Data Architecture,
401, 408
perform gap analysis, 401, 405
resolve impacts across
Architecture Landscape,
401, 406, 407
select reference models,
viewpoints and tools,
400, 402
steps, 400
update Architecture
Definition Document,
401, 409
to identify potential Data
Architecture
components, 400
Initiative, 71
Initiatives or work packages, 124
Interface Catalog, 161, 428,
432–433, 435
Interoperability, 32, 171, 253, 254,
283, 319, 429, 441, 623

Invoice Management, 35
ISO/IEC/IEEE 42010:2022
standard, 58, 62
IT investment and acquisition
strategy, 673

J

Job, 248

K

Key components of Architecture
Repository
Architecture Capability, 135
Architecture Landscape,
130, 131
Architecture Metamodel, 129
Architecture Requirements
Repository, 132–134
Governance Repository, 135
Reference Library, 134
Solutions Landscape, 131, 132
Standards Library, 134, 135
Key performance indicators (KPIs),
78, 223, 765, 770, 776
KPIs, *see* Key performance
indicators (KPIs)

L

LeanIX, 716
Location Catalog, 367, 374, 375, 456
Logical application component,
96–98, 160, 161, 163, 411

863

INDEX

Logical data components, 93–95, 411, 415
Logical data model, 364
Logical technology component entity, 100

M

Mapping existing frameworks and methods, 701–703
Mastering the TOGAF Standard (book), 9, 10
Matrices, 153, 154, 184
Maturity models, 288, 628, 698–700
Measure, 75, 78–80, 84, 121
Measurement data, 686
Metrics, 78, 234, 243, 348
Microservices Architecture (MSA), 45
Migration planning, 495, 496, 748
    artifacts, 508
        Business Value and Risk Matrix, 511–513
        Transition Architecture State Evolution Table, 508–511
    assign business value to work packages, 496, 499, 500
    complete architecture development cycle, 497, 506
    complete Implementation and Migration Plan, 497, 505
    confirm Architecture Roadmap, 497, 504, 505
    confirm management framework interactions, 496–498
    business transformation readiness assessment, 802
    business value, 803, 804, 814
    constraints, 800
    cost savings, 800
    estimate resource requirements, project timings, and availability/delivery vehicle, 496, 501
    HR business partner, 803
    and implementation, 800, 801
    implementation factor catalog, 798, 799
    meetings, 802, 803
    migration plan, 798
    new planning application, 799
    optimized discharge planning, 803
    patient feedback application, 799
    plateau planning diagram, 801
    prioritize migration projects through cost/benefit assessment and risk validation, 496, 501–503
    project manager, 803, 814
    projects, 804
    risk analysis, 800
    risk levels, 805
    risk matrix, 803, 804

risk/process/application realization diagram, 804, 805
staff training, 800
standardization, 800
steps, 813
work package portfolio map, 801
work packages, 802
Minimum Viable Architecture, 728, 736, 737, 744, 746, 752, 755
MSA, *see* Microservices Architecture (MSA)
Motivation elements, 108
Multiple views, 177
MySQL database system, 100

N

Netflix, 35
Network and Communications Diagram, 458-459
Networked Computing/Hardware Diagram, 458

O

Objectives, 77
Open Group, 17, 23, 32
Open Group Architecture Forum, 41
Opportunities and solutions, 459-460, 748
architecture definition document, 793, 797, 798
architecture roadmap, 795, 796
architecture work, 797
artifacts, 475
  Architecture Definition Increments Table, 483, 484
  Benefits Catalog, 477, 478
  Consolidated Gaps, Solutions and Dependencies Matrix, 481-483
  Implementation Factor Catalog, 478-480
  Project Context Diagram, 476, 477
confirm readiness and risk for business transformation, 461, 468-469
consolidate and reconcile interoperability requirements, 461, 466-467
create Architecture Roadmap, 491, 493, 494
  architecture tool, 492
  roadmap's iterative development, 485
  TOGAF Framework, 485
  Work Package Portfolio Map, 488-490
  work packages, 486-488, 491-493
create Architecture Roadmap and Implementation and Migration Plan, 462, 473-475

INDEX

Opportunities and solutions (*cont.*)
  building blocks/
    capabilities, 797
  determine business constraints
    for implementation,
    461, 464
  determine/confirm key
    corporate change
    attributes, 461, 463
  formulate Implementation and
    Migration Strategy, 462, 470
  group sessions, 797
  HR business partner, 796
  identify and group major work
    packages, 462, 471
  identify Transition
    Architecture(s), 462, 472
  to identify work packages, 460
  refine and validate
    dependencies,
    461, 467–468
  review and consolidate gap
    analysis results, 461,
    465, 466
  steps, 461–462
  implementation factor
    catalog, 813
  plateau planning diagram, 794
  requirements, 797
  steps, 812
  transition architecture, 813
  work package portfolio map,
    794, 795
  work packages, 794–796

Organization Decomposition
  Diagram, 369, 394, 395
Organization Diagram, 384, 385
Organization Map, 153, 154, 339,
  362, 363, 368, 383, 384
Organizations, 219, 236, 267
Organization-Specific
  Architectures, 144, 145
Organization-Specific
  Solutions, 149

P

Patient Feedback Application,
  391–393, 414, 417, 430, 433,
  434, 799
Physical application component
  entity, 98, 174
Physical data component, 93, 95,
  160, 411
Physical technology component
  entity, 100
Platform Decomposition
  Diagram, 456–457
Preliminary Framework and
  Principles, 22
Preliminary phase, ADM, 112, 305
  Academic Diagnostic Medicine,
    783, 784, 809
  artifacts, 320
  assess Enterprise Architecture
    maturity, 306, 312, 313
  choose and tailor architecture
    framework, 314, 317–319

INDEX

define enterprise, 306, 307
determine Architecture
    Capability, 305
driver/goal/objective diagram,
    309, 809
establish Enterprise
    Architecture team, 314, 315
follow-up meeting, 783
identify key drivers and
    organizational context,
    306, 308
implementation of strategy, 784
manage relationships between
    frameworks, 306, 310, 311
organization-wide, 306
Principles Catalog, 321, 322
projects, 783
requirements catalog, 783, 784
set Architecture Principles,
    314, 316
steps, 306, 809
strategy meetings, 782
Principle, requirement and
    constraint relationship, 69
Principles, 68
Process/Application Realization
    Diagram, 438–439
Process/Event/Control/Product
    Catalog, 367, 375, 376
Process Flow Diagram, 153, 369,
    395, 396, 483
Processing Diagram, 457
Product Lifecycle Diagram, 368,
    391, 392
Project and portfolio
    management, 682
Project Context Diagram,
    124, 163, 476, 477

Q

Q3 Legal Department Report, 78

R

RACI matrix, 315, 316, 324, 349, 371
Reference Library, 44, 134
Reference Library of the
    Architecture Repository,
    177, 186, 352, 353, 403
Reference materials, 39, 43, 44, 123
Relationship between architecture
    deliverables, artifacts, and
    building blocks, 150
Relationship between views and
    viewpoints, 177, 178
Repository functionality, 181
Request for Architecture Work
    document, 300, 302, 305
Requirement, 68
Requirements Catalog, 546
Requirements Management,
    539–540
    artifacts, 546
        Requirements Catalog, 546
    assess and revise gap
        analysis, 545
    assess impact of changed
        requirements, 544

INDEX

Requirements Management (*cont.*)
  capture, manage and refine requirements, 540
  establish Baseline Requirements, 542–543
  excerpt of Architecture Repository, 542
  identify and document requirements, 542
  identify changed requirements and record priorities, 543–544
  identify new and changed requirements, 543
  implement change in current phase, 545
  implement requirements, 544
  manage and maintain requirements, 540
  monitor Baseline Requirements, 543
  update Architecture Requirements Repository, 544

Review gap analysis, 746

Risk management, 590, 591, 632, 667
  ADM phases, 634
  architecture project risk, 635
  enterprise risk management, 635
  initial/residual risks, 635, 636
  initial risk assessment, 637, 644, 668
  agreement, 647
  architecture work, 644
  phases, 644
  risk classification scheme, 646
  risk frequency level, 644, 645
  risk impact level, 644, 645
  mitigating factors, 635, 636
  phases, 632–634
  risk categorization, 668
  risk classification, 637, 644
    categories, 638, 639
    domain/segment, 640
    implementation factor catalog, 640, 641
    organizations, 642
    potential impact, 638
  risk identification, 637, 642, 643
  risk mitigation, 637, 647, 648, 650, 668
  risk monitoring, 637, 650
  risk types, 667
  risk visualization, 651
    risk calculation, 653, 654
    risk heat map, 651, 652
    risk to diagrams, 655
  steps, 636

Risk Management and Stakeholder Management, 716, 718

Risk Management Plan, 496, 503, 642–644

Risk mitigation, 647, 648, 650

Risk monitoring, 636, 650–651

INDEX

Robotic Process Automation (RPA), 775
Role, definition, 248
Role/Application Matrix, 428, 434
Role Catalog, 367, 371, 372
Role descriptions, 196, 249, 255

S

SBA Administration, 28, 29
Segment, 132, 249, 250, 252, 256
Segment Architects, 191, 253, 254, 263, 279, 291, 294, 819
    architecture capabilities, 259, 260
    architecture domains, 256
    architecture roles, 259
    Business Architecture, 255
    Business domains/segments *vs.* architecture domains, 256, 257
    horizontal perspective, 258–263, 278
    job, 256
    job assignment, 255
    job description states, 255
    organization, 258
    organizational domain, 261
    organizational perspective, 256
    role, 259, 260, 278
    technical architecture domain, 255
    use of, 255
    vertical perspective, 257, 278
    vs. Solution Architecture, 261, 262
    specializations, 260
Segment Architecture Requirements, 133
Segment architectures, 130, 731, 733, 734, 740, 755
Senior management involvement, 671, 672, 677, 699
Senior management training, 683
Series Guides, 51, 58
    ADM application methods, 42
    Enterprise Architecture, 40
    Enterprise Architecture Capability, 40, 41
    guidance material, 41
    industry trends and market demands, 41
    reference models, 42
Service Level Agreement (SLA), 377
SIB, *see* Standards Information Base (SIB)
SLA, *see* Service Level Agreement (SLA)
Simple, Measurable, Actionable, Realistic, and Timebound (SMART), 77, 491, 765
Skills frameworks
    competency levels, 264
    consultancy and/or project management assignment, 264
    proficiency levels, 264, 266

INDEX

Skills frameworks (*cont.*)
    skill categories, alphabetical order, 265
    TOGAF Standard, 278
SMART, *see* Simple, Measurable, Actionable, Realistic, and Timebound (SMART)
Software Distribution Diagram, 162, 429, 440
Software Engineering Diagram, 162, 439
Solution Architects, 254–255, 259, 263, 264, 277, 279, 291, 292, 294
Solution Architecture, 91, 238, 261, 262, 287, 293
Solution Building Blocks, 132, 164, 170, 185, 186
    actual implementation choices representation, 173
    Architecture Repository
        central location, 174, 175
        Solutions Landscape, 175
    business service, 173
    functionality and attributes, 174
    HR Management capability, 174
    physical/concrete components, 174
    real-world implementation, 174
Solution Concept Diagram, 325, 328, 332, 336, 343, 344, 360
Solutions Landscape, 131, 132, 147, 175, 182, 183
Stackable LEGO bricks, 165

Stakeholder management, 590, 591
    architecture team, 603
    communication plan, 603, 605, 611, 613, 614
        architect, 612
        components, 612
        scope, 611
        stakeholder group, 611
        stakeholder power grid, 611
    definition, 601
    practical approach, steps, 604
    stakeholder analysis, 603–605, 608
        architecture work, 607
        classification, 609
        stakeholder power grid, 607, 608
    stakeholder catalog, 609–611
    stakeholder concept, 601, 602
    stakeholder engagement, 603, 604, 615
    structured approach, 601
Stakeholders, 68, 104, 602, 685, 752
    and architects, 602, 665
    architecture decisions, 602
    architecture development, 656, 668
    architecture development process, 601
    classification, 607, 609, 611
    communication, 611
    concerns and constraints, 786
    concept, 601, 602
    cross-mapping, 624, 625

element, 104
engagement, 329, 335, 494
engaging, 591, 603, 604
enterprise architecture, 601
groups, 605, 613, 614
identification, 603-605
interoperability, 623
involvement, 682
key concerns, 609, 668
management, 665
power grid, 607, 608
project success, 603
readiness, 746
Standards Information Base (SIB), 19
Standards Library, 134, 135
Statement of Architecture Work document, 301, 303, 326, 330, 333, 356, 359, 360, 407, 410, 420, 425, 426, 449, 452
Strategic Architecture Requirements, 132-133
Strategic architectures, 130, 731, 733, 734, 740, 754
Strategic decision-making, 49, 269, 272, 696, 697
Strategic decisions, 168, 205, 206, 208
Strategy/Capability Matrix, 159, 368, 382

T

TAFIM framework, 20, 28

Tailoring TOGAF Standard
design, customization
tailoring for organizational context, 714, 715, 722
tailoring governance and compliance, 716, 723
tailoring the architecture development method, 715, 716, 722
tailoring the content framework and deliverables, 716, 723
tailoring tools and techniques, 716, 723
healthcare industry (*see* Healthcare industry, Tailoring TOGAF Standard)
Target Architecture, 28, 38, 67, 70-72, 237, 244, 298
Target values, 78
Technical Architect, 249
Technical Architecture domain, 255, 291, 294
Technical Corrigendum, 25, 26, 30, 54, 62
Technical Reference Model (TRM), 19
Technology Architect, 256
Technology architecture, 257, 260, 441
artifacts, 453
Application/Technology Matrix, 455
Environments and Locations Diagram, 456

INDEX

Technology architecture (*cont.*)
    Network and
        Communications
        Diagram, 458–459
    Networked Computing/
        Hardware Diagram, 458
    Platform Decomposition
        Diagram, 456–457
    Processing Diagram, 457
    Technology Portfolio
        Catalog, 454–455
    Technology Standards
        Catalog, 454
    conduct formal stakeholder
        review, 444, 449
    to craft target, 442
    define candidate roadmap
        components, 443, 447
    develop Baseline and Target
        Technology Architecture
        description, 443, 445–446
    domain, 263
    finalize, 444, 451
    to identify potential Technology
        Architecture
        components, 442
    perform gap analysis,
        443, 446–447
    resolve impacts across
        Architecture Landscape,
        444, 448–449
    select reference models,
        viewpoints and tools,
        443, 445
    steps, 443
    update Architecture Definition
        Document, 444, 451–452
Technology Architecture Entities
    definition, 99
    logical technology
        component, 100
    overview, 98, 99
    physical technology
        component, 100
    technology service, 99
Technology Portfolio Catalog, 162,
    339, 454–455
Technology service, 99
Technology Standards Catalog,
    162, 454
The Contract Management
    capability, 35
The Resource Base, 20
The Solutions Continuum, 138, 184
    addressing business needs, 148
    aim of, 147
    architectures specification and
        construction, 147
    Common Systems Solutions,
        148, 149
    definition, 147
    Foundation Solutions, 148
    Industry Solutions, 149
    Organization-Specific
        Solutions, 149
    perspectives, 148
    solution types, 147
    solution value, 147

INDEX

The TOGAF Standard, 10th Edition (book), 24
Three continua
    Architecture Continuum, 138, 141 (*see also* Architecture Continuum)
    Enterprise Continuum, 138–140
    Solutions Continuum (*see* The Solutions Continuum)
TOGAF 2.0, 17
TOGAF 4, 17
TOGAF 5, 17, 18
TOGAF 7, 18, 21, 55, 282
TOGAF 8, 21–24, 282
TOGAF documentation, 3, 43
TOGAF framework
    components and/or Series Guides, 4
    scaffolding, 3
    The Series Guides, 3
TOGAF Library, 308, 310
    Architecture Repository, 44
    definition, 43
    different styles and methodologies, 44, 45
    empowers organizations, 52
    foundational nature of framework, 33
    Fundamental Content, 44
    other standards and frameworks, 45
    practical applications, 45
    Reference Library, 44
    Series Guides, 34
    specific verticals/use cases, 45
    support for architecture areas/domains, 45
    The Open Group Library, 43, 51
TOGAF Standard, 8, 307, 823
    ADM, 828
    aims, 25
    alternative routes, 9
    applicable framework, 2
    ArchiMate/modeling ethics, 12
    architectural styles, 563, 564
    Architecture Capability, 4, 33
    architecture frameworks, 824
    Architecture Repository (*see* Architecture Repository)
    BIZBOK Guide, 826
    components/aspects, 818–820
    concepts, 817, 825
    create Segment Architect profile, 249
    criticism, 5, 6
    de facto standard, 5
    definition, 2, 33
    design the architecture, 4
    different architecture styles support, 36
    documentation, 827, 828
    domains, 256
    early versions, 17, 18
    Enterprise Architecture establishment, 50
    Enterprise Architecture framework, 32
    flexible by design, 40

INDEX

TOGAF Standard (*cont.*)
   flexible implementation roadmap, 4
   vs. frameworks, 826
   general approach, 32
   governance and execution, 4
   independent documents, 39
   IT Architecture, 825
   jobs, roles, and assignments, 248
   myths, 824
   one-to-one description, 2
   organizations, 824
   overview, 32
   practical and pragmatic, 7
   practical application, 6
   practical approach, 828
   practical translation, 9
   recent versions, 18, 19
   rigid/prescriptive framework, 825
   skills, 264
   structure, 33, 34
   The Open Group, 32
   updation, 827
   what and how, 6
TOGAF *vs.* TAFIM framework, 26–28
Transition Planning, 567, 568, 575–578, 585, 586
Tribe Architect, 258

TRM, *see* Technical Reference Model (TRM)
Types of entities, 112

U

UML, 716, 718
Use-case models, 364

V

Value Chain Diagram, 336, 339–341, 343
Value realization process, 529, 530, 554, 750
Value Stream/Capability Diagram, 347, 381–382
Value Stream/Capability Matrix, 368, 381
Value Stream Map, 336, 346, 347, 368, 380, 381
Value streams
   acquire talent, 81, 83
   business capabilities, 80
   business processes, 80, 82
   definition, 80
   key principle, 82
   signed employee agreement, 82
   stages, 80–82
Viewpoints, 176, 177, 186
Views, 176, 177, 186

Virtual assistants, 774
Virtual Reality (VR), 776
Visual Tool, 472
VR, *see* Virtual Reality (VR)

W, X, Y

Work package, 71, 72

Work Package Portfolio Map, 124, 461, 462, 468, 471, 476, 485, 488–490, 492, 499, 514–516

Z

Zachman Framework, 14–16, 30, 716

GPSR Compliance

The European Union's (EU) General Product Safety Regulation (GPSR) is a set of rules that requires consumer products to be safe and our obligations to ensure this.

If you have any concerns about our products, you can contact us on

ProductSafety@springernature.com

In case Publisher is established outside the EU, the EU authorized representative is:

Springer Nature Customer Service Center GmbH
Europaplatz 3
69115 Heidelberg, Germany

www.ingramcontent.com/pod-product-compliance
Lightning Source LLC
LaVergne TN
LVHW021954060526
838201LV00048B/1569